LEARNING GUIDE WITH INTEGRATED REVIEW WORKSHEETS

C. BRAD DAVIS

PAMELA TRIM

Southwest Tennessee Community College

PRECALCULUS

SIXTH EDITION

Robert Blitzer

Miami Dade College

Reproduced by Pearson from electronic files supplied by the author.

Publishing as Pearson, 330 Hudson Street, NY NY 10013

2 18

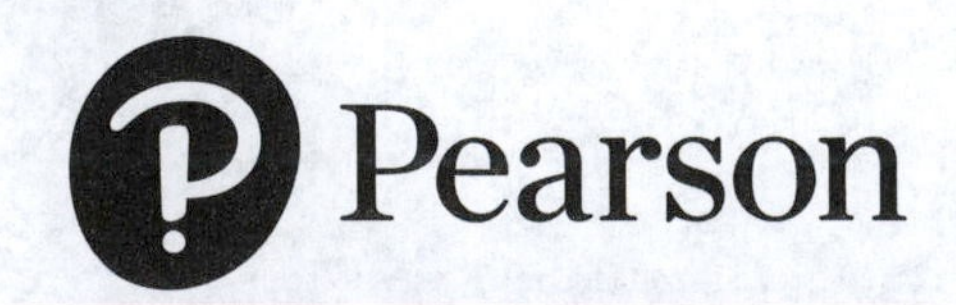

ISBN-13: 978-0-13-447010-8
ISBN-10: 0-13-447010-9

Learning Guide
Precalculus 6th edition

Table of Contents

About this *Learning Guide* ii

Chapter P Fundamental Concepts of Algebra 1

Chapter 1 Functions and Graphs 55

Chapter 2 Polynomial and Rational Functions 107

Chapter 3 Exponential and Logarithmic Functions 143

Chapter 4 Trigonometric Functions 167

Chapter 5 Analytic Trigonometry 221

Chapter 6 Additional Topics in Trigonometry 247

Chapter 7 Systems of Equations and Inequalities 293

Chapter 8 Matrices and Determinants 319

Chapter 9 Conic Sections and Analytic Geometry 345

Chapter 10 Sequences, Induction, and Probability 379

Chapter 11 Introduction to Calculus 409

Classroom Activities A-1

Integrated Review Worksheets 1

About this *Learning Guide*

Dear Student,

We are glad you have decided to read this introduction to your *Learning Guide*. By looking at the following great questions that your classmates have asked, you will discover how to get the most out of this *Learning Guide* as you progress through your algebra course.

We hope you have a great semester!

Bob Blitzer, *Miami-Dade College, FL*
C. Brad Davis

Great Questions that Students are Asking!

What is the relationship between this Learning Guide and the textbook?

Each section of the textbook begins with a list of learning objectives that focus on the section's most important ideas. Your *Learning Guide* is organized around the textbook's learning objectives. These objectives are the headers for the *Solved Problems* and *Pencil Problems* that form the essence of this guide.

What am I supposed to do with this Learning Guide? Why is it 3-hole punched?

- It is 3-hole punched so you can use it as the basis of your **course notebook**.
- You should **insert your lecture notes** at the beginning of each section.
- The main component of the *Learning Guide* is the Guided Practice. This is a series of ***Solved Problems*** that are paired with unsolved ***Pencil Problems*** for you to attempt. Answers to these problems are given at the end of each section.

What are the benefits of using this Learning Guide?

- It will help you become better organized. This includes organizing your class notes, assigned homework, quizzes, and tests.
- It will enable you to use your textbook more efficiently.
- It will help increase your study skills.
- It will help you to prepare for the chapter tests.

About this *Learning Guide* (continued)

What are the Classroom Activities? What should I do with them?

Some of the Classroom Activities are designed to introduce a topic or review important material before a section of the textbook is covered in class; others are designed to reinforce the concepts or give an application of the mathematics in the section just covered. Your instructor may use these in class or assign them as group projects for homework.

Integrated Review Worksheets

The Integrated Review worksheets offer additional practice exercises of relevant intermediate algebra topics with ample space for students to show their work. The Integrated Review Worksheets perfectly complement the Integrated Review MyLab Math course for College Algebra, seventh edition.

A complete list of the prerequisite topics covered by the worksheets can be found in the Table of Contents immediately preceding the worksheets.

What else should I know about Precalculus?

As you have noticed, math textbooks can be lengthy. It is unlikely that your course will cover all the material in the entire book, so be sure to pay attention to the syllabus provided by your instructor.

You will notice that this *Learning Guide* uses the same subdivisions as the main textbook. Each Objective is written out so you will know exactly what concept is presented in the various examples. This will allow you to easily find the specific help you need as you move from resource to resource, and will allow you to easily skip the topics that your instructor is not planning to cover.

To take this course you should have the following:

- ***Precalculus 6th edition*** textbook (or eBook)
- ***Learning Guide***

Additional resources available for this course (optional, unless specified by your instructor):

- ***Student Solution Manual*** (contains solutions to selected exercises)
- ***MyLabMath*** (online homework, exercise tutorials, and more)

Note to Instructors:

Thank you for choosing this *Learning Guide* for your class. Its design and purpose is to facilitate the teaching and learning process by keeping your students organized. Please encourage your students to read these introductory pages.

We hope you and your class have a rewarding semester.

Bob Blitzer, *Miami-Dade College, FL*
C. Brad Davis

Section P.1
Algebraic Expressions, Mathematical Models, and Real Numbers

It costs how much?

You are looking ahead to the next school year and wondering how much money you will need.
Is there any way that you can use trends for college costs over the past few years to predict how much college will cost next year?
In the Exercise Set for this section, you will use a model that will allow you to project average costs at private U.S. colleges in the near future.

Objective #1: Evaluate algebraic expressions.

✔ ***Solved Problem #1***

1. Evaluate $8+6(x-3)^2$ for $x=13$.

$$\begin{aligned} 8+6(x-3)^2 &= 8+6(13-3)^2 \\ &= 8+6(10)^2 \\ &= 8+6(100) \\ &= 8+600 \\ &= 608 \end{aligned}$$

Pencil Problem #1

1. Evaluate $4+5(x-7)^3$ for $x=9$.

Objective #2: Use mathematical models.

✔ ***Solved Problem #2***

2. The formula $T = 4x^2 + 330x + 3310$ models the average cost of tuition and fees, T, for public U.S. colleges for the school year ending x years after 2000. Use this formula to project the average cost of tuition and fees at public U.S. colleges for the school year ending in 2014.

Because 2014 is 14 years after 2000, we substitute 14 for x in the formula.

$T = 4x^2 + 330x + 3310$

$T = 4(14)^2 + 330(14) + 3310$

$T = 4(196)^2 + 330(14) + 3310$

$T = 784 + 4620 + 3310$

$T = 8714$

The formula indicates that for the school year ending in 2014, the average cost of tuition and fees at public U.S. colleges will be \$8714.

Pencil Problem #2

2. The formula $T = 21x^2 + 862x + 15,552$ models the average cost of tuition and fees, T, for private U.S. colleges for the school year ending x years after 2000. Use this formula to project the average cost of tuition and fees at private U.S. colleges for the school year ending in 2020.

Objective #3: Find the intersection of two sets.

✔ *Solved Problem #3*

3. Find the intersection: $\{3, 4, 5, 6, 7\} \cap \{3, 7, 8, 9\}$.

The elements common to $\{3, 4, 5, 6, 7\}$ and $\{3, 7, 8, 9\}$ are 3 and 7.

$\{3, 4, 5, 6, 7\} \cap \{3, 7, 8, 9\} = \{3, 7\}$

Pencil Problem #3

3. Find the intersection: $\{1, 2, 3, 4\} \cap \{2, 4, 5\}$.

Objective #4: Find the union of two sets.

✔ *Solved Problem #4*

4. Find the union: $\{3, 4, 5, 6, 7\} \cup \{3, 7, 8, 9\}$.

List the elements from the first set: 3, 4, 5, 6, and 7. Now list any elements from the second set not in the first: 8 and 9.

$\{3, 4, 5, 6, 7\} \cup \{3, 7, 8, 9\} = \{3, 4, 5, 6, 7, 8, 9\}$

Pencil Problem #4

4. Find the union: $\{1, 2, 3, 4\} \cup \{2, 4, 5\}$.

Objective #5: Recognize subsets of the real numbers..

✔ *Solved Problem #5*

5. Consider the following set of numbers:
$\left\{-9, -1.3, 0, 0.\overline{3}, \frac{\pi}{2}, \sqrt{9}, \sqrt{10}\right\}$.

5a. List the natural numbers.

The natural numbers are used for counting. The only natural number is $\sqrt{9}$ because $\sqrt{9} = 3$.

5b. List the rational numbers.

All numbers that can be expressed as quotients of integers are rational numbers: $-9\left(-9 = \frac{-9}{1}\right)$, $0\left(0 = \frac{0}{1}\right)$, and $\sqrt{9}\left(\sqrt{9} = \frac{3}{1}\right)$. All numbers that are terminating or repeating decimals are rational numbers: -1.3 and $0.\overline{3}$.

Pencil Problem #5

5. Consider the following set of numbers:
$\left\{-11, -\frac{5}{6}, 0, 0.75, \sqrt{5}, \pi, \sqrt{64}\right\}$.

5a. List the natural numbers.

5b. List the rational numbers.

Objective #6: Use inequality symbols.

✔ *Solved Problem #6*

6. Indicate whether each statement is true or false.

6a. $-8 > -3$

This statement is false. Because −8 lies to the left of −3 on a number line, −8 is less than −3. So, $-8 < -3$.

6b. $9 \le 9$

This statement is true because $9 = 9$.

Pencil Problem #6

6. Indicate whether each statement is true or false.

6a. $-7 < -2$

6b. $-5 \ge 2$

Objective #7: Evaluate absolute value.

✔ *Solved Problem #7*

7. Rewrite each expression without absolute value bars.

7a. $|1-\sqrt{2}|$

Because $\sqrt{2} \approx 1.4$, the number $1-\sqrt{2}$ is negative.
Thus, $|1-\sqrt{2}| = -(1-\sqrt{2}) = \sqrt{2}-1$.

7b. $|\pi-3|$

Because $\pi \approx 3.14$, the number $\pi - 3$ is positive.
Thus, $|\pi-3| = \pi-3$.

7c. $\frac{|x|}{x}$ if $x > 0$

If $x > 0$, then $|x| = x$. Thus, $\frac{|x|}{x} = \frac{x}{x} = 1$.

Pencil Problem #7

7. Rewrite each expression without absolute value bars.

7a. $|12-\pi|$

7b. $|\sqrt{2}-5|$

7c. $\frac{-3}{|-3|}$

Objective #8: Use absolute value to express distance.

✔ *Solved Problem #8*

8. Find the distance between −4 and 5 on the real number line.

$|-4-5| = |-9| = 9$

Pencil Problem #8

8. Find the distance between −19 and −4 on the real number line.

Objective #9: Identify properties of the real numbers.

✔ *Solved Problem #9*

9. State the name of the property illustrated.

9a. $2+\sqrt{5}=\sqrt{5}+2$

The order of the numbers in the addition has changed. This illustrates the commutative property of addition.

9b. $1\cdot(x+3)=x+3$

One has been deleted from a product. This illustrates the identity property of multiplication.

Pencil Problem #9

9. State the name of the property illustrated.

9a. $6+(2+7)=(6+2)+7$

9b. $2(-8+6)=-16+12$

Objective #10: Simplify algebraic expressions.

✔ *Solved Problem #10*

10. Simplify: $6+4[7-(x-2)]$.

$$\begin{aligned} 6+4[7-(x-2)] &= 6+4[7-x+2] \\ &= 6+4[9-x] \\ &= 6+36-4x \\ &= (6+36)-4x \\ &= 42-4x \end{aligned}$$

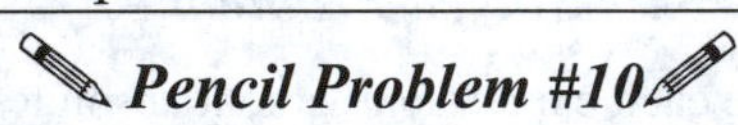

10. Simplify: $7-4[3-(4y-5)]$.

Answers for Pencil Problems *(Textbook Exercise references in parentheses)*:

1. 44 *(P.1 #9)* **2.** $41,192 *(P.1 #131c)* **3.** $\{2, 4\}$ *(P.1 #21)* **4.** $\{1, 2, 3, 4, 5\}$ *(P.1 #29)*
5. a. $\sqrt{64}$ **b.** $-11, -\frac{5}{6}, 0, 0.75, \sqrt{64}$ *(P.1 #37)* **6. a.** true **b.** false **7. a.** $12-\pi$ *(P.1 #53)*
b. $5-\sqrt{2}$ *(P.1 #55)* **c.** -1 *(P.1 #57)* **8.** 15 *(P.1 #71)* **9. a.** associative property of addition *(P.1 #77)*
b. distributive property *(P.1 #81)* **10.** $16y-25$ *(P.1 #93)*

Section P.2
Exponents and Scientific Notation

WOW, THAT'S BIG!

Did you know that in the summer of 2012 the national debt passed \$16,000,000,000,000 or \$16 trillion? Yes, that's 12 zeros you count. In this section, you will express the national debt in a form called *scientific notation* and use this form to calculate your share of the debt.

Objective #1: Use the properties of exponents.

✔ Solved Problem #1

1a. Multiply using the product rule:
$3^3 \cdot 3^2$

$3^3 \cdot 3^2 = 3^{3+2} = 3^5$ or 243

1b. Multiply using the product rule:
$(4x^3y^4)(10x^2y^6)$

$(4x^3y^4)(10x^2y^6) = 4 \cdot 10 \cdot x^{3+2} \cdot y^{4+6} = 40x^5y^{10}$

1c. Divide using the quotient rule.
$\frac{(-3)^6}{(-3)^3}$

$\frac{(-3)^6}{(-3)^3} = (-3)^{6-3} = (-3)^3$ or -27

1d. Divide using the quotient rule.
$\frac{27x^{14}y^8}{3x^3y^5}$

$\frac{27x^{14}y^8}{3x^3y^5} = \frac{27}{3} \cdot x^{14-3} \cdot y^{8-5} = 9x^{11}y^3$

Pencil Problem #1

1a. Multiply using the product rule:
$x^3 \cdot x^7$

1b. Multiply using the product rule:
$(-9x^3y)(-2x^6y^4)$

1c. Divide using the quotient rule.
$\frac{2^8}{2^4}$

1d. Divide using the quotient rule.
$\frac{25a^{13}b^4}{-5a^2b^3}$

1e. Evaluate -8^0.

Because there are no parentheses only 8 is raised to the 0 power.

$-8^0 = -(8^0) = -1$

1e. Evaluate $(-3)^0$.

1f. Write with a positive exponent. Simplify, if possible.

5^{-2}

$5^{-2} = \frac{1}{5^2} = \frac{1}{25}$

1f. Write with a positive exponent. Simplify, if possible.

4^{-3}

1g. Write with a positive exponent. Simplify, if possible.

$3x^{-6}y^4$

$3x^{-6}y^4 = 3 \cdot \frac{1}{x^6} \cdot y^4 = \frac{3y^4}{x^6}$

1g. Write with a positive exponent. Simplify, if possible.

$(4x^3)^{-2}$

1h. Simplify using the power rule.

$(3^3)^2$

$(3^3)^2 = 3^{3 \cdot 2} = 3^6$ or 729

1h. Simplify using the power rule.

$(2^2)^3$

1i. Simplify using the power rule.

$(y^7)^{-2}$

$(y^7)^{-2} = y^{7(-2)} = y^{-14} = \frac{1}{y^{14}}$

1i. Simplify using the power rule.

$(x^{-5})^3$

1j. Simplify: $(-4x)^3$.

$(-4x)^3 = (-4)^3(x)^3 = -64x^3$

1j. Simplify: $(8x^3)^2$.

1k. Simplify: $\left(-\frac{2}{y}\right)^5$.

$\left(-\frac{2}{y}\right)^5 = \frac{(-2)^5}{y^5} = \frac{-32}{y^5} = -\frac{32}{y^5}$

1k. Simplify: $\left(-\frac{4}{x}\right)^3$.

Objective #2: Simplify exponential expressions.

Solved Problem #2

2. Simplify.

2a. $(2x^3y^6)^4$

$(2x^3y^6)^4 = (2)^4(x^3)^4(y^6)^4$

$= 2^4x^{3\cdot 4}y^{6\cdot 4}$

$= 16x^{12}y^{24}$

2b. $(-6x^2y^5)(3xy^3)$

$(-6x^2y^5)(3xy^3) = (-6)(3)x^2xy^5y^3$

$= -18x^{2+1}y^{5+3}$

$= -18x^3y^8$

Pencil Problem #2

2. Simplify.

2a. $(-3x^2y^5)^2$

2b. $(3x^4)(2x^7)$

2c. $\dfrac{100x^{12}y^2}{20x^{16}y^{-4}}$

$$\frac{100x^{12}y^2}{20x^{16}y^{-4}} = \left(\frac{100}{20}\right)\left(\frac{x^{12}}{x^{16}}\right)\left(\frac{y^2}{y^{-4}}\right)$$
$$= 5x^{12-16}y^{2-(-4)}$$
$$= 5x^{-4}y^6$$
$$= \frac{5y^6}{x^4}$$

2c. $\dfrac{24x^3y^5}{32x^7y^{-9}}$

2d. $\left(\dfrac{5x}{y^4}\right)^{-2}$

$$\left(\frac{5x}{y^4}\right)^{-2} = \frac{(5x)^{-2}}{(y^4)^{-2}}$$
$$= \frac{5^{-2}x^{-2}}{y^{-6}}$$
$$= \frac{y^6}{5^2x^2}$$
$$= \frac{y^6}{25x^2}$$

2d. $\left(\dfrac{5x^3}{y}\right)^{-2}$

Objective #3: Use scientific notation.

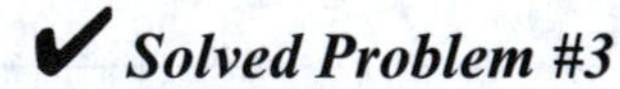

Solved Problem #3

3a. Write in decimal notation:
-2.6×10^9

Move the decimal point 9 places to the right.
$-2.6\times10^9 = -2,600,000,000$

Pencil Problem #3

3a. Write in decimal notation:
-7.16×10^6

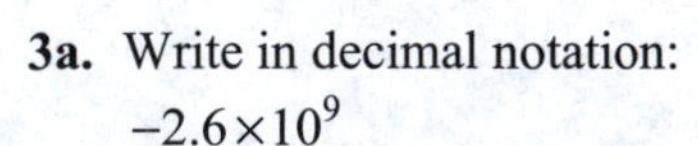

3b. Write in decimal notation:
3.017×10^{-6}

Move the decimal point 6 places to the left.
$3.017\times10^{-6}=0.000003017$

3b. Write in decimal notation:
7.9×10^{-1}

3c. Write in scientific notation: 5,210,000,000

The decimal point needs to be moved 9 places to the left.
$5{,}210{,}000{,}000=5.21\times10^{9}$

3c. Write in scientific notation: 32,000

3d. Write in scientific notation: –0.000000006893

The decimal point needs to be moved 8 places to the right.
$-0.00000006893=-6.893\times10^{-8}$

3d. Write in scientific notation: –0.00000000504

3e. Perform the indicated computation. Write the answer in scientific notation.

$(7.1\times10^{5})(5\times10^{-7})$

$$\begin{aligned}(7.1\times10^{5})(5\times10^{-7})&=(7.1\times5)\times10^{5+(-7)}\\&=35.5\times10^{-2}\\&=(3.55\times10^{1})\times10^{-2}\\&=3.55\times10^{-1}\end{aligned}$$

3e. Perform the indicated computation. Write the answer in scientific notation.

$(1.6\times10^{15})(4\times10^{-11})$

3f. Perform the indicated computation. Write the answer in scientific notation.

$$\frac{1.2\times10^{6}}{3\times10^{-3}}$$

$$\frac{1.2\times10^{6}}{3\times10^{-3}}=\frac{1.2}{3}\times10^{6-(-3)}$$

$$=0.4\times10^{9}$$

$$=(4\times10^{-1})\times10^{9}$$

$$=4\times10^{8}$$

3f. Perform the indicated computation. Write the answer in scientific notation.

$$\frac{2.4\times10^{-2}}{4.8\times10^{-6}}$$

Answers for Pencil Problems *(Textbook Exercise references in parentheses)*:

1a. x^{10} *(P.2 #27)* **1b.** $18x^{9}y^{5}$ *(P.2 #47)* **1c.** 16 *(P.2 #17)* **1d.** $-5a^{11}b$ *(P.2 #51)*

1e. 1 *(P.2 #7)* **1f.** $\frac{1}{64}$ *(P.2 #11)* **1g.** $\frac{1}{16x^{6}}$ *(P.2 #55)*

1h. 64 *(P.2 #15)* **1i.** $\frac{1}{x^{15}}$ *(P.2 #33)* **1j.** $64x^{6}$ *(P.2 #39)* **1k.** $-\frac{64}{x^{3}}$ *(P.2 #41)*

2a. $9x^{4}y^{10}$ *(P.2 #43)* **2b.** $6x^{11}$ *(P.2 #45)* **2c.** $\frac{3y^{14}}{4x^{4}}$ *(P.2 #57)* **2d.** $\frac{y^{2}}{25x^{6}}$ *(P.2 #59)*

3a. −7,160,000 *(P.2 #69)* **3b.** 0.79 *(P.2 #71)* **3c.** 3.2×10^{4} *(P.2 #77)* **3d.** -5.04×10^{-9} *(P.2 #85)*

3e. 6.4×10^{4} *(P.2 #89)* **3f.** 5×10^{3} *(P.2 #101)*

Section P.3
Radicals and Rational Exponents

Radicals in Space?

What does space travel have to do with radicals?

Imagine that in the future we will be able to travel at velocities approaching the speed of light
(approximately 186,000 miles per second). According to Einstein's theory of special relativity, time would pass more quickly on Earth than it would in the moving spaceship.

Objective #1: Evaluate square roots.

✔ ***Solved Problem #1***

1a. Evaluate $\sqrt{81}$.

$\sqrt{81} = 9$ Check: $9^2 = 81$

1b. Evaluate $-\sqrt{9}$.

$-\sqrt{9} = -3$ Check: $(-3)^2 = 9$

1c. Evaluate $\sqrt{\frac{1}{25}}$.

$\sqrt{\frac{1}{25}} = \frac{1}{5}$ Check: $\left(\frac{1}{5}\right)^2 = \frac{1}{25}$

1d. Evaluate $\sqrt{36+64}$.

$\sqrt{36+64} = \sqrt{100} = 10$

1a. Evaluate $\sqrt{36}$.

1b. Evaluate $-\sqrt{36}$.

1c. Evaluate $\sqrt{\frac{1}{81}}$.

1d. Evaluate $\sqrt{25-16}$.

1e. Evaluate $\sqrt{36}+\sqrt{64}$.

$\sqrt{36}+\sqrt{64}=6+8=14$

1e. Evaluate $\sqrt{25}-\sqrt{16}$.

Objective #2: Simplify expressions of the form $\sqrt{a^2}$.

Solved Problem #2

2. Evaluate $\sqrt{(-6)^2}$.

$\sqrt{(-6)^2}=|-6|=6$

Pencil Problem #2

2. Evaluate $\sqrt{(-13)^2}$.

Objective #3: Use the product rule to simplify square roots.

Solved Problem #3

3a. Simplify $\sqrt{75}$.

$\sqrt{75}=\sqrt{25\cdot 3}=\sqrt{25}\cdot\sqrt{3}=5\sqrt{3}$

3b. Simplify $\sqrt{5x}\cdot\sqrt{10x}$.

$$\begin{aligned}\sqrt{5x}\cdot\sqrt{10x}&=\sqrt{5x\cdot 10x}\\&=\sqrt{50x^2}\\&=\sqrt{25x^2\cdot 2}\\&=\sqrt{25x^2}\sqrt{2}\\&=5x\sqrt{2}\end{aligned}$$

Pencil Problem #3

3a. Simplify $\sqrt{50}$.

3b. Simplify $\sqrt{2x}\cdot\sqrt{6x}$.

Objective #4: Use the quotient rule to simplify square roots.

✔ Solved Problem #4

4a. Simplify $\sqrt{\frac{25}{16}}$.

$\sqrt{\frac{25}{16}} = \frac{\sqrt{25}}{\sqrt{16}} = \frac{5}{4}$

4b. Simplify $\frac{\sqrt{150x^3}}{\sqrt{2x}}$.

$$\begin{aligned}\frac{\sqrt{150x^3}}{\sqrt{2x}} &= \sqrt{\frac{150x^3}{2x}}\\ &= \sqrt{75x^2}\\ &= \sqrt{25x^2 \cdot 3}\\ &= \sqrt{25x^2}\sqrt{3}\\ &= 5x\sqrt{3}\end{aligned}$$

Pencil Problem #4

4a. Simplify $\sqrt{\frac{49}{16}}$.

4b. Simplify $\frac{\sqrt{48x^3}}{\sqrt{3x}}$.

Objective #5: Add and subtract square roots.

✔ Solved Problem #5

5a. Add: $8\sqrt{13}+9\sqrt{13}$.

$8\sqrt{13}+9\sqrt{13} = (8+9)\sqrt{13} = 17\sqrt{13}$

Pencil Problem #5

5a. Subtract: $6\sqrt{17x}-8\sqrt{17x}$.

5b. Subtract: $6\sqrt{18x}-4\sqrt{8x}$.

$$\begin{aligned}6\sqrt{18x}-4\sqrt{8x}&=6\sqrt{9\cdot 2x}-4\sqrt{4\cdot 2x}\\&=6\cdot 3\sqrt{2x}-4\cdot 2\sqrt{2x}\\&=18\sqrt{2x}-8\sqrt{2x}\\&=(18-8)\sqrt{2x}\\&=10\sqrt{2x}\end{aligned}$$

5b. Add: $3\sqrt{18}+5\sqrt{50}$.

Objective #6: Rationalize denominators.

Solved Problem #6

6a. Rationalize the denominator : $\dfrac{6}{\sqrt{12}}$.

Multiply by $\sqrt{3}$ to obtain the square root of a perfect square, $\sqrt{12}\cdot\sqrt{3}=\sqrt{36}$.

$$\frac{6}{\sqrt{12}}\cdot\frac{\sqrt{3}}{\sqrt{3}}=\frac{6\sqrt{3}}{\sqrt{36}}=\frac{6\sqrt{3}}{6}=\sqrt{3}$$

6b. Rationalize the denominator : $\dfrac{8}{4+\sqrt{5}}$.

Multiply by $4-\sqrt{5}$, the conjugate of $4+\sqrt{5}$.

$$\begin{aligned}\frac{8}{4+\sqrt{5}}\cdot\frac{4-\sqrt{5}}{4-\sqrt{5}}&=\frac{8(4-\sqrt{5})}{4^2-(\sqrt{5})^2}\\&=\frac{8(4-\sqrt{5})}{16-5}\\&=\frac{8(4-\sqrt{5})}{11}\text{ or }\frac{32-8\sqrt{5}}{11}\end{aligned}$$

Pencil Problem #6

6a. Rationalize the denominator : $\dfrac{\sqrt{2}}{\sqrt{5}}$.

6b. Rationalize the denominator : $\dfrac{7}{\sqrt{5}-2}$.

Objective #7: Evaluate and perform operations with higher roots.

✔ *Solved Problem #7*

7a. Simplify $\sqrt[5]{8}\cdot\sqrt[5]{8}$.

$$\begin{aligned}\sqrt[5]{8}\cdot\sqrt[5]{8} &= \sqrt[5]{8\cdot 8}\\ &= \sqrt[5]{64}\\ &= \sqrt[5]{32\cdot 2}\\ &= \sqrt[5]{32}\cdot\sqrt[5]{2}\\ &= 2\sqrt[5]{2}\end{aligned}$$

7b. Simplify $\sqrt[3]{\frac{125}{27}}$.

$$\sqrt[3]{\frac{125}{27}} = \frac{\sqrt[3]{125}}{\sqrt[3]{27}} = \frac{5}{3}$$

7c. Subtract: $3\sqrt[3]{81}-4\sqrt[3]{3}$.

$$\begin{aligned}3\sqrt[3]{81}-4\sqrt[3]{3} &= 3\sqrt[3]{27\cdot 3}-4\sqrt[3]{3}\\ &= 3\cdot 3\sqrt[3]{3}-4\sqrt[3]{3}\\ &= 9\sqrt[3]{3}-4\sqrt[3]{3}\\ &= (9-4)\sqrt[3]{3}\\ &= 5\sqrt[3]{3}\end{aligned}$$

Pencil Problem #7

7a. Simplify $\sqrt[3]{9}\cdot\sqrt[3]{6}$.

7b. Simplify $\frac{\sqrt[5]{64x^6}}{\sqrt[5]{2x}}$.

7c. Add: $5\sqrt[3]{16}+\sqrt[3]{54}$.

Objective #8: Understand and use rational exponents..

✔ *Solved Problem #8*	*Pencil Problem #8*
8a. Simplify $(-8)^{\frac{1}{3}}$. $(-8)^{\frac{1}{3}} = \sqrt[3]{-8} = -2$	**8a.** Simplify $36^{\frac{1}{2}}$.
8b. Simplify $32^{-\frac{2}{5}}$. $32^{-\frac{2}{5}} = \dfrac{1}{32^{\frac{2}{5}}} = \dfrac{1}{(\sqrt[5]{32})^2} = \dfrac{1}{2^2} = \dfrac{1}{4}$	**8b.** Simplify $125^{\frac{2}{3}}$.
8c. Simplify $\dfrac{20x^4}{5x^{\frac{3}{2}}}$. $\dfrac{20x^4}{5x^{\frac{3}{2}}} = \dfrac{20}{5} \cdot x^{4-\frac{3}{2}} = 4x^{\frac{8}{2}-\frac{3}{2}} = 4x^{\frac{5}{2}}$	**8c.** Simplify $(7x^{\frac{1}{3}})(2x^{\frac{1}{4}})$.
8d. Simplify $\sqrt[6]{x^3}$. $\sqrt[6]{x^3} = x^{\frac{3}{6}} = x^{\frac{1}{2}} = \sqrt{x}$	**8d.** Simplify $\sqrt[6]{x^4}$.

Answers for Pencil Problems *(Textbook Exercise references in parentheses)*:

1a. 6 *(P.3 #1)* **1b.** -6 *(P.3 #3)* **1c.** $\frac{1}{9}$ *(P.3 #23)* **1d.** 3 *(P.3 #7)* **1e.** 1 *(P.3 #9)*

2. 13 *(P.3 #11)* **3a.** $5\sqrt{2}$ *(P.3 #13)* **3b.** $2x\sqrt{3}$ *(P.3 #17)*

4a. $\frac{7}{4}$ *(P.3 #25)* **4b.** $4x$ *(P.3 #27)* **5a.** $-2\sqrt{17x}$ *(P.3 #35)* **5b.** $34\sqrt{2}$ *(P.3 #41)*

6a. $\frac{\sqrt{10}}{5}$ *(P.3 #47)* **6b.** $7(\sqrt{5}+2)$ *(P.3 #51)*

7a. $3\sqrt[3]{2}$ *(P.3 #71)* **7b.** $2x$ *(P.3 #73)* **7c.** $13\sqrt[3]{2}$ *(P.3 #77)*

8a. 6 *(P.3 #83)* **8b.** 25 *(P.3 #87)* **8c.** $14x^{\frac{7}{12}}$ *(P.3 #91)* **8d.** $\sqrt[3]{x^2}$ *(P.3 #105)*

Section P.4
Polynomials

What Are the Best Dimensions for a Box?

Many children get excited about gift boxes of all shapes and sizes, with the possible *exception* of clothing-sized boxes. (I must confess I dreaded boxes of that size.)

While completing the application exercises in this section of the textbook, we will use polynomials to model the dimensions of a box. We will then apply the concepts of this section to model the area of the box's base and its volume.

Objective #1: Understand the vocabulary of polynomials.

✔ *Solved Problem #1*

1. True or false: $7x^5-3x^3+8$ is a polynomial of degree 7 with three terms.

False. The expression $7x^5-3x^3+8$ is a polynomial with three terms, but its degree is 5, not 7.

1. True or false: $x^2-4x^3+9x-12x^4+63$ is a polynomial of degree 2 with five terms.

Objective #2: Add and subtract polynomials.

✔ *Solved Problem #2*

2a. Add:
$(-17x^3+4x^2-11x-5)+(16x^3-3x^2+3x-15)$.

$$(-17x^3+4x^2-11x-5)+(16x^3-3x^2+3x-15)$$
$$=(-17x^3+16x^3)+(4x^2-3x^2)+(-11x+3x)+(-5-15)$$
$$=-x^3+x^2+(-8x)+(-20)$$
$$=-x^3+x^2-8x-20$$

2a. Add: $(-6x^3+5x^2-8x+9)+(17x^3+2x^2-4x-13)$.

2b. Subtract:
$(13x^3 - 9x^2 - 7x + 1) - (-7x^3 + 2x^2 - 5x + 9)$.

$(13x^3 - 9x^2 - 7x + 1) - (-7x^3 + 2x^2 - 5x + 9)$
$= (13x^3 - 9x^2 - 7x + 1) + (7x^3 - 2x^2 + 5x - 9)$
$= (13x^3 + 7x^3) + (-9x^2 - 2x^2) + (-7x + 5x) + (1 - 9)$
$= 20x^3 + (-11x^2) + (-2x) + (-8)$
$= 20x^3 - 11x^2 - 2x - 8$

2b. Subtract:
$(17x^3 - 5x^2 + 4x - 3) - (5x^3 - 9x^2 - 8x + 11)$.

Objective #3: Multiply polynomials.

✔ *Solved Problem #3*

3. Multiply: $(5x - 2)(3x^2 - 5x + 4)$.

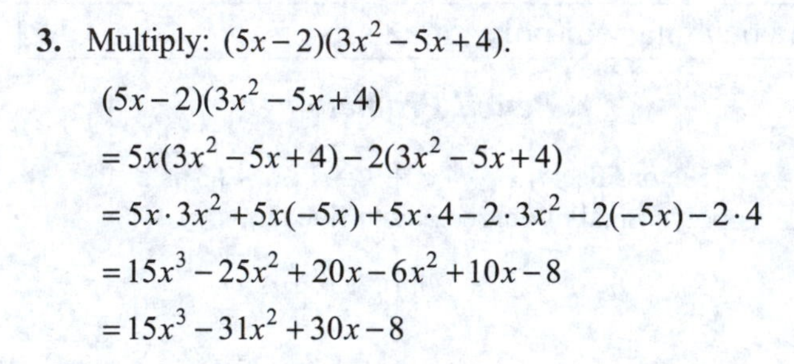

$(5x - 2)(3x^2 - 5x + 4)$
$= 5x(3x^2 - 5x + 4) - 2(3x^2 - 5x + 4)$
$= 5x \cdot 3x^2 + 5x(-5x) + 5x \cdot 4 - 2 \cdot 3x^2 - 2(-5x) - 2 \cdot 4$
$= 15x^3 - 25x^2 + 20x - 6x^2 + 10x - 8$
$= 15x^3 - 31x^2 + 30x - 8$

Pencil Problem #3

3. Multiply: $(2x - 3)(x^2 - 3x + 5)$.

Objective #4: Use FOIL in polynomial multiplication.

✔ *Solved Problem #4*

4. Multiply: $(7x - 5)(4x - 3)$.
Use FOIL.

First: $7x \cdot 4x$ Outside: $7x(-3)$
Inside: $-5 \cdot 4x$ Last: $-5(-3)$

$(7x - 5)(4x - 3)$
$= 7x \cdot 4x + 7x(-3) - 5 \cdot 4x - 5(-3)$
$= 28x^2 - 21x - 20x + 15$
$= 28x^2 - 41x + 15$

Pencil Problem #4

4. Multiply: $(3x + 5)(2x + 1)$.

Objective #5: Use special products in polynomial multiplication.

✔ *Solved Problem #5*

5a. Multiply: $(7x+8)(7x-8)$.

Use $(A+B)(A-B)=A^2-B^2$.

$$(7x+8)(7x-8)=(7x)^2-8^2$$
$$=49x^2-64$$

5b. Multiply: $(5x+4)^2$.

Use $(A+B)^2=A^2+2AB+B^2$.

$$(5x+4)^2=(5x)^2+2(5x)(4)+4^2$$
$$=25x^2+40x+16$$

5c. Multiply: $(x-9)^2$.

Use $(A-B)^2=A^2-2AB+B^2$.

$$(x-9)^2=x^2-2\cdot x\cdot 9+9^2$$
$$=x^2-18x+81$$

Pencil Problem #5

5a. Multiply: $(5-7x)(5+7x)$.

5b. Multiply: $(2x+3)^2$.

5c. Multiply: $(x-3)^2$.

Objective #6: Perform operations with polynomials in several variables.

✔ *Solved Problem #6*

6a. Multiply: $(7x-6y)(3x-y)$.

Use FOIL.

$$(7x-6y)(3x-y)$$
$$=7x\cdot 3x+7x(-y)-6y\cdot 3x-6y(-y)$$
$$=21x^2-7xy-18xy+6y^2$$
$$=21x^2-25xy+6y^2$$

Pencil Problem #6

6a. Multiply: $(x+5y)(7x+3y)$.

6b. Multiply: $(2x+4y)^2$. Use $(A+B)^2 = A^2 + 2AB + B^2$. $(2x+4y)^2 = (2x)^2 + 2(2x)(4y) + (4y)^2$ $= 4x^2 + 16xy + 16y^2$	**6b.** Multiply: $(7x+5y)^2$.
6c. Multiply: $(3x+2+5y)(3x+2-5y)$. Group terms and use $(A+B)(A-B) = A^2 - B^2$. $(3x+2+5y)(3x+2-5y)$ $= [(3x+2)+5y][(3x+2)-5y]$ $= (3x+2)^2 - (5y)^2$ $= (3x)^2 + 2(3x)(2) + 2^2 - 25y^2$ $= 9x^2 + 12x + 4 - 25y^2$	**6c.** Multiply: $(x+y+3)(x+y-3)$.
6d. Multiply: $(2x+y+3)^2$. Group terms and use $(A+B)^2 = A^2 + 2AB + B^2$. $(2x+y+3)^2 = [(2x+y)+3]^2$ $(2x+y)^2 + 2(2x+y)(3) + 3^2$ $= (2x)^2 + 2(2x)(y) + y^2 + 12x + 6y + 9$ $= 4x^2 + 4xy + y^2 + 12x + 6y + 9$	**6d.** Multiply: $(x+y+1)^2$.

Answers for Pencil Problems *(Textbook Exercise references in parentheses)*:

1. False *(P.4 #7)* **2a.** $11x^3 + 7x^2 - 12x - 4$ *(P.4 #9)* **2b.** $12x^3 + 4x^2 + 12x - 14$ *(P.4 #11)*
3. $2x^3 - 9x^2 + 19x - 15$ *(P.4 #17)* **4.** $6x^2 + 13x + 5$ *(P.4 #23)*
5a. $25 - 49x^2$ *(P.4 #35)* **5b.** $4x^2 + 12x + 9$ *(P.4 #43)* **5c.** $x^2 - 6x + 9$ *(P.4 #45)*
6a. $7x^2 + 38xy + 15y^2$ *(P.4 #59)* **6b.** $49x^2 + 70xy + 25y^2$ *(P.4 #65)*
6c. $x^2 + 2xy + y^2 - 9$ *(P.4 #73)* **6d.** $x^2 + 2xy + y^2 + 2x + 2y + 1$ *(P.4 #79)*

Section P.5
Factoring Polynomials

What's the sales price?

Many times retailers advertise their discounts in terms of percentages by which the price is reduced, such as 30% off. If a product still doesn't sell, the retailer may offer an additional 30% off the price that has already been reduced by 30%.

In this section's Exercise Set, you will see how the 30% discount followed by another 30% discount can be expressed as a polynomial. By factoring the polynomial and simplifying, you will see that our double discount means that we pay 49% of the original price.

Objective #1: Factor out the greatest common factor.

✔ ***Solved Problem #1***

1a. Factor $10x^3 - 4x^2$.

2 is the greatest integer that divides 10 and 4. x^2 is the greatest expression that divides x^3 and x^2. The GCF is $2x^2$.

$$\begin{aligned} 10x^3 - 4x^2 &= 2x^2(5x) - 2x^2(2) \\ &= 2x^2(5x - 2) \end{aligned}$$

1b. Factor $2x(x-7)+3(x-7)$.

The GCF is the binomial factor $(x - 7)$.

$$2x(x-7)+3(x-7)=(x-7)(2x+3)$$

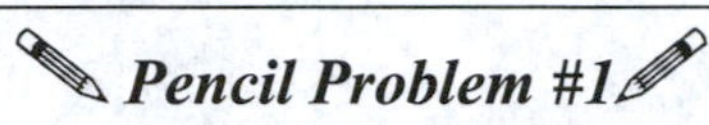

1a. Factor $3x^2 + 6x$.

1b. Factor $x(x+5)+3(x+5)$.

Objective #2: Factor by grouping.

✔ *Solved Problem #2*

2. Factor $x^3+5x^2-2x-10$.

The GCF of the first two terms is x^2, and the GCF of the last two terms is −2. After factoring out these GCFs, factor out the common binomial factor.

$$\begin{aligned} x^3+5x^2-2x-10 &= (x^3+5x^2)+(-2x-10) \\ &= x^2(x+5)-2(x+5) \\ &= (x+5)(x^2-2) \end{aligned}$$

Pencil Problem #2

2. Factor $x^3-2x^2+5x-10$.

Objective #3: Factor trinomials.

✔ *Solved Problem #3*

3a. Factor $x^2-5x-14$.

The leading coefficient is 1. We look for factors of −14 that sum to −5.

−7(2) = −14 and −7 + 2 = −5
The numbers are −7 and 2.
$x^2-5x-14=(x-7)(x+2)$

Pencil Problem #3

3a. Factor $x^2-8x+15$.

3b. Factor $6x^2 + 19x - 7$.

The leading coefficient is 6, not 1. $6x^2$ factors as $6x(x)$ or $3x(2x)$. -7 factors as $-7(1)$ or $7(-1)$.

The possible factorizations are

$(6x-7)(x+1)$	$(6x+1)(x-7)$
$(6x+7)(x-1)$	$(6x-1)(x+7)$
$(3x-7)(2x+1)$	$(3x+1)(2x-7)$
$(3x+7)(2x-1)$	$(3x-1)(2x+7)$

We want the combination, if there is one, that results in a sum of Outside and Inside terms of $19x$. Compute the sums of the Outside and Inside terms in the possible factorizations until you find one that results in $19x$.

For $(3x-1)(2x+7)$:
Outside: $3x(7) = 21x$
Inside: $-1(2x) = -2x$
Sum: $21x + (-2x) = 19x$
So, $6x^2 + 19x - 7 = (3x-1)(2x+7)$.

3b. Factor $9x^2 - 9x + 2$.

Objective #4: Factor the difference of squares.

✔ *Solved Problem #4*

4. Factor: $36x^2 - 25$.

Note that $36x^2 = (6x)^2$ and $25 = 5^2$ can both be expressed as squares.

Use $A^2 - B^2 = (A+B)(A-B)$.

$$36x^2 - 25 = (6x)^2 - 5^2$$
$$= (6x+5)(6x-5)$$

Pencil Problem #4

4. Factor $9x^2 - 25y^2$.

Objective #5: Factor perfect square trinomials.

✔ *Solved Problem #5*

5a. Factor $x^2 + 14x + 49$.

Note that the first term is the square of x, the last term is the square of 7, and the middle term is twice the product of x and 7.

Factor using $A^2 + 2AB + B^2 = (A+B)^2$.

$$x^2 + 14x + 49 = x^2 + 2 \cdot x \cdot 7 + 7^2$$
$$= (x+7)^2$$

5b. Factor $16x^2 - 56x + 49$.

Note that the first term is the square of $4x$, the last term is the square of 7, and the middle term is twice the product of $4x$ and 7.
Factor using $A^2 - 2AB + B^2 = (A-B)^2$.

$$16x^2 - 56x + 49 = (4x)^2 - 2 \cdot 4x \cdot 7 + 7^2$$
$$= (4x-7)^2$$

Pencil Problem #5

5a. Factor $x^2 + 2x + 1$.

5b. Factor $9x^2 - 6x + 1$.

Objective #6: Factor the sum or difference of two cubes.

✔ *Solved Problem #6*

6a. Factor $x^3 + 1$.

Note that both terms can be expressed as cubes.
Factor using $A^3 + B^3 = (A+B)(A^2 - AB + B^2)$.

$$x^3 + 1 = x^3 + 1^3$$
$$= (x+1)(x^2 - x \cdot 1 + 1^2)$$
$$= (x+1)(x^2 - x + 1)$$

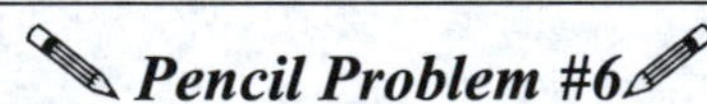

Pencil Problem #6

6a. Factor $x^3 + 27$.

6b. Factor $125x^3 - 8$.

Note that the first term is the cube of $5x$ and the second term is the cube of 2.

Factor using $A^3 - B^3 = (A-B)(A^2 + AB + B^2)$.

$$\begin{aligned} 125x^3 - 8 &= (5x)^3 - 2^3 \\ &= (5x-2)[(5x)^2 + 5x \cdot 2 + 2^2] \\ &= (5x-2)(25x^2 + 10x + 4) \end{aligned}$$

6b. Factor $8x^3 - 1$.

Objective #7: Use a general strategy for factoring polynomials..

✔ ***Solved Problem #7***

7a. Factor $3x^3 - 30x^2 + 75x$.

First, factor out the GCF, $3x$.

$3x^3 - 30x^2 + 75x = 3x(x^2 - 10x + 25)$

Now factor the trinomial. Find factors of 25 that sum to −10, or use the formula for a perfect square trinomial, $A^2 - 2AB + B^2 = (A-B)^2$.

$$\begin{aligned} 3x^3 - 30x^2 + 75x &= 3x(x^2 - 10x + 25) \\ &= 3x(x^2 - 2 \cdot x \cdot 5 + 5^2) \\ &= 3x(x-5)^2 \end{aligned}$$

7b. Factor $x^2 - 36a^2 + 20x + 100$.

Regroup factors and look for opportunities to factor within groupings.

$x^2 - 36a^2 + 20x + 100 = (x^2 + 20x + 100) - 36a^2$

Factor the expression in parentheses using $A^2 + 2AB + B^2 = (A+B)^2$.

$$\begin{aligned} (x^2 + 20x + 100) - 36a^2 &= (x^2 + 2 \cdot x \cdot 10 + 10^2) - 36a^2 \\ &= (x+10)^2 - 36a^2 \end{aligned}$$

Pencil Problem #7

7a. Factor $20y^4 - 45y^2$.

7b. Factor $x^2 - 12x + 36 - 49y^2$.

This last form is the difference of squares. Factor using $A^2 - B^2 = (A+B)(A-B)$.

$$\begin{aligned}(x+10)^2 - 36a^2 &= (x+10)^2 - (6a)^2\\ &= [(x+10)+6a][(x+10)-6a]\\ &= (x+10+6a)(x+10-6a)\end{aligned}$$

So, $x^2 - 36a^2 + 20x + 100 = (x+10+6a)(x+10-6a)$.

Objective #8: Factor algebraic expressions containing fractional and negative exponents.

✔ *Solved Problem #8*

8. Factor and simplify: $x(x-1)^{-\frac{1}{2}} + (x-1)^{\frac{1}{2}}$.

The GCF is $(x-1)$ with the smaller exponent. Thus, the GCF is $(x-1)^{-\frac{1}{2}}$.

$$\begin{aligned}x(x-1)^{-\frac{1}{2}} + (x-1)^{\frac{1}{2}} &= (x-1)^{-\frac{1}{2}}\cdot x + (x-1)^{-\frac{1}{2}}\cdot(x-1)\\ &= (x-1)^{-\frac{1}{2}}[x+(x-1)]\\ &= (x-1)^{-\frac{1}{2}}(2x-1)\\ &= \frac{2x-1}{(x-1)^{1/2}}\end{aligned}$$

Pencil Problem #8

8. Factor and simplify: $(x+3)^{\frac{1}{2}} - (x+3)^{\frac{3}{2}}$.

Answers for Pencil Problems *(Textbook Exercise references in parentheses)*:

1a. $3x(x+2)$ *(P.5 #3)* **1b.** $(x+5)(x+3)$ *(P.5 #7)* **2.** $(x-2)(x^2+5)$ *(P.5 #11)*
3a. $(x-5)(x-3)$ *(P.5 #21)* **3b.** $(3x-2)(3x-1)$ *(P.5 #31)*
4. $(3x-5y)(3x+5y)$ *(P.5 #43)* **5a.** $(x+1)^2$ *(P.5 #49)* **5b.** $(3x-1)^2$ *(P.5 #55)*
6a. $(x+3)(x^2-3x+9)$ *(P.5 #57)* **6b.** $(2x-1)(4x^2+2x+1)$ *(P.5 #61)*
7a. $5y^2(2y+3)(2y-3)$ *(P.5 #83)* **7b.** $(x-6+7y)(x-6-7y)$ *(P.5 #85)* **8.** $-(x+3)^{\frac{1}{2}}(x+2)$ *(P.5 #97)*

Section P.6
Rational Expressions

Ouch! That Hurts!!

Though it may not be fun to get a flu shot, it is a great way to protect yourself from getting sick!

In this section of the textbook, one of the application problems will explore the costs for inoculating various percentages of the population.

Objective #1: Specify numbers that must be excluded from the domain of a rational expression.

✔ Solved Problem #1

1. Find all real numbers that must be excluded from the domain of each rational expression.

$$\frac{7x}{x^2-5x-14}$$

Factor the denominator.

$$x^2-5x-14=(x-7)(x+2)$$

The first factor would be 0 if $x=7$. The second factor would be 0 if $x=-2$. We must exclude -2 and 7 from the domain.

Pencil Problem #1

1. Find all real numbers that must be excluded from the domain of each rational expression.

$$\frac{x+5}{x^2-25}$$

Objective #2: Simplify rational expressions.

✔ Solved Problem #2

2a. Simplify $\frac{x^3+3x^2}{x+3}$.

Note that $x \neq -3$ since -3 would make the denominator 0.

Factor the numerator and divide out common factors.

$$\frac{x^3+3x^2}{x+3}=\frac{x^2(x+3)}{x+3}=\frac{x^2\cancel{(x+3)}}{\cancel{x+3}}$$

$$=x^2,\ x \neq -3$$

Pencil Problem #2

2a. Simplify $\frac{3x-9}{x^2-6x+9}$.

2b. Simplify $\dfrac{x^2-1}{x^2+2x+1}$.

By factoring the denominator, $x^2+2x+1=(x+1)^2$, we see that $x \neq -1$.

Factor the numerator and denominator and divide out common factors.

$$\frac{x^2-1}{x^2+2x+1}=\frac{(x+1)(x-1)}{(x+1)(x+1)}=\frac{\cancel{(x+1)}(x-1)}{\cancel{(x+1)}(x+1)}$$

$$=\frac{x-1}{x+1},\ x \neq -1$$

2b. Simplify $\dfrac{y^2+7y-18}{y^2-3y+2}$.

Objective #3: Multiply rational expressions.

✔ *Solved Problem #3*

3. Multiply: $\dfrac{x+3}{x^2-4}\cdot\dfrac{x^2-x-6}{x^2+6x+9}$.

Factor and divide by common factors.

$$\frac{x+3}{x^2-4}\cdot\frac{x^2-x-6}{x^2+6x+9}=\frac{x+3}{(x+2)(x-2)}\cdot\frac{(x-3)(x+2)}{(x+3)(x+3)}$$

$$=\frac{\cancel{x+3}}{\cancel{(x+2)}(x-2)}\cdot\frac{(x-3)\cancel{(x+2)}}{\cancel{(x+3)}(x+3)}$$

$$=\frac{x-3}{(x-2)(x+3)},\ x \neq -3,-2,2$$

To see which values must be excluded from the domain, look at the factored forms of the denominators in the second step.

Pencil Problem #3

3. Multiply: $\dfrac{x^2-5x+6}{x^2-2x-3}\cdot\dfrac{x^2-1}{x^2-4}$.

Objective #4: Divide rational expressions.

✔ *Solved Problem #4*

4. Divide: $\dfrac{x^2-2x+1}{x^3+x}\div\dfrac{x^2+x-2}{3x^2+3}$.

Invert the divisor and multiply.

$$\frac{x^2-2x+1}{x^3+x}\div\frac{x^2+x-2}{3x^2+3}=\frac{x^2-2x+1}{x^3+x}\cdot\frac{3x^2+3}{x^2+x-2}$$

$$=\frac{(x-1)\cancel{(x-1)}}{x\cancel{(x^2+1)}}\cdot\frac{3\cancel{(x^2+1)}}{(x+2)\cancel{(x-1)}}$$

$$=\frac{3(x-1)}{x(x+2)},\ x \neq -2,0,1$$

Pencil Problem #4

4. Divide: $\dfrac{x^2-25}{2x-2}\div\dfrac{x^2+10x+25}{x^2+4x-5}$.

Objective #5: Add and subtract rational expressions.

✔ *Solved Problem #5*

5a. Subtract: $\frac{x}{x+1}-\frac{3x+2}{x+1}$.

The expressions have the same denominator. Subtract numerators.

$$\frac{x}{x+1}-\frac{3x+2}{x+1}=\frac{x-(3x+2)}{x+1}$$
$$=\frac{x-3x-2}{x+1}$$
$$=\frac{-2x-2}{x+1}$$
$$=\frac{-2(x+1)}{x+1}$$
$$=-2,\ x\neq -1$$

5b. Subtract: $\frac{x}{x^2-10x+25}-\frac{x-4}{2x-10}$.

Factor the denominators.

$x^2-10x+25=(x-5)(x-5)$

$2x-10=2(x-5)$

LCD $=2(x-5)(x-5)$

$$\frac{x}{x^2-10x+25}-\frac{x-4}{2x-10}$$
$$=\frac{x}{(x-5)(x-5)}-\frac{x-4}{2(x-5)}$$
$$=\frac{2x}{2(x-5)(x-5)}-\frac{(x-4)(x-5)}{2(x-5)(x-5)}$$
$$=\frac{2x-(x-4)(x-5)}{2(x-5)^2}$$
$$=\frac{2x-(x^2-9x+20)}{2(x-5)^2}$$
$$=\frac{2x-x^2+9x-20}{2(x-5)^2}$$
$$=\frac{-x^2+11x-20}{2(x-5)^2},\ x\neq 5$$

✎ *Pencil Problem #5* ✎

5a. Add: $\frac{4x+1}{6x+5}+\frac{8x+9}{6x+5}$.

5b. Subtract: $\frac{3x}{x^2+3x-10}-\frac{2x}{x^2+x-6}$.

5c. Add: $\frac{x}{x-3}+\frac{x-1}{x+3}$.

The denominators are not equal and have no common factor.

$LCD=(x-3)(x+3)$

$$\frac{x}{x-3}+\frac{x-1}{x+3}=\frac{x(x+3)}{(x-3)(x+3)}+\frac{(x-1)(x-3)}{(x+3)(x-3)}$$

$$=\frac{x(x+3)+(x-1)(x-3)}{(x-3)(x+3)}$$

$$=\frac{x^2+3x+x^2-4x+3}{(x-3)(x+3)}$$

$$=\frac{2x^2-x+3}{(x-3)(x+3)},\ x\neq 3,-3$$

5c. Add: $\frac{3}{x+4}+\frac{6}{x+5}$.

Objective #6: Simplify complex rational expressions.

✔ *Solved Problem #6*

6a. Simplify: $\dfrac{\frac{1}{x}-\frac{3}{2}}{\frac{1}{x}+\frac{3}{4}}$.

Subtract and add in the numerator and denominator to obtain a single rational expression in each.

$$\frac{1}{x}-\frac{3}{2}=\frac{1\cdot 2}{x\cdot 2}-\frac{3\cdot x}{2\cdot x}=\frac{2}{2x}-\frac{3x}{2x}=\frac{2-3x}{2x}$$

$$\frac{1}{x}+\frac{3}{4}=\frac{1\cdot 4}{x\cdot 4}+\frac{3\cdot x}{4\cdot x}=\frac{4}{4x}+\frac{3x}{4x}=\frac{4+3x}{4x}$$

Now return to the original complex fraction.

$$\frac{\frac{1}{x}-\frac{3}{2}}{\frac{1}{x}+\frac{3}{4}}=\frac{\frac{2-3x}{2x}}{\frac{4+3x}{4x}}$$

$$=\frac{2-3x}{2x}\cdot\frac{4x}{4+3x}$$

$$=\frac{2-3x}{\cancel{2x}}\cdot\frac{2\cdot\cancel{2x}}{4+3x}$$

$$=\frac{2(2-3x)}{4+3x},\ x\neq 0,-\frac{4}{3}$$

✎ *Pencil Problem #6* ✎

6a. Simplify: $\dfrac{1+\frac{1}{x}}{3-\frac{1}{x}}$.

6b. Simplify: $\dfrac{\frac{1}{x+7}-\frac{1}{x}}{7}$.

The LCD of the fractions within the complex fraction is $x(x+7)$. Multiply the numerator and the denominator of the complex fraction by the LCD.

$$\frac{\frac{1}{x+7}-\frac{1}{x}}{7}=\frac{\left(\frac{1}{x+7}-\frac{1}{x}\right)x(x+7)}{7x(x+7)}$$

$$=\frac{\frac{1}{x+7}\cdot x(x+7)-\frac{1}{x}\cdot x(x+7)}{7x(x+7)}$$

$$=\frac{x-(x+7)}{7x(x+7)}$$

$$=\frac{-7}{7x(x+7)}$$

$$=\frac{-1}{x(x+7)},\ x\neq -7,0$$

6b. Simplify: $\dfrac{\frac{3}{x-2}-\frac{4}{x+2}}{\frac{7}{x^2-4}}$.

Objective #7: Simplify rational expressions that occur in calculus.

✔ *Solved Problem #7*

7. Simplify: $\dfrac{\sqrt{x}+\frac{1}{\sqrt{x}}}{x}$.

Multiply the numerator and denominator by $\sqrt{x}$.

$$\frac{\sqrt{x}+\frac{1}{\sqrt{x}}}{x}=\frac{\left(\sqrt{x}+\frac{1}{\sqrt{x}}\right)}{x}\cdot\frac{\sqrt{x}}{\sqrt{x}}$$

$$=\frac{\sqrt{x}\sqrt{x}+\frac{1}{\sqrt{x}}\sqrt{x}}{x\sqrt{x}}$$

$$=\frac{x+1}{\sqrt{x^3}}$$

Note that the simplification in the denominator above can be done as follows:

$$x\sqrt{x}=x\cdot x^{\frac{1}{2}}=x^{1+\frac{1}{2}}=x^{\frac{3}{2}}=\sqrt{x^3}.$$

✎ *Pencil Problem #7* ✎

7. Simplify: $\dfrac{\sqrt{x}-\frac{1}{3\sqrt{x}}}{\sqrt{x}}$.

Objective #8: Rationalize numerators.

✔ *Solved Problem #8*

8. Rationalize the numerator: $\dfrac{\sqrt{x+3}-\sqrt{x}}{3}$.

Multiply the numerator and the denominator by the conjugate of the numerator.

$$\frac{\sqrt{x+3}-\sqrt{x}}{3}=\frac{\sqrt{x+3}-\sqrt{x}}{3}\cdot\frac{\sqrt{x+3}+\sqrt{x}}{\sqrt{x+3}+\sqrt{x}}$$
$$=\frac{(\sqrt{x+3})^2-(\sqrt{x})^2}{3(\sqrt{x+3}+\sqrt{x})}$$
$$=\frac{x+3-x}{3(\sqrt{x+3}+\sqrt{x})}$$
$$=\frac{3}{3(\sqrt{x+3}+\sqrt{x})}$$
$$=\frac{1}{\sqrt{x+3}+\sqrt{x}}$$

✎ *Pencil Problem #8* ✎

8. Rationalize the numerator: $\dfrac{\sqrt{x+5}-\sqrt{x}}{5}$.

Answers for Pencil Problems *(Textbook Exercise references in parentheses)*:

1. $-5, 5$ *(P.6 #3)* **2a.** $\dfrac{3}{x-3},\ x\neq 3$ *(P.6 #7)* **2b.** $\dfrac{y+9}{y-1},\ y\neq 1,2$ *(P.6 #11)*

3. $\dfrac{x-1}{x+2},\ x\neq -2,-1,2,3$ *(P.6 #19)* **4.** $\dfrac{x-5}{2},\ x\neq -5,1$ *(P.6 #29)*

5a. $2,\ x\neq -\dfrac{5}{6}$ *(P.6 #33)* **5b.** $\dfrac{x^2-x}{(x+5)(x-2)(x+3)},\ x\neq -5,-3,2$ *(P.6 #53)*

5c. $\dfrac{9x+39}{(x+4)(x+5)},\ x\neq -5,-4$ *(P.6 #41)*

6a. $\dfrac{x+1}{3x-1},\ x\neq 0,\dfrac{1}{3}$ *(P.6 #61)* **6b.** $-\dfrac{x-14}{7},\ x\neq -2,2$ *(P.6 #67)*

7. $1-\dfrac{1}{3x}, x>0$ *(P.6 #73)* **8.** $\dfrac{1}{\sqrt{x+5}+\sqrt{x}}$ *(P.6 #79)*

Section P.7
Equations

Maybe I Should Ride the Bus Instead

Did you know that the likelihood that a driver will be involved in a fatal crash decreases with age until about age 45 and then increases after that? Formulas that model data that first decrease and then increase contain a variable squared. When we use these models to answer questions about the data, we often need to find the solutions of a *quadratic equation*.

Quadratic equations may have exactly two distinct solutions. Thus, when we find the age at which drivers are involved in 3 fatal crashes per 100 million miles driven, we will find two different ages, one less 45 and the other greater than 45.

Objective #1: Solve linear equations in one variable

Solved Problem #1

1. Solve and check: $4(2x+1)=29+3(2x-5)$

Simplify the algebraic expression on each side.

$$4(2x+1)=29+3(2x-5)$$
$$8x+4=29+6x-15$$
$$8x+4=6x+14$$

Collect variable terms on one side and constant terms on the other side.

$$8x-6x+4=6x-6x+14$$
$$2x+4=14$$
$$2x+4-4=14-4$$

Isolate the variable and solve.

$$\frac{2x}{2}=\frac{10}{2}$$
$$x=5$$

Check:

$$4(2x+1)=29+3(2x-5)$$
$$4(2\cdot 5+1)=29+3(2\cdot 5-5)$$
$$4(11)=29+3(5)$$
$$44=44$$

The solution set is $\{5\}$.

Pencil Problem #1

1. Solve and check: $3(x-2)+7=2(x+5)$

Objective #2: Solve linear equations containing fractions.

✔ *Solved Problem #2*

2. Solve and check: $\frac{x-3}{4}=\frac{5}{14}-\frac{x+5}{7}$

The LCD is 28.

$$\frac{x-3}{4}=\frac{5}{14}-\frac{x+5}{7}$$

$$28\left(\frac{x-3}{4}\right)=28\left(\frac{5}{14}-\frac{x+5}{7}\right)$$

$$\frac{28}{1}\left(\frac{x-3}{4}\right)=\frac{28}{1}\left(\frac{5}{14}\right)-\frac{28}{1}\left(\frac{x+5}{7}\right)$$

$$7(x-3)=2(5)-4(x+5)$$

$$7x-21=10-4x-20$$

$$7x-21=-4x-10$$

$$7x+4x-21=-4x+4x-10$$

$$11x-21=-10$$

$$11x-21+21=-10+21$$

$$11x=11$$

$$\frac{11x}{11}=\frac{11}{11}$$

$$x=1$$

Check:

$$\frac{x-3}{4}=\frac{5}{14}-\frac{x+5}{7}$$

$$\frac{1-3}{4}=\frac{5}{14}-\frac{1+5}{7}$$

$$\frac{-2}{4}=\frac{5}{14}-\frac{6}{7}$$

$$-\frac{1}{2}=-\frac{1}{2}$$

The solution set is $\{1\}$.

Pencil Problem #2

2. Solve and check: $\frac{x+3}{6}=\frac{3}{8}+\frac{x-5}{4}$

Objective #3: Solve rational equations with variables in denominators.

✔ *Solved Problem #3a*

3a. Solve: $\frac{6}{x+3}-\frac{5}{x-2}=\frac{-20}{x^2+x-6}$

By factoring $x^2+x-6=(x+3)(x-2)$, we see that the LCD is $(x+3)(x-2)$ and that $x \neq -3$ and $x \neq 2$.

$$\frac{6}{x+3}-\frac{5}{x-2}=\frac{-20}{x^2+x-6},\ x \neq -3,\ x \neq 2$$

$$(x+3)(x-2)\left(\frac{6}{x+3}-\frac{5}{x-2}\right)=(x+3)(x-2)\cdot\frac{-20}{(x+3)(x-2)}$$

$$\cancel{(x+3)}(x-2)\cdot\frac{6}{\cancel{x+3}}-(x+3)\cancel{(x-2)}\cdot\frac{5}{\cancel{x-2}}=\cancel{(x+3)}\cancel{(x-2)}\cdot\frac{-20}{\cancel{(x+3)}\cancel{(x-2)}}$$

$$6(x-2)-5(x+3)=-20$$

$$6x-12-5x-15=-20$$

$$x-27=-20$$

$$x-27+27=-20+27$$

$$x=7$$

You can check the proposed solution x = 7 in the original equation. Since $x=7$ is not part of the restriction $x \neq -3$ and $x \neq 2$, the solution set is $\{7\}$.

✎ *Pencil Problem #3a* ✎

3a. Solve: $\frac{2}{x+1}-\frac{1}{x-1}=\frac{2x}{x^2-1}$

✔ *Solved Problem #3b*

3b. Solve: $\dfrac{1}{x+2}=\dfrac{4}{x^2-4}-\dfrac{1}{x-2}$

By factoring $x^2-4=(x+2)(x-2)$, we see that the LCD is $(x+2)(x-2)$ and that $x\neq -2$ and $x\neq 2$.

$$\frac{1}{x+2}=\frac{4}{x^2-4}-\frac{1}{x-2},\ x\neq -2,\ x\neq 2$$

$$(x+2)(x-2)\cdot\frac{1}{x+2}=(x+2)(x-2)\left(\frac{4}{(x+2)(x-2)}-\frac{1}{x-2}\right)$$

$$\cancel{(x+2)}(x-2)\cdot\frac{1}{\cancel{x+2}}=\cancel{(x+2)}\,\cancel{(x-2)}\cdot\frac{4}{\cancel{(x+2)}\,\cancel{(x-2)}}-(x+2)\cancel{(x-2)}\cdot\frac{1}{\cancel{x-2}}$$

$$\begin{aligned}
1(x-2)&=4-1(x+2)\\
x-2&=4-x-2\\
x-2&=-x+2\\
x-2+x&=-x+2+x\\
2x-2&=2\\
2x-2+2&=2+2\\
2x&=4\\
\frac{2x}{2}&=\frac{4}{2}\\
x&=2
\end{aligned}$$

Since $x=2$ is part of the restriction $x\neq -2$ and $x\neq 2$, the solution set is $\varnothing$, the empty set.

Pencil Problem #3b

3b. Solve: $\dfrac{1}{x-4}-\dfrac{5}{x+2}=\dfrac{6}{x^2-2x-8}$

Objective #4: Solve a formula for a variable.

✔ *Solved Problem #4*

4. Solve for q: $\frac{1}{p}+\frac{1}{q}=\frac{1}{f}$

The LCD is pqf.

$$\frac{1}{p}+\frac{1}{q}=\frac{1}{f}$$

$$pqf\left(\frac{1}{p}+\frac{1}{q}\right)=pqf\cdot\frac{1}{f}$$

$$\cancel{p}qf\cdot\frac{1}{\cancel{p}}+p\cancel{q}f\cdot\frac{1}{\cancel{q}}=pq\cancel{f}\cdot\frac{1}{\cancel{f}}$$

$$qf+pf=pq$$

$$qf+pf-qf=pq-qf$$

$$pf=q(p-f)$$

$$\frac{pf}{p-f}=\frac{q(p-f)}{p-f}$$

$$\frac{pf}{p-f}=q \text{ or } q=\frac{pf}{p-f}$$

Pencil Problem #4

4. Solve for f: $\frac{1}{p}+\frac{1}{q}=\frac{1}{f}$

Objective #5: Solve equations involving absolute value.

✔ *Solved Problem #5*

5. Solve: $4|1-2x|-20=0$

First isolate the absolute value expression.

$$4|1-2x|-20=0$$

$$4|1-2x|=20$$

$$|1-2x|=5$$

Now rewrite $|u|=c$ as $u=c$ or $u=-c$.

$1-2x=5$ or $1-2x=-5$

$-2x=4$ $\quad -2x=-6$

$x=-2$ $\quad x=3$

You can check -2 and 3 in the original equation. The solution set is $\{-2, 3\}$.

Pencil Problem #5

5. Solve: $2|3x-2|=14$

Objective #6: Solve quadratic equations by factoring.

✔ Solved Problem #6

6. Solve by factoring.

6a. $3x^2 - 9x = 0$

$$3x^2 - 9x = 0$$
$$3x(x-3) = 0$$
$$3x = 0 \text{ or } x - 3 = 0$$
$$x = 0 \text{ or } \quad x = 3$$

The solution set is $\{0, 3\}$.

6b. $2x^2 + x = 1$

$$2x^2 + x = 1$$
$$2x^2 + x - 1 = 0$$
$$(2x-1)(x+1) = 0$$
$$2x - 1 = 0 \text{ or } x + 1 = 0$$
$$2x = 1 \text{ or } \quad x = -1$$
$$x = \frac{1}{2}$$

The solution set is $\left\{-1, \frac{1}{2}\right\}$.

Pencil Problem #6

6. Solve by factoring.

6a. $5x^2 = 20x$

6b. $x^2 = 8x - 15$

Objective #7: Solve quadratic equations by the square root property.

✔ Solved Problem #7

7. Solve by the square root property.

7a. $3x^2 - 21 = 0$

$$3x^2 - 21 = 0$$
$$3x^2 = 21$$
$$x^2 = 7$$
$$x = \pm\sqrt{7}$$

The solution set is $\left\{-\sqrt{7}, \sqrt{7}\right\}$.

Pencil Problem #7

7. Solve by the square root property.

7a. $5x^2 + 1 = 51$

7b. $(x+5)^2 = 11$

$$(x+5)^2 = 11$$
$$x+5 = \pm\sqrt{11}$$
$$x = -5 \pm \sqrt{11}$$

The solution set is $\{-5+\sqrt{11},\ -5-\sqrt{11}\}$.

7b. $3(x-4)^2 = 15$

Objective #8: Solve quadratic equations by completing the square.

Solved Problem #8

8. Solve by completing the square: $x^2+4x-1=0$

$$x^2+4x-1=0$$
$$x^2+4x \quad = 1$$

Half of 4 is 2, and 2^2 is 4, which should be added to both sides.

$$x^2+4x+4=1+4$$
$$x^2+4x+4=5$$
$$(x+2)^2=5$$
$$x+2=\sqrt{5} \quad \text{or} \quad x+2=-\sqrt{5}$$
$$x=-2+\sqrt{5} \qquad x=-2-\sqrt{5}$$

The solution set is $\{-2 \pm \sqrt{5}\}$.

Pencil Problem #8

8. Solve by completing the square: $x^2+6x-11=0$

Objective #9: Solve quadratic equations using the quadratic formula.

✔ *Solved Problem #9*

9. Solve using the quadratic formula: $2x^2 + 2x - 1 = 0$

The equation is in the form $ax^2 + bx + c = 0,$ where $a = 2$, $b = 2$, and $c = -1$.

$$x = \frac{-b \pm \sqrt{b^2 - 4ac}}{2a}$$
$$= \frac{-2 \pm \sqrt{2^2 - 4(2)(-1)}}{2(2)}$$
$$= \frac{-2 \pm \sqrt{4+8}}{4}$$
$$= \frac{-2 \pm \sqrt{12}}{4}$$
$$= \frac{-2 \pm 2\sqrt{3}}{4}$$
$$= \frac{2(-1 \pm \sqrt{3})}{4}$$
$$= \frac{-1 \pm \sqrt{3}}{2}$$

The solution set is $\left\{\frac{-1+\sqrt{3}}{2}, \frac{-1-\sqrt{3}}{2}\right\}$.

Pencil Problem #9

9. Solve using the quadratic formula: $3x^2 - 3x - 4 = 0$

Objective #10: Use the discriminant to determine the number and type of solutions.

✔ *Solved Problem #10*

10. Compute the discriminant and determine the number and type of solutions: $3x^2 - 2x + 5 = 0$

$$b^2 - 4ac = (-2)^2 - 4(3)(-5)$$
$$= -56$$

Since the discriminant is negative, there is no real solution.

Pencil Problem #10

10. Compute the discriminant and determine the number and type of solutions: $2x^2 - 11x + 3 = 0$

Objective #11: Determine the most efficient method to use when solving a quadratic equation.

✔ *Solved Problem #11*

11. What is the most efficient method for solving a quadratic equation of the form $ax^2 + c = 0$?

The most efficient method is to solve for x^2 and apply the square root property.

Pencil Problem #11

11. What is the most efficient method for solving a quadratic equation of the form $u^2 = d$, where u is a first-degree polynomial?

Objective #12: Solve radical equations.

✔ *Solved Problem #12*

12. Solve: $\sqrt{x+3}+3=x$

$$\sqrt{x+3}+3=x$$
$$\sqrt{x+3}=x-3$$
$$(\sqrt{x+3})^2=(x-3)^2$$
$$x+3=x^2-6x+9$$
$$0=x^2-7x+6$$
$$0=(x-6)(x-1)$$
$$x-6=0 \quad \text{or} \quad x-1=0$$
$$x=6 \qquad\qquad x=1$$

Check 6: $\sqrt{x+3}+3=x$

$$\sqrt{6+3}+3=6$$
$$6=6$$

Check 1: $\sqrt{x+3}+3=x$

$$\sqrt{1+3}+3=1$$
$$5=1$$

The solution set is $\{6\}$.

Pencil Problem #12

12. Solve: $\sqrt{2x+13}=x+7$

Answers for Pencil Problems *(Textbook Exercise references in parentheses)*:

1. $\{9\}$ *(P.7 #9)*
2. $\left\{\frac{33}{2}\right\}$ *(P.7 #11)*

3a. $\{-3\}$ *(P.7 #23)* **3b.** $\varnothing$ *(P.7 #25)*

4. $f = \frac{pq}{p+q}$ *(P.7 #39)*
5. $\left\{-\frac{5}{3}, 3\right\}$ *(P.7 #47)*

6a. $\{0, 4\}$ *(P.7 #59)* **6b.** $\{3, 5\}$ *(P.7 #57)*

7a. $\{-\sqrt{10}, \sqrt{10}\}$ *(P.7 #63)* **7b.** $\{4+\sqrt{5}, 4-\sqrt{5}\}$ *(P.7 #65)*

8. $\{3+\sqrt{11}, 3-\sqrt{11}\}$ *(P.7 #71)*
9. $\left\{\frac{3+\sqrt{57}}{6}, \frac{3-\sqrt{57}}{6}\right\}$ *(P.7 #79)*
10. 97; two unequal real solutions *(P.7 #85)*
11. The square root property *(P.7 #101)*
12. $\{-6\}$ *(P.7 #119)*

Section P.8
Modeling with Equations

Counting Your Money!

From how much you can expect to earn at your first job after college to how much you need to save each month for retirement, mathematical models can help you plan your finances. In this section, you will see applications that involve starting salaries with a college degree based on major, discounts on electronics, and sharing the cost of a yacht.

Objective #1: Use equations to solve problems.

✔ Solved Problem #1

1a. The average yearly salary of a woman with a bachelor's degree exceeds that of a woman with an associate's degree by \$14 thousand. The average yearly salary of a woman with a master's degree exceeds that of a woman with an associate's degree by \$26 thousand. Combined, three women with each of these educational attainments earn \$139 thousand. Find the average yearly salary of women with each of these levels of education.

Since the salaries for women with bachelor's and master's degrees are compared to salaries of women with associate's degrees, we let x = the average salary of a woman with an associate's degree. The salaries for the other two degrees exceed this salary by a specified amount, so we add that amount to x.

$x + 14$ = the average salary of a woman with a bachelor's degree
$x + 26$ = the average salary of a woman with a master's degree

Since the combined salary is \$139 thousand, we add the three salaries and set the sum equal to 139. Then we solve for x.

$$x+(x+14)+(x+26)=139$$
$$3x+40=139$$
$$3x=99$$
$$x=33$$

$x + 14 = 33 + 14 = 47$
$x + 26 = 33 + 26 = 59$

The average salaries are \$33 thousand for an associate's degree, \$47 thousand for a bachelor's degree, and \$59 thousand for a master's degree.
You should verify that these salaries are \$139 thousand combined.

Pencil Problem #1

1a. According to the American Bureau of Labor Statistics, you will devote 37 years to sleeping and watching TV. The number of years sleeping will exceed the number of years watching TV by 19. Over your lifetime, how many years will you spend on each of these activities?

1b. After a 30% price reduction, you purchase a new computer for $840. What was the computer's price before the reduction?

Let x = the computer's price before the reduction.

Since the price reduction is 30% of the original price, the discount can be expressed as $0.30x$. The price is reduced by this amount, so the reduced price can be found by subtracting the discount from the original price: $x - 0.30x$.

The reduced price is $840, so $x - 0.30x = 840$.

$$x - 0.30x = 840$$
$$1x - 0.30x = 840$$
$$(1 - 0.30)x = 840$$
$$0.70x = 840$$
$$\frac{0.70x}{0.70} = \frac{840}{0.70}$$
$$x = 1200$$

Before the reduction, the computer's price was $1200.

1b. After a 20% reduction, you purchase a television for $336. What was the television's price before the reduction?

1c. A rectangular garden measures 16 feet by 12 feet. A path of uniform width is to be added so as to surround the entire garden. The landscape artist doing the work wants the garden and path to cover an area of 320 square feet. How wide should the path be?

Let x = the width of the path.

Since the path is added on all sides of the garden, the length of the larger rectangle including the garden and the path is $16 + 2x$, and the width is $12 + 2x$. See the figure. Since the area of the larger rectangle is to be 320 square feet, we write the equation using $A = lw$.

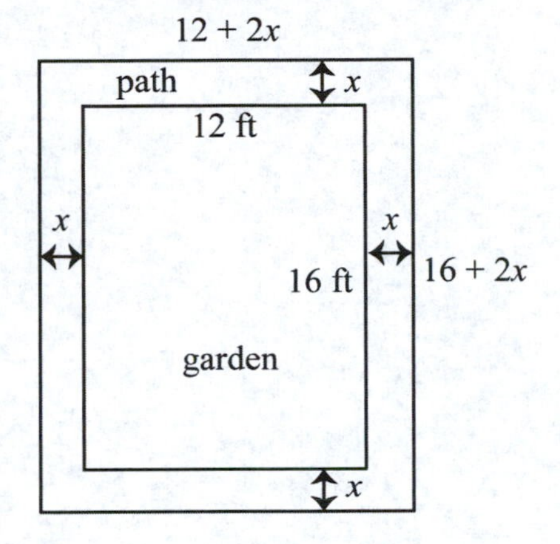

$$(16+2x)(12+2x) = 320$$
$$192+56x+4x^2 = 320$$
$$4x^2+56x-128 = 0$$
$$4(x^2+14x-32) = 0$$
$$4(x+16)(x-2) = 0$$
$$x+16 = 0 \quad \text{or} \quad x-2 = 0$$
$$x = -16 \qquad x = 2$$

Since the width of the path cannot be negative, discard −16.

If the path is 2 feet wide, then the length of the larger rectangle is 20 feet and the width is 16 feet, resulting in an area of 320 square feet. The width of the path is 2 feet.

1c. A pool measuring 10 meters by 20 meters is surrounded by a path of uniform width. If the area of the pool and the path combined is 600 square meters, what is the width of the path?

1d. A radio tower is supported by two wires that are each 130 yards long and attached to the ground 50 yards from the base of the tower. How tall is the tower?

Let h = the height of the tower. The situation is illustrated by the right triangle in the figure.

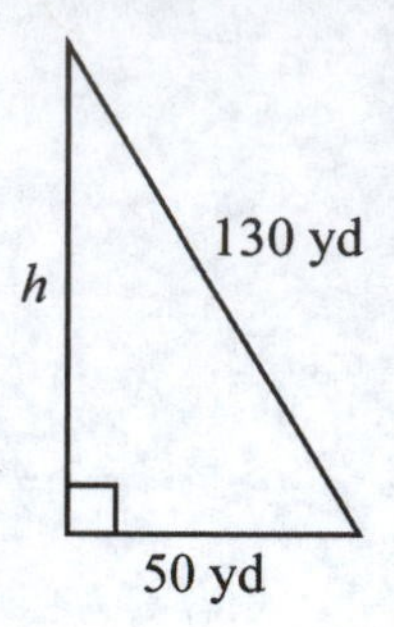

We write an equation using the Pythagorean Theorem, where the lengths of the legs are h and 50 and the length of the hypotenuse is 130.

$h^2 + 50^2 = 130^2$

Solve the equation using the square root property.

$$h^2 + 50^2 = 130^2$$
$$h^2 + 2500 = 16,900$$
$$h^2 = 14,400$$
$$h = \pm\sqrt{14,400} = \pm 120$$

Disregard the negative value. You should check that a height of 120 yards satisfies the conditions of the problem. The radio tower is 120 yards tall.

1d. A 20-foot ladder is 15 feet from a house. How far up the house, to the nearest tenth of a foot, does the ladder reach?

✔ *Solved Problem #1e*

1e. A group of people equally share in a $5,000,000 lottery. Before the money is divided, three more winning ticket holders are declared. As a result, each person's share is reduced by $375,000. How many people were in the original group of winners?

Let x = the number of people in the original group.

Then $x + 3$ = the number of people in the new group.

The share for each person in the original group, $\frac{5,000,000}{x}$, reduced by 375,000 is the share for each person in the new group, $\frac{5,000,000}{x+3}$. The equation is $\frac{5,000,000}{x} - 375,000 = \frac{5,000,000}{x+3}$.

The LCD is $x(x + 3)$.

$$\frac{5,000,000}{x} - 375,000 = \frac{5,000,000}{x+3}$$

$$x(x+3)\left(\frac{5,000,000}{x} - 375,000\right) = x(x+3)\cdot\frac{5,000,000}{x+3}$$

$$\cancel{x}(x+3)\cdot\frac{5,000,000}{\cancel{x}} - x(x+3)\cdot 375,000 = x\cancel{(x+3)}\cdot\frac{5,000,000}{\cancel{x+3}}$$

$$5,000,000(x+3) - 375,000x(x+3) = 5,000,000x$$

$$5,000,000x + 15,000,000 - 375,000x^2 - 1,125,000x = 5,000,000x$$

$$-375,000x^2 - 1,125,000x + 15,000,000 = 0$$

$$-375,000(x^2 + 3x - 40) = 0$$

$$-375,000(x+8)(x-5) = 0$$

$$x + 8 = 0 \quad \text{or} \quad x - 5 = 0$$

$$x = -8 \qquad\qquad x = 5$$

Disregard the negative value. If 5 people share $5,000,000, each person receives $1,000,000. If 3 more people join the group, then each receives $625,000, which is $375,000 less than $1,000,000.
There were 5 people in the original group.

Pencil Problem #1e

1e. A group of people equally share in a \$20,000,000 lottery. Before the money is divided, two more winning ticket holders are declared. As a result, each person's share is reduced by \$500,000. How many people were in the original group of winners?

Answers for Pencil Problems *(Textbook Exercise references in parentheses)*:

1a. TV: 9 years; sleeping: 28 years *(P.8 #1)*
1b. \$420 *(P.8 #15)*
1c. 5 meters *(P.8 #35)*
1d. 13.2 feet *(P.8 #39)*
1e. 8 people *(P.8 #45)*

Section P.9
Linear Inequalities and Absolute Values Inequalities

Are You in LOVE?

As the years go by in a relationship, three key components of love…

passion

commitment

intimacy

…progress differently over time.

Passion peaks early in a relationship and then declines.
By contrast, intimacy and commitment build gradually.

In the applications of this section of the textbook, we will use mathematics to explore the relationships among these three variables of love.

Objective #1: Use interval notation.

✔ *Solved Problem #1*

1a. Express $[-2,5)$ in set-builder notation and graph.

$\{x|-2 \leq x < 5\}$

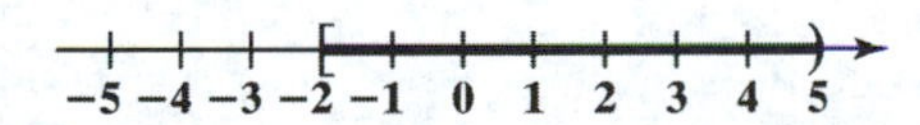

1b. Express $(-\infty,-1)$ in set-builder notation and graph.

$\{x|x < -1\}$

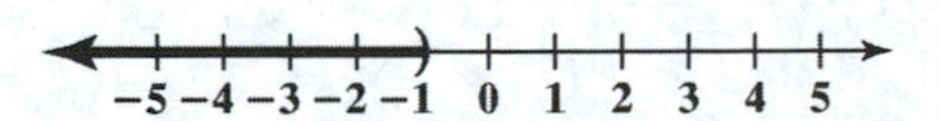

Pencil Problem #1

1a. Express $(1,6]$ in set-builder notation and graph.

1b. Express $[-3,\infty)$ in set-builder notation and graph.

Objective #2: Find intersections and unions of intervals.

✔ *Solved Problem #2*

2. Use graphs to find each set:

2a. $[1, 3] \cap (2, 6)$

Graph each interval. The intersection consists of the portion of the number line that the two graphs have in common.

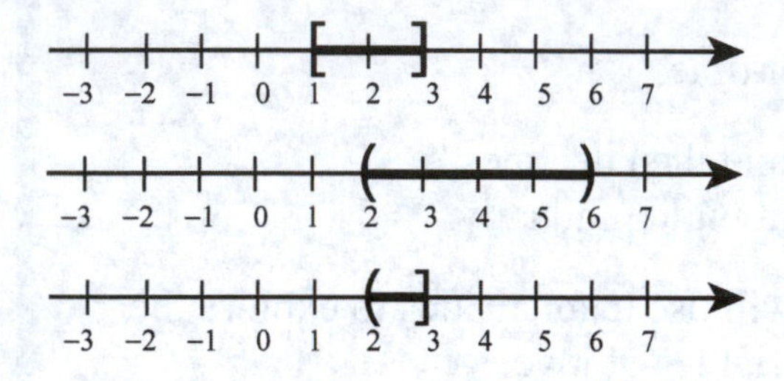

$[1, 3] \cap (2, 6) = (2, 3]$

2b. $[1, 3] \cup (2, 6)$

Graph each interval. The union consists of the portion of the number line in either one of the intervals or the other or both.

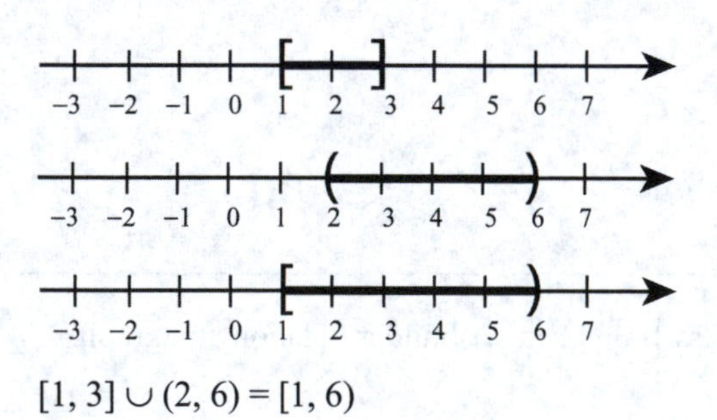

$[1, 3] \cup (2, 6) = [1, 6)$

Pencil Problem #2

2. Use graphs to find each set:

2a. $(-3, 0) \cap [-1, 2]$

2b. $(-3, 0) \cup [-1, 2]$

Objective #3: Solve linear inequalities.

✔ *Solved Problem #3*

3a. Solve and graph the solution set on a number line:
$$3x+1>7x-15$$

$$3x+1>7x-15$$
$$-4x>-16$$
$$\frac{-4x}{-4}<\frac{-16}{-4}$$
$$x<4$$

The solution set is $(-\infty,4)$.

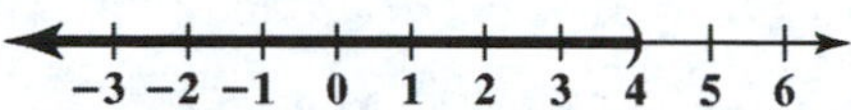

3b. Solve and graph the solution set on a number line:
$$\frac{x-4}{2}\geq\frac{x-2}{3}+\frac{5}{6}$$

$$\frac{x-4}{2}\geq\frac{x-2}{3}+\frac{5}{6}$$
$$6\left(\frac{x-4}{2}\right)\geq6\left(\frac{x-2}{3}+\frac{5}{6}\right)$$
$$3(x-4)\geq2(x-2)+5$$
$$3x-12\geq2x-4+5$$
$$3x-12\geq2x+1$$
$$x\geq13$$

The solution set is $[13,\infty)$.

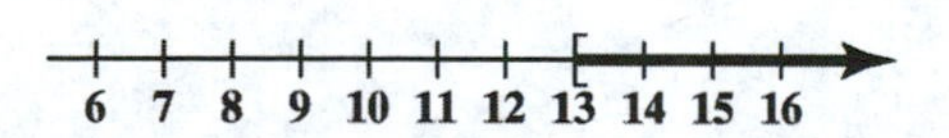

Pencil Problem #2

3a. Solve and graph the solution set on a number line:
$$-9x\geq36$$

3b. Solve and graph the solution set on a number line:
$$\frac{x}{4}-\frac{3}{2}=\frac{x}{2}+1$$

3c. A car can be rented from Basic Rental for $260 per week with no extra charge for mileage. Continental charges $80 per week plus 25 cents for each mile driven to rent the same car. How many miles must be driven in a week to make the rental cost for Basic Rental a better deal than Continental's?

Let $x =$ number of miles driven in a week.

$$\underbrace{260}_{\text{Cost for Basic Rental}} < \underbrace{80+0.25x}_{\text{Cost for Continental}}$$

$$260 < 80+0.25x$$
$$180 < 0.25x$$
$$\frac{180}{0.25} < \frac{0.25x}{0.25}$$
$$720 < x$$
$$x > 720$$

Driving more than 720 miles per week makes Basic Rental a better deal.

3c. An elevator at a construction site has a maximum capacity of 3000 pounds. If the elevator operator weighs 245 pounds and each cement bag weighs 95 pounds, how many bags of cement can be safely lifted on the elevator in one trip?

Objective #4: Solve compound inequalities.

✔ *Solved Problem #4*

4. Solve the compound inequality: $1 \le 2x+3 < 11$

$$1 \le 2x+3 < 11$$
$$1-3 \le 2x+3-3 < 11-3$$
$$-2 \le 2x < 8$$
$$\frac{-2}{2} \le \frac{2x}{2} < \frac{8}{2}$$
$$-1 \le x < 4$$

The solution set is $[-1,4)$.

Pencil Problem #4

4. Solve the compound inequality: $-11 < 2x-1 \le -5$

Objective #5: Solve absolute value inequalities.

✔ *Solved Problem #5*

5a. Solve the inequality:

$$|x-2|<5$$

Rewrite without absolute value bars.
$|u|<c$ means $-c<u<c$.

$$-5<x-2<5$$
$$-5+2<x-2+2<5+2$$
$$-3<x<7$$

The solution set is $(-3,7)$.

5b. Solve the inequality:

$$-3|5x-2|+20\geq-19$$

First, isolate the absolute value expression on one side of the inequality.

$$-3|5x-2|+20\geq-19$$
$$-3|5x-2|\geq-39$$
$$\frac{-3|5x-2|}{-3}\leq\frac{-39}{-3}$$
$$|5x-2|\leq13$$

Rewrite $|5x-2|\leq13$ without absolute value bars.
$|u|\leq c$ means $-c\leq u\leq c$.

$$-13\leq5x-2\leq13$$
$$-13+2\leq5x-2+2\leq13+2$$
$$-11\leq5x\leq15$$
$$\frac{-11}{5}\leq\frac{5x}{5}\leq\frac{15}{5}$$
$$-\frac{11}{5}\leq x\leq3$$

The solution set is $\left[\frac{-11}{5},3\right]$.

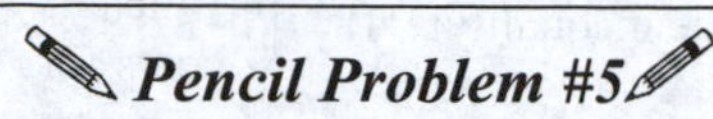

Pencil Problem #5

5a. Solve the inequality:

$$|2x-6|<8$$

5b. Solve the inequality:

$$|2(x-1)+4|\leq8$$

5c. Solve the inequality: $18 < |6-3x|$

Rewrite with the absolute value expression on the left.

$|6-3x| > 18$

This means the same as $6-3x < -18$ or $6-3x > 18$.

$$6-3x < -18 \quad \text{or} \quad 6-3x > 18$$
$$-3x < -24 \qquad -3x > 12$$
$$\frac{-3x}{-3} > \frac{-24}{-3} \qquad \frac{-3x}{-3} < \frac{12}{-3}$$
$$x > 8 \qquad x < -4$$

The solution set is $\{x \mid x < -4 \text{ or } x > 8\}$ or $(-\infty, -4) \cup (8, \infty)$.

5c. Solve the inequality: $1 < |2-3x|$

Answers for Pencil Problems *(Textbook Exercise references in parentheses)*:

1a. $\{x \mid 1 < x \leq 6\}$; 1 6 *(P.9 #1)*

1b. $\{x \mid x \geq -3\}$; −3 *(P.9 #9)*

2a. $[-1, 0)$ *(P.9 #15)* **2b.** $(-3, 2]$ *(P.9 #17)*

3a. $(-\infty, -4]$; −4 *(P.9 #31)*

3b. $[-10, \infty)$; −10 *(P.9 #41)* **3c.** at most 29 bags *(P.9 #119)*

4. $(-5, -2]$ *(P.9 #53)*

5a. $(-1, 7)$ *(P.9 #61)* **5b.** $[-5, 3]$ *(P.9 #63)* **5c.** $\left(-\infty, \frac{1}{3}\right) \cup (1, \infty)$ *(P.9 #87)*

Section 1.1
Graphs and Graphing Utilities

Let it snow! Let it snow! Let it snow!

The arrival of snow can range from light flurries to a full-fledged blizzard. Snow can be welcomed as a beautiful backdrop to outdoor activities or it can be a nuisance and endanger drivers.

We will look at how graphs can be used to explain both mathematical concepts and everyday situations. Specifically, in the application exercises of this section of the textbook, you will match stories of varying snowfalls to the graphs that explain them.

Objective #1: Plot points in the rectangular coordinate system.

✔ *Solved Problem #1*

1a. Plot the points:
$A(-2, 4)$, $B(4, -2)$, $C(-3, 0)$, and $D(0, -3)$.

From the origin, point A is left 2 units and up 4 units.

From the origin, point B is right 4 units and down 2 units.

From the origin, point C is left 3 units.

From the origin, point D is down 3 units.

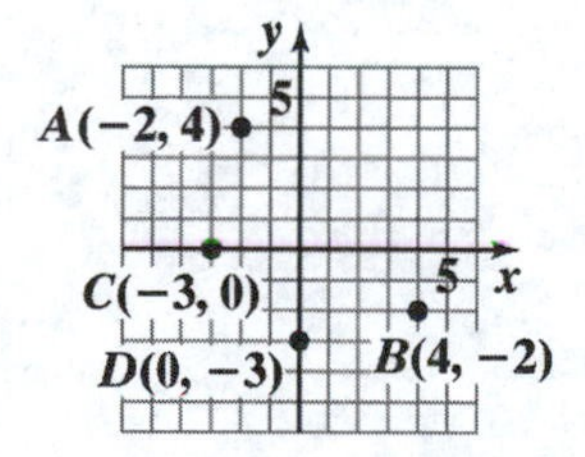

1b. If a point is on the x-axis it is neither up nor down, so $x = 0$.

False. The y-coordinate gives the distance up or down, so $y = 0$.

Pencil Problem #1

1a. Plot the points:
$A(1, 4)$, $B(-2, 3)$, $C(-3, -5)$, and $D(-4, 0)$.

1b. True or false: If a point is on the y-axis, its x-coordinate must be 0.

Objective #2: Graph equations in the rectangular coordinate system.

✔ Solved Problem #2

2a. Graph $y = 4 - x$.

x	$y = 4 - x$	(x, y)
-3	$y = 4-(-3) = 7$	$(-3,7)$
-2	$y = 4-(-2) = 6$	$(-2,6)$
-1	$y = 4-(-1) = 5$	$(-1,5)$
0	$y = 4-(0) = 4$	$(0,4)$
1	$y = 4-(1) = 3$	$(1,3)$
2	$y = 4-(2) = 2$	$(2,2)$
3	$y = 4-(3) = 1$	$(3,1)$

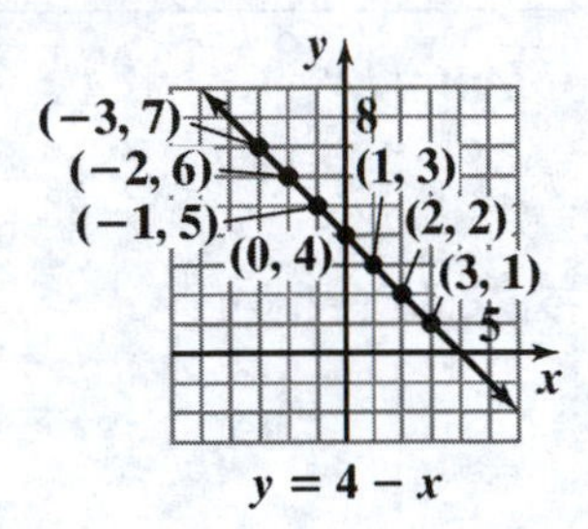

2b. Graph $y = |x+1|$.

x	$y = \|x+1\|$	(x, y)
-4	$y = \|-4+1\| = \|-3\| = 3$	$(-4,3)$
-3	$y = \|-3+1\| = \|-2\| = 2$	$(-3,2)$
-2	$y = \|-2+1\| = \|-1\| = 1$	$(-2,1)$
-1	$y = \|-1+1\| = \|0\| = 0$	$(-1,0)$
0	$y = \|0+1\| = \|1\| = 1$	$(0,1)$
1	$y = \|1+1\| = \|2\| = 2$	$(1,2)$
2	$y = \|2+1\| = \|3\| = 3$	$(2,3)$

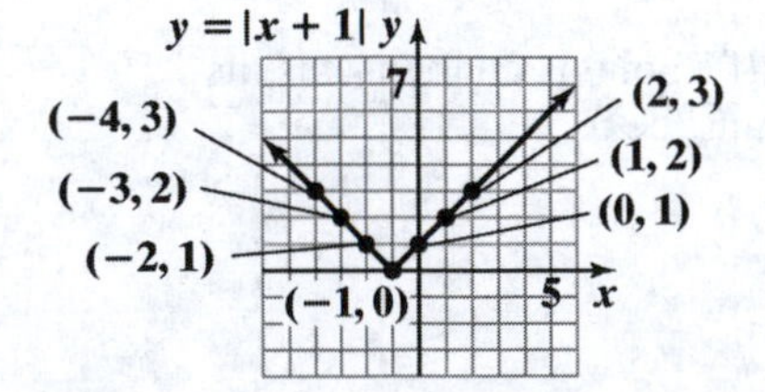

✎ Pencil Problem #2 ✎

2a. Graph $y = x^2 - 2$. Let $x = -3, -2, -1, 0, 1, 2,$ and 3.

2b. Graph $y = 2|x|$. Let $x = -3, -2, -1, 0, 1, 2,$ and 3.

Objective #3: Interpret information about a graphing utility's viewing rectangle or table.

✔ Solved Problem #3

3. What is the meaning of a
[–100, 100, 50] by [–100, 100, 10]
viewing rectangle?

The minimum x-value is –100, the maximum x-value is 100, and the distance between consecutive tick marks is 50.

The minimum y-value is –100, the maximum y-value is 100, and the distance between consecutive tick marks is 10.

Pencil Problem #3

3. What is the meaning of a
[–20, 80, 10] by [–30, 70, 10]
viewing rectangle?

Objective #4: Use a graph to determine intercepts.

✔ Solved Problem #4

4a. Identify the x- and y- intercepts:

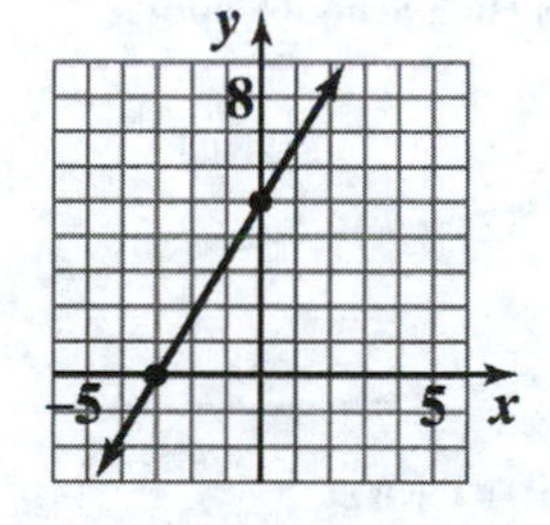

The graph crosses the x-axis at (–3,0).
Thus, the x-intercept is –3.

The graph crosses the y-axis at (0,5).
Thus, the y-intercept is 5.

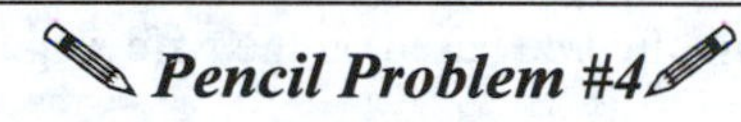

Pencil Problem #4

4a. Identify the x- and y- intercepts:

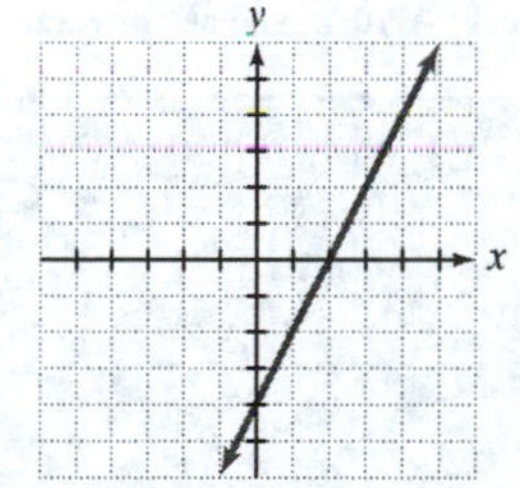

4b. Identify the x- and y- intercepts:

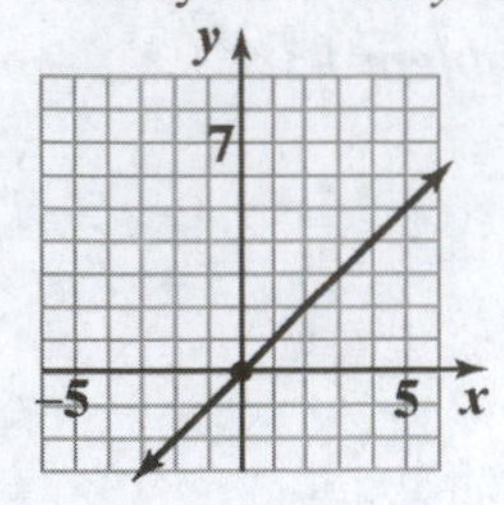

The graph crosses the x-axis at (0,0).
Thus, the x-intercept is 0.

The graph crosses the y-axis at (0,0).
Thus, the y-intercept is 0.

4b. Identify the x- and y- intercepts:

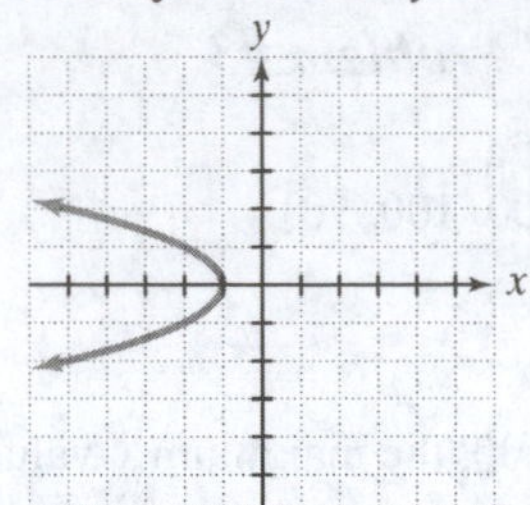

Objective #5: Interpret information given by graphs.

✔ *Solved Problem #5*

5. The line graphs show the percentage of marriages ending in divorce based on the wife's age at marriage.

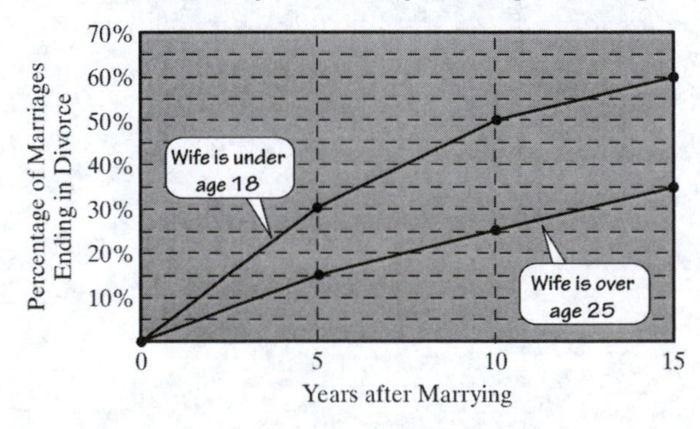

The model $d = 4n + 5$ approximates the data in the graph when the wife is under 18 at the time of marriage. In the model, n is the number of years after marriage and d is the percentage of marriages ending in divorce.

(continued on next page)

Pencil Problem #5

5. The graphs show the percentage of high school seniors who used alcohol or marijuana.

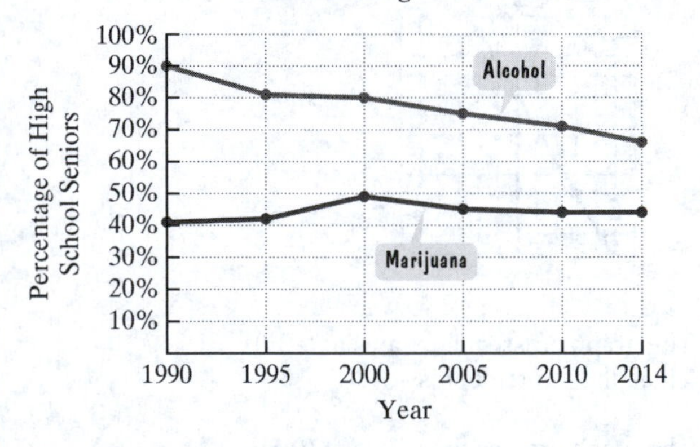

Source: University of Michigan Institute for Social Research

The data for seniors who used marijuana can be modeled by $M = 0.1n + 43$, where M is the percentage of seniors who used marijuana n years after 1990.

(continued on next page)

5a. Use the formula to determine the percentage of marriages ending in divorce after 15 years when the wife is under 18 at the time of marriage.

$d = 4n + 5$
$d = 4(15) + 5$
$d = 60 + 5$
$d = 65$

According to the formula, 65% of marriages end in divorce after 15 years when the wife is under 18 at the time of marriage.

5a. Use the formula to determine the percentage of seniors who used marijuana in 2010.

5b. Use the appropriate line graph to determine the percentage of marriages ending in divorce after 15 years when the wife is under 18 at the time of marriage.

Locate 15 on the horizontal axis and locate the point above it on the graph. Read across to the corresponding percentage on the vertical axis. This percentage is 60. According to the line graph, 60% of marriages end in divorce after 15 years when the wife is under 18 at the time of marriage.

5b. Use the appropriate line graph to determine the percentage of seniors who used marijuana in 2010.

5c. Does the value given by the model underestimate or overestimate the value shown by the graph? By how much?

The value given by the model, 65%, is greater than the value shown by the graph, 60%, so the model overestimates the percentage by 65 – 5, or 5.

5c. Does the formula underestimate or overestimate the percentage of seniors who used marijuana in 2010 as shown by the graph.

Answers for Pencil Problems *(Textbook Exercise references in parentheses)*:

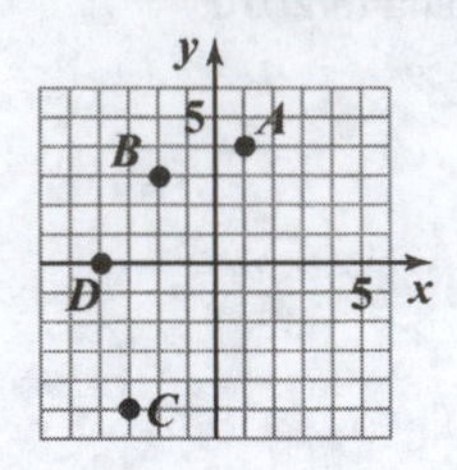

1a. *(1.1 #1-9)* **1b.** True *(1.1.#73)*

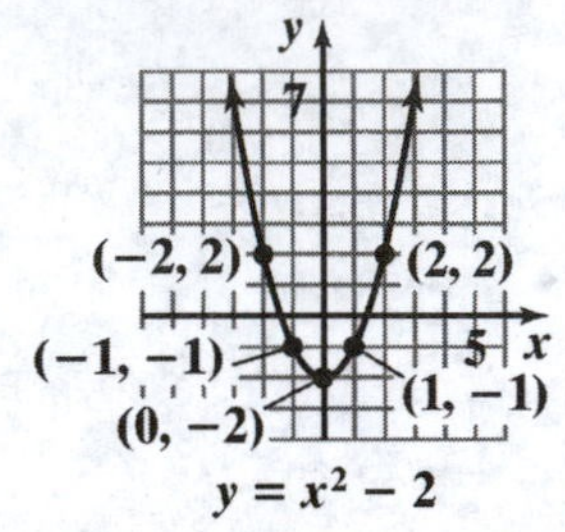

2a. *(1.1 #13)* **2b.**

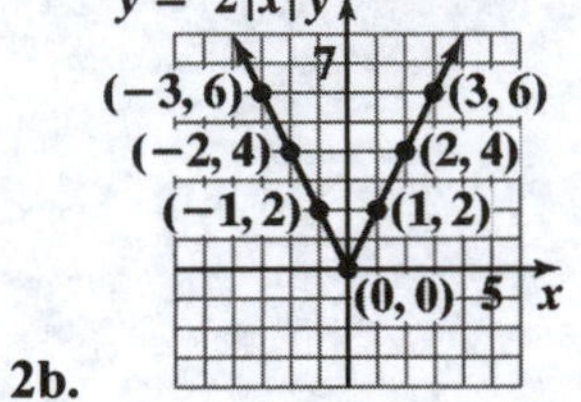

(1.1 #21)

3. The minimum x-value is –20, the maximum x-value is 80, and the distance between consecutive tick marks is 10. The minimum y-value is –30, the maximum y-value is 70, and the distance between consecutive tick marks is 10. *(1.1 #31)*

4a. x-intercept: 2; y-intercept: –4 *(1.1 #41)* **4b.** x-intercept: –1 ; y-intercept: none *(1.1 #45)*

5a. 45% *(1.1 #55b)* **5b.** ≈44% *(1.1 #55a)* **5c.** overestimates by 1, although answers vary *(1.1 #55b)*

Section 1.2
Basics of Functions and Their Graphs

Say *WHAT???*

You may have noticed that mathematical notation occasionally can have more than one meaning depending on the context.

For example, $(-3,6)$ could refer to the ordered pair where $x=-3$ and $y=6$, or it could refer to the open interval $-3<x<6$.

Similarly, in this section of the textbook, we will use the notation, $f(x)$. It may surprise you to find out that it does *not* mean to multiply "f times x."

It will be important for you to gain an understanding of what this notation *does* mean as you work through this essential concept of "functions."

Objective #1: Find the domain and range of a relation.

✔ *Solved Problem #1*

1. Find the domain and range of the relation:
$\{(0, 9.1), (10, 6.7), (20, 10.7), (30, 13.2), (42, 21.7)\}$

The domain is the set of all first components.

Domain:
$\{0, 10, 20, 30, 42\}$.

The range is the set of all second components.

Range:
$\{9.1, 6.7, 10.7, 13.2, 21.7\}$.

✎ *Pencil Problem #1* ✎

1. Find the domain and range of the relation:
$\{(3, 4), (3, 5), (4, 4), (4, 5)\}$

Objective #2: Determine whether a relation is a function.

✔ *Solved Problem #2*

2a. Determine whether the relation is a function:
$$\{(1,2),(3,4),(5,6),(5,8)\}$$

5 corresponds to both 6 and 8. If any element in the domain corresponds to more than one element in the range, the relation is not a function.

Thus, the relation is not a function.

✎ *Pencil Problem #2* ✎

2a. Determine whether the relation is a function:

$\{(3, 4), (3, 5), (4, 4), (4, 5)\}$

2b. Determine whether the relation is a function:

$$\{(1,2),(3,4),(6,5),(8,5)\}$$

Every element in the domain corresponds to exactly one element in the range. No two ordered pairs in the given relation have the same first component and different second components.

Thus, the relation is a function.

2b. Determine whether the relation is a function:

$$\{(-3,-3),(-2,-2),(-1,-1),(0,0)\}$$

Objective #3: Determine whether an equation represents a function.

✔ *Solved Problem #3*

3. Solve each equation for *y* and then determine whether the equation defines *y* as a function of *x*.

3a. $2x+y=6$

Subtract $2x$ from both sides to solve for *y*.

$$2x+y=6$$
$$2x-2x+y=6-2x$$
$$y=6-2x$$

For each value of *x*, there is only one value of *y*, so the equation defines *y* as a function of *x*.

3b. $x^2+y^2=1$

Subtract x^2 from both sides and then use the square root property to solve for *y*.

$$x^2+y^2=1$$
$$x^2-x^2+y^2=1-x^2$$
$$y^2=1-x^2$$
$$y=\pm\sqrt{1-x^2}$$

For values of *x* between −1 and 1, there are two values of *y*. For example, if $x = 0$, then $y = \pm 1$. Thus, the equation does not define *y* as a function of *x*.

Pencil Problem #3

3. Solve each equation for *y* and then determine whether the equation defines *y* as a function of *x*.

3a. $x^2+y=16$

3b. $x=y^2$

Objective #4: Evaluate a function.

✔ *Solved Problem #4*

4. If $f(x) = x^2 - 2x + 7$, evaluate each of the following.

4a. $f(-5)$

Substitute –5 for *x*. Place parentheses around –5 when making the substitution.

$$f(-5) = (-5)^2 - 2(-5) + 7$$
$$= 25 + 10 + 7 = 42$$

4b. $f(x + 4)$

Substitute $x + 4$ for x and then simplify. Place parentheses around $x + 4$ when making the substitution.

Use $(A + B)^2 = A^2 + 2AB + B^2$ to expand $(x + 4)^2$ and the distributive property to multiply $-2(x + 4)$. Then combine like terms.

$$f(x + 4) = (x + 4)^2 - 2(x + 4) + 7$$
$$= x^2 + 8x + 16 - 2x - 8 + 7$$
$$= x^2 + 6x + 15$$

4c. $f(-x)$

Substitute $-x$ for x. Place parentheses around $-x$ when making the substitution.

$$f(-x) = (-x)^2 - 2(-x) + 7$$
$$= x^2 + 2x + 7$$

Pencil Problem #4

4. If $g(x) = x^2 + 2x + 3$, evaluate each of the following.

4a. $g(-1)$

4b. $g(x + 5)$

4c. $g(-x)$

Objective #5: Graph functions by plotting points.

✔ Solved Problem #5

5. Graph the functions $f(x) = 2x$ and $g(x) = 2x - 3$ in the same rectangular coordinate system. Select integers for x, starting with -2 and ending with 2. How is the graph of g related to the graph of f?

Make a table for $f(x) = 2x$:

x	$f(x) = 2x$	(x, y)
-2	$f(-2) = 2(-2) = -4$	$(-2, -4)$
-1	$f(-1) = 2(-1) = -2$	$(-1, -2)$
0	$f(0) = 2(0) = 0$	$(0, 0)$
1	$f(1) = 2(1) = 2$	$(1, 2)$
2	$f(2) = 2(2) = 4$	$(2, 4)$

Make a table for $g(x) = 2x - 3$:

x	$g(x) = 2x - 3$	(x, y)
-2	$g(-2) = 2(-2) - 3 = -7$	$(-2, -7)$
-1	$g(-1) = 2(-1) - 3 = -5$	$(-1, -5)$
0	$g(0) = 2(0) - 3 = -3$	$(0, -3)$
1	$g(1) = 2(1) - 3 = -1$	$(1, -1)$
2	$g(2) = 2(2) - 3 = 1$	$(2, 1)$

Plot the points and draw the lines that pass through them.

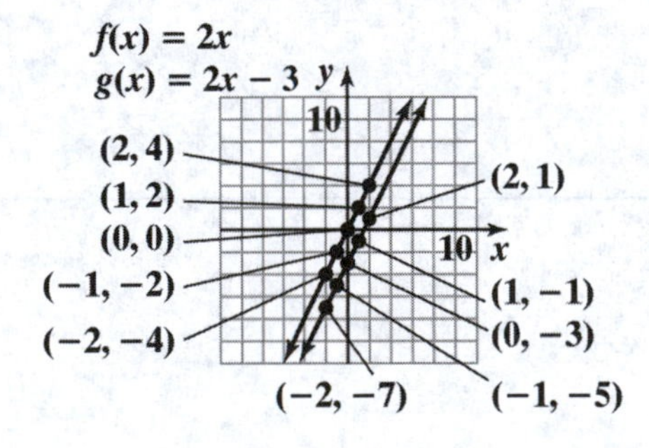

The graph of g is the graph of f shifted down by 3 units.

Pencil Problem #5

5. Graph the functions $f(x) = |x|$ and $g(x) = |x| - 2$ in the same rectangular coordinate system. Select integers for x, starting with -2 and ending with 2. How is the graph of g related to the graph of f?

Objective #6: Use the vertical line test to identify functions.

✔ *Solved Problem #6*

6. Use the vertical line test to determine if the graph represents y as a function of x.

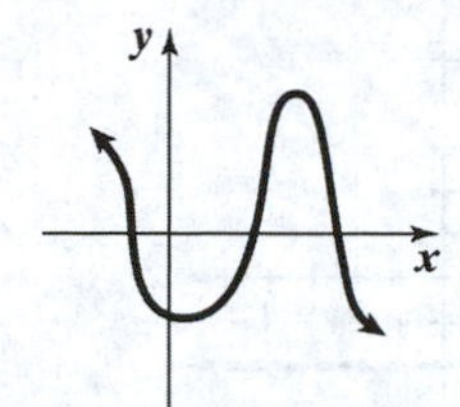

The graph passes the vertical line test and thus y is a function of x.

Pencil Problem #6

6. Use the vertical line test to determine if the graph represents y as a function of x.

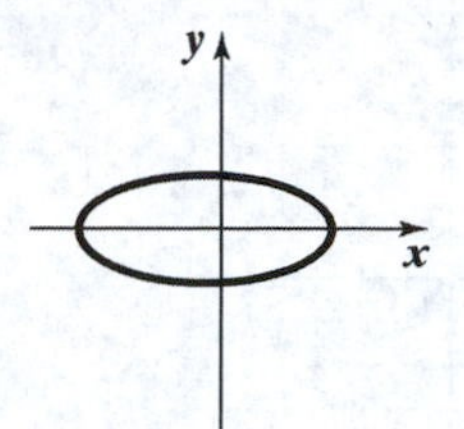

Objective #7: Obtain information about a function from its graph.

✔ *Solved Problem #7*

7a. The following is the graph of g.

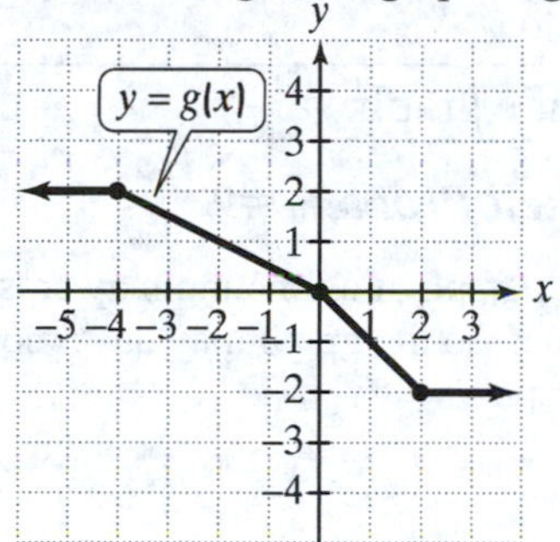

Use the graph to find $g(-20)$.

The graph indicates that to the left of $x = -4$, the graph is at a constant height of 2.

Thus, $g(-20) = 2$.

7b. Use the graph from *Problem 7a* above to find the value of x for which $g(x) = -1$.

$g(1) = -1$

The height of the graph is -1 when $x = 1$.

Pencil Problem #7

7a. The following is the graph of g.

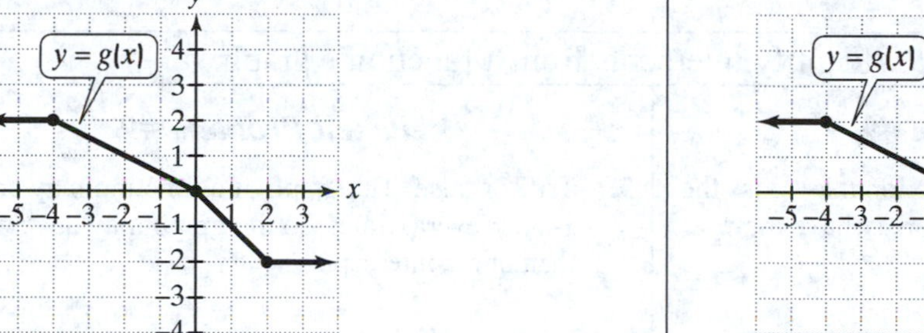

Use the graph to find $g(-4)$.

7b. Use the graph from *Problem 7a* above to find the value of x for which $g(x) = 1$.

Objective #8: Identify the domain and range of a function from its graph.

✔ *Solved Problem #8*

8. Use the graph of the function to identify its domain and its range.

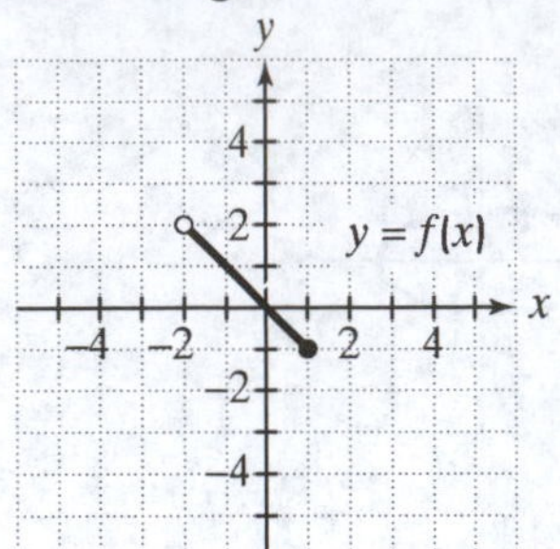

Inputs on the x-axis extend from −2, excluding −2, to 1, including 1.
The domain is $(-2,1]$.

Outputs on the y-axis extend from −1, including −1, to 2, excluding 2.
The range is $[-1,2)$.

Pencil Problem #8

8. Use the graph of the function to identify its domain and its range.

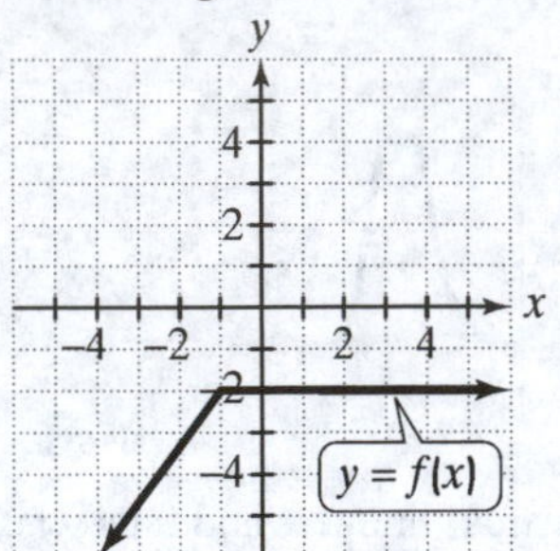

Objective #9: Identify intercepts from a function's graph.

✔ *Solved Problem #9*

9. True or false: The graph of a function may cross the y-axis several times, so the graph may have more than one y-intercept.

False. Since each point on the y-axis has x-coordinate 0 and a function may have only one y-value for each x-value, the graph of a function has at most one y-coordinate.

Pencil Problem #9

9. True or false: The graph of a function may cross the x-axis several times, so the graph may have more than one x-intercept.

Answers for Pencil Problems *(Textbook Exercise references in parentheses)*:

1. Domain: {3, 4}. Range: {4, 5}. *(1.2 #3)* **2a.** not a function *(1.2 #3)* **2b.** function *(1.2 #7)*

3a. $y = 16 - x^2$; y is a function of x. *(1.2 #13)* **3b.** $y = \pm\sqrt{x}$; y is not a function of x. *(1.2 #17)*

4a. 2 *(1.2 #29a)* **4b.** $x^2 + 12x + 38$ *(1.2 #29b)* **4c.** $x^2 - 2x + 3$ *(1.2 #29c)*

5.

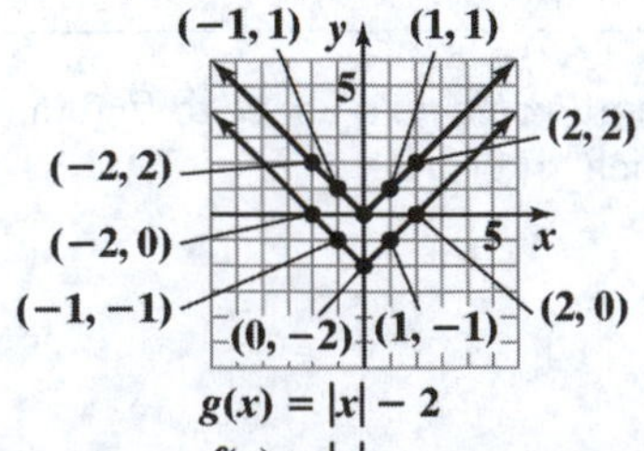

The graph of g is the graph of f shifted down 2 units. *(1.2 #45)*

6. not a function *(1.2 #59)* **7a.** 2 *(1.2 #71)* **7b.** −2 *(1.2 #75)*

8. Domain: $(-\infty,\infty)$. Range: $(-\infty,-2]$ *(1.2 #87)* **9.** True *(1.2 #77)*

Section 1.3
More on Functions and Their Graphs

Can I Really Tell That from a Graph?

Graphs provide a visual representation of how a function changes over time. Many characteristics of a function are much more evident from the function's graph than from the equation that defines the function.

We can use graphs to determine for what years the fuel efficiency of cars was increasing and decreasing and at what age men and women attain their maximum percent body fat. A graph can even help you understand your cellphone bill better.

Objective #1: Identify intervals on which a function increases, decreases, or is constant.

✔ *Solved Problem #1*

1. State the intervals on which the given function is increasing, decreasing, or constant.

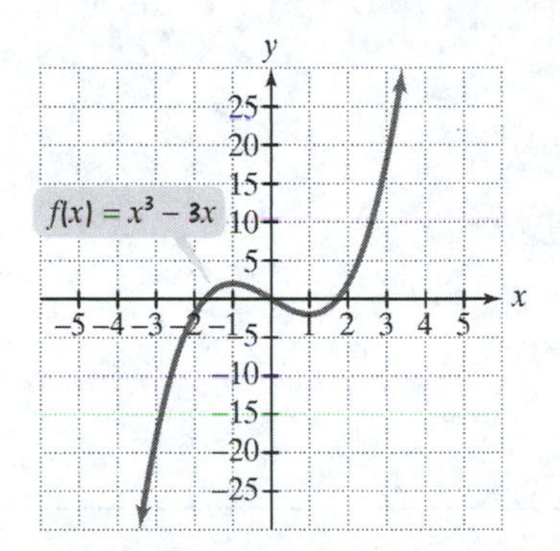

The intervals are stated in terms of x-values.
When we start at the left and follow along the graph, at first the graph is going up. This continues until $x = -1$. The function is increasing on the interval $(-\infty, -1)$.

At $x = -1$, the graph turns and moves downward until we get to $x = 1$. The function is decreasing on the interval $(-1, 1)$.

At $x = 1$, the graph turns again and continues in an upward direction. The function is increasing on the interval $(1, \infty)$.

Pencil Problem #1

1. State the intervals on which the given function is increasing, decreasing, or constant.

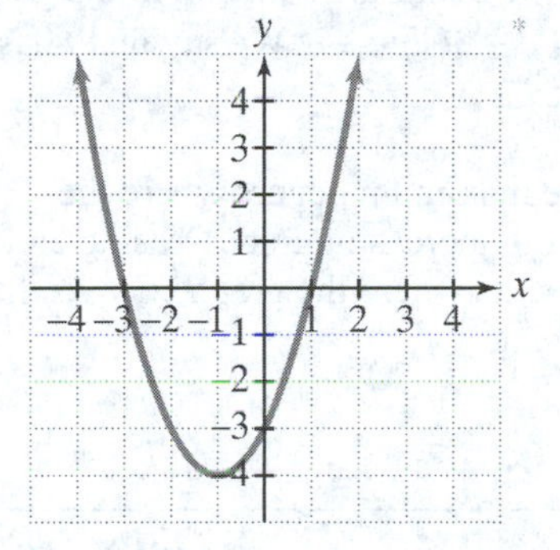

Objective #2: Use graphs to locate relative maxima or minima.

✔ Solved Problem #2

2. Look at the graph in Solved Problem #1. Locate values at which the function f has any relative maxima or minima. What are these relative maxima or minima?

The graph has a turning point at $x = -1$. The value of $f(x)$ or y at $x = -1$ is greater than the values of $f(x)$ for values of x near -1 (for values of x between -2 and 0, for example). Thus, f has a relative maximum at $x = -1$. The relative maximum is the value of $f(x)$ or y corresponding to $x = -1$. Using the equation in the graph, we find that $f(-1) = (-1)^3 - 3(-1) = 2$. We say that f has a relative maximum of 2 at $x = -1$.

The graph has a second turning point at $x = 1$. The value of $f(x)$ or y at $x = 1$ is less than the values of $f(x)$ for values of x near 1 (for values of x between 0 and 2, for example). Thus, f has a relative minimum at $x = 1$. The relative minimum is the value of $f(x)$ or y corresponding to $x = 1$. Using the equation in the graph, we find that $f(1) = (1)^3 - 3(1) = -2$. We say that f has a relative minimum of -2 at $x = 1$.

Note that the relative maximum occurs where the functions changes from increasing to decreasing and the relative minimum occurs where the graph changes from decreasing to increasing.

✎ Pencil Problem #2

2. The graph of a function f is given below. Locate values at which the function f has any relative maxima or minima. What are these relative maxima or minima? Read y-values from the graph, as needed, since the equation is not given.

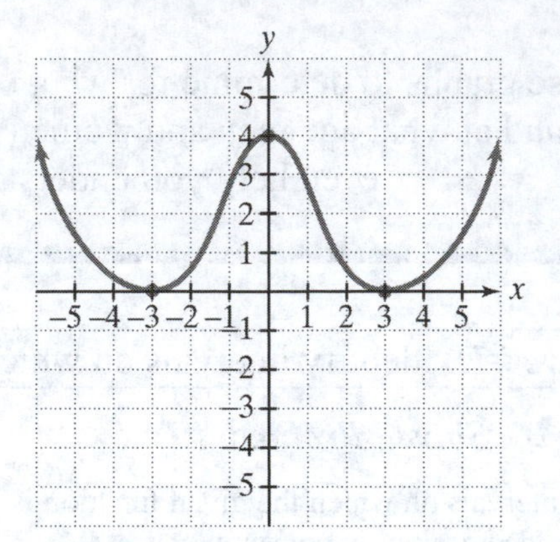

Objective #3: Test for symmetry.

✔ Solved Problem #3

3. Determine whether the graph of each equation is symmetric with respect to the y-axis, the x-axis, or the origin.

3a. $y = x^2 - 1$

To test for y-axis symmetry, replace x with $-x$.

$y = (-x)^2 - 1$

$y = x^2 - 1$

The result is the original equation, so the graph is symmetric with respect to the y-axis.

✎ Pencil Problem #3

3. Determine whether the graph of each equation is symmetric with respect to the y-axis, the x-axis, or the origin.

3a. $y = x^2 + 6$

To test for x-axis symmetry, replace y with $-y$.

$$-y = x^2 - 1$$
$$y = -x^2 + 1$$

The result is not equivalent to the original equation, so the graph is not symmetric with respect to the x-axis.
To test for origin symmetry, replace x with $-x$ and y with $-y$.

$$-y = (-x)^2 - 1$$
$$-y = x^2 - 1$$
$$y = -x^2 + 1$$

The result is not equivalent to the original equation, so the graph is not symmetric with respect to the origin.

The graph of $y = x^2 - 1$ is symmetric with respect to the y-axis only.

3b. $y^5 = x^3$

To test for y-axis symmetry, replace x with $-x$.

$$y^5 = (-x)^3$$
$$y^5 = -x^3$$

The result is not equivalent to the original equation, so the graph is not symmetric with respect to the y-axis.

To test for x-axis symmetry, replace y with $-y$.

$$(-y)^5 = x^3$$
$$-y^5 = x^3$$
$$y^5 = -x^3$$

The result is not equivalent to the original equation, so the graph is not symmetric with respect to the x-axis.

To test for origin symmetry, replace x with $-x$ and y with $-y$.

$$(-y)^5 = (-x)^3$$
$$-y^5 = -x^3$$
$$y^5 = x^3$$

The result is the original equation, so the graph is symmetric with respect to the origin.

The graph of $y^5 = x^3$ is symmetric with respect to the origin only.

3b. $x^2 + y^2 = 100$

Objective #4: Identify even or odd functions and recognize their symmetries.

Solved Problem #4

4. Determine whether each of the following functions is even, odd, or neither.

4a. $f(x) = x^2 + 6$

Replace x with $-x$.
$f(-x) = (-x)^2 + 6 = x^2 + 6 = f(x)$

The function did not change when we replaced x with $-x$. The function is even.

4b. $g(x) = 7x^3 - x$

Replace x with $-x$.
$g(-x) = 7(-x)^3 - (-x) = -7x^3 + x = -g(x)$

Each term of the equation defining the function changed sign when we replaced x with $-x$. The function is odd.

4c. $h(x) = x^5 + 1$

Replace x with $-x$.
$h(x) = (-x)^5 + 1 = -x^5 + 1$

The resulting function is not equal to the original function, so the function is not even. Only the sign of one term changed, so the function is not odd. The function is neither even nor odd.

Pencil Problem #4

4. Determine whether each of the following functions is even, odd, or neither.

4a. $f(x) = x^3 + x$

4b. $g(x) = x^2 + x$

4c. $h(x) = x^2 - x^4$

Objective #5: Understand and use piecewise functions.

✔ *Solved Problem #4*

5. Graph the piecewise function defined by

$$f(x)=\begin{cases}3 & \text{if} \quad x\le -1\\ x-2 & \text{if} \quad x>-1\end{cases}$$

For $x \le -1$, the function value is always 3, so (–4, 3) and (–1, 3) are examples of points on the first piece of the graph.

For $x > -1$, we use $f(x) = x - 2$. We have points such as (0, –2) and (2, 0) on the graph. Note that this piece of the graph will approach the point (–1, –3) but this point is not part of the graph. We use an open dot at (–1, –3).

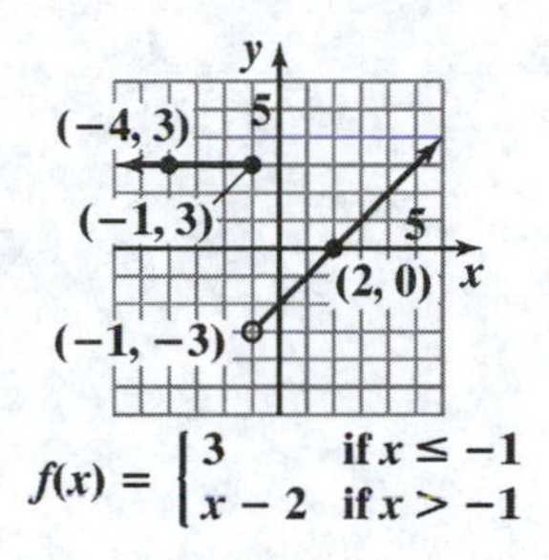

Pencil Problem #4

5. Graph the piecewise function defined by

$$f(x)=\begin{cases}2x & \text{if} \quad x\le 0\\ 2 & \text{if} \quad x>0\end{cases}$$

Objective #6: Find and simplify a function's difference quotient.

✔ *Solved Problem #6*

6. Find and simplify the difference quotient for $f(x) = -2x^2 + x + 5$.

$$\frac{f(x+h)-f(x)}{h}$$

$$=\frac{[-2(x+h)^2+(x+h)+5]-(-2x^2+x+5)}{h}$$

$$=\frac{-2(x^2+2xh+h^2)+x+h+5+2x^2-x-5}{h}$$

$$=\frac{-2x^2-4xh-2h^2+x+h+5+2x^2-x-5}{h}$$

$$=\frac{-4xh-2h^2+h}{h}$$

$$=\frac{h(-4x-2h+1)}{h}$$

$$=-4x-2h+1,\ h\neq 0$$

Pencil Problem #6

6. Find and simplify the difference quotient for $f(x) = x^2 - 4x + 3$.

Answers for Pencil Problems *(Textbook Exercise references in parentheses)*:

1. decreasing on $(-\infty, -1)$; increasing on $(-1, \infty)$ *(1.3 #1)*

2. relative minimum of 0 at $x = -3$; relative maximum of 4 at $x = 0$; relative minimum of 0 at $x = 3$ *(1.3 #13)*

3a. y-axis *(1.3 #17)* **3b.** x-axis, y-axis, and the origin *(1.3 #27)*

4a. odd *(1.3 #37)* **4b.** neither *(1.3 #39)* **4c.** even *(1.3 #41)*

5.

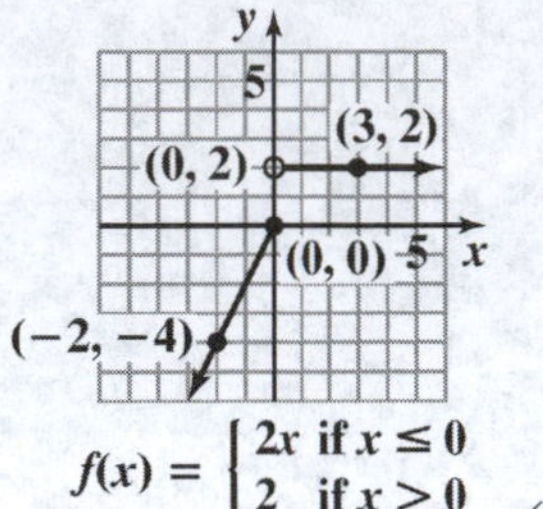

$$f(x) = \begin{cases} 2x & \text{if } x \le 0 \\ 2 & \text{if } x > 0 \end{cases}$$ *(1.3 #61)*

6. $2x + h - 4, h \neq 0$ *(1.3 #77)*

Section 1.4
Linear Functions and Slope

READ FOR LIFE!

Is there a relationship between literacy and child mortality?

As the percentage of adult females who are literate increases, does the mortality of children under age five decrease? Data from the United Nations indicates that this is, indeed, the case.

In this section of the textbook, you will be given a graph for which each point represents one country. You will use the concept of slope to see how much the mortality rate decreases for each 1% increase in the literacy rate of adult females in a country.

Objective #1: Calculate a line's slope.

✔ *Solved Problem #1*

1. Find the slope of the line passing through $(4,-2)$ and $(-1,5)$.

$$m=\frac{y_2-y_1}{x_2-x_1}$$

$$m=\frac{5-(-2)}{-1-4}$$

$$=\frac{7}{-5}$$

$$=-\frac{7}{5}$$

Pencil Problem #1

1. Find the slope of the line passing through $(-2,1)$ and $(2,2)$.

Objective #2: Write the point-slope form of the equation of a line.

✔ *Solved Problem #2*

2a. Write the point-slope form of the equation of the line with slope 6 that passes through the point $(2,-5)$. Then solve the equation for y.

Begin by finding the point-slope equation of a line.

$$y-y_1=m(x-x_1)$$

$$y-(-5)=6(x-2)$$

$$y+5=6(x-2)$$

Now solve this equation for y.

$$y+5=6(x-2)$$

$$y+5=6x-12$$

$$y=6x-17$$

Pencil Problem #2

2a. Write the point-slope form of the equation of the line with slope -3 that passes through the point $(-2,-3)$. Then solve the equation for y.

2b. A line passes through the points $(-2,-1)$ and $(-1,-6)$. Find the equation of the line in point-slope form and then solve the equation for y.

Begin by finding the slope: $m=\dfrac{-6-(-1)}{-1-(-2)}=\dfrac{-5}{1}=-5$

Using the slope and either point, find the point-slope equation of a line.

$y-y_1=m(x-x_1)$ or $y-y_1=m(x-x_1)$

$y-(-1)=-5(x-(-2))$ $\quad$ $y-(-6)=-5(x-(-1))$

$y+1=-5(x+2)$ $\quad$ $y+6=-5(x+1)$

To obtain slope-intercept form, solve the above equation for y:

$y+1=-5(x+2)$ or $y+6=-5(x+1)$

$y+1=-5x-10$ $\quad$ $y+6=-5x-5$

$y=-5x-11$ $\quad$ $y=-5x-11$

2b. A line passes through the points $(-3,-1)$ and $(2,4)$. Find the equation of the line in point-slope form and then solve the equation for y.

Objective #3: Write and graph the slope-intercept form of the equation of a line.

Solved Problem #3

3. Graph: $f(x)=\dfrac{3}{5}x+1$

The y-intercept is 1, so plot the point $(0,1)$.

The slope is $m=\dfrac{3}{5}$.

Find another point by going up 3 units and to the right 5 units.

Use a straightedge to draw a line through the two points.

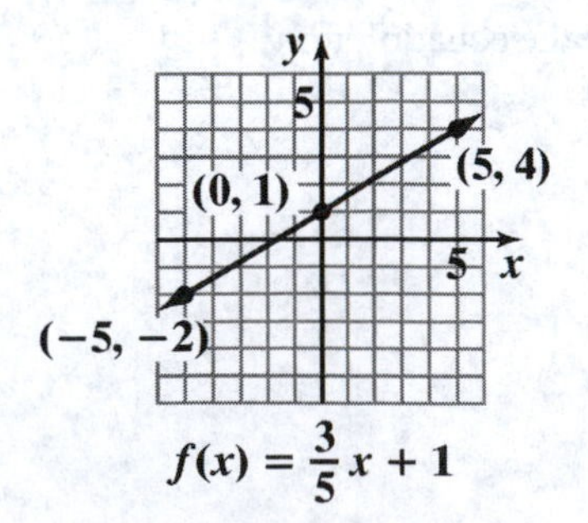

Pencil Problem #3

3. Graph: $f(x)=\dfrac{3}{4}x-2$

Objective #4: Graph horizontal or vertical lines.

✔ *Solved Problem #4*

4a. Graph $y = 3$ in the rectangular coordinate system.

$y = 3$ is a horizontal line.

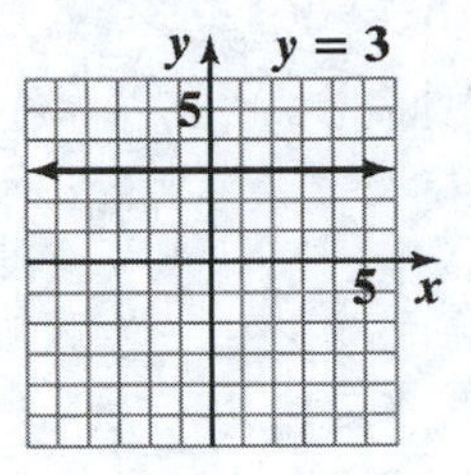

4b. Graph $x = -3$ in the rectangular coordinate system.

$x = -3$ is a vertical line.

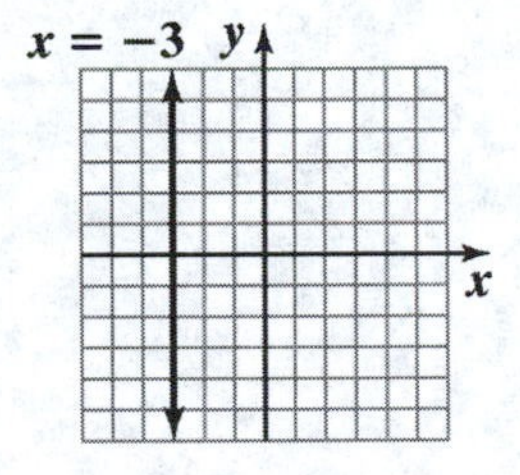

Pencil Problem #4

4a. Graph $y = -2$ in the rectangular coordinate system.

4b. Graph $x = 5$ in the rectangular coordinate system.

Objective #5: Recognize and use the general form of a line's equation.

✔ Solved Problem #5

5. Find the slope and y-intercept of the line whose equation is $3x+6y-12=0$.

Solve for y.

$$3x+6y-12=0$$
$$6y=-3x+12$$
$$\frac{6y}{6}=\frac{-3x+12}{6}$$
$$y=\frac{-3}{6}x+\frac{12}{6}$$
$$y=-\frac{1}{2}x+2$$

The coefficient of x, $-\frac{1}{2}$, is the slope, and the constant term, 2, is the y-intercept.

Pencil Problem #5

5. Find the slope and y-intercept of the line whose equation is $2x+3y-18=0$.

Objective #6: Use intercepts to graph a linear function in standard form.

✔ Solved Problem #6

6. Graph: $3x-2y-6=0$

Find the x–intercept by setting $y=0$.

$$3x-2y-6=0$$
$$3x-2(0)-6=0$$
$$3x=6$$
$$x=2$$

Find the y–intercept by setting $x=0$.

$$3x-2y-6=0$$
$$3(0)-2y-6=0$$
$$-2y=6$$
$$y=-3$$

Plot the points and draw the line that passes through them.

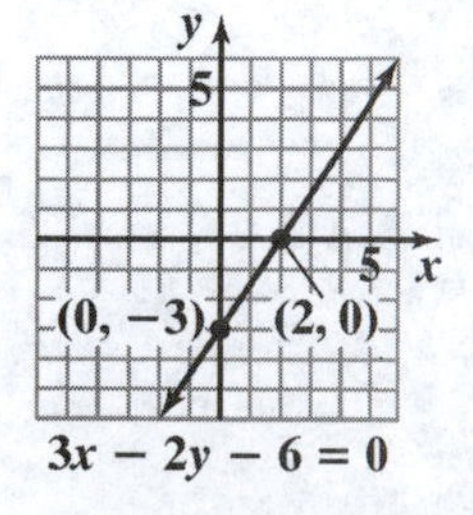

Pencil Problem #6

6. Graph: $6x-2y-12=0$

Objective #7: Model data with linear functions and make predictions.

✔ Solved Problem #7

7. The amount of carbon dioxide in the atmosphere, measured in parts per million, has been increasing as a result of the burning of oil and coal. The buildup of gases and particles is believed to trap heat and raise the planet's temperature. When the atmospheric concentration of carbon dioxide is 317 parts per million, the average global temperature is 57.04°F. When the atmospheric concentration of carbon dioxide is 354 parts per million, the average global temperature is 57.64°F.

Write a linear function that models average global temperature, $f(x)$, for an atmospheric concentration of carbon dioxide of x parts per million. Use the function to project the average global temperature when the atmospheric concentration of carbon dioxide is 600 parts per million.

Write the equation of the line through the points (317, 57.04) and (354, 57.64). First find the slope.

$$m = \frac{y_2 - y_1}{x_2 - x_1} = \frac{57.64 - 57.04}{354 - 317} = \frac{0.6}{37} \approx 0.016$$

Use this slope and the point (317, 57.04) in the point-slope form.

$$\begin{aligned} y - y_1 &= m(x - x_1) \\ y - 57.04 &= 0.016(x - 317) \\ y - 57.04 &= 0.016x - 5.072 \\ y &= 0.016x + 51.968 \end{aligned}$$

Using function notation and rounding the constant, we have

$$f(x) = 0.016x + 52.0$$

To predict the temperature when the atmospheric concentration of carbon dioxide is 600 parts per million, find $f(600)$.

$$f(600) = 0.016(600) + 52.0 = 61.6$$

The model predicts an average global temperature of 61.6°F when the atmospheric concentration of carbon dioxide is 600 parts per million.

Pencil Problem #7

7. The life expectancy for men born in 1980 is 70.0 years, and the life expectancy for men born in 2000 is 74.3 years. Let x represent the number of birth years after 1960 and y male life expectancy. Write a linear function that models life expectancy, $E(x)$, for American men born x years after 1960. Use the function to project the life expectancy of American men born in 2020.

Answers for Pencil Problems *(Textbook Exercise references in parentheses)*:

1. $\frac{1}{4}$ *(1.4 #3)*

2a. $y+3=-3(x+2); y=-3x-9$ *(1.4 #15)* **2b.** $y+1=1(x+3)$ or $y-4=1(x-2); y=x+2$ *(1.4 #29)*

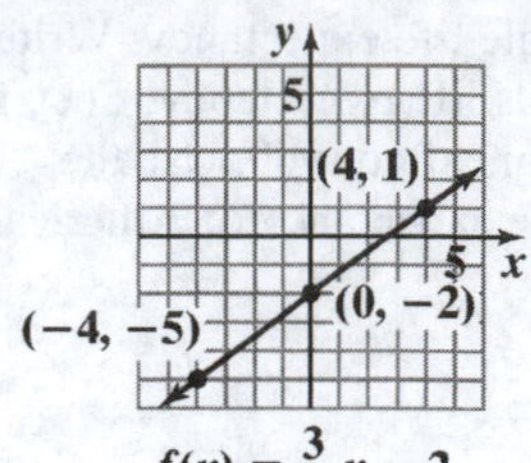

3. *(1.4 #43)*

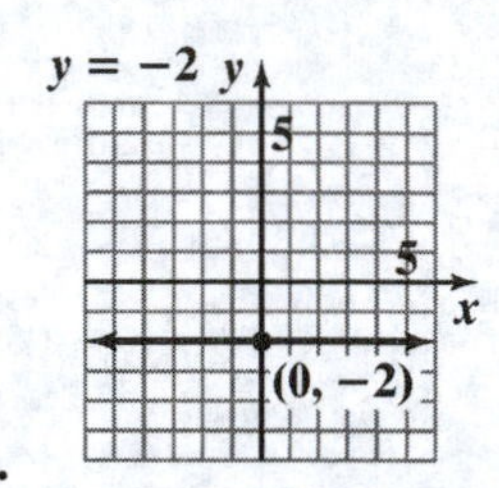

4a. *(1.4 #49)*

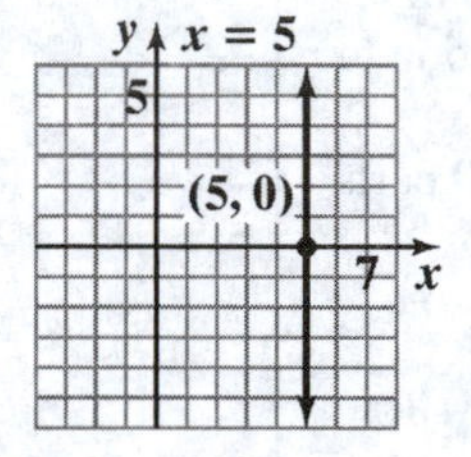

4b. *(1.4 #52)*

5. slope: $-\frac{2}{3}$; y-intercept: 6 *(1.4 #61)*

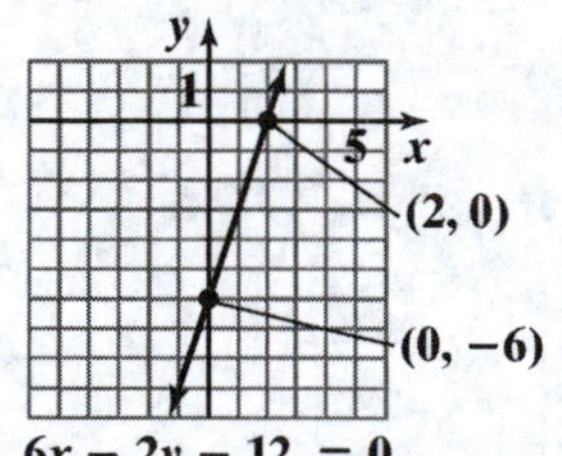

6. *(1.4 #67)*

7. $E(x)=0.215x+65.7$; 78.6 years *(1.4 #89)*

Section 1.5
More on Slope

Will They Ever Catch Up?

Many quantities, such as the number of men and the number of women living alone, are increasing over time. We can use slope to indicate how fast such quantities are growing on average.

If the slopes are the same, the quantities are growing at the same rate. However, if the slopes are different, then one quantity is growing faster than the other.
There were 9.0 million men and 14.0 million women living alone in 1990. Since then the number of men living alone has increased faster than the number of women living alone. If this trend continues, eventually, the number of men living alone will catch up to the number of women living alone.

Objective #1: Find slopes and equations of parallel and perpendicular lines.

✔ *Solved Problem #1*

1a. Write an equation of the line passing through $(-2,5)$ and parallel to the line whose equation is $y=3x+1$. Express the equation in point-slope form and slope-intercept form.

Since the line is parallel to $y=3x+1$, we know it will have slope $m=3$.

We are given that it passes through $(-2,5)$. We use the slope and point to write the equation in point-slope form.

$$y-y_1=m(x-x_1)$$
$$y-5=3(x-(-2))$$
$$y-5=3(x+2)$$

Point-Slope form: $y-5=3(x+2)$

Solve for y to obtain slope-intercept form.

$$y-5=3(x+2)$$
$$y-5=3x+6$$
$$y=3x+11$$
$$f(x)=3x+11$$

Slope-Intercept form: $y=3x+11$

Pencil Problem #1

1a. Write an equation of the line passing through $(-8,-10)$ and parallel to the line whose equation is $y=-4x+3$. Express the equation in point-slope form and slope-intercept form.

1b. Write an equation of the line passing through $(-2,-6)$ and perpendicular to the line whose equation is $x+3y-12=0$. Express the equation in point-slope form and general form.

First, find the slope of the line $x+3y-12=0$.
Solve the given equation for y to obtain slope-intercept form.

$$x+3y-12=0$$
$$3y=-x+12$$
$$y=-\frac{1}{3}x+4$$

Since the slope of the given line is $-\frac{1}{3}$, the slope of any line perpendicular to the given line is 3.

We use the slope of 3 and the point $(-2,-6)$ to write the equation in point-slope form. Then gather the variable and constant terms on one side with zero on the other side.

$$y-y_1=m(x-x_1)$$
$$y-(-6)=3(x-(-2))$$
$$y+6=3(x+2)$$
$$y+6=3x+6$$
$$0=3x-y \text{ or } 3x-y=0$$

1b. Write an equation of the line passing through $(4,-7)$ and perpendicular to the line whose equation is $x-2y-3=0$. Express the equation in point-slope form and general form.

Objective #2: Interpret slope as rate of change.

✔ *Solved Problem #2*

2. In 2000, there 11.2 million men living alone and in 2013, there were 15.0 million men living alone. Use the ordered pairs (2000, 11.2) and (2013, 15.0) to find the slope of the line through the points. Express the slope correct to two decimal places and describe what it represents.

$$m=\frac{\text{Change in } y}{\text{Change in } x}=\frac{15.0-11.2}{2013-2000}$$
$$=\frac{3.8}{13}\approx 0.29$$

The number of men living alone increased at an average rate of approximately 0.29 million men per year.

Pencil Problem #2

2. In 1994, 617 active-duty servicemembers were discharged under the “don’t ask, don’t tell” policy. In 1998, 1163 were discharged under the policy. Use the ordered pairs (1994, 617) and (1998, 1163) to find the slope of the line through the points. Express the slope correct to the nearest whole number and describe what it represents.

Objective #3: Find a function's average rate of change..

✔ *Solved Problem #3*

3a. Find the average rate of change of $f(x)=x^3$ from $x_1 = 0$ to $x_2 = 1$.

$$\frac{f(x_2)-f(x_1)}{x_2-x_1}=\frac{f(1)-f(0)}{1-0}$$
$$=\frac{1^3-0^3}{1}$$
$$=1$$

The average rate of change is 1.

3b. Find the average rate of change of $f(x)=x^3$ from $x_1 = 1$ to $x_2 = 2$.

$$\frac{f(x_2)-f(x_1)}{x_2-x_1}=\frac{f(2)-f(1)}{2-1}$$
$$=\frac{2^3-1^3}{1}$$
$$=\frac{8-1}{1}$$
$$=7$$

The average rate of change is 7.

3c. Find the average rate of change of $f(x)=x^3$ from $x_1 = -2$ to $x_2 = 0$.

$$\frac{f(x_2)-f(x_1)}{x_2-x_1}=\frac{f(0)-f(-2)}{0-(-2)}$$
$$=\frac{0^3-(-2)^3}{0+2}$$
$$=\frac{0-(-8)}{2}$$
$$=\frac{8}{2}$$
$$=4$$

The average rate of change is 4.

✎ *Pencil Problem #3* ✎

3a. Find the average rate of change of $f(x)=3x$ from $x_1 = 0$ to $x_2 = 5$.

3b. Find the average rate of change of $f(x)=x^2+2x$ from $x_1 = 3$ to $x_2 = 5$

3c. Find the average rate of change of $f(x)=\sqrt{x}$ from $x_1 = 4$ to $x_2 = 9$.

3d. The distance, $s(t)$, in feet, traveled by a ball rolling down a ramp is given by the function $s(t) = 4t^2$, where t is the time, in seconds, after the ball is released. Find the ball's average velocity from $t_1 = 1$ second to $t_2 = 2$ seconds.

$$\frac{\Delta s}{\Delta t} = \frac{s(2) - s(1)}{2 - 1}$$
$$= \frac{4 \cdot 2^2 - 4 \cdot 1^2}{1}$$
$$= \frac{16 - 4}{1}$$
$$= 12 \text{ ft/sec}$$

3d. The distance, $s(t)$, in feet, traveled by a ball rolling down a ramp is given by the function $s(t) = 10t^2$, where t is the time, in seconds, after the ball is released. Find the ball's average velocity from $t_1 = 3$ second to $t_2 = 4$ seconds.

3e. The distance, $s(t)$, in feet, traveled by a ball rolling down a ramp is given by the function $s(t) = 4t^2$, where t is the time, in seconds, after the ball is released. Find the ball's average velocity from $t_1 = 1$ second to $t_2 = 1.5$ seconds.

$$\frac{\Delta s}{\Delta t} = \frac{s(1.5) - s(1)}{1.5 - 1}$$
$$= \frac{4 \cdot 1.5^2 - 4 \cdot 1^2}{0.5}$$
$$= \frac{9 - 4}{0.5}$$
$$= 10 \text{ ft/sec}$$

3e. The distance, $s(t)$, in feet, traveled by a ball rolling down a ramp is given by the function $s(t) = 10t^2$, where t is the time, in seconds, after the ball is released. Find the ball's average velocity from $t_1 = 3$ second to $t_2 = 3.5$ seconds.

Answers for Pencil Problems *(Textbook Exercise references in parentheses)*:

1a. Point-Slope form: $y + 10 = -4(x + 8)$, Slope-Intercept form: $y = -4x - 42$ *(1.5 #5)*

1b. Point-Slope form: $y + 7 = -2(x - 4)$, General form: $2x + y - 1 = 0$ *(1.5 #11)*

2. 137; There was an average increase of approximately 137 discharges per year. *(1.5 #29)*

3a. 3 *(1.5 #13)* **3b.** 10 *(1.5 #15)* **3c.** $\frac{1}{5}$ *(1.5 #17)*

3d. 70 ft/sec *(1.5 #19a)* **3e.** 60 ft/sec *(1.5 #19b)*

Section 1.6
Transformations of Functions

Movies and Mathematics

Have you ever seen special effects in a movie where a person or object is continuously transformed into something different? This is called morphing.
In mathematics, we can use transformations of a known graph to graph a function with a similar equation. This is achieved through horizontal and vertical shifts, reflections, and stretching and shrinking of the known graph.

Objective #1: Recognize graphs of common functions.

✔ *Solved Problem #1*

1. True or false: The graphs of the standard quadratic function $f(x)=x^2$ and the absolute value function $g(x)=|x|$ have the same type of symmetry.

True. Both functions are even and their graphs are symmetric with respect to the *y*-axis.

Pencil Problem #1

1. True or false: The graphs of the identity function $f(x)=x$ and the standard cubic function $g(x)=x^3$ have the same type of symmetry.

Objective #2: Use vertical shifts to graph functions.

✔ *Solved Problem #2*

2. Use the graph of $f(x)=|x|$ to obtain the graph of $g(x)=|x|+3$.

The graph of *g* is the graph of *f* shifted vertically up by 3 units. Add 3 to each *y*-coordinate.

Since the points (–3, 3), (0, 0), and (3, 3) are on the graph of *f*, the points (–3, 6), (0, 3), and (3, 6) are on the graph of *g*.

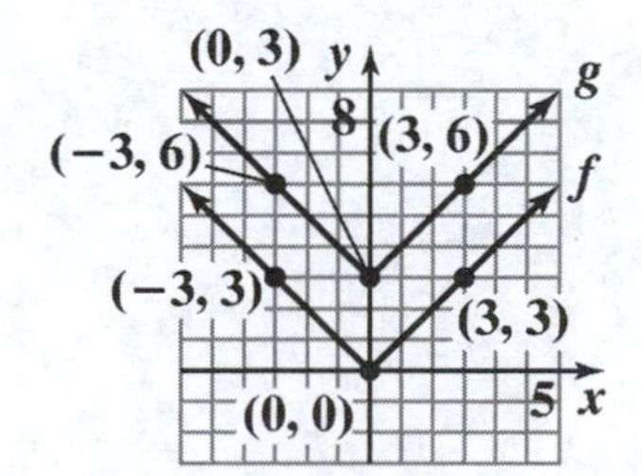

Pencil Problem #2

2. Use the graph of $f(x)=x^2$ to obtain the graph of $g(x)=x^2-2$.

Objective #3: Use horizontal shifts to graph functions.

✔ *Solved Problem #3*

3. Use the graph of $f(x)=\sqrt{x}$ to obtain the graph of $g(x)=\sqrt{x-4}$.

$g(x)=\sqrt{x-4}=f(x-4)$

The graph of g is the graph of f shifted horizontally to the right by 4 units. Add 4 to each x-coordinate.

Since the points (0, 0), (1, 1), and (4, 2) are on the graph of f, the points (4, 0), (5, 1), and (8, 2) are on the graph of g.

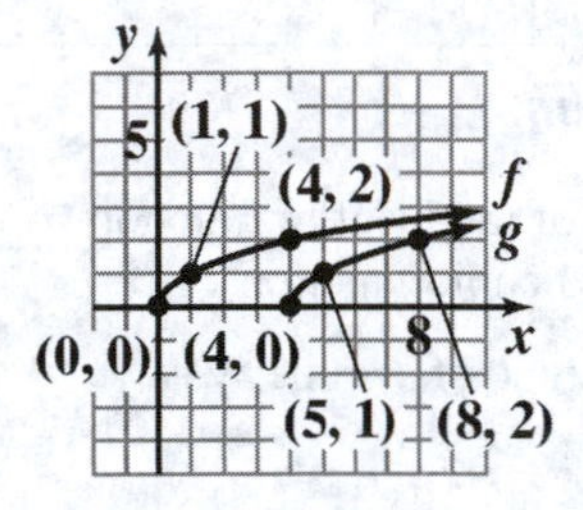

Pencil Problem #3

3. Use the graph of $f(x)=|x|$ to obtain the graph of $g(x)=|x+4|$.

Objective #4: Use reflections to graph functions.

✔ *Solved Problem #4*

4a. Use the graph of $f(x)=|x|$ to obtain the graph of $g(x)=-|x|$.

The graph of g is a reflection of the graph of f about the x-axis because $g(x)=-f(x)$. Replace each y-coordinate with its opposite.

Since the points (–3, 3), (0, 0), and (3, 3) are on the graph of f, the points (–3, –3), (0, 0), and (3, –3) are on the graph of g.

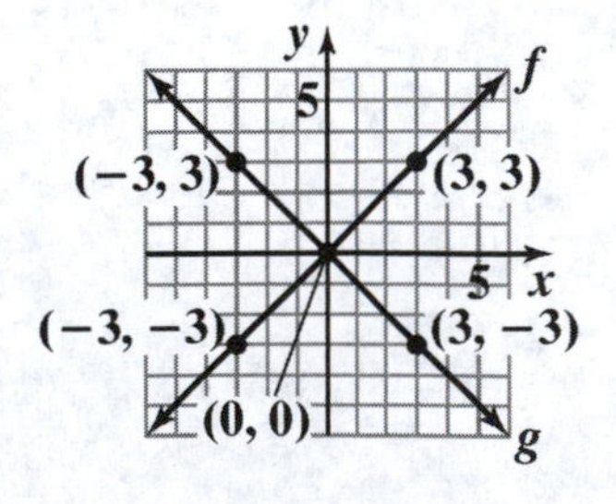

Pencil Problem #4

4a. Use the graph of $f(x)=x^3$ to obtain the graph of $h(x)=-x^3$.

4b. Use the graph of $f(x) = \sqrt[3]{x}$ to obtain the graph of $h(x) = \sqrt[3]{-x}$.

The graph of h is a reflection of the graph of f about the y-axis because $h(x) = f(-x)$. Replace each x-coordinate with its opposite.

Since the points (–1, –1), (0, 0), and (1, 1) are on the graph of f, the points (1, –1), (0, 0), and (–1, 1) are on the graph of h.

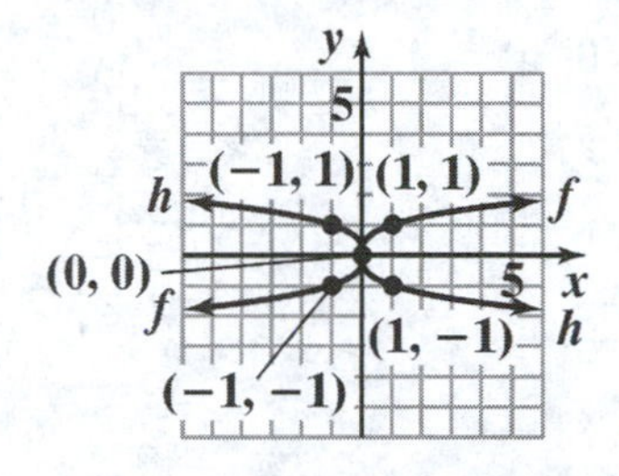

4b. Use the graph of f shown below to obtain the graph of $g(x) = f(-x)$.

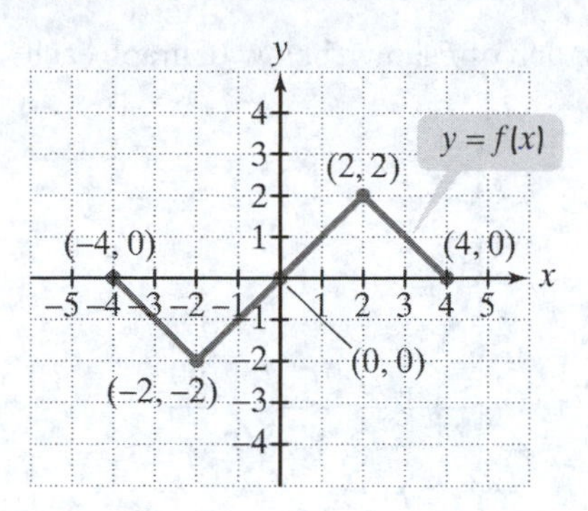

Objective #5: Use vertical stretching and shrinking to graph functions.

✔ Solved Problem #5

5. Use the graph of $f(x) = |x|$ to obtain the graph of $g(x) = 2|x|$.

The graph of g is obtained by vertically stretching the graph of f because $g(x) = 2f(x)$. Multiply each y-coordinate by 2.

Since the points (–2, 2), (0, 0), and (2, 2) are on the graph of f, the points (–2, 4), (0, 0), and (2, 4) are on the graph of g.

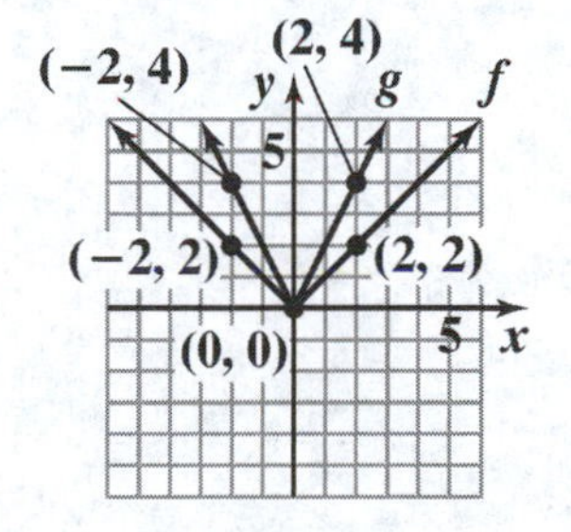

Pencil Problem #5

5. Use the graph of $f(x) = x^3$ to obtain the graph of $h(x) = \frac{1}{2}x^3$.

Objective #6: Use horizontal stretching and shrinking to graph functions.

✔ Solved Problem #6

6. Use the graph of f shown below to graph each function.

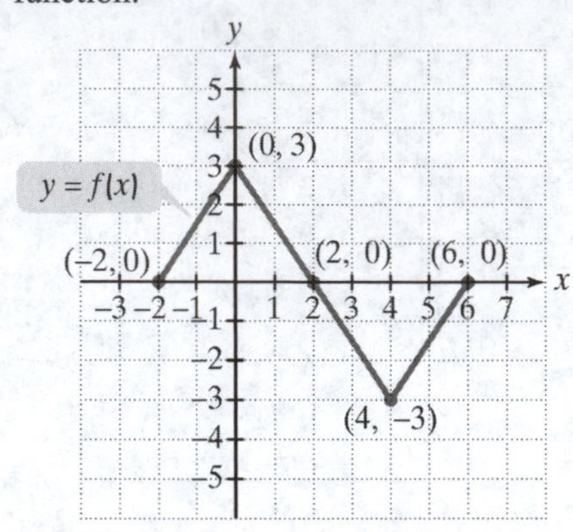

6a. $g(x) = f(2x)$

Divide the x-coordinate of each point on the graph of f by 2. The points (–1, 0), (0, 3), (1, 0), (2, –3), and (3, 0) are on the graph of g.

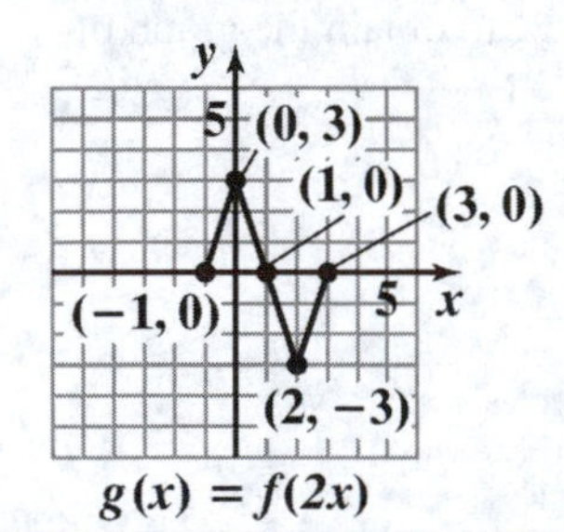

6b. $h(x) = f(\frac{1}{2}x)$

Multiply the x-coordinate of each point on the graph of f by 2. The points (–4, 0), (0, 3), (4, 0), (8, –3), and (12, 0) are on the graph of h.

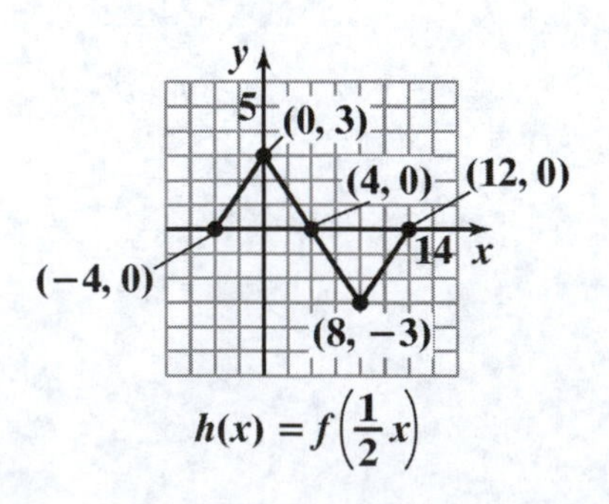

✎ Pencil Problem #6 ✎

6. Use the graph of f shown below to graph each function.

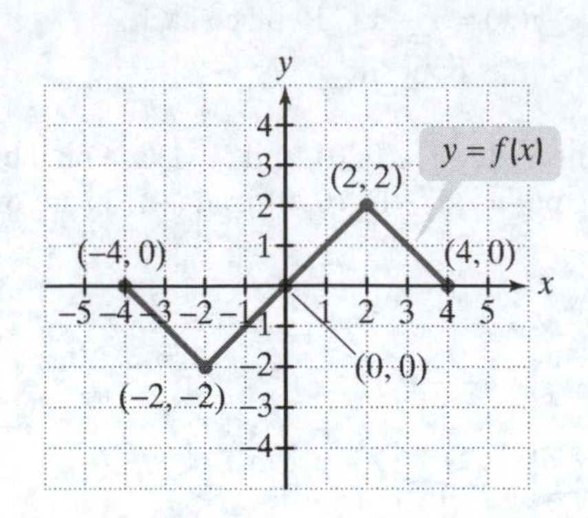

6a. $g(x) = f(2x)$

6b. $g(x) = f(\frac{1}{2}x)$

Objective #7: Graph functions involving a sequence of transformations.

✔ *Solved Problem #7*

7. Use the graph of $f(x) = x^2$ to graph

$g(x) = 2(x-1)^2 + 3.$

The graph of g is the graph of f horizontally shifted to the right 1 unit, vertically stretched by a factor of 2, and vertically shifted up 3 units. Beginning with a point on the graph of f, add 1 to each x-coordinate, then multiply each y-coordinate by 2, and finally add 3 to each y-coordinate.

$(-1, 1) \rightarrow (0, 1) \rightarrow (0, 2) \rightarrow (0, 5)$
$(0, 0) \rightarrow (1, 0) \rightarrow (1, 0) \rightarrow (1, 3)$
$(1, 1) \rightarrow (2, 1) \rightarrow (2, 2) \rightarrow (2, 5)$

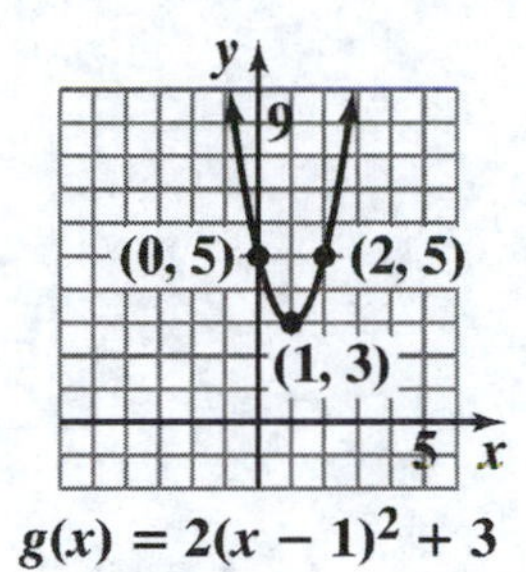

Pencil Problem #7

7. Use the graph of $f(x) = x^3$ to graph

$h(x) = \frac{1}{2}(x-3)^2 - 2.$

Answers for Pencil Problems *(Textbook Exercise references in parentheses)*:

1. True *(1.6 #99)*

2.

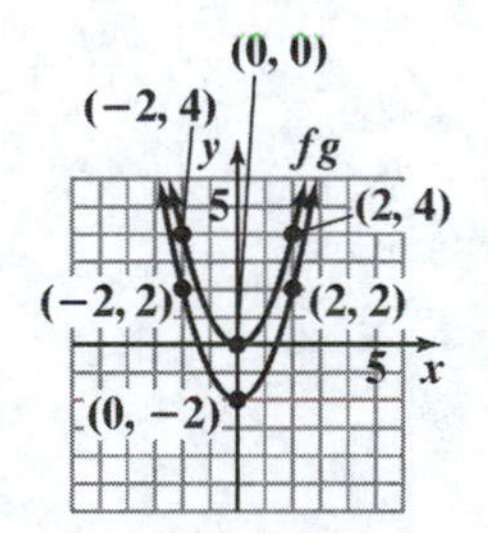

(1.6 #53)

3.

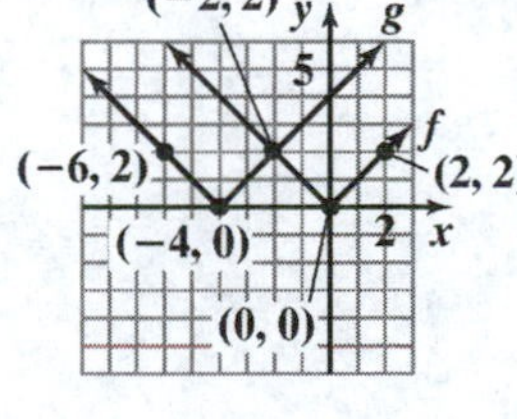

(1.6 #83)

4a.

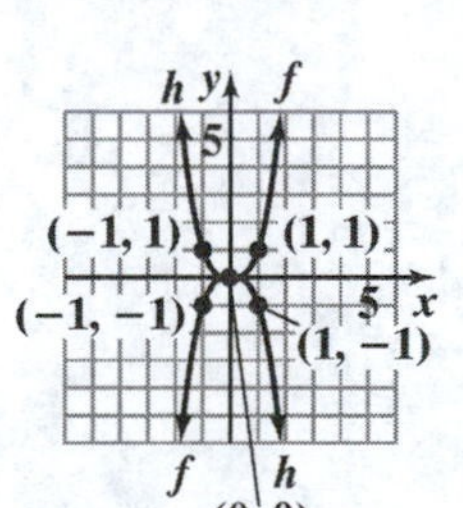

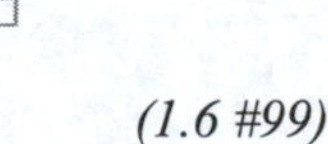

(1.6 #99)

4b.

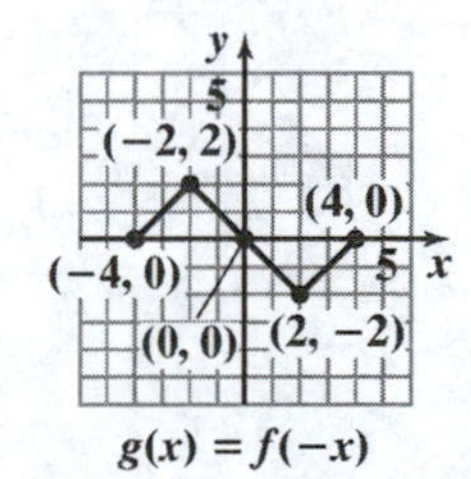

(1.6 #24)

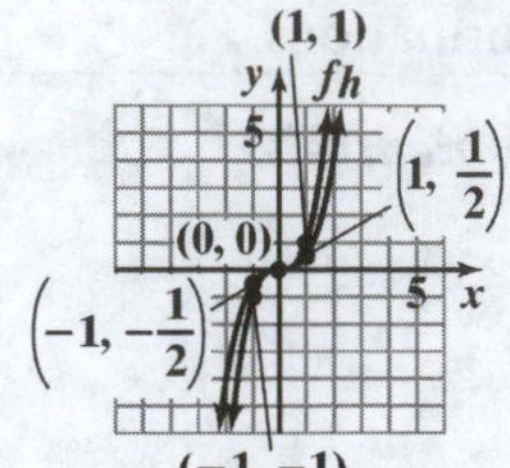

5. *(1.6 #101)*

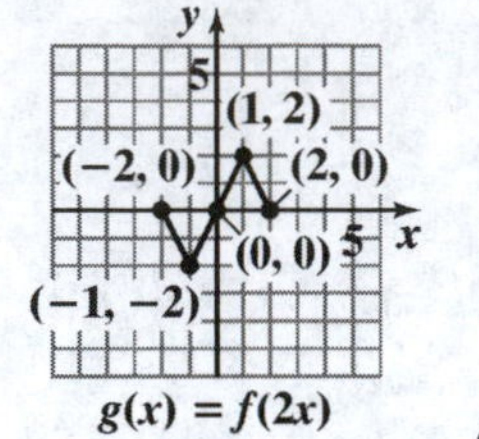

6a. *(1.6 #29)*

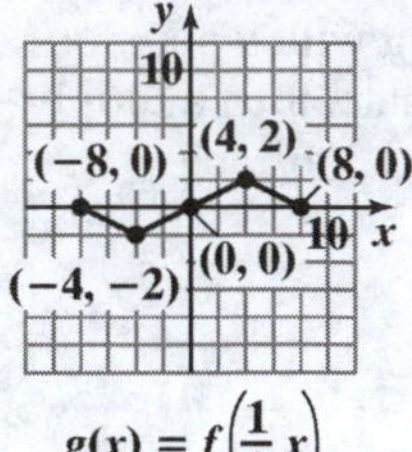

6b. *(1.6 #30)*

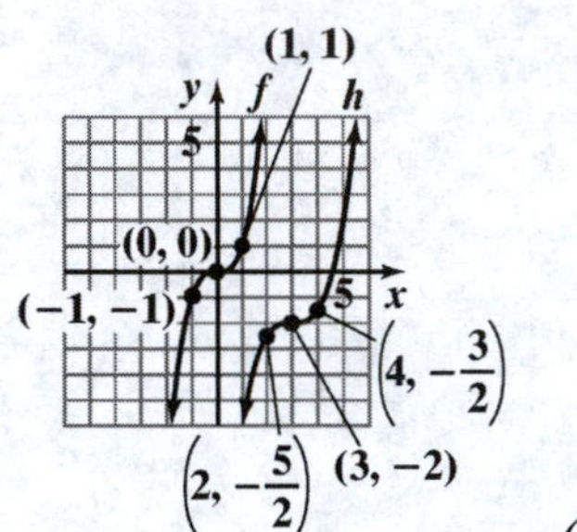

7. *(1.6 #105)*

Section 1.7
Combinations of Functions; Composite Functions

We're Born. We Die.

The figure below quantifies these statements by showing the number of births and deaths in the United States from 2000 through 2011.

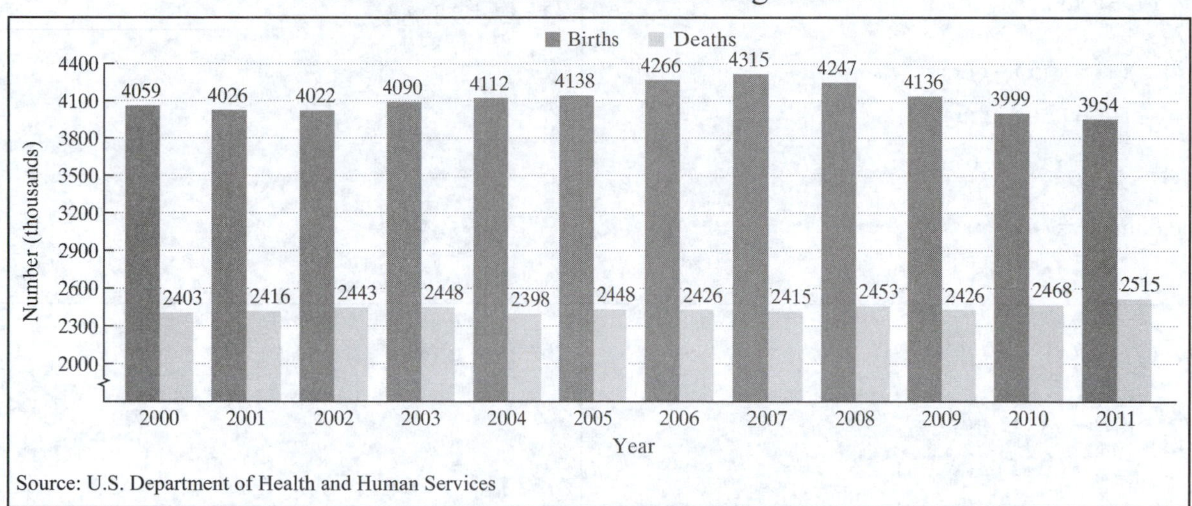

Source: U.S. Department of Health and Human Services

In this section, we look at these data from the perspective of functions. By considering the yearly change in the U.S. population, you will see that functions can be subtracted using procedures that will remind you of combining algebraic expressions.

Objective #1: Find the domain of a function.

Solved Problem #1

1a. Find the domain of $f(x) = x^2 + 3x - 17$.

The function contains neither division nor an even root. It is defined for all real numbers. The domain is $(-\infty, \infty)$.

1b. Find the domain of $j(x) = \dfrac{5x}{\sqrt{24-3x}}$.

The function contains both an even root and division. The expression under the radical must be nonnegative and the denominator cannot equal 0. Thus, $24 - 3x$ must be greater than 0.

$$24 - 3x > 0$$
$$24 > 3x$$
$$8 > x \quad \text{or} \quad x < 8$$

The domain is $(-\infty, 8)$.

Pencil Problem #1

1a. Find the domain of $f(x) = 3(x-4)$.

1b. Find the domain of $g(x) = \dfrac{\sqrt{x-2}}{x-5}$.

Objective #2: Combine functions using the algebra of functions, specifying domains.

✔ Solved Problem #2

2. Let $f(x)=x-5$ and $g(x)=x^2-1$. Find each function and determine its domain.

2a. $(f+g)(x)$

$$(f+g)(x)=f(x)+g(x)$$
$$=(x-5)+(x^2-1)$$
$$=x^2+x-6$$

The domain is $(-\infty, \infty)$.

2b. $(f-g)(x)$

$$(f-g)(x)=f(x)-g(x)$$
$$=(x-5)-(x^2-1)$$
$$=x-5-x^2+1$$
$$=-x^2+x-4$$

The domain is $(-\infty, \infty)$.

2c. $(fg)(x)$

$$(fg)(x)=f(x)\cdot g(x)$$
$$=(x-5)(x^2-1)$$
$$=x^3-5x^2-x+5$$

The domain is $(-\infty, \infty)$.

2d. $\left(\frac{f}{g}\right)(x)$

$$\left(\frac{f}{g}\right)(x)=\frac{f(x)}{g(x)}=\frac{x-5}{x^2-1}$$

The function contains division; it is undefined when $x^2-1=0$ or $x^2=1$ or $x=\pm 1$. The domain is $(-\infty,-1)\cup(-1,1)\cup(1,\infty)$.

✎ Pencil Problem #2 ✎

2. Let $f(x)=2x^2-x-3$ and $g(x)=x+1$. Find each function and determine its domain.

2a. $(f+g)(x)$

2b. $(f-g)(x)$

2c. $(fg)(x)$

2d. $\left(\frac{f}{g}\right)(x)$

Objective #3: Form composite functions.

✔ *Solved Problem #3*

3. Given $f(x)=5x+6$ and $g(x)=2x^2-x-1$, find each of the following.

3a. $(f\circ g)(x)$

$$
\begin{aligned}
(f\circ g)(x) &= f(g(x))\\
&= 5g(x)+6\\
&= 5(2x^2-x-1)+6\\
&= 10x^2-5x-5+6\\
&= 10x^2-5x+1
\end{aligned}
$$

3b. $(g\circ f)(x)$

$$
\begin{aligned}
(g\circ f)(x) &= g(f(x))\\
&= 2[f(x)]^2-f(x)-1\\
&= 2(5x+6)^2-(5x+6)-1\\
&= 2(25x^2+60x+36)-5x-6-1\\
&= 50x^2+120x+72-5x-7\\
&= 50x^2+115x+65
\end{aligned}
$$

3c. $(f\circ g)(-1)$

$$
\begin{aligned}
(f\circ g)(-1) &= 10(-1)^2-5(-1)+1\\
&= 10+5+1\\
&= 16
\end{aligned}
$$

✎ *Pencil Problem #3* ✎

3. Given $f(x)=4x-3$ and $g(x)=5x^2-2$, find each of the following.

3a. $(f\circ g)(x)$

3b. $(g\circ f)(x)$

3c. $(f\circ g)(2)$

Objective #4: Determine domains for composite functions.

✔ *Solved Problem #4*

4. Given $f(x)=\dfrac{4}{x+2}$ and $g(x)=\dfrac{1}{x}$, find each of the following.

4a. $(f\circ g)(x)$

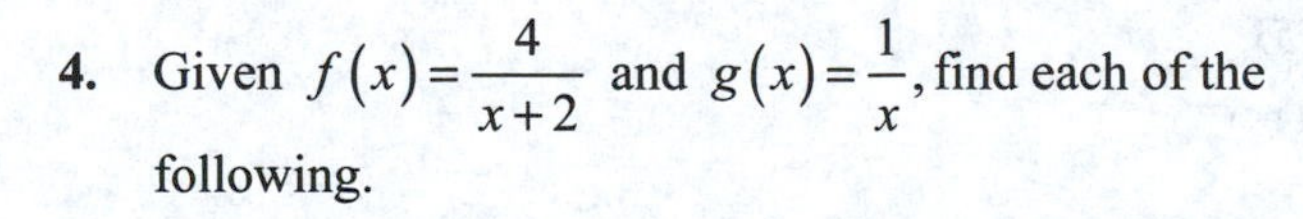

$$(f\circ g)(x)=\frac{4}{g(x)+2}=\frac{4}{\frac{1}{x}+2}\cdot\frac{x}{x}=\frac{4x}{1+2x}$$

✎ *Pencil Problem #4* ✎

4. Given $f(x)=\dfrac{2}{x+3}$ and $g(x)=\dfrac{1}{x}$, find each of the following.

4a. $(f\circ g)(x)$

4b. The domain of $f \circ g$

The function g is undefined when $x = 0$, so 0 is not in the domain of $f \circ g$. The function f is undefined for $x = -2$, so any values of x for which $g(x) = -2$ are not in the domain of $f \circ g$. Solving $\frac{1}{x} = -2$, we find that $x = -\frac{1}{2}$.

The domain is $\left(-\infty, -\frac{1}{2}\right) \cup \left(-\frac{1}{2}, 0\right) \cup (0, \infty)$.

4b. The domain of $f \circ g$

Objective #5: Write functions as compositions.

✔ *Solved Problem #5*

5. Express $h(x) = \sqrt{x^2 + 5}$ as the composition of two functions.

A natural way to write h as the composition of two functions is to take the square root of $g(x) = x^2 + 5$.

Let $f(x) = \sqrt{x}$ and $g(x) = x^2 + 5$.

Then $(f \circ g)(x) = \sqrt{g(x)} = \sqrt{x^2 + 5} = h(x)$.

Pencil Problem #5

5. Express $h(x) = (3x - 1)^4$ as the composition of two functions.

Answers for Pencil Problems *(Textbook Exercise references in parentheses)*:

1a. $(-\infty, \infty)$ *(1.7 #1)* **1b.** $[2, 5) \cup (5, \infty)$ *(1.7 #27)*

2a. $(f + g)(x) = 2x^2 - 2$; domain: $(-\infty, \infty)$ *(1.7 #35)*

2b. $(f - g)(x) = 2x^2 - 2x - 4$; domain: $(-\infty, \infty)$ *(1.7 #35)*

2c. $(fg)(x) = 2x^3 + x^2 - 4x - 3$; domain: $(-\infty, \infty)$ *(1.7 #35)*

2d. $\left(\frac{f}{g}\right)(x) = \frac{2x^2 - x - 3}{x + 1}$; domain: $(-\infty, -1) \cup (1, \infty)$ *(1.7 #35)*

3a. $(f \circ g)(x) = 20x^2 - 11$ *(1.7 #55a)* **3b.** $(g \circ f)(x) = 80x^2 - 120x + 43$ *(1.7 #55 b)* **3c.** 69 *(1.7 #55c)*

4a. $(f \circ g)(x) = \frac{2x}{1 + 3x}$ *(1.7 #67a)* **4b.** $\left(-\infty, -\frac{1}{3}\right) \cup \left(-\frac{1}{3}, 0\right) \cup (0, \infty)$ *(1.7 #67b)*

5. Let $f(x) = x^4$ and $g(x) = 3x - 1$. Then $(f \circ g)(x) = h(x)$. *(1.7 #75)*

Section 1.8
Inverse Functions

Hey! That's My Birthday Too !

What is the probability that two people in the same room share a birthday?

It might be higher than you think.

In this section we will explore the graph of the function that represents this probability.

Objective #1: Verify inverse functions.

✔ *Solved Problem #1*

1. Show that each function is the inverse of the other:

$f(x) = 4x - 7$ and $g(x) = \dfrac{x+7}{4}$.

First, show that $f(g(x)) = x$.

$$f(g(x)) = 4\left(\frac{x+7}{4}\right) - 7$$
$$= x + 7 - 7$$
$$= x$$

Next, show that $g(f(x)) = x$.

$$g(f(x)) = \frac{(4x-7)+7}{4}$$
$$= \frac{4x}{4}$$
$$= x$$

Pencil Problem #1

1. Determine whether $f(x) = \dfrac{3}{x-4}$ and $g(x) = \dfrac{3}{x} + 4$ are inverses of each other.

Objective #2: Find the inverse of a function.

✔ *Solved Problem #2*

2a. Find the inverse of $f(x) = 2x + 7$.

Replace $f(x)$ with y.

$y = 2x + 7$

Interchange x and y and solve for y.

$$x = 2y + 7$$
$$x - 7 = 2y$$
$$\frac{x-7}{2} = y$$

Replace y with $f^{-1}(x)$.

$$f^{-1}(x) = \frac{x-7}{2}$$

Pencil Problem #2

2a. Find the inverse of $f(x) = x + 3$.

2b. Find the inverse of $f(x) = 4x^3 - 1$.

Replace *f*(*x*) with *y*.

$$y = 4x^3 - 1$$

Interchange *x* and *y* and solve for *y*.

$$x = 4y^3 - 1$$
$$x + 1 = 4y^3$$
$$\frac{x+1}{4} = y^3$$
$$\sqrt[3]{\frac{x+1}{4}} = y$$

Replace *y* with $f^{-1}(x)$.

$$f^{-1}(x) = \sqrt[3]{\frac{x+1}{4}}$$

2b. Find the inverse of $f(x) = (x+2)^3$.

2c. Find the inverse of $f(x) = \frac{x+1}{x-5}$.

Replace *f*(*x*) with *y*.

$$y = \frac{x+1}{x-5}$$

Interchange *x* and *y* and solve for *y*.

$$x = \frac{y+1}{y-5}$$
$$x(y-5) = y+1$$
$$xy - 5x = y+1$$
$$xy - y = 5x+1$$
$$y(x-1) = 5x+1$$
$$y = \frac{5x+1}{x-1}$$

Replace *y* with $f^{-1}(x)$.

$$f^{-1}(x) = \frac{5x+1}{x-1}$$

2c. Find the inverse of $f(x) = \frac{x+4}{x-2}$.

Objective #3: Use the horizontal line test to determine if a function has an inverse function.

✔ *Solved Problem #3*

3. Use the horizontal line test to determine if the following graph represents a function that has an inverse function.

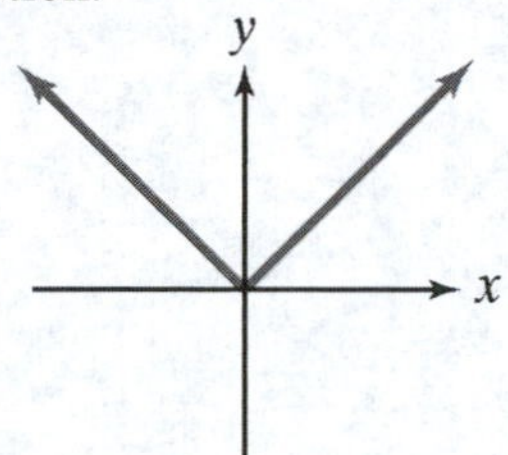

Since a horizontal line can be drawn that intersects the graph more than once, it fails the horizontal line test.

Thus, this graph does not represent a function that has an inverse function.

✎ *Pencil Problem #3* ✎

3. Use the horizontal line test to determine if the following graph represents a function that has an inverse function.

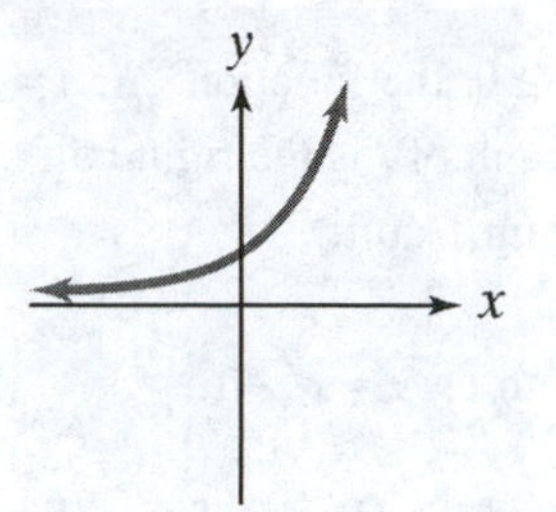

Objective #4: Use the graph of a one-to-one function to graph its inverse function.

✔ *Solved Problem #4*

4. The graph of function f consists of two line segments, one segment from $(-2,-2)$ to $(-1,0)$, and a second segment from $(-1,0)$ to $(1,2)$. Graph f and use the graph to draw the graph of its inverse function.

Since f has a line segment from $(-2,-2)$ to $(-1,0)$, then f^{-1} has a line segment from $(-2,-2)$ to $(0,-1)$.

Since f has a line segment from $(-1,0)$ to $(1,2)$, then f^{-1} has a line segment from $(0,-1)$ to $(2,1)$.

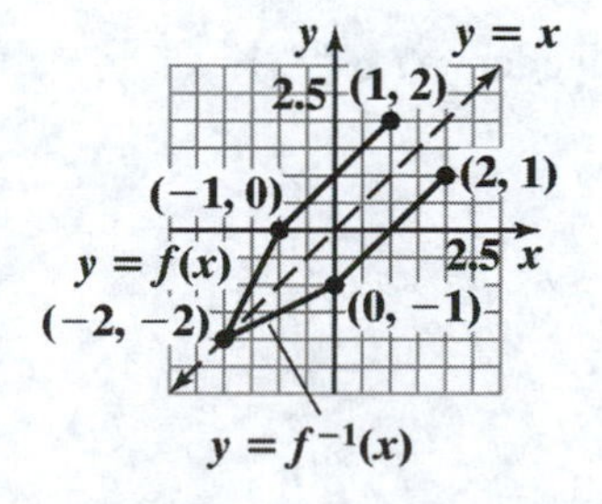

✎ *Pencil Problem #4* ✎

4. The graph of a linear function f contains the points $(0,-4)$, $(2,0)$, $(3,2)$, and $(4,4)$. Draw the graph of the inverse function.

Objective #5: Find the inverse of a function and graph both functions on the same axes.

✔ Solved Problem #5

5. Find the inverse of $f(x)=x^2+1$ if $x \ge 0$. Graph f and f^{-1} in the same rectangular coordinate system.

Restricted to $x \ge 0$, the function $f(x)=x^2+1$ has an inverse. The graph of f is the right half of the graph of $y=x^2$ shifted up 1 unit.

Replace $f(x)$ with y: $y=x^2+1$.

Interchange x and y and solve for y. Since the values of x are nonnegative in the original function, the values of y must be nonnegative in the inverse function. We choose the positive square root in the third step below.

$$x=y^2+1$$
$$x-1=y^2$$
$$\sqrt{x-1}=y$$

Replace y with $f^{-1}(x)$: $f^{-1}(x)=\sqrt{x-1}$.

The graph of f^{-1} is the graph of the square root function shifted 1 unit to the left. The graph of f^{-1} is also the reflection of the graph of f about the line $y=x$.

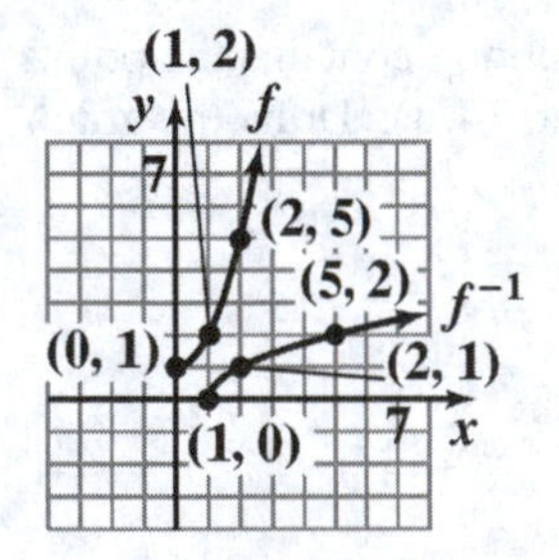

✎ Pencil Problem #5 ✎

5. Find the inverse of $f(x)=(x-1)^2$ if $x \le 1$. Graph f and f^{-1} in the same rectangular coordinate system.

Answers for Pencil Problems *(Textbook Exercise references in parentheses)*:

1. The functions are inverses of each other. *(1.8 #7)* **2a.** $f^{-1}(x)=x-3$ *(1.8 #11)*

2b. $f^{-1}(x)=\sqrt[3]{x}-2$ *(1.8 #19)* **2c.** $f^{-1}(x)=\dfrac{2x+4}{x-1}$ *(1.8 #25)* **3.** has inverse function *(1.8 #33)*

4.

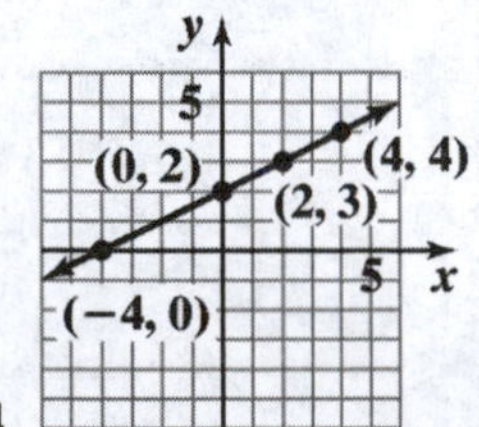

(1.8 #35)

5.

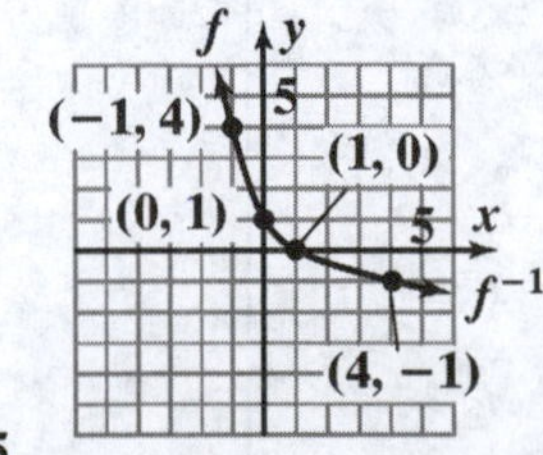

(1.8 #43)

Section 1.9
Distance and Midpoint Formulas; Circles

Round and Round !

In 1893, George Washington Gale Ferris, Jr. designed and built the first Ferris wheel as the centerpiece for the World's Columbian Exposition in Chicago.

The rectangular coordinate system gives us a unique way of knowing a circle. It enables us to translate a circle's geometric definition into an algebraic equation. In this section, we will learn, and then apply, these algebraic techniques.

Objective #1: Find the distance between two points.

✔ *Solved Problem #1*

1. Find the distance between $(-1,-3)$ and $(2,3)$. Express the answer in simplified radical form and then round to two decimal places.

$$d=\sqrt{(x_2-x_1)^2+(y_2-y_1)^2}$$
$$d=\sqrt{(2-(-1))^2+(3-(-3))^2}$$
$$=\sqrt{3^2+6^2}$$
$$=\sqrt{9+36}$$
$$=\sqrt{45}$$
$$=3\sqrt{5}$$
$$\approx 6.71 \text{ units}$$

Pencil Problem #1

1. Find the distance between $(4,-1)$ and $(-6,3)$. Express the answer in simplified radical form and then round to two decimal places.

Objective #2: Find the midpoint of a line segment.

✔ *Solved Problem #2*

2. Find the midpoint of the line segment with endpoints $(1,2)$ and $(7,-3)$.

$$\text{Midpoint}=\left(\frac{x_1+x_2}{2},\frac{y_1+y_2}{2}\right)$$
$$=\left(\frac{1+7}{2},\frac{2+(-3)}{2}\right)$$
$$=\left(\frac{8}{2},\frac{-1}{2}\right)$$
$$=\left(4,-\frac{1}{2}\right)$$

Pencil Problem #2

2. Find the midpoint of the line segment with endpoints $(6,8)$ and $(2,4)$.

Objective #3: Write the standard form of a circle's equation.

✔ *Solved Problem #3*

3a. Write the standard form of the equation of the circle with center $(0,0)$ and radius 4.

$$(x-h)^2+(y-k)^2=r^2$$
$$(x-0)^2+(y-0)^2=4^2$$
$$x^2+y^2=16$$

Pencil Problem #3

3a. Write the standard form of the equation of the circle with center $(0,0)$ and radius 7.

✔ *Solved Problem #3*

3b. Write the standard form of the equation of the circle with center $(0,-6)$ and radius 10.

$$(x-h)^2+(y-k)^2=r^2$$
$$(x-0)^2+(y-(-6))^2=10^2$$
$$x^2+(y+6)^2=100$$

Pencil Problem #3

3b. Write the standard form of the equation of the circle with center $(-1,4)$ and radius 2.

Objective #4: Give the center and radius of a circle whose equation is in standard form.

✔ *Solved Problem #4*

4a. Find the center and radius of the circle whose equation is $(x+3)^2+(y-1)^2=4$.

$$(x+3)^2+(y-1)^2=4$$
$$(x-(-3))^2+(y-1)^2=2^2$$

The center is $(-3,1)$ and the radius is 2 units.

Pencil Problem #4

4a. Find the center and radius of the circle whose equation is $(x-3)^2+(y-1)^2=36$.

4b. Graph the equation in Solved Problem 4a.

Plot points 2 units above and below and to the left and right of the center, (–3,1). Draw a circle through these points.

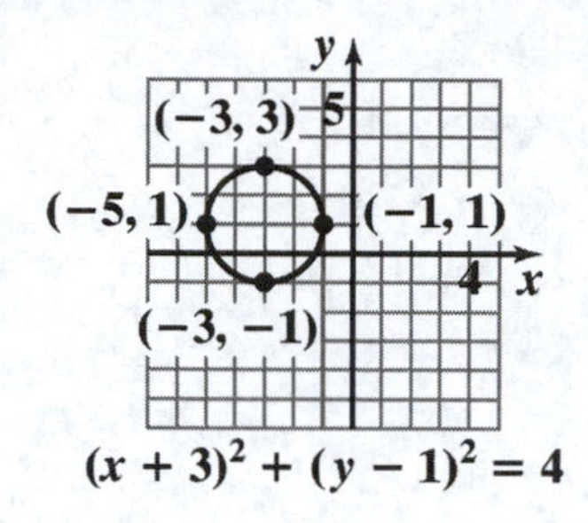

$(x + 3)^2 + (y - 1)^2 = 4$

4b. Graph the equation in Pencil Problem 4a.

4c. Use the graph in Solved Problem 4b to identify the relation's domain and range.

The leftmost point on the circle has an x-coordinate of –5, and the rightmost point has an x-coordinate of –1. The domain is [–5, –1].

The lowest point on the graph has a y-coordinate of –1, and the highest point on the graph has a y-coordinate of 3. The range is [–1, 3].

4c. Use the graph in Pencil Problem 4b to identify the relation's domain and range.

Objective #5: Convert the general form of a circle's equation to standard form.

✔ *Solved Problem #5*

5. Write in standard form and graph:

$$x^2 + y^2 + 4x - 4y - 1 = 0$$

$$x^2 + y^2 + 4x - 4y - 1 = 0$$

$$\left(x^2 + 4x \quad\right) + \left(y^2 - 4y \quad\right) = 1$$

Complete the squares.

For x: $\left(\frac{b}{2}\right)^2 = \left(\frac{4}{2}\right)^2 = (2)^2 = 4$

For y: $\left(\frac{b}{2}\right)^2 = \left(\frac{-4}{2}\right)^2 = (-2)^2 = 4$

✎ *Pencil Problem #5* ✎

5. Write in standard form and graph:

$$x^2 + y^2 + 8x - 2y - 8 = 0$$

Add these values to both sides of the equation.

$$x^2+y^2+4x-4y-1=0$$
$$\left(x^2+4x\quad\right)+\left(y^2-4y\quad\right)=1$$
$$\left(x^2+4x+4\right)+\left(y^2-4y+4\right)=1+4+4$$
$$(x+2)^2+(y-2)^2=9$$
$$(x-(-2))^2+(y-2)^2=3^2$$

The center is $(-2,2)$ and the radius is 3 units.

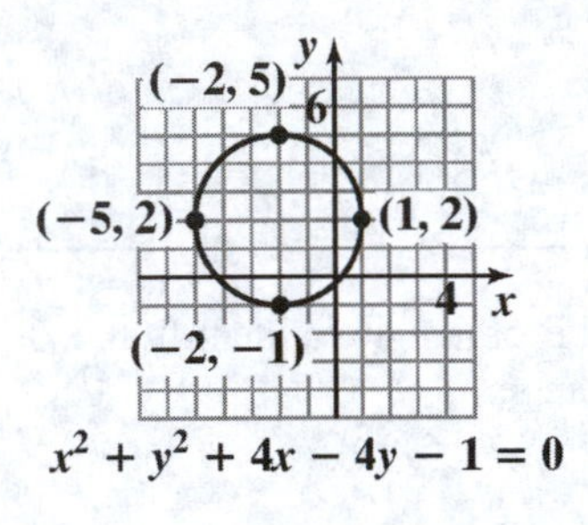

Answers for Pencil Problems *(Textbook Exercise references in parentheses)*:

1. $2\sqrt{29}\approx 10.77$ *(1.9 #3)* **2.** $(4,6)$ *(1.9 #19)*

3a. $x^2+y^2=49$ *(1.9 #31)*

3b. $(x+1)^2+(y-4)^2=4$ *(1.9 #35)*

4a. center: $(3,1)$; radius: 6 units *(1.9 #43)*

4b.

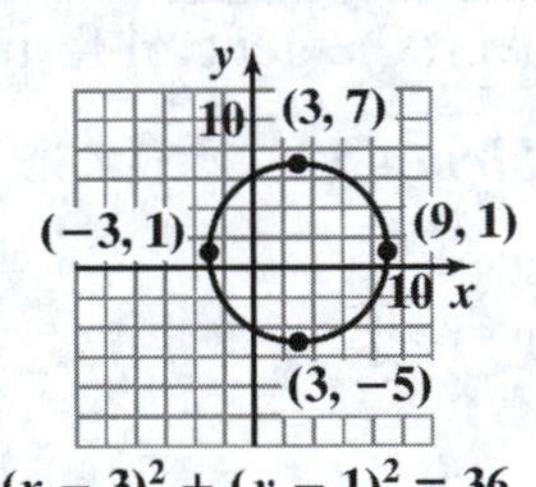

$(x-3)^2+(y-1)^2=36$ *(1.9 #43)*

4c. domain: $[-3,9]$; range: $[-5,7]$ *(1.9 #43)*

5. $(x+4)^2+(y-1)^2=25$;

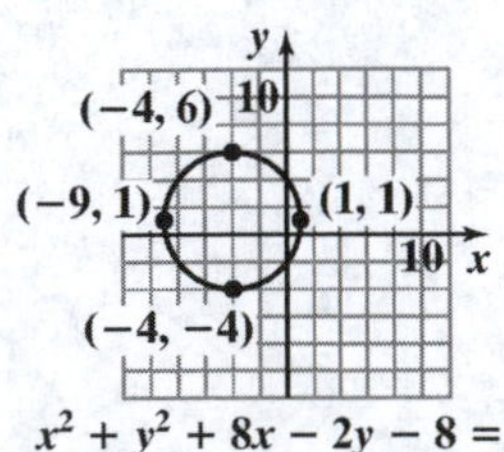

$x^2+y^2+8x-2y-8=0$ *(1.9 #57)*

Section 1.10
Modeling with Functions

The Mathematics of Soda?

Although we have concerns about the effects of soft drinks on your health, we do want to explore the connection between mathematics and the size and shape of your soda can. Given that the can is supposed to hold 12 ounces of soda, how can we determine the dimensions of the can and its surface area?

In this section, you will learn how to express the surface area of a can with a fixed volume as a function of its radius, In calculus, you will take the problem a step further and find the radius that minimizes the surface area of the soda can.

Objective #1: Construct functions from verbal conditions.

✔ *Solved Problem #1*

1a. You are choosing between two texting plans. Plan A has a monthly fee of $15 with a charge of $0.08 per text. Plan B has a monthly fee of $3 with a charge of $0.12 per text. Express the monthly cost for each plan as a function of the numbers of text messages in a month, x. For how many text messages will the costs for the two plans be the same?

We start with plan A. There is a fee of $15 plus a charge of $0.08 per text for x text messages. Multiply the number of texts, x, by the cost per text, $0.08, and add to the monthly fee, $15.

$f(x) = 15 + 0.08x$ or $f(x) = 0.08x + 15$

For plan B, there is a fee of $3 plus a charge of $0.12 per text for x text messages. Multiply the number of texts, x, by the cost per text, $0.12, and add to the monthly fee, $3.

$g(x) = 3 + 0.12x$ or $g(x) = 0.12x + 3$

To find when the costs of the two plans are the same, we solve $f(x) = g(x)$.

$$\begin{aligned} f(x) &= g(x) \\ 0.08x + 15 &= 0.12x + 3 \\ 15 &= 0.04x + 3 \\ 12 &= 0.04x \\ 300 &= x \end{aligned}$$

The two plans have the same cost for 300 text messages. To check this, note that the cost of plan A for 300 text messages is

$f(300) = 0.08(300) + 15 = 24 + 15 = \$39,$

and the cost of plan B is

$g(300) = 0.12(300) + 3 = 36 + 3 = \$39.$

✎ *Pencil Problem #1* ✎

1a. You are choosing between two plans at a discount warehouse. Plan A has an annual membership of $100 and you pay 80% of the manufacturer's recommended list price. Plan B has an annual membership of $40 and you pay 90% of the manufacturer's recommended list price. Express the total yearly amount paid to the warehouse for each plan as a function of the dollars of merchandise purchased, x. For how many dollars of merchandise will the amount paid under the two plans be the same?

1b. On a certain route, an airline carries 8000 passengers per month, each paying \$100. For each \$1 increase in ticket price, the airline will lose 100 passengers. Express the number of passengers per month, N, as a function of the ticket price, x. Then express the monthly revenue for the route, R, as a function of the ticket price, x.

If the new ticket price is x dollars, then the amount the ticket price has increased over the original price, \$100, is $x - 100$.

For each \$1 increase, the airline will lose 100 passengers. So for an increase of $(x - 100)$ dollars, the airline will lose $100(x - 100)$ passengers.

The airline now has 8000 passengers. If it loses $100(x - 100)$ passengers, it will have $8000 - 100(x - 100)$ passengers.

$$\begin{aligned} N(x) &= 8000 - 100(x - 100) \\ &= 8000 - 100x + 10{,}000 \\ &= -100x + 18{,}000 \end{aligned}$$

So, $N(x) = -100x + 18{,}000$ models the number of passengers per month.

Since the ticket price is x dollars and there are $N(x)$ passengers, multiply $N(x)$ by x to find the revenue.

$$\begin{aligned} R(x) &= (-100x + 18{,}000) \cdot x \\ &= -100x^2 + 18{,}000x \end{aligned}$$

So, $R(x) = -100x^2 + 18{,}000x$ models the monthly revenue for the route.

1b. With a ticket price of \$20, the average attendance at football games is 30,000. For each \$1 increase in ticket price, attendance decreases by 500. Express the average attendance at a football game, N, as a function of the ticket price, x. Then express the revenue from a football game, R, as a function of the ticket price, x.

Objective #2: Construct functions from formulas.

✔ *Solved Problem #2*

2a. You have 200 feet of fencing to enclose a rectangular garden. Express the area, A, of the garden as function of one of its dimensions, x.

Let x and y be the length and the width of the garden, respectively. Since the garden is rectangular, it area is then $A = xy$. However, we want to express the area in terms of x only. We need to substitute an expression for y in terms of x.

We know that the perimeter of the garden will be 200 feet, since this is the amount of fencing available. We can also express the perimeter in terms of its length, x, and width, y.

Perimeter = 200 feet
Perimeter = $2x + 2y$

Solve the equation $2x + 2y = 200$ for y.

$$2x + 2y = 200$$
$$2y = 200 - 2x$$
$$\frac{2y}{2} = \frac{200 - 2x}{2}$$
$$y = 100 - x$$

Now replace y with $100 - x$ in $A = xy$ and write the result as a function.

$A(x) = x(100 - x)$ or $A(x) = 100x - x^2$

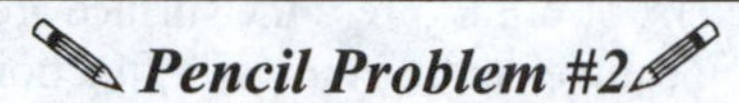

Pencil Problem #2

2a. You have 800 feet of fencing to enclose a rectangular field. Express the area, A, of the field as function of one of its dimensions, x.

2b. A cylindrical can is to hold 1 liter, or 1000 cubic centimeters, of oil. Express the surface area of the can, A, in square centimeters, as a function of its radius, r, in centimeters.

$$A = 2\pi r^2 + 2\pi rh$$

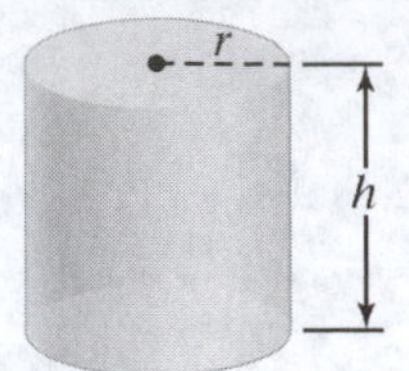

The surface area of a cylindrical can is given by the formula $A = 2\pi r^2 + 2\pi rh$, where r is the radius and h is the height. We need to express this in terms of r only. We need to replace h with an expression in terms of r.

The volume of the can is 1000 cubic centimeters, and volume is given by the formula $V = \pi r^2 h$, where r is the radius and h is the height. Replace V with 1000 in the formula and solve for h.

$$V = \pi r^2 h$$

$$1000 = \pi r^2 h$$

$$\frac{1000}{\pi r^2} = \frac{\pi r^2 h}{\pi r^2}$$

$$\frac{1000}{\pi r^2} = h$$

Now replace h in the surface area formula with $\frac{1000}{\pi r^2}$.

$$A = 2\pi r^2 + 2\pi rh$$

$$= 2\pi r^2 + 2\pi r\left(\frac{1000}{\pi r^2}\right)$$

$$= 2\pi r^2 + \frac{2000}{r}$$

Now write as a function.

$$A(r) = 2\pi r^2 + \frac{2000}{r}$$

2b. The figure shows an open box with a square base. The box is to have a volume of 10 cubic feet. Express the amount of material needed to construct the box, A, as a function of the length of its square base, x. [Hint: How is this problem similar to Solved Problem #2b? Both involve finding a function for the surface area of a figure when the volume is given.]

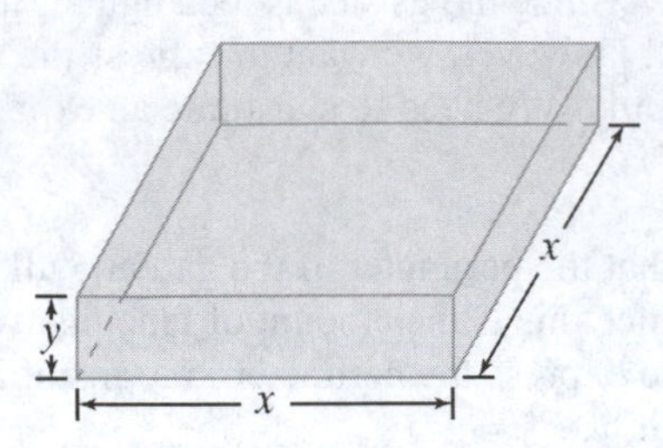

2c. Let $P = (x, y)$ be a point on the graph of $y = x^3$. Express the distance, d, from P to the origin as a function of the point's x-coordinate.

Since P is on the graph of $y = x^3$, the y-coordinate of P can be replaced by x^3. Thus,
$P = (x, y) = (x, x^3)$

Recall that the origin is the point with coordinates (0, 0). We use the distance formula to find the distance between (x, x^3) and (0, 0).

$$d = \sqrt{(x-0)^2 + (x^3-0)^2} = \sqrt{(x)^2 + (x^3)^2} = \sqrt{x^2 + x^6}$$

Now write as a function.
$d(x) = \sqrt{x^2 + x^6}$ or $d(x) = \sqrt{x^6 + x^2}$

2c. Let $P = (x, y)$ be a point on the graph of $y = x^2 - 4$. Express the distance, d, from P to the origin as a function of the point's x-coordinate.

Answers for Pencil Problems *(Textbook Exercise references in parentheses)*:

1a. Plan A: $f(x) = 100 + 0.8x$; plan B: $g(x) = 40 + 0.9x$; \$600 of merchandise *(1.10 #7)*

1b. Attendance: $N(x) = -500x + 40{,}000$; revenue: $R(x) = -500x^2 + 40{,}000x$ *(1.10 #9)*

2a. $A(x) = x(400 - x)$ or $A(x) = 400x - x^2$ *(1.10 #21)*

2b. $A(x) = \dfrac{40}{x} + x^2$ *(1.10 #31)*

2c. $d(x) = \sqrt{x^4 - 7x^2 + 16}$ *(1.10 #35)*

Section 2.1
Complex Numbers

Why Study Something if it is IMAGINARY???

Great Question!
The numbers that we study in this section were given the name "*imaginary*" at a time when mathematicians believed such numbers to be useless.

Since that time, many *real-life* applications for so-called imaginary numbers have been discovered, but the name they were originally given has endured.

Objective #1: Add and subtract complex numbers.

Solved Problem #1

1a. Add: $(5-2i)+(3+3i)$

$$\begin{aligned}(5-2i)+(3+3i) &= 5-2i+3+3i\\ &= 8+i\end{aligned}$$

1b. Subtract: $(2+6i)-(12-4i)$

$$\begin{aligned}(2+6i)-(12-4i) &= 2+6i-12+4i\\ &= -10+10i\end{aligned}$$

Pencil Problem #1

1a. Add: $(7+2i)+(1-4i)$

1b. Subtract: $(3+2i)-(5-7i)$

Objective #2: Multiply complex numbers.

✔ *Solved Problem #2*

2a. Multiply: $(5+4i)(6-7i)$

$$
\begin{aligned}
(5+4i)(6-7i) &= 30-35i+24i-28i^2 \\
&= 30-35i+24i-28(-1) \\
&= 30+28-35i+24i \\
&= 58-11i
\end{aligned}
$$

2b. Multiply: $7i(2-9i)$

$$
\begin{aligned}
7i(2-9i) &= 7i\cdot 2-7i\cdot 9i \\
&= 14i-63i^2 \\
&= 14i-63(-1) \\
&= 63+14i
\end{aligned}
$$

Pencil Problem #2

2a. Multiply: $(-5+4i)(3+i)$

2b. Multiply: $-3i(7i-5)$

Objective #3: Divide complex numbers.

✔ *Solved Problem #3*

3. Divide and express the result in standard form: $\dfrac{5+4i}{4-i}$

Multiply the numerator and the denominator by the conjugate of the denominator, $4+i$.

$$
\begin{aligned}
\frac{5+4i}{4-i} &= \frac{(5+4i)}{(4-i)}\cdot\frac{(4+i)}{(4+i)} \\
&= \frac{20+5i+16i+4i^2}{16+1} \\
&= \frac{20+21i+4(-1)}{16+1} \\
&= \frac{16+21i}{17} \\
&= \frac{16}{17}+\frac{21}{17}i
\end{aligned}
$$

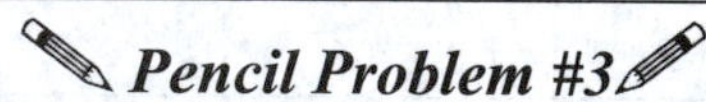

Pencil Problem #3

3. Divide and express the result in standard form: $\dfrac{2+3i}{2+i}$

Objective #4: Operations with square roots of negative numbers.

✔ *Solved Problem #4*

4. Perform the indicated operations and write the result in standard form.

4a. $\sqrt{-27}+\sqrt{-48}$

$$\begin{aligned}\sqrt{-27}+\sqrt{-48} &= i\sqrt{27}+i\sqrt{48}\\ &= i\sqrt{9\cdot 3}+i\sqrt{16\cdot 3}\\ &= 3i\sqrt{3}+4i\sqrt{3}\\ &= 7i\sqrt{3}\end{aligned}$$

4b. $(-2+\sqrt{-3})^2$

$$\begin{aligned}(-2+\sqrt{-3})^2 &= (-2+i\sqrt{3})^2\\ &= (-2)^2+2(-2)(i\sqrt{3})+(i\sqrt{3})^2\\ &= 4-4i\sqrt{3}+3i^2\\ &= 4-4i\sqrt{3}+3(-1)\\ &= 1-4i\sqrt{3}\end{aligned}$$

4c. $\dfrac{-14+\sqrt{-12}}{2}$

$$\begin{aligned}\frac{-14+\sqrt{-12}}{2} &= \frac{-14+i\sqrt{12}}{2}\\ &= \frac{-14+2i\sqrt{3}}{2}\\ &= \frac{-14}{2}+\frac{2i\sqrt{3}}{2}\\ &= -7+i\sqrt{3}\end{aligned}$$

Pencil Problem #4

4. Perform the indicated operations and write the result in standard form.

4a. $\sqrt{-64}-\sqrt{-25}$

4b. $(-3-\sqrt{-7})^2$

4c. $\dfrac{-8+\sqrt{-32}}{24}$

Objective #5: Solve quadratic equations with complex imaginary solutions.

Solved Problem #5

5. Solve using the quadratic formula: $x^2 - 2x + 2 = 0$

The equation is in the form $ax^2 + bx + c = 0$, where $a = 1$, $b = -2$, and $c = 2$.

$$x = \frac{-b \pm \sqrt{b^2 - 4ac}}{2a}$$
$$= \frac{-(-2) \pm \sqrt{(-2)^2 - 4(1)(2)}}{2(1)}$$
$$= \frac{2 \pm \sqrt{4-8}}{2}$$
$$= \frac{2 \pm \sqrt{-4}}{2}$$
$$= \frac{2 \pm 2i}{2}$$
$$= \frac{2(1 \pm i)}{2}$$
$$= 1 \pm i$$

The solution set is $\{1 + i, 1 - i\}$.

Pencil Problem #5

5. Solve using the quadratic formula: $x^2 - 6x + 10 = 0$

Answers for Pencil Problems *(Textbook Exercise references in parentheses)*:

1a. $8 - 2i$ *(2.1 #1)* **1b.** $-2 + 9i$ *(2.1 #3)*

2a. $-19 + 7i$ *(2.1 #11)* **2b.** $21 + 15i$ *(2.1 #9)*

3. $\frac{7}{5} + \frac{4}{5}i$ *(2.1 #27)*

4a. $3i$ *(2.1 #29)* **4b.** $2 + 6i\sqrt{7}$ *(2.1 #35)* **4c.** $-\frac{1}{3} + i\frac{\sqrt{2}}{6}$ *(2.1 #37)*

5. $\{3 + i, 3 - i\}$ *(2.1 #45)*

Section 2.2
Quadratic Functions and Their Graphs

Heads UP!!!

Many sports involve objects that are thrown, kicked, or hit, and then proceed with no additional force of their own.
Such objects are called projectiles.

In this section of your textbook, you will learn to use graphs of quadratic functions to gain a visual understanding of various projectile sports.

Objective #1: Recognize characteristics of parabolas.

✔ *Solved Problem #1*

1. True or false: The *vertex* of a parabola is also called the *turning point.*

True

Pencil Problem #1

1. True or false: The *vertex* of a parabola is always the minimum point of the parabola.

Objective #2: Graph parabolas.

✔ *Solved Problem #2*

2a. Graph the quadratic function: $f(x) = -(x-1)^2 + 4$

Since $a = -1$ is negative, the parabola opens downward.
The vertex of the parabola is $(h,k) = (1,4)$.

Replace $f(x)$ with 0 to find x–intercepts.

$$0 = -(x-1)^2 + 4$$
$$(x-1)^2 = 4$$
$$x - 1 = \pm\sqrt{4}$$
$$x - 1 = \pm 2$$
$$x - 1 = 2 \quad \text{or} \quad x - 1 = -2$$
$$x = 3 \qquad\qquad x = -1$$

The x–intercepts are -1 and 3.

Pencil Problem #2

2a. Graph the quadratic function: $f(x) = (x-4)^2 - 1$

Set $x = 0$ and solve for y to obtain the y–intercept.

$y = -(0-1)^2 + 4 = 3$

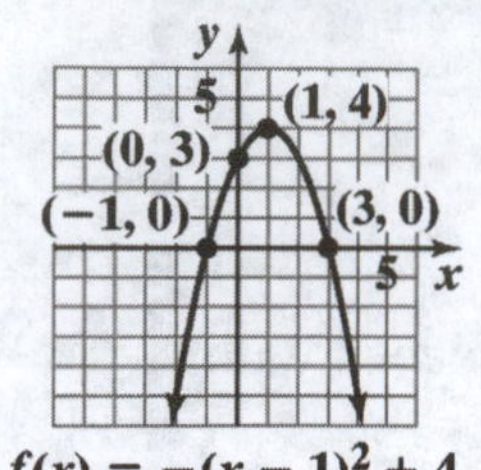

$f(x) = -(x-1)^2 + 4$

2b. Graph the quadratic function $f(x) = -x^2 + 4x + 1$. Use the graph to identify the function's domain and its range.

Since $a = -1$ is negative, the parabola opens downward.

The x–coordinate of the vertex of the parabola is

$$-\frac{b}{2a} = -\frac{4}{2(-1)} = -\frac{4}{-2} = 2.$$

The y–coordinate of the vertex of the parabola is

$$f\left(-\frac{b}{2a}\right) = f(2) = -(2)^2 + 4(2) + 1 = 5.$$

The vertex is $(2,5)$.

Replace $f(x)$ with 0 to find x–intercepts.

$$0 = -x^2 + 4x + 1$$

$$x = \frac{-b \pm \sqrt{b^2 - 4ac}}{2a}$$

$$x = \frac{-4 \pm \sqrt{4^2 - 4(-1)(1)}}{2(-1)}$$

$$x = 2 \pm \sqrt{5}$$

$x \approx -0.2$ or $x \approx 4.2$

The x–intercepts are -0.2 and 4.2.

Set $x = 0$ and solve for y to obtain the y–intercept.

$y = -0^2 + 4 \cdot 0 + 1 = 1$

$f(x) = -x^2 + 4x + 1$

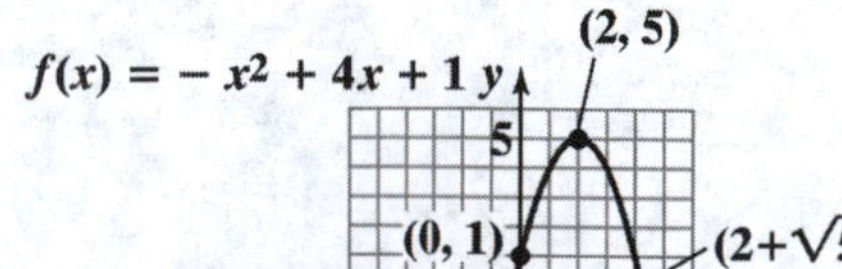

Domain: $(-\infty, \infty)$ Range: $(-\infty, 5]$

2b. Graph the quadratic function $f(x) = x^2 + 3x - 10$. Use the graph to identify the function's range.

Objective #3: Determine a quadratic function's minimum or maximum value.

✔ *Solved Problem #3*

3. Consider the quadratic function $f(x)=4x^2-16x+1000$.

3a. Determine, without graphing, whether the function has a minimum value or a maximum value.

Because $a>0$, the function has a minimum value.

3b. Find the minimum or maximum value and determine where it occurs.

The minimum value occurs at $-\frac{b}{2a}=-\frac{-16}{2(4)}=2$.

The minimum of $f(x)$ is $f(2)=4\cdot 2^2-16\cdot 2+1000$
$=984$.

3c. Identify the function's domain and its range.

Like all quadratic functions, the domain is $(-\infty,\infty)$.

Because the minimum is 984, the range includes all real numbers at or above 984. The range is $[984,\infty)$.

Pencil Problem #3

3. Consider the quadratic function $f(x)=-4x^2+8x-3$.

3a. Determine, without graphing, whether the function has a minimum value or a maximum value.

3b. Find the minimum or maximum value and determine where it occurs.

3c. Identify the function's domain and its range.

Objective #4: Solve problems involving a quadratic function's minimum or maximum value.

✔ *Solved Problem #4*

4. Among all pairs of numbers whose difference is 8, find a pair whose product is as small as possible. What is the minimum product?

Let the two numbers be represented by x and y, and let the product be represented by P.

We must minimize $P=xy$.

Because the difference of the two numbers is 8, then $x-y=8$.

Solve for y in terms of x.

$$x-y=8$$
$$-y=-x+8$$
$$y=x-8$$

Write P as a function of x.

Pencil Problem #4

4. Among all pairs of numbers whose sum is 16, find a pair whose product is as large as possible. What is the maximum product?

$$P = xy$$
$$P(x) = x(x-8)$$
$$P(x) = x^2 - 8x$$

Because $a > 0$, the function has a minimum value that

occurs at $x = -\frac{b}{2a}$

$$= -\frac{-8}{2(1)}$$

$$= 4.$$

Substitute to find the other number.

$$y = x - 8$$
$$y = 4 - 8$$
$$= -4$$

The two numbers are 4 and −4.

The minimum product is $P = xy = (4)(-4) = -16$.

Answers for Pencil Problems *(Textbook Exercise references in parentheses)*:

1. false *(2.2 #41)*

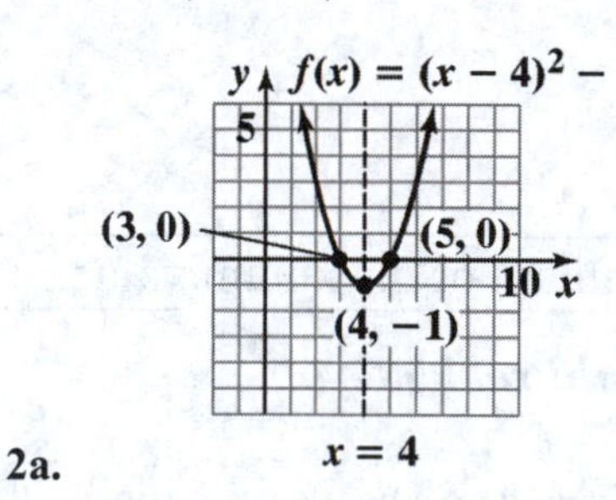

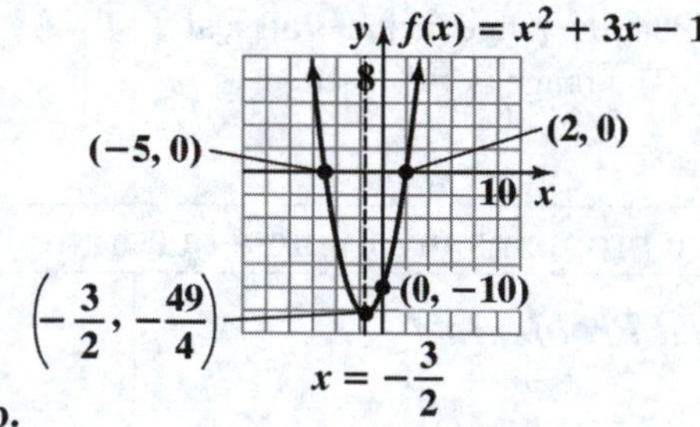

2a. *(2.2 #17)* **2b.**

Range: $\left[-\frac{49}{4}, \infty\right)$ *(2.2 #29)*

3a. maximum *(2.2 #41a)* **3b.** The maximum is 1 at $x = 1$. *(2.2 #41b)*

3c. Domain: $(-\infty, \infty)$; Range: $(-\infty, 1]$ *(2.2 #41c)*

4. The maximum product is 64 when the numbers are 8 and 8. *(2.2 #61)*

Section 2.3
Polynomial Functions and Their Graphs

Pay at the Pump !

Other than outrage, what is going on at the gas pumps?
Is surging demand creating the increasing oil prices?
Like all things in a free market economy, the price of a commodity is based on supply and demand.

In the Exercise Set for this section, we will explore the volatility of gas prices over the past several years.

Objective #1: Identify polynomial functions.

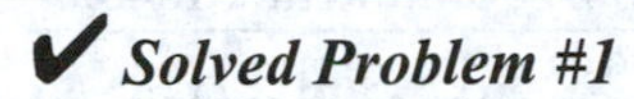

1. The exponents on the variables in a polynomial function must be nonnegative integers.

True

1. The coefficients of the variables in a polynomial function must be nonnegative integers.

Objective #2: Recognize characteristics of graphs of polynomial functions.

2. The graph of a polynomial function may have a sharp corner.

False. The graphs of polynomial functions are smooth, meaning that they have rounded curves and no sharp corners.

2. The graph of a polynomial function may have a gap or break.

Objective #3: Determine end behavior.

✔ ***Solved Problem #3***

3. Use the Leading Coefficient Test to determine the end behavior of the graph of each function.

3a. $f(x) = x^4 - 4x^2$

The term with the greater exponent is x^4, or $1x^4$. The leading coefficient is 1, which is positive. The degree of the function is 4, which is even. Even-degree polynomial functions have the same behavior at each end. Since the leading coefficient is positive, the graph rises to the left and rises to the right.

3. Use the Leading Coefficient Test to determine the end behavior of the graph of each function.

3a. $f(x) = 5x^3 + 7x^2 - x + 9$

3b. $f(x) = 2x^3(x-1)(x+5)$

The function is in factored form, but we can determine the degree and the leading coefficient without multiplying it out. The factors $2x^3$, $x - 1$, and $x + 5$ are of degree 3, 1, and 1, respectively. When we multiply expressions with the same base, we add exponents, so the degree of the function is 3 + 1 + 1, or 5, which is odd. Without multiplying out, you should be able to see that the leading coefficient is 2, which is positive.

Odd-degree polynomial functions have graphs with opposite behavior at each end. Since the leading coefficient is positive, the graph falls to the left and rises to the right.

3b. $f(x) = -x^2(x-1)(x+3)$

Objective #4: Use factoring to find zeros of polynomial functions.

✔ *Solved Problem #4*

4. Find all zeros of $f(x) = x^3 + 2x^2 - 4x - 8$.

Set $f(x)$ equal to zero.

$$x^3 + 2x^2 - 4x - 8 = 0$$
$$x^2(x+2) - 4(x+2) = 0$$
$$(x+2)(x^2 - 4) = 0$$
$$(x+2)(x+2)(x-2) = 0$$

Apply the zero-product principle.

$x + 2 = 0$ or $x + 2 = 0$ or $x - 2 = 0$

$x = -2$ $\quad$ $x = -2$ $\quad$ $x = 2$

The zeros are -2 and 2.

Pencil Problem #4

4. Find all zeros of $f(x) = x^3 + 2x^2 - x - 2$.

Objective #5: Identify zeros and their multiplicities.

✔ *Solved Problem #5*

5. Find the zeros of $f(x) = -4\left(x + \frac{1}{2}\right)^2 (x-5)^3$ and give the multiplicity of each zero. State whether the graph crosses the x-axis or touches the x-axis and turns around at each zero.

Set each factor equal to zero.

$x + \frac{1}{2} = 0$ or $x - 5 = 0$

$x = -\frac{1}{2}$ $\quad$ $x = 5$

$-\frac{1}{2}$ is a zero of multiplicity 2, and 5 is a zero of multiplicity 3.

Because the multiplicity of $-\frac{1}{2}$ is even, the graph touches the x-axis and turns around at this zero.

Because the multiplicity of 5 is odd, the graph crosses the x-axis at this zero.

Pencil Problem #5

5. Find the zeros of $f(x) = 4(x-3)(x+6)^3$ and give the multiplicity of each zero. State whether the graph crosses the x-axis or touches the x-axis and turns around at each zero.

Objective #6: Use the Intermediate Value Theorem.

✔ *Solved Problem #6*

6. Show that the polynomial function $f(x) = 3x^3 - 10x + 9$ has a real zero between −3 and −2.

Evaluate *f* at −3 and −2.

$f(-3) = 3(-3)^3 - 10(-3) + 9 = -42$

$f(-2) = 3(-2)^3 - 10(-2) + 9 = 5$

The sign change between *f*(−3) and *f*(−2) shows that *f* has a real zero between −3 and −2.

Pencil Problem #6

6. Show that the polynomial function $f(x) = 2x^4 - 4x^2 + 1$ has a real zero between −1 and 0.

Objective #7: Understand the relationship between degree and turning points.

✔ *Solved Problem #7*

7. If a polynomial function, *f*, is of degree 5, what is the greatest number of turning points on its graph?

The greatest number of turning points on the graph of a polynomial of degree 5 is $5-1$, or 4.

Pencil Problem #7

7. If a polynomial function, *f*, is of degree 4, what is the greatest number of turning points on its graph?

Objective #8: Graph polynomial functions.

✔ *Solved Problem #5*

8. Use the five-step strategy to graph $f(x) = x^3 - 3x^2$.

Step 1: Determine end behavior.

Since $f(x) = x^3 - 3x^2$ is an odd-degree polynomial and since the leading coefficient, 1, is positive, the graph falls to the left and rises to the right.

Step 2: Find *x*-intercepts by setting $f(x) = 0$.

$x^3 - 3x^2 = 0$

$x^2(x-3) = 0$

Apply the zero-product principle.

$x^2 = 0$ or $x - 3 = 0$

$x = 0$ $\quad$ $x = 3$

The zeros of *f* are 0 and 3. The graph touches the *x*-axis at 0 since it has multiplicity 2. The graph crosses the *x*-axis at 3 since it has multiplicity 1.

Pencil Problem #5

8. Use the five-step strategy to graph $f(x) = x^4 - 9x^2$.

Step 3: Find the y-intercept by computing $f(0)$.

$$f(x) = x^3 - 3x^2$$
$$f(0) = 0^3 - 3(0)^2$$
$$= 0$$

There is a y-intercept at 0, so the graph passes through (0, 0).

Step 4: Use possible symmetry to help draw the graph.

$$f(x) = x^3 - 3x^2$$
$$f(-x) = (-x)^3 - 3(-x)^2$$
$$= -x^3 - 3x^2$$

Since $f(-x) \neq f(x)$ and since $f(-x) \neq -f(x)$, the function is neither even nor odd, and the graph is neither symmetric with respect to the y-axis nor the origin.

Step 5: Draw the graph.

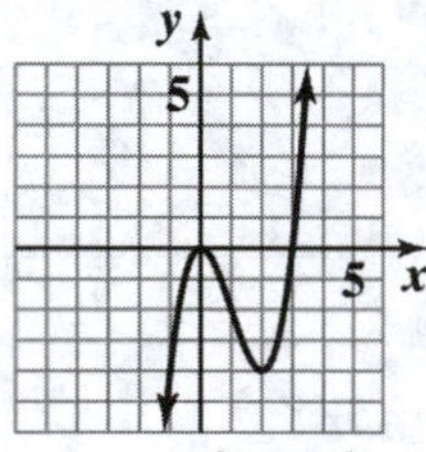

$f(x) = x^3 - 3x^2$

Answers for Pencil Problems *(Textbook Exercise references in parentheses)*:

1. False *(2.3 #3)* **2.** False *(2.3 #13)*

3a. The graph falls to the left and rises to the right. *(2.3 #19)*

3b. The graph falls to the left and falls to the right. *(2.3 #59a)*

4. -2, -1, and 1 *(2.3 #41b)*

5. zeros: 3 (multiplicity 1) and -6 (multiplicity 3); The graph crosses the x-axis at 3 and at -6. *(2.3 #27)*

6. $f(-1) = -1$ and $f(0) = 1$; The sign change between $f(-1)$ and $f(0)$ shows that f has a real zero between -1 and 0. *(2.3 #35)*

7. 3 *(2.3 #47e)*

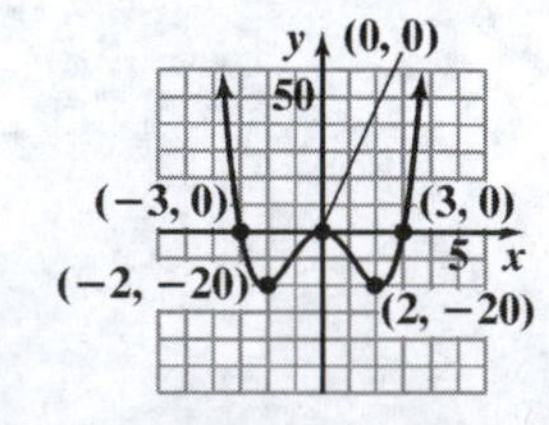

8. $f(x) = x^4 - 9x^2$ *(2.3 #43)*

Section 2.4
Dividing Polynomials; Remainder and Factor Theorems

What Happened to My Sweater?

It's that first brisk morning in autumn and you go to the closet for your favorite sweater. But what's that? There's a hole. No. There are dozens of holes.
In this section's Exercise Set, you will work with a polynomial function that models the number of eggs in a female moth based on her abdominal width. The techniques of this section provide a new way of evaluating the function to find out how many moths were eating your sweater.

Objective #1: Use long division to divide polynomials.

✔ *Solved Problem #1*

1. Divide $2x^4+3x^3-7x-10$ by x^2-2x.

Rewrite the dividend with the missing power of x and divide.

$$\begin{array}{r} 2x^2+7x+14 \\ x^2-2x\overline{)\,2x^4+3x^3+0x^2-7x-10} \\ \underline{2x^4-4x^3} \\ 7x^3+0x^2 \\ \underline{7x^3-14x^2} \\ 14x^2-7x \\ \underline{14x^2-28x} \\ 21x-10 \end{array}$$

Thus, $\dfrac{2x^4+3x^3-7x-10}{x^2-2x}=2x^2+7x+14+\dfrac{21x-10}{x^2-2x}$

Pencil Problem #1

1. Divide $4x^4-4x^2+6x$ by $x-4$ using long division.

Objective #2: Use synthetic division to divide polynomials.

✔ *Solved Problem #2*

2. Use synthetic division: $(x^3-7x-6)\div(x+2)$

$$\begin{array}{r|rrrr} -2 & 1 & 0 & -7 & -6 \\ & & -2 & 4 & 6 \\ \hline & 1 & -2 & -3 & 0 \end{array}$$

Thus, $(x^3-7x-6)\div(x+2)=x^2-2x-3$

Pencil Problem #2

2. Use synthetic division: $(3x^2+7x-20)\div(x+5)$

Objective #3: Evaluate a polynomial function using the Remainder Theorem.

✔ *Solved Problem #3*

3. Given $f(x)=3x^3+4x^2-5x+3$, use the Remainder Theorem to find $f(-4)$.

$$\begin{array}{r|rrrr}
-4 & 3 & 4 & -5 & 3 \\
 & & -12 & 32 & -108 \\
\hline
 & 3 & -8 & 27 & -105 \leftarrow f(-4)=-105
\end{array}$$

Pencil Problem #3

3. Given $f(x)=2x^3-11x^2+7x-5$, use the Remainder Theorem to find $f(4)$.

Objective #4: Use the Factor Theorem to solve a polynomial equation.

✔ *Solved Problem #4*

4. Solve the equation $15x^3+14x^2-3x-2=0$ given that -1 is a zero of $f(x)=15x^2+14x^2-3x-2$.

Synthetic division verifies that $x+1$ is a factor.

$$\begin{array}{r|rrrr}
-1 & 15 & 14 & -3 & -2 \\
 & & -15 & 1 & 2 \\
\hline
 & 15 & -1 & -2 & 0
\end{array}$$

Next, continue factoring to find all solutions.

$$15x^3+14x^2-3x-2=0$$

$$(x+1)(15x^2-x-2)=0$$

$$(x+1)(5x-2)(3x+1)=0$$

$$x+1=0 \quad \text{or} \quad 5x-2=0 \quad \text{or} \quad 3x+1=0$$

$$x=-1 \qquad 5x=2 \qquad 3x=-1$$

$$x=\frac{2}{5} \qquad x=-\frac{1}{3}$$

The solution set is $\left\{-1,-\frac{1}{3},\frac{2}{5}\right\}$

Pencil Problem #4

4. Solve the equation $2x^3-5x^2+x+2=0$ given that 2 is a zero of $f(x)=2x^3-5x^2+x+2$.

Answers for Pencil Problems *(Textbook Exercise references in parentheses)*:

1. $4x^3+16x^2+60x+246+\frac{984}{x-4}$ *(2.4 #11)*

2. $3x-8+\frac{20}{x+5}$ *(2.4 #19)*

3. -25 *(2.4 #33)*

4. $\left\{-\frac{1}{2},\ 1,\ 2\right\}$ *(2.4 #43)*

Section 2.5
Zeros of Polynomials

Do I Have to Check My Bag?

Airlines have regulations on the sizes of carry-on luggage that are allowed. As a passenger, you are interested in the volume of your luggage, but the airline is concerned about the sum of bag's length, width, and depth.

In this section's Exercise Set, you will work with a polynomial function that relates the two quantities and allows you to find dimensions of a carry-on bag that meet both your volume requirement and the airline's regulations.

Objective #1: Use the Rational Zero Theorem to find possible rational zeros.

✔ *Solved Problem #1*

1. List all possible rational zeros of $f(x)=4x^5+12x^4-x-3$.

Factors of the constant term -3: $\pm1,\ \pm3$

Factors of the leading coefficient 4: $\pm1,\ \pm2,\ \pm4$

The possible rational zeros are:

$$\frac{\text{Factors of } -3}{\text{Factors of } 4}=\frac{\pm1,\ \pm3}{\pm1,\ \pm2,\ \pm4}$$

$$=\pm1,\ \pm3,\ \pm\frac{1}{2},\ \pm\frac{3}{2},\ \pm\frac{1}{4},\ \pm\frac{3}{4}$$

Pencil Problem #1

1. List all possible rational zeros of $f(x)=3x^4-11x^3-x^2+19x+6$.

Objective #2: Find zeros of a polynomial function.

✔ *Solved Problem #2*

2. Find all zeros of $f(x)=x^3+x^2-5x-2$.

First, list the possible rational zeros:

$$\frac{\text{Factors of } -2}{\text{Factors of } 1}=\frac{\pm1,\ \pm2}{\pm1}=\pm1,\ \pm2$$

Now use synthetic division to find a rational zero from among the list of possible rational zeros. Try 2:

```
2| 1  1  -5  -2
      2   6   2
   -----------
   1  3   1   0
```

The last number in the bottom row is 0. Thus 2 is a zero and $x-2$ is a factor.

The first three numbers in the bottom row of the synthetic division give the coefficients of the other factor. This factor is x^2+3x+1.

Pencil Problem #2

2. Find all zeros of $f(x)=x^3+4x^2-3x-6$.

Factor completely: $x^3 + x^2 - 5x - 2 = 0$

$(x-2)(x^2 + 3x + 1) = 0$

Since $x^2 + 3x + 1$ is not factorable, use the quadratic formula to find the remaining zeros.

$$x = \frac{-b \pm \sqrt{b^2 - 4ac}}{2a}$$

$$x = \frac{-3 \pm \sqrt{3^2 - 4(1)(1)}}{2(1)} = \frac{-3 \pm \sqrt{5}}{2}$$

The zeros are 2 and $\frac{-3 \pm \sqrt{5}}{2}$.

Objective #3: Solve polynomial equations.

✔ *Solved Problem #3*

3. Solve: $x^4 - 6x^3 + 22x^2 - 30x + 13 = 0$

First, list the possible rational roots:

$$\frac{\text{Factors of 13}}{\text{Factors of 1}} = \frac{\pm 1,\ \pm 13}{\pm 1} = \pm 1,\ \pm 13$$

Now use synthetic division to find a rational root from among the list of possible rational roots. Try 1.

$$\begin{array}{r|rrrrr} 1 & 1 & -6 & 22 & -30 & 13 \\ & & 1 & -5 & 17 & -13 \\ \hline & 1 & -5 & 17 & -13 & 0 \end{array}$$

The last number in the bottom row is 0.
Thus, 1 is a root.

Rewrite the equation in factored form using the bottom row of the synthetic division to obtain the coefficients of the other factor.

$x^4 - 6x^3 + 22x^2 - 30x + 13 = 0$

$(x-1)(x^3 - 5x^2 + 17x - 13) = 0$

Use the same approach to find another root. Try 1 again.

$$\begin{array}{r|rrrr} 1 & 1 & -5 & 17 & -13 \\ & & 1 & -4 & 13 \\ \hline & 1 & -4 & 13 & 0 \end{array}$$

The last number in the bottom row is 0.
Thus, 1 is a root (of multiplicity 2).

The first three numbers in the bottom row of the synthetic division give the coefficients of the factor $x^2 - 4x + 13$.

Pencil Problem #3

3. Solve: $x^3 - 2x^2 - 11x + 12 = 0$

Since $x^2 - 4x + 13$ is not factorable, use the quadratic formula to find the remaining roots.

$$x = \frac{-b \pm \sqrt{b^2 - 4ac}}{2a}$$

$$x = \frac{-(-4) \pm \sqrt{(-4)^2 - 4(1)(13)}}{2(1)}$$

$$x = \frac{4 \pm \sqrt{-36}}{2}$$

$$x = \frac{4 \pm 6i}{2}$$

$$x = 2 \pm 3i$$

The roots are 1 and $2 \pm 3i$.

Objective #4: Use the Linear Factorization Theorem to find polynomials with given zeros.

✔ *Solved Problem #4*

4. Find a third-degree polynomial function $f(x)$ with real coefficients that has -3 and i as zeros such that $f(1) = 8$.

Because i is a zero and the polynomial has real coefficients, the conjugate, $-i$, must also be a zero. We can now use the Linear Factorization Theorem.

$$\begin{aligned} f(x) &= a_n(x - c_1)(x - c_2)(x - c_3) \\ &= a_n(x - (-3))(x - i)(x - (-i)) \\ &= a_n(x+3)(x-i)(x+i) \\ &= a_n(x+3)(x^2 - i^2) \\ &= a_n(x+3)(x^2 - (-1)) \\ &= a_n(x+3)(x^2 + 1) \\ &= a_n(x^3 + 3x^2 + x + 3) \end{aligned}$$

Now we use $f(1) = 8$ to find a_n.

$$f(1) = a_n(1^3 + 3 \cdot 1^2 + 1 + 3) = 8$$

$$8a_n = 8$$

$$a_n = 1$$

Now substitute 1 for a_n in the formula for $f(x)$.

$$f(x) = 1(x^3 + 3x^2 + x + 3)$$

or $f(x) = x^3 + 3x^2 + x + 3$

Pencil Problem #4

4. Find a fourth-degree polynomial function $f(x)$ with real coefficients that has i and $3i$ as zeros such that $f(-1) = 20$.

Objective #5: Use Descartes's Rule of Signs.

✔ *Solved Problem #5*

5. Determine the possible numbers of positive and negative real zeros of $f(x) = x^4 - 14x^3 + 71x^2 - 154x + 120$.

Count the number of sign changes in $f(x)$.

$f(x) = x^4 - 14x^3 + 71x^2 - 154x + 120$

Since $f(x)$ has four sign changes, it has 4, 2, or 0 positive real zeros.

Count the number of sign changes in $f(-x)$.

$$\begin{aligned} f(-x) &= (-x)^4 - 14(-x)^3 + 71(-x)^2 - 154(-x) + 120 \\ &= x^4 + 14x^3 + 71x^2 + 154x + 120 \end{aligned}$$

Since $f(-x)$ has no sign changes, $f(x)$ has 0 negative real zeros.

Pencil Problem #5

5. Determine the possible numbers of positive and negative real zeros of $f(x) = x^3 + 2x^2 + 5x + 4$.

Answers for Pencil Problems *(Textbook Exercise references in parentheses)*:

1. $\pm 1, \pm 2, \pm 3, \pm 6, \pm\frac{1}{3}, \pm\frac{2}{3}$ *(2.5 #3)*

2. $-1, \frac{-3-\sqrt{33}}{3}$, and $\frac{-3+\sqrt{33}}{3}$ *(2.5 #13)*

3. $\{-3, 1, 4\}$ *(2.5 #17)*

4. $f(x) = x^4 + 10x^2 + 9$ *(2.5 #29)*

5. f has no positive real zeros and either 3 or 1 negative real zeros *(2.5 #33)*

Section 2.6
Rational Functions and Their Graphs

Decreasing Costs with Increased Production?

In a simple business model, the cost, $C(x)$, to produce x units of a product is the sum of the fixed and variable costs and can be expressed in a form similar to $C(x) = \$500{,}000 + \$400x$. In this model, the cost increases by \$400 for each additional unit.

If we divide the cost, $C(x)$, by x, the number of units produced, we obtain the function $\overline{C}(x)$, which represents the average cost of each item. By studying the rational function $\overline{C}(x)$, we'll see that the average cost per item decreases for each additional unit.

Objective #1: Find the domains of rational functions.

✔ *Solved Problem #1*

1a. Find the domain of $g(x) = \dfrac{x}{x^2 - 25}$.

The denominator of $g(x) = \dfrac{x}{x^2 - 25}$ is 0 when $x = -5$ or $x = 5$. The domain of g consists of all real numbers except -5 and 5. This can be expressed as

$\{x \mid x \neq -5, x \neq 5\}$ or $(-\infty, -5) \cup (-5, 5) \cup (5, \infty)$

1b. Find the domain of $h(x) = \dfrac{x+5}{x^2 + 25}$.

No real numbers cause the denominator of $h(x) = \dfrac{x+5}{x^2 + 25}$ to equal 0. The domain of h consists of all real numbers, or $(-\infty, \infty)$.

Pencil Problem #1

1a. Find the domain of $h(x) = \dfrac{x+7}{x^2 - 49}$.

1b. Find the domain of $f(x) = \dfrac{x+7}{x^2 + 49}$.

Objective #2: Use arrow notation.

✔ *Solved Problem #2*

2. True or false: The notation "$x \to a^+$" means that the values of x are increasing without bound.

False. "$x \to a^+$" means that x is approaching a from the right.

2. True or false: If $f(x) \to 0$ as $x \to \infty$, then the graph of f approaches the x-axis to the right.

Objective #3: Solve polynomial equations.

✔ *Solved Problem #3*

3a. Find the vertical asymptotes, if any, of the graph of the rational function: $g(x)=\dfrac{x-1}{x^2-1}$.

The numerator and denominator have a factor in common. Therefore, simplify *g*.

$$g(x)=\frac{x-1}{x^2-1}=\frac{x-1}{(x+1)(x-1)}=\frac{1}{x+1}$$

The only zero of the denominator of the simplified function is -1.

Thus, the line $x=-1$ is a vertical asymptote for the graph of *g*.

3b. Find the vertical asymptotes, if any, of the graph of the rational function: $h(x)=\dfrac{x-1}{x^2+1}$.

The denominator cannot be factored.
The denominator has no real zeros.

Thus, the graph of *h* has no vertical asymptotes.

Pencil Problem #3

3a. Find the vertical asymptotes, if any, of the graph of the rational function: $h(x)=\dfrac{x}{x(x+4)}$.

3b. Find the vertical asymptotes, if any, of the graph of the rational function: $r(x)=\dfrac{x}{x^2+4}$.

Objective #4: Identify horizontal asymptotes.

✔ *Solved Problem #4*

4a. Find the horizontal asymptotes, if any, of the graph of the rational function: $f(x)=\dfrac{9x^2}{3x^2+1}$.

The degree of the numerator, 2, is equal to the degree of the denominator, 2.
The leading coefficients of the numerator and denominator are 9 and 3, respectively.

Thus, the equation of the horizontal asymptote is $y=\dfrac{9}{3}$ or $y=3$.

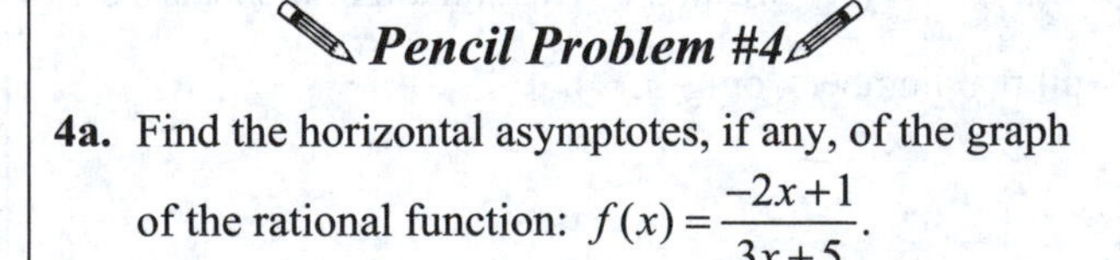

Pencil Problem #4

4a. Find the horizontal asymptotes, if any, of the graph of the rational function: $f(x)=\dfrac{-2x+1}{3x+5}$.

4b. Find the horizontal asymptotes, if any, of the graph of the rational function: $h(x)=\dfrac{9x^3}{3x^2+1}$.

The degree of the numerator, 3, is greater than the degree of the denominator, 2.

Thus, the graph of h has no horizontal asymptote.

4b. Find the horizontal asymptotes, if any, of the graph of the rational function: $f(x)=\dfrac{12x}{3x^2+1}$.

Objective #5: Use transformations to graph rational functions.

✔ *Solved Problem #5*

5. Use the graph of $f(x)=\dfrac{1}{x}$ to graph

$g(x)=\dfrac{1}{x+2}-1$.

Start with the graph of $f(x)=\dfrac{1}{x}$ and two points on its graph, such as $(-1,-1)$ and $(1,1)$.

First move the graph two units to the left to graph $y=\dfrac{1}{x+2}$; the indicated points end up at $(-3,-1)$ and $(-1,1)$. The vertical asymptote is now $x=-2$.

Next move the graph down one unit to graph $g(x)=\dfrac{1}{x+2}-1$; the indicated points end up at $(-3,-2)$ and $(-1,0)$. The horizontal asymptote is now $y=-1$.

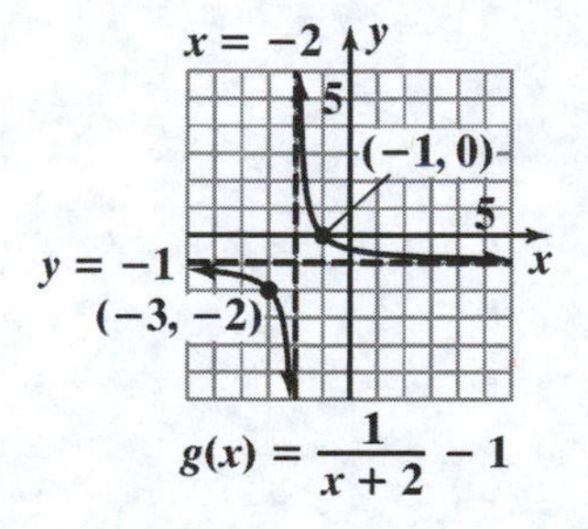

Pencil Problem #5

5. Use the graph of $f(x)=\dfrac{1}{x}$ to graph

$g(x)=\dfrac{1}{x+1}-2$.

Objective #6: Graph rational functions.

✔ *Solved Problem #6*

6a. Graph: $f(x)=\dfrac{3x-3}{x-2}$

Step 1: $f(-x)=\dfrac{3(-x)-3}{-x-2}=\dfrac{-3x-3}{-x-2}=\dfrac{3x+3}{x+2}$

Because $f(-x)$ does not equal $f(x)$ or $-f(x)$, the graph has neither y-axis symmetry nor origin symmetry.

Step 2: $f(0)=\dfrac{3(0)-3}{0-2}=\dfrac{3}{2}$

The y-intercept is $\dfrac{3}{2}$.

Step 3: $3x-3=0$

$3x=3$

$x=1$

The x-intercept is 1.

Step 4: $x-2=0$

$x=2$

The line $x=2$ is the only vertical asymptote for the graph of f.

Step 5: The numerator and denominator have the same degree, 1. The leading coefficients of the numerator and denominator are 3 and 1, respectively. Thus, the equation of the horizontal asymptote is $y=\dfrac{3}{1}$ or $y=3$.

Step 6: Plot points between and beyond each x-intercept and vertical asymptote:

x	-1	$\frac{3}{2}$	3	5
$f(x)$	2	-3	6	4

Step 7: Use the preceding information to graph the function.

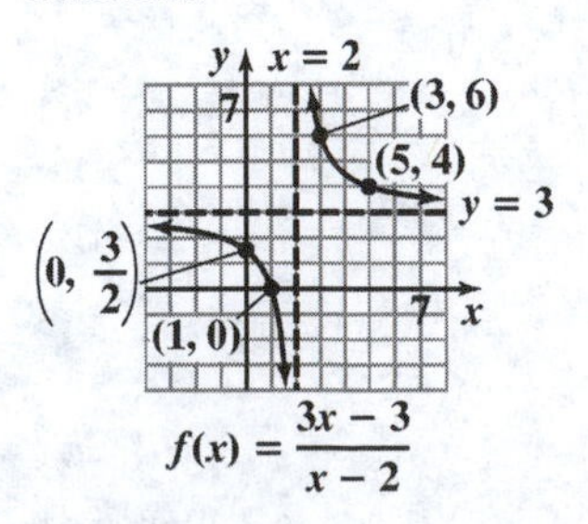

✎ *Pencil Problem #6* ✎

6a. Graph: $f(x)=\dfrac{-x}{x+1}$

6b. Graph: $f(x)=\dfrac{x^4}{x^2+2}$

Step 1: $f(-x)=\dfrac{(-x)^4}{(-x)^2+2}=\dfrac{x^4}{x^2+2}=f(x)$

Because $f(-x)=f(x)$, the graph has y-axis symmetry.

Step 2: $f(0)=\dfrac{0^4}{0^2+2}=0$

The y-intercept is 0, so the graph passes through the origin.

Step 3: $x^4=0$

$x=0$

There is only one x-intercept. This verifies that the graph passes through the origin.

Step 4: $x^2+2=0$

$x^2=-2$

$x=\pm i\sqrt{2}$

Since these solutions are not real, the graph of f will not have any vertical asymptotes.

Step 5: The degree of the numerator, 4, is greater than the degree of the denominator, 2, so the graph will not have a horizontal asymptote.

Step 6: Plot some points other than the intercepts:

x	-2	-1	1	2
$f(x)$	$\frac{8}{3}$	$\frac{1}{3}$	$\frac{8}{3}$	$\frac{1}{3}$

Step 7: Use the preceding information to graph the function.

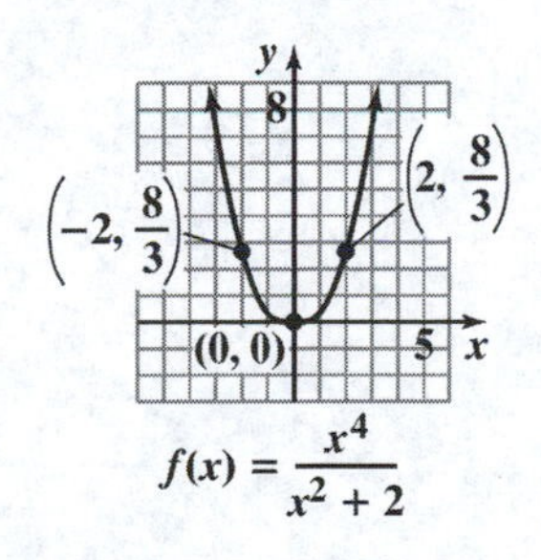

6b. Graph: $f(x)=-\dfrac{1}{x^2-4}$

Objective #7: Identify slant asymptotes.

✔ *Solved Problem #7*

7. Find the slant asymptote of $f(x)=\dfrac{2x^2-5x+7}{x-2}$.

Note that the graph of *f* has a slant asymptote because the degree of the numerator is exactly one more than the degree of the denominator and the denominator is not a factor of the numerator.

Divide $2x^2-5x+7$ by $x-2$.

$$\begin{array}{r|rrr} 2 & 2 & -5 & 7 \\ & & 4 & -2 \\ \hline & 2 & -1 & 5 \end{array}$$

So, $\dfrac{2x^2-5x+7}{x-2}=2x-1+\dfrac{5}{x-2}$.

The equation of the slant asymptote is $y = 2x - 1$.

Pencil Problem #7

7. Find the slant asymptote of $f(x)=\dfrac{x^2+x-6}{x-3}$.

Objective #8: Solve applied problems involving rational functions.

✔ *Solved Problem #8*

8. A company is planning to manufacture wheelchairs. The cost, *C*, in dollars, of producing *x* wheelchairs is $C(x) = 500{,}000 + 400x$.

8a. Write the average cost function, $\overline{C}$.

The average cost is the cost divided by the number of wheelchairs produced.

$$\overline{C}(x)=\frac{500{,}000+400x}{x}$$

Pencil Problem #8

8. A company is planning to manufacture mountain bikes. The cost, *C*, in dollars, of producing *x* mountain bikes is $C(x) = 100{,}000 + 100x$.

8a. Write the average cost function, $\overline{C}$.

8b. Find and interpret $\overline{C}(1000)$ and $\overline{C}(10,000)$.

$$\overline{C}(1000) = \frac{500,000 + 400(1000)}{1000} = 900$$

The average cost per wheelchair of producing 1000 wheelchairs is $900.

$$\overline{C}(10,000) = \frac{500,000 + 400(10,000)}{10,000} = 405$$

The average cost per wheelchair of producing 10,000 wheelchairs is $405.

8c. What is the horizontal asymptote for the graph of $\overline{C}$? Describe what this means for the company.

The horizontal asymptote is $y = \frac{400}{1}$ or $y = 400$.

The cost per wheelchair approaches $400 as more wheelchairs are produced.

8b. Find and interpret $\overline{C}(1000)$ and $\overline{C}(4000)$.

8c. What is the horizontal asymptote for the graph of $\overline{C}$? Describe what this means for the company.

Answers for Pencil Problems *(Textbook Exercise references in parentheses)*:

1a. $\{x|x \neq -7, x \neq 7\}$ or $(-\infty, -7)\cup(-7, 7)\cup(7, \infty)$ *(2.6 #5)*

1b. all real numbers or $(-\infty, \infty)$ *(2.6 #7)*

2. True *(2.6 #14)*

3a. vertical asymptote: $x = -4$ *(2.6 #25)*

3b. no vertical asymptotes *(2.6 #27)*

4a. horizontal asymptote: $y = \frac{-2}{3}$ *(2.6 #43)*

4b. horizontal asymptote: $y = 0$ *(2.6 #37)*

5.

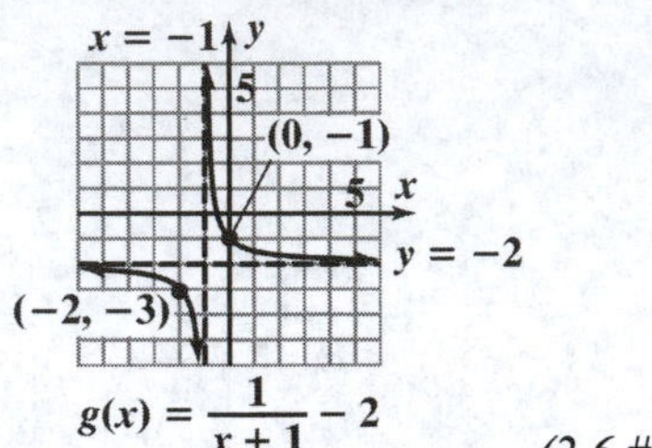

(2.6 #49)

6a.

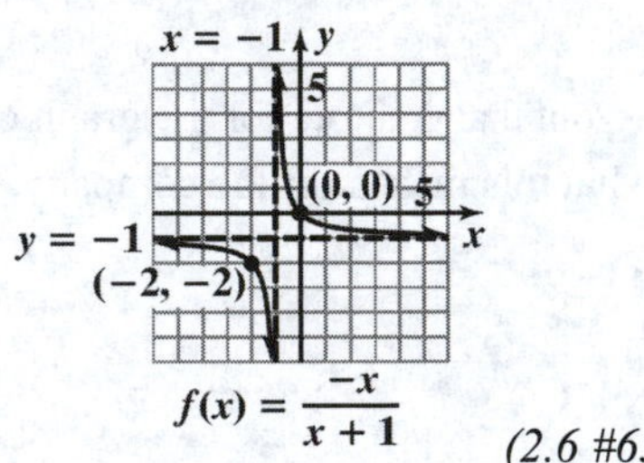

(2.6 #63)

6b.

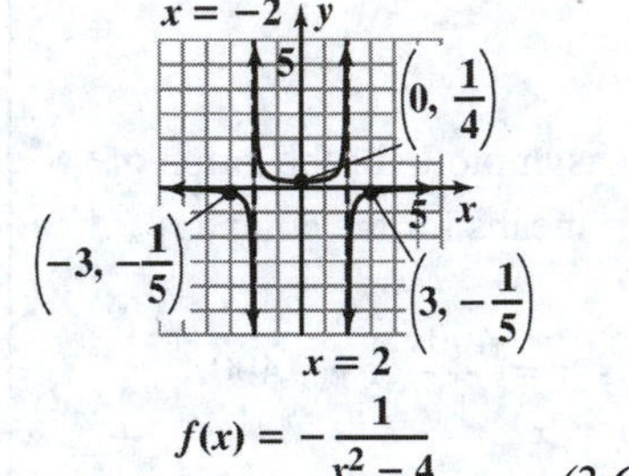

(2.6 #65)

7. $y = x + 4$ *(2.6 #85a)*

8a. $\bar{C}(x) = \frac{100{,}000 + 100x}{x}$ *(2.6 #99b)*

8b. $\bar{C}(1000) = 200$; The average cost per mountain bike of producing 1000 mountain bikes is \$200; $\bar{C}(4000) = 125$; The average cost per mountain bike of producing 4000 mountain bikes is \$125. *(2.6 #99c)*

8c. $y = 100$; The cost per mountain bike approaches \$100 as more mountain bikes are produced. *(2.6 #99d)*

Section 2.7
Polynomial and Rational Inequalities

Tailgaters Beware!

It is never a good idea to follow too closely behind the car in front of you. But when the roads are wet it can be even more dangerous.

In this section, we apply the mathematical concepts we learn to explore the different stopping distances required for a car driving on wet pavement and a car driving on dry pavement.

Objective #1: Solve polynomial inequalities.

Solved Problem #1

1. Solve and graph the solution set on a real number line: $x^2 - x > 20$

$$x^2 - x > 20$$
$$x^2 - x - 20 > 0$$

Solve the related quadratic equation to find the boundary points.

$$x^2 - x - 20 = 0$$
$$(x+4)(x-5) = 0$$

Apply the zero-product principle.

$x + 4 = 0$ or $x - 5 = 0$

$x = -4$ $\quad$ $x = 5$

The boundary points are -4 and 5.

Interval	Test Value	Test	Conclusion
$(-\infty,-4)$	-5	$(-5)^2 - (-5) > 20$ $30 > 20$, true	$(-\infty,-4)$ belongs to the solution set.
$(-4,5)$	0	$(0)^2 - (0) > 20$ $0 > 20$, false	$(-4,5)$ does not belong to the solution set.
$(5,\infty)$	10	$(10)^2 - (10) > 20$ $90 > 20$, true	$(5,\infty)$ belongs to the solution set.

The solution set is $(-\infty,-4) \cup (5,\infty)$.

−8 −6 −4 −2 0 2 4 6 8 10 12

Pencil Problem #1

1. Solve and graph the solution set on a real number line: $4x^2 + 7x < -3$

Objective #2: Solve rational inequalities.

Solved Problem #2

2. Solve and graph the solution set on a real number line: $\frac{2x}{x+1} \geq 1$

$$\frac{2x}{x+1} \geq 1$$
$$\frac{2x}{x+1} - 1 \geq 0$$
$$\frac{2x}{x+1} - \frac{x+1}{x+1} \geq 0$$
$$\frac{2x - x - 1}{x+1} \geq 0$$
$$\frac{x-1}{x+1} \geq 0$$

Find the values of x that make the numerator and denominator zero.

$x - 1 = 0$ and $x + 1 = 0$

$x = 1$ $\qquad$ $x = -1$

The boundary points are -1 and 1.

Interval	Test Value	Test	Conclusion
$(-\infty, -1)$	-2	$\frac{2(-2)}{-2+1} \geq 1$ $4 \geq 1$, true	$(-\infty, -1)$ belongs to the solution set.
$(-1, 1)$	0	$\frac{2(0)}{0+1} \geq 1$ $0 \geq 1$, false	$(-1, 1)$ does not belong to the solution set.
$(1, \infty)$	2	$\frac{2(2)}{2+1} \geq 1$ $\frac{4}{3} \geq 1$, true	$(1, \infty)$ belongs to the solution set.

Exclude -1 from the solution set because it would make the denominator zero. The solution set is $(-\infty, -1) \cup [1, \infty)$.

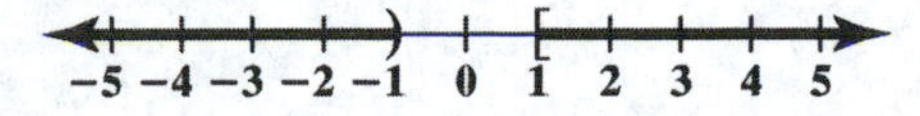

Pencil Problem #2

2. Solve and graph the solution set on a real number line: $\dfrac{x+1}{x+3}<2$

Objective #3: Solve problems modeled by polynomial or rational inequalities.

✔ *Solved Problem #3*

3. An object is propelled straight up from ground level with an initial velocity of 80 feet per second. Its height at time t is modeled by $s(t) = -16t^2 + 80t$ where the height, $s(t)$, is measured in feet and the time, t, is measured in seconds. In which time interval will the object be more than 64 feet above the ground?

To find when the object will be more than 64 feet above the ground, solve the inequality $-16t^2 + 80t > 64$.
Solve the related quadratic equation.

$$-16t^2 + 80t = 64$$
$$-16t^2 + 80t - 64 = 0$$
$$t^2 - 5t + 4 = 0$$
$$(t-4)(t-1) = 0$$
$$t - 4 = 0 \quad \text{or} \quad t - 1 = 0$$
$$t = 4 \qquad\qquad t = 1$$

The boundary points are 1 and 4.

Interval	Test Value	Test	Conclusion
$(0,1)$	0.5	$-16(0.5)^2 + 80(0.5) > 64$ $36 > 64$, false	$(0,1)$ does not belong to the solution set.
$(1,4)$	2	$-16(2)^2 + 80(2) > 64$ $96 > 64$, true	$(1,4)$ belongs to the solution set.
$(4,\infty)$	5	$-16(5)^2 + 80(5) > 64$ $0 > 64$, false	$(4,\infty)$ does not belong to the solution set.

The solution set is $(1,4)$.

This means that the object will be more than 64 feet above the ground between 1 and 4 seconds excluding $t = 1$ and $t = 4$.

Pencil Problem #3

3. You throw a ball straight up from a rooftop 160 feet high with an initial speed of 48 feet per second. The function $s(t) = -16t^2 + 48t + 160$ models the ball's height above the ground, $s(t)$, in feet, t seconds after it was thrown. During which time period will the ball's height exceed that of the rooftop?

Answers for Pencil Problems *(Textbook Exercise references in parentheses)*:

1. $\left(-1,-\frac{3}{4}\right)$

$-1 \quad 0 \quad 1$ *(2.7 #15)*

2. $(-\infty,-5)\cup(-3,\infty)$

$-6\ -5\ -4\ -3\ -2\ -1\ 0\ 1\ 2\ 3\ 4$ *(2.7 #55)*

3. The ball exceeds the height of the building between 0 and 3 seconds. *(2.7 #76)*

Section 2.8
Modeling Using Variation

How Far Would You Go To Lose Weight?

On the moon your weight would be significantly less.

To find out how much less,
be sure to work on the application problems
in this section of your textbook!

Objective #1: Solve direct variation problems.

✔ Solved Problem #1

1. The number of gallons of water, W, used when taking a shower varies directly as the time, t, in minutes, in the shower. A shower lasting 5 minutes uses 30 gallons of water. How much water is used in a shower lasting 11 minutes?

Since W varies directly with t, we have $W = kt$.
Use the given values to find k.

$$W = kt$$
$$30 = k \cdot 5$$
$$\frac{30}{5} = \frac{k \cdot 5}{5}$$
$$6 = k$$

The equation becomes $W = 6t$. Find W when $t = 11$.

$$W = 6t$$
$$W = 6 \cdot 11$$
$$= 66$$

An 11 minute shower will use 66 gallons of water.

Pencil Problem #1

1. An alligator's tail length, T, varies directly as its body length, B. An alligator with a body length of 4 feet has a tail length of 3.6 feet. What is the tail length of an alligator whose body length is 6 feet?

Objective #2: Solve inverse variation problems.

Solved Problem #2

2. The length of a violin string varies inversely as the frequency of its vibrations. A violin string 8 inches long vibrates at a frequency of 640 cycles per second. What is the frequency of a 10-inch string?

Beginning with $y = \frac{k}{x}$, we will use l for the length of the string and f for the frequency.

Use the given values to find k.

$$f = \frac{k}{l}$$
$$640 = \frac{k}{8}$$
$$8 \cdot 640 = 8 \cdot \frac{k}{8}$$
$$5120 = k$$

The equation becomes $f = \frac{k}{l}$

$$f = \frac{5120}{l}$$

Find f when $l = 10$.

$$f = \frac{5120}{l}$$
$$f = \frac{5120}{10}$$
$$f = 512$$

A string length of 10 inches will vibrate at 512 cycles per second.

Pencil Problem #2

2. A bicyclist tips his bicycle when making a turn. The angle B, formed by the vertical direction and the bicycle, is called the banking angle. The banking angle varies inversely as the cycle's turning radius. When the turning radius is 4 feet, the banking angle is $28°$. What is the banking angle when the turning radius is 3.5 feet?

Objective #3: Solve combined variation problems.

✔ *Solved Problem #3*

3. The number of minutes needed to solve an Exercise Set of variation problems varies directly as the number of problems and inversely as the number of people working to solve the problems. It takes 4 people 32 minutes to solve 16 problems. How many minutes will it take 8 people to solve 24 problems?

Let $m =$ the number of minutes needed to solve an exercise set.
Let $p =$ the number of people working on the problems.
Let $x =$ the number of problems in the exercise set.

Use $m = \frac{kx}{p}$ to find *k*.

$$m = \frac{kx}{p}$$
$$32 = \frac{k16}{4}$$
$$32 = 4k$$
$$k = 8$$

Thus, $m = \frac{8x}{p}$.

Find *m* when $p = 8$ and $x = 24$.

$$m = \frac{8 \cdot 24}{8}$$
$$m = 24$$

It will take 24 minutes for 8 people to solve 24 problems.

Pencil Problem #3

3. Body-mass index, or BMI, varies directly as one's weight, in pounds, and inversely as the square of one's height, in inches. A person who weighs 180 pounds and is 5 feet, or 60 inches, tall has a BMI of 35.15. What is the BMI, to the nearest tenth, for a 170 pound person who is 5 feet 10 inches tall?

Objective #4: Solve problems involving joint variation.

✔ Solved Problem #4

4. The volume of a cone, V, varies jointly as its height, h, and the square of its radius, r. A cone with a radius measuring 6 feet and a height measuring 10 feet has a volume of 120π cubic feet. Find the volume of a cone having a radius of 12 feet and a height of 2 feet.

Find k: $V = khr^2$

$$120\pi = k \cdot 10 \cdot 6^2$$

$$120\pi = k \cdot 360$$

$$\frac{120\pi}{360} = \frac{k \cdot 360}{360}$$

$$\frac{\pi}{3} = k$$

Thus, $V = \frac{\pi}{3}hr^2 = \frac{\pi hr^2}{3}$.

$$V = \frac{\pi hr^2}{3}$$

$$V = \frac{\pi \cdot 2 \cdot 12^2}{3} = 96\pi$$

The volume of a cone having a radius of 12 feet and a height of 2 feet is 96π cubic feet.

Pencil Problem #4

4. The heat loss of a glass window varies jointly as the window's area and the difference between the outside and inside temperatures. A window 3 feet wide by 6 feet long loses 1200 Btu per hour when the temperature outside is $20°$ colder than the temperature inside. Find the heat loss through a glass window that is 6 feet wide by 9 feet long when the temperature outside is $10°$ colder than the temperature inside.

Answers for Pencil Problems *(Textbook Exercise references in parentheses)*:

1. 5.4 feet *(2.8 #21)* **2.** 32° *(2.8 #27)* **3.** BMI: 24.4 *(2.8 #31)* **4.** 1800 Btu *(2.8 #33)*

Section 3.1
Exponential Functions

Shop 'til You Drop!

Are you just browsing? Take your time.
Researchers know, to the dollar, the average amount the typical consumer spends at the shopping mall. And the longer you stay, the more you spend. So if you say you're just browsing, that's just fine with the mall merchants. Browsing is time and, as we will explore in this section, time is money.

Objective #1: Evaluate exponential functions.

✔ *Solved Problem #1*

1. The exponential function $f(x)=42.2(1.56)^x$ models the average amount spent, $f(x)$, in dollars, at a shopping mall after x hours. What is the average amount spent, to the nearest dollar, after three hours?

$$f(x)=42.2(1.56)^x$$
$$f(3)=42.2(1.56)^3$$
$$\approx 160.20876$$
$$\approx 160$$

The average amount spent after three hours at a mall is \$160.

Pencil Problem #1

1. The exponential function $f(x)=574(1.026)^x$ models the population of India, $f(x)$, in millions, x years after 1974. Find India's population, to the nearest million, in the year 2028.

Objective #2: Graph exponential functions.

✔ *Solved Problem #2*

2a. Graph: $f(x)=3^x$

Make a table of values:

x	$f(x)=3^x$	(x,y)
-3	$3^{-3}=\frac{1}{27}$	$\left(-3,\frac{1}{27}\right)$
-2	$3^{-2}=\frac{1}{9}$	$\left(-2,\frac{1}{9}\right)$
-1	$3^{-1}=\frac{1}{3}$	$\left(-1,\frac{1}{3}\right)$
0	$3^0=1$	$(0,1)$
1	$3^1=3$	$(1,3)$
2	$3^2=9$	$(2,9)$
3	$3^3=27$	$(3,27)$

Pencil Problem #2

2a. Graph: $f(x)=4^x$

Plot the points in the table and connect with a smooth curve.

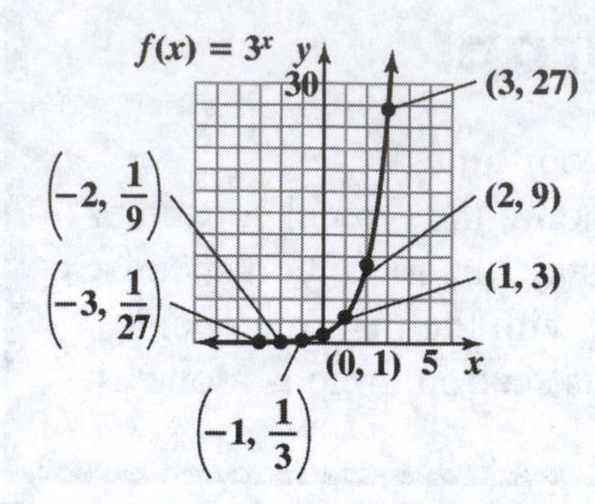

2b. Graph: $f(x)=\left(\frac{1}{3}\right)^x$

Make a table of values:

x	$f(x)=\left(\frac{1}{3}\right)^x$	(x, y)
−3	$\left(\frac{1}{3}\right)^{-3}=27$	$(-3, 27)$
−2	$\left(\frac{1}{3}\right)^{-2}=9$	$(-2, 9)$
−1	$\left(\frac{1}{3}\right)^{-1}=3$	$(-1, 3)$
0	$\left(\frac{1}{3}\right)^{0}=1$	$(0, 1)$
1	$\left(\frac{1}{3}\right)^{1}=\frac{1}{3}$	$\left(1, \frac{1}{3}\right)$
2	$\left(\frac{1}{3}\right)^{2}=\frac{1}{9}$	$\left(2, \frac{1}{9}\right)$
3	$\left(\frac{1}{3}\right)^{3}=\frac{1}{27}$	$\left(3, \frac{1}{27}\right)$

Plot the points in the table and connect with a smooth curve.

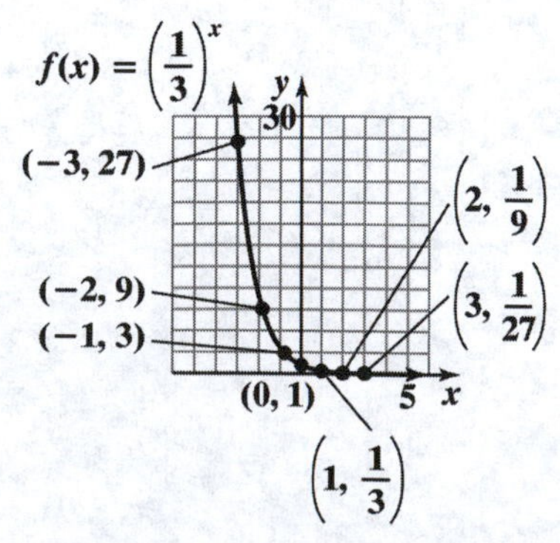

2b. Graph: $g(x)=\left(\frac{3}{2}\right)^x$

2c. Use the graph of $f(x) = 3^x$ to obtain the graph of $g(x) = 3^{x-1}$.

The graph of $g(x) = 3^{x-1}$ is the graph of $f(x) = 3^x$ shifted 1 unit to the right. We identified two points on the graph of *f*, which we graphed in Solved Problem 2a, and added 1 to each of the *x*-coordinates.

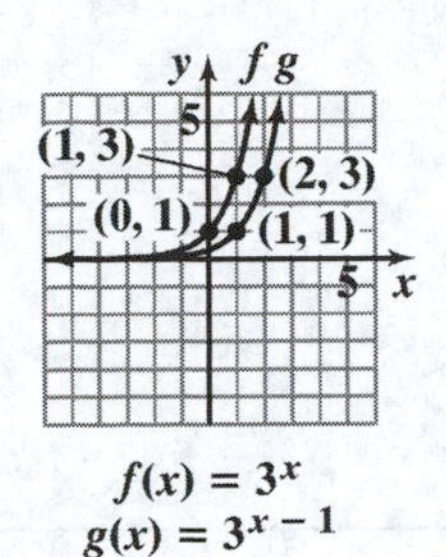

$f(x) = 3^x$
$g(x) = 3^{x-1}$

2c. Use the graph of $f(x) = 2^x$ to obtain the graph of $g(x) = 2^{x+1}$.

2d. Use the graph of $f(x) = 2^x$ to obtain the graph of $g(x) = 2^x + 1$.

The graph of $g(x) = 2^x + 1$ is the graph of $f(x) = 2^x$ shifted up 1 unit. We identified two points on the graph of *f* and added 1 to each of the *y*-coordinates. The asymptote is also shifted up 1 unit.

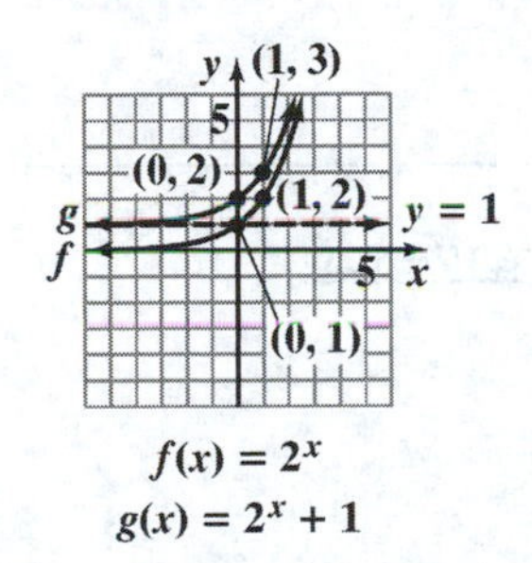

$f(x) = 2^x$
$g(x) = 2^x + 1$

2d. Use the graph of $f(x) = 2^x$ to obtain the graph of $g(x) = 2^x - 1$.

Objective #3: Evaluate functions with base *e*.

✔ *Solved Problem #3*

3. The exponential function $f(x) = 1145e^{0.0325x}$ models the gray wolf population of the Western Great Lakes, $f(x)$, x years after 1978. If trends continue, project the gray wolf's population in the recovery area in 2017.

2017 is 39 years after 1978.

$f(x) = 1145e^{0.0325x}$

$f(39) = 1145e^{0.0325(39)} \approx 4067$

In 2017 the gray wolf population of the Western Great Lakes was projected to be about 4067.

Pencil Problem #3

3. The exponential function $g(x) = 32.7e^{0.0217x}$ models the percentage of high school seniors who applied to more than three colleges *x* years after 1980. What percentage of high school seniors applied to more than three colleges in 2013?

Objective #4: Use compound interest formulas.

✔ *Solved Problem #4*

4a. A sum of $10,000 is invested at an annual rate of 8%. Find the balance in the account after 5 years subject to quarterly compounding.

$$A = P\left(1+\frac{r}{n}\right)^{nt}$$
$$= \$10{,}000\left(1+\frac{0.08}{4}\right)^{4\cdot 5}$$
$$\approx \$14{,}859.47$$

4b. A sum of $10,000 is invested at an annual rate of 8%. Find the balance in the account after 5 years subject to continuous compounding.

$$A = Pe^{rt}$$
$$= \$10{,}000e^{0.08(5)}$$
$$\approx \$14{,}918.25$$

Pencil Problem #4

4a. A sum of $10,000 is invested at an annual rate of 5.5%. Find the balance in the account after 5 years subject to monthly compounding.

4b. A sum of $10,000 is invested at an annual rate of 5.5%. Find the balance in the account after 5 years subject to continuous compounding.

Answers for Pencil Problems *(Textbook Exercise references in parentheses)*:

1. 2295 million *(3.1 #65c)*

2a.
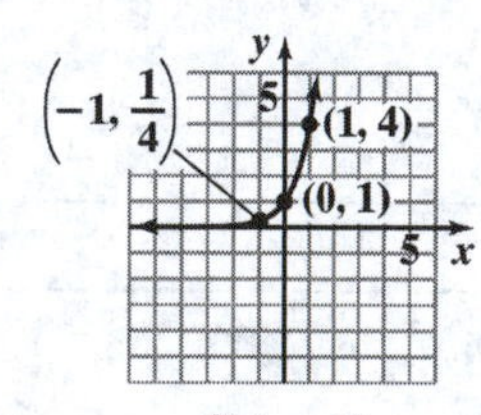

$f(x) = 4^x$ *(3.1 #11)*

2b.
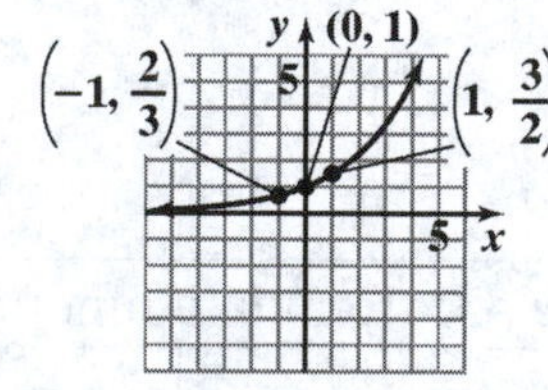

$g(x) = \left(\frac{3}{2}\right)^x$ *(3.1 #13)*

2c.
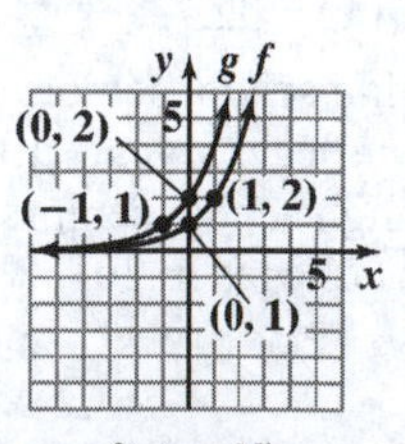

$f(x) = 2^x$
$g(x) = 2^{x+1}$ *(3.1 #25)*

2d.
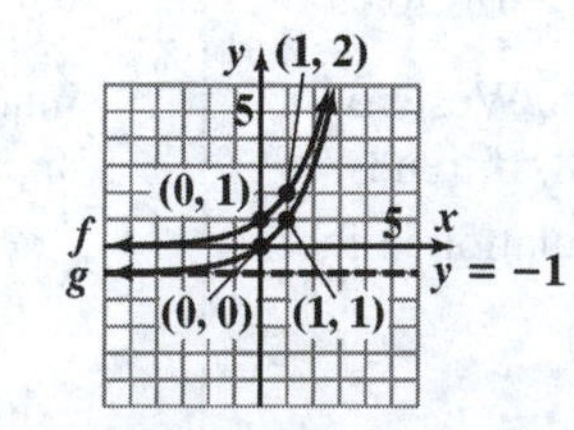

$f(x) = 2^x$
$g(x) = 2^x - 1$ *(3.1 #27)*

3. 67% *(3.1 #71b)*

4a. $13,157.04 *(3.1 #53c)* **4b.** $13,165.31 *(3.1 #53d)*

Section 3.2
Logarithmic Functions

Speak Up!

The loudness level of a sound is measured in decibels. Decibel levels range from 0, a barely audible sound, to 160, a sound resulting in a ruptured eardrum.

We will see that decibels can be modeled by a logarithmic function, the topic of this section.

Objective #1: Change from logarithmic to exponential form.

1. Write the equation $2 = \log_b 25$ in its equivalent exponential form.

$$2 = \log_b 25$$
$$b^2 = 25$$

1. Write the equation $\log_6 216 = y$ in its equivalent exponential form.

Objective #2: Change from exponential to logarithmic form.

Solved Problem #2

2. Write the equation $2^5 = x$ in its equivalent logarithmic form.

$$2^5 = x$$
$$5 = \log_2 x$$

2. Write the equation $7^y = 200$ in its equivalent logarithmic form.

Objective #3: Evaluate logarithms.

Solved Problem #3

3a. Evaluate: $\log_5 \frac{1}{125}$.

$\log_5 \frac{1}{125} = -3$ because

$$5^{-3} = \frac{1}{5^3} = \frac{1}{125}$$

Pencil Problem #3

3a. Evaluate: $\log_5 \frac{1}{5}$.

3b. Evaluate: $\log_{36} 6$.

$\log_{36} 6 = \frac{1}{2}$ because $36^{\frac{1}{2}} = \sqrt{36} = 6.$

3b. Evaluate: $\log_4 16$.

Objective #4: Use basic logarithmic properties.

✔ *Solved Problem #4*

4. Evaluate $\log_7 7^8$.

Because $\log_b b^x = x$, we conclude $\log_7 7^8 = 8$.

Pencil Problem #4

4. Evaluate $\log_4 1$.

Objective #5: Graph logarithmic functions.

✔ *Solved Problem #5*

5. Graph $f(x) = 3^x$ and $g(x) = \log_3 x$ in the same rectangular coordinate system.

Set up a table of coordinates for $f(x) = 3^x$.

x	-2	-1	0	1	2	3
$f(x) = 3^x$	$\frac{1}{9}$	$\frac{1}{3}$	1	3	9	27

Reverse these coordinates to obtain the coordinates of $g(x) = \log_3 x$.

x	$\frac{1}{9}$	$\frac{1}{3}$	1	3	9	27
$g(x) = \log_3 x$	-2	-1	0	1	2	3

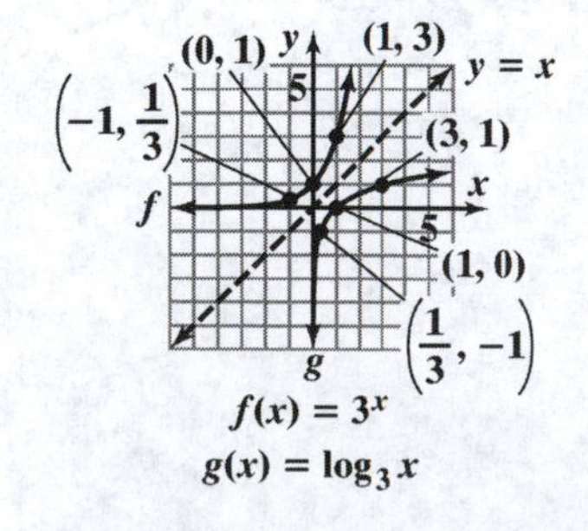

Pencil Problem #5

5. Graph $f(x) = \left(\frac{1}{2}\right)^x$ and $g(x) = \log_{\frac{1}{2}} x$ in the same rectangular coordinate system.

Objective #6: Find the domain of a logarithmic function.

✔ *Solved Problem #6*

6. Find the domain of $f(x)=\log_4(x-5)$.

$$x-5>0$$
$$x>5$$

The domain of *f* is $(5,\infty)$.

Pencil Problem #6

6. Find the domain of $f(x)=\log(2-x)$.

Objective #7: Use common logarithms.

✔ *Solved Problem #7*

7. The percentage of adult height attained by a boy who is *x* years old can be modeled by $f(x)=29+48.8\log(x+1)$, where *x* represents the boy's age and $f(x)$ represents the percentage of his adult height. Approximately what percentage of his adult height has a boy attained at age ten?

$$f(x)=29+48.8\log(x+1)$$
$$f(10)=29+48.8\log(10+1)$$
$$=29+48.8\log 11$$
$$\approx 80$$

A 10-year-old boy has attained approximately 80% of his adult height.

Pencil Problem #7

7. The percentage of adult height attained by a girl who is *x* years old can be modeled by $f(x)=62+35\log(x-4)$, where *x* represents the girl's age and $f(x)$ represents the percentage of her adult height. Approximately what percentage of her adult height has a girl attained at age 13?

Objective #8: Use natural logarithms.

✔ *Solved Problem #8*

8a. Find the domain of $f(x)=\ln(4-x)$.

The domain of *f* consists of all *x* for which $4-x>0$.

$$4-x>0$$
$$-x>-4$$
$$x<4$$

The domain of *f* is $(-\infty,4)$.

Pencil Problem #8

8a. Find the domain of $g(x)=\ln(x+2)$.

8b. The function $f(x)=13.4\ln x-11.6$ models the temperature increase, $f(x)$, in an enclosed vehicle after x minutes when the outside air temperature is between 72°F and 96°F. Use the function to find the temperature increase, to the nearest degree, after 30 minutes.

Evaluate the function at 30.

$$f(x)=13.4\ln x-11.6$$
$$f(30)=13.4\ln 30-11.6$$
$$\approx 34$$

The temperature will increase by about 34° after 30 minutes.

8b. The function $f(x)=-35.2\ln x+34.5$ models wives' weekly housework hours, $f(x)$, x years after 1964. Use the function to find the number of weekly housework hours for wives in 2010. Round to the nearest hour.

Answers for Pencil Problems *(Textbook Exercise references in parentheses)*:

1. $6^y=216$ *(3.2 #7)*

2. $\log_7 200=y$ *(3.2 #19)*

3a. -1 *(3.2 #25)* **3b.** 2 *(3.2 #21)*

4. 0 *(3.2 #37)*

5. 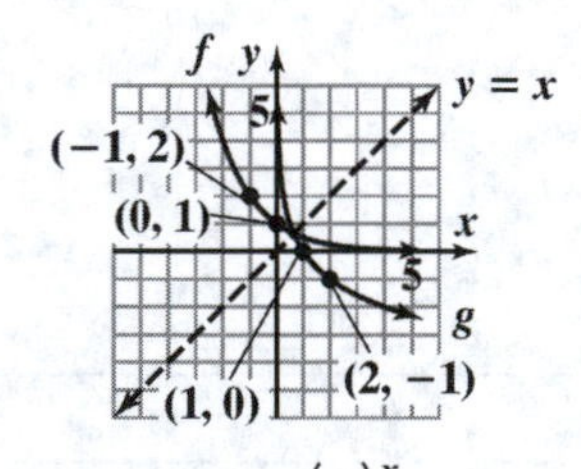

$f(x)=\left(\frac{1}{2}\right)^x$

$g(x)=\log_{1/2}x$ *(3.2 #45)*

6. $(-\infty,2)$ *(3.2 #77)*

7. 95.4% *(3.2 #113)*

8a. $(-2,\infty)$ *(3.2 #65)*

8b. 21 hours *(3.2 #115)*

Section 3.3
Properties of Logarithms

How Smart Is This Chimp?

Scientists are often amazed at what a chimpanzee can learn. These scientists are not simply interested in *what* a chimp can learn, but also in a particular chimp's *maximum capacity* for learning, and *how long* the learning takes.

A typical chimpanzee learning sign language can master a maximum of 65 signs. In the Exercise Set of the textbook, one application will take a look at the number of weeks
it will take for a chimp to master 30 signs.

Objective #1: Use the product rule.

✔ *Solved Problem #1*

1. Use the product rule to expand: $\log(100x)$.

$$\log(100x) = \log 100 + \log x$$
$$= \log 10^2 + \log x$$
$$= 2 + \log x$$

Pencil Problem #1

1. Use the product rule to expand: $\log_5(7 \cdot 3)$.

Objective #2: Use the quotient rule.

✔ *Solved Problem #2*

2. Use the quotient rule to expand: $\ln\left(\frac{e^5}{11}\right)$.

$$\ln\left(\frac{e^5}{11}\right) = \ln e^5 - \ln 11$$
$$= 5 - \ln 11$$

Pencil Problem #2

2. Use the quotient rule to expand: $\log_4\left(\frac{64}{y}\right)$.

Objective #3: Use the power rule.

✔ *Solved Problem #3*

3a. Use the power rule to expand: $\log_6 3^9$

$$\log_6 3^9 = 9\log_6 3$$

Pencil Problem #3

3a. Use the power rule to expand: $\log_b x^3$.

3b. Use the power rule to expand: $\log(x+4)^2$

$$\log(x+4)^2 = 2\log(x+4)$$

3b. Use the power rule to expand: $\log N^{-6}$

Objective #4: Expand logarithmic expressions.

✔ *Solved Problem #4*

4a. Use logarithmic properties to expand: $\ln\sqrt[3]{x}$.

$$\begin{aligned}\ln\sqrt[3]{x} &= \ln x^{\frac{1}{3}} \\ &= \frac{1}{3}\ln x\end{aligned}$$

4b. Use logarithmic properties to expand: $\log_b\left(x^4\sqrt[3]{y}\right)$.

$$\begin{aligned}\log_b\left(x^4\sqrt[3]{y}\right) &= \log_b x^4 + \log_b\sqrt[3]{y} \\ &= \log_b x^4 + \log_b y^{\frac{1}{3}} \\ &= 4\log_b x + \frac{1}{3}\log_b y\end{aligned}$$

4c. Use logarithmic properties to expand: $\log_5\left(\frac{\sqrt{x}}{25y^3}\right)$.

$$\begin{aligned}\log_5\left(\frac{\sqrt{x}}{25y^3}\right) &= \log_5\left(\frac{x^{\frac{1}{2}}}{25y^3}\right) \\ &= \log_5 x^{\frac{1}{2}} - \log_5\left(25y^3\right) \\ &= \log_5 x^{\frac{1}{2}} - \left(\log_5 25 + \log_5 y^3\right) \\ &= \log_5 x^{\frac{1}{2}} - \log_5 25 - \log_5 y^3 \\ &= \frac{1}{2}\log_5 x - 2 - 3\log_5 y\end{aligned}$$

✎ *Pencil Problem #4* ✎

4a. Use logarithmic properties to expand: $\ln\sqrt[5]{x}$.

4b. Use logarithmic properties to expand: $\log_b(x^2 y)$.

4c. Use logarithmic properties to expand: $\log_6\left(\frac{36}{\sqrt{x+1}}\right)$.

Objective #5: Condense logarithmic expressions.

✔ *Solved Problem #5*

5a. Write as a single logarithm: $\log 25 + \log 4$.

$$\begin{aligned} \log 25 + \log 4 &= \log(25 \cdot 4) \\ &= \log 100 \\ &= 2 \end{aligned}$$

5b. Write as a single logarithm: $2\ln x + \frac{1}{3}\ln(x+5)$.

$$\begin{aligned} 2\ln x + \frac{1}{3}\ln(x+5) &= \ln x^2 + \ln(x+5)^{\frac{1}{3}} \\ &= \ln x^2 + \ln \sqrt[3]{x+5} \\ &= \ln\left(x^2 \sqrt[3]{x+5}\right) \end{aligned}$$

5c. Write as a single logarithm:
$\frac{1}{4}\log_b x - 2\log_b 5 - 10\log_b y$.

$$\begin{aligned} &\tfrac{1}{4}\log_b x - 2\log_b 5 - 10\log_b y \\ &= \log_b x^{\frac{1}{4}} - \log_b 5^2 - \log_b y^{10} \\ &= \log_b x^{\frac{1}{4}} - \left(\log_b 5^2 + \log_b y^{10}\right) \\ &= \log_b \sqrt[4]{x} - \left(\log_b 25 + \log_b y^{10}\right) \\ &= \log_b \sqrt[4]{x} - \left(\log_b 25y^{10}\right) \\ &= \log_b\left(\frac{\sqrt[4]{x}}{25y^{10}}\right) \end{aligned}$$

Pencil Problem #5

5a. Write as a single logarithm: $\log 5 + \log 2$.

5b. Write as a single logarithm: $4\ln(x+6) - 3\ln x$.

5c. Write as a single logarithm:
$3\ln x + 5\ln y - 6\ln z$.

Objective #6: Use the change-of-base property.

✔ Solved Problem #6

6a. Use common logarithms to evaluate $\log_7 2506$.

$$\log_7 2506 = \frac{\log 2506}{\log 7} \approx 4.02$$

6b. Use natural logarithms to evaluate $\log_7 2506$.

$$\log_7 2506 = \frac{\ln 2506}{\ln 7} \approx 4.02$$

Pencil Problem #6

6a. Use common logarithms to evaluate $\log_{0.1} 17$.

6b. Use natural logarithms to evaluate $\log_{0.1} 17$.

Answers for Pencil Problems *(Textbook Exercise references in parentheses)*:

1. $\log_5 7 + \log_5 3$ *(3.3 #1)*

2. $3 - \log_4 y$ *(3.3 #11)*

3a. $3\log_b x$ *(3.3 #15)* **3b.** $-6\log N$ *(3.3 #17)*

4a. $\frac{1}{5}\ln x$ *(3.3 #19)* **4b.** $2\log_b x + \log_b y$ *(3.3 #21)* **4c.** $2 - \frac{1}{2}\log_6(x+1)$ *(3.3 #25)*

5a. 1 *(3.3 #41)* **5b.** $\ln\left[\frac{(x+6)^4}{x^3}\right]$ *(3.3 #59)* **5c.** $\ln\left(\frac{x^3 y^5}{z^6}\right)$ *(3.3 #61)*

6a. -1.2304 *(3.3 #75)* **6b.** -1.2304 *(3.3 #75)*

Section 3.4
Exponential and Logarithmic Equations

Under the Sea !

Though it can be pitch black in the depths of the ocean, sunlight is visible as you get closer to the surface. About 12% of the surface sunlight reaches a depth of 20 feet and about 1.6% reaches to a depth of 100 feet.

In the applications of this section, you will use an exponential function to determine the depths that correspond to various percentages of light.

Objective #1: Use like bases to solve exponential equations.

✔ *Solved Problem #1*

1a. Solve: $5^{3x-6} = 125$.

$$5^{3x-6} = 125$$
$$5^{3x-6} = 5^3$$
$$3x - 6 = 3$$
$$3x = 9$$
$$x = 3$$

The solution set is $\{3\}$.

1b. Solve: $8^{x+2} = 4^{x-3}$.

$$8^{x+2} = 4^{x-3}$$
$$(2^3)^{x+2} = (2^2)^{x-3}$$
$$2^{3(x+2)} = 2^{2(x-3)}$$
$$3(x+2) = 2(x-3)$$
$$3x + 6 = 2x - 6$$
$$x + 6 = -6$$
$$x = -12$$

The solution set is $\{-12\}$.

Pencil Problem #1

1a. Solve: $4^{2x-1} = 64$.

1b. Solve: $8^{x+3} = 16^{x-1}$.

Objective #2: Use logarithms to solve exponential equations.

✔ *Solved Problem #2*

2a. Solve: $10^x = 8000$.

Take the common log of both sides of the equation.

$$\begin{aligned} 10^x &= 8000 \\ \log 10^x &= \log 8000 \\ x &= \log 8000 \\ x &\approx 3.90 \end{aligned}$$

The solution set is $\{\log 8000 \approx 3.90\}$.

2b. Solve: $7e^{2x} - 5 = 58$.

Isolate the exponential expression, then take the natural log of both sides of the equation.

$$\begin{aligned} 7e^{2x} - 5 &= 58 \\ 7e^{2x} &= 63 \\ e^{2x} &= 9 \\ \ln e^{2x} &= \ln 9 \\ 2x &= \ln 9 \\ x &= \frac{\ln 9}{2} \\ x &= \frac{\ln 3^2}{2} \\ x &= \frac{2\ln 3}{2} \\ x &= \ln 3 \\ x &\approx 1.10 \end{aligned}$$

The solution set is $\{\ln 3 \approx 1.10\}$.

✎ *Pencil Problem #2* ✎

2a. Solve: $10^x = 3.91$.

2b. Solve: $7^{x+2} = 410$.

Objective #3: Use exponential form to solve logarithmic equations.

✔ *Solved Problem #3*

3a. Solve: $4\ln(3x) = 8$.

$$4\ln(3x) = 8$$
$$\ln(3x) = 2$$
$$e^2 = 3x$$
$$\frac{e^2}{3} = x$$

Check:

$$4\ln(3x) = 8$$
$$4\ln\left[3\left(\frac{e^2}{3}\right)\right] = 8$$
$$4\ln\left(e^2\right) = 8$$
$$4 \cdot 2 = 8$$
$$8 = 8, \text{ true}$$

The solution set is $\left\{\frac{e^2}{3}\right\}$.

3b. Solve: $\log x + \log(x-3) = 1$.

$$\log x + \log(x-3) = 1$$
$$\log(x^2 - 3x) = 1$$
$$10^1 = x^2 - 3x$$
$$0 = x^2 - 3x - 10$$
$$0 = (x+2)(x-5)$$

$x + 2 = 0$ or $x - 5 = 0$

$x = -2$ $\quad x = 5$

Check -2:

$$\log x + \log(x-3) = 1$$
$$\log(-2) + \log(-2-3) = 1$$

-2 does not check.

Check 5:

$$\log x + \log(x-3) = 1$$
$$\log 5 + \log(5-3) = 1$$
$$\log 5 + \log 2 = 1$$
$$\log 10 = 1$$
$$1 = 1, \text{ true}$$

The solution set is $\{5\}$.

✎ *Pencil Problem #3* ✎

3a. Solve: $\log_4(x+5) = 3$.

3b. Solve: $\log_5 x + \log_5(4x-1) = 1$.

Objective #4: Use the one-to-one property of logarithms to solve logarithmic equations.

✔ Solved Problem #4

4. Solve: $\ln(x-3) = \ln(7x-23) - \ln(x+1)$

$$\ln(x-3) = \ln(7x-23) - \ln(x+1)$$
$$\ln(x-3) = \ln\left(\frac{7x-23}{x+1}\right)$$
$$x-3 = \frac{7x-23}{x+1}$$
$$(x+1)(x-3) = (x+1)\frac{7x-23}{x+1}$$
$$x^2 - 2x - 3 = 7x - 23$$
$$x^2 - 9x + 20 = 0$$
$$(x-4)(x-5) = 0$$
$$x-4=0 \quad \text{or} \quad x-5=0$$
$$x=4 \qquad\qquad x=5$$

Check 4:

$$\ln(x-3) = \ln(7x-23) - \ln(x+1)$$
$$\ln(4-3) = \ln(7\cdot 4-23) - \ln(4+1)$$
$$\ln 1 = \ln 5 - \ln 5$$
$$0 = 0, \text{ true}$$

Check 5:

$$\ln(x-3) = \ln(7x-23) - \ln(x+1)$$
$$\ln(5-3) = \ln(7\cdot 5-23) - \ln(5+1)$$
$$\ln 2 = \ln 12 - \ln 6$$
$$\ln 2 = \ln\left(\frac{12}{6}\right)$$
$$\ln 2 = \ln 2, \text{ true}$$

The solution set is $\{4, 5\}$.

Pencil Problem #4

4. Solve: $\log(x+4) - \log 2 = \log(5x+1)$

Objective #5: Solve applied problems involving exponential and logarithmic equations.

✔ *Solved Problem #5*

5a. Medical research indicates that the risk of having a car accident increases exponentially as the concentration of alcohol in the blood increases. The risk is modeled by $R = 6e^{12.77x}$ where x is the blood alcohol concentration and R, given as a percent, is the risk of having a car accident. What blood alcohol concentration corresponds to a 7% risk of a car accident?

$$R = 6e^{12.77x}$$
$$6e^{12.77x} = 7$$
$$e^{12.77x} = \frac{7}{6}$$
$$\ln e^{12.77x} = \ln\frac{7}{6}$$
$$12.77x = \ln\frac{7}{6}$$
$$x = \frac{\ln\frac{7}{6}}{12.77}$$
$$x \approx 0.01$$

For a blood alcohol concentration of 0.01, the risk of a car accident is 7%.

5b. How long, to the nearest tenth of a year, will it take \$1000 to grow to \$3600 at 8% annual interest compounded quarterly?

$$A = P\left(1+\frac{r}{n}\right)^{nt}$$
$$3600 = 1000\left(1+\frac{0.08}{4}\right)^{4t}$$
$$3.6 = 1.02^{4t}$$
$$1.02^{4t} = 3.6$$
$$\ln 1.02^{4t} = \ln 3.6$$
$$4t\ln 1.02 = \ln 3.6$$
$$t = \frac{\ln 3.6}{4\ln 1.02}$$
$$t \approx 16.2$$

After approximately 16.2 years, the \$1000 will grow to \$3600.

Pencil Problem #5

5a. The formula $A = 37.3e^{0.0095e}$ models the population of California, A, in millions, t years after 2010. When will the population of California reach 40 million?

5b. How long, to the nearest tenth of a year, will it take \$8000 to grow to \$16,000 at 8% annual interest compounded continuously?

Answers for Pencil Problems *(Textbook Exercise references in parentheses)*:

1a. $\{2\}$ *(3.4 #7)* **1b.** $\{13\}$ *(3.4 #19)*

2a. $\{\log 3.91 \approx 0.59\}$ *(3.4 #23)*

2b. $\left\{\dfrac{\ln 410}{\ln 7} - 2 \approx 1.09\right\}$ *(3.4 #37)*

3a. {59} *(3.4 #53)*

3b. $\left\{\dfrac{5}{4}\right\}$; note: reject -1 *(3.4 #67)*

4. $\left\{\dfrac{2}{9}\right\}$ *(3.4 #83)*

5a. 2017 *(3.4 #103b)*

5b. 8.7 years *(3.4 #111)*

Section 3.5
Exponential Growth and Decay; Modeling Data

NOT Tooth Decay!

One of algebra's many applications is to predict the behavior of variables. This can be done with *exponential growth* and *decay models*. With exponential growth or decay, quantities grow or decay at a rate directly proportional to their size. Populations that are growing exponentially grow extremely rapidly as they get larger because there are more adults to have offspring.

In this section we explore how to create such functions and how to use them to make predictions.

Objective #1: Model exponential growth and decay.

✔ *Solved Problem #1*

1. In 2000, the population of Africa was 807 million and by 2011 it had grown to 1052 million.

1a. Use the exponential growth model $A = A_0 e^{kt}$, in which t is the number of years after 2000, to find the exponential growth function that models the data.

2011 is 11 years after 2000.
Thus, when $t = 11$, $A = 1052$.
$A_0 = 807$

Substitute these values to find k.

$$A = A_0 e^{kt}$$
$$1052 = 807e^{k(11)}$$

Solve for k.

$$1052 = 807e^{k(11)}$$
$$\frac{1052}{807} = e^{11k}$$
$$\ln \frac{1052}{807} = \ln e^{11k}$$
$$k = \frac{\ln \frac{1052}{807}}{11}$$
$$k \approx 0.024$$

Thus, the growth function is $A = 807e^{0.024t}$.

Pencil Problem #1

1. In 2000, the population of Israel was approximately 6.04 million and by 2050 it is projected to grow to 10 million.

1a. Use the exponential growth model $A = A_0 e^{kt}$, in which t is the number of years after 2000, to find an exponential growth function that models the data.

1b. By which year will Africa's population reach 2000 million, or two billion?

$$A = 807e^{0.024t}$$
$$2000 = 807e^{0.024t}$$
$$\frac{2000}{807} = e^{0.024t}$$
$$\ln\frac{2000}{807} = \ln e^{0.024t}$$
$$\ln\frac{2000}{807} = 0.024t$$
$$t = \frac{\ln\frac{2000}{807}}{0.024} \approx 38$$

Africa's population will reach 2000 million approximately 38 years after 2000, or 2038.

1b. In which year will Israel's population be 9 million?

Objective #2: Use logistic growth models.

✔ ***Solved Problem #2***

2. In a learning theory project, psychologists discovered that

$$f(t) = \frac{0.8}{1+e^{-0.2t}}$$

is a model for describing the proportion of correct responses, $f(t)$, after t learning trials.

2a. Find the proportion of correct responses after 10 learning trials.

We substitute 10 for t in the logistic growth function.

$$f(10) = \frac{0.8}{1+e^{-0.2(10)}} \approx 0.7$$

The proportion of correct responses after 10 trials is approximately 0.7.

 ***Pencil Problem #2***

2. The logistic growth function

$$f(t) = \frac{100{,}000}{1+5000e^{-t}}$$

describes the number of people, $f(t)$, who have become ill with influenza t weeks after its initial outbreak in a particular community.

2a. How many people were ill by the end of the fourth week?

2b. What is the limiting size of $f(t)$, the proportion of correct responses, as continued learning trials take place?

The number in the numerator of the logistic growth function, 0.8, is the limiting size of the proportion of correct responses.

2b. What is the limiting size of the population that becomes ill?

Objective #3: Use Newton's Law of Cooling.

✔ *Solved Problem #3*

3. An object is heated to 100°C. It is left to cool in a room that has a temperature of 30°C. After 5 minutes, the temperature of the object is 80°C.

3a. Use Newton's Law of Cooling, $T = C + (T_0 - C)e^{kt}$, to find a model for the temperature of the object, T, after t minutes.

The initial temperature is 100°C: $T_0 = 100$.

The constant temperature of the room is 30°C: $C = 30$.

$$
\begin{aligned}
T &= C + (T_0 - C)e^{kt} \\
&= 30 + (100 - 30)e^{kt} \\
&= 30 + 70e^{kt}
\end{aligned}
$$

After 5 minutes, the temperature of the object is 80°C: $T = 80$ when $t = 5$. Substitute these values into $T = 30 + 70e^{kt}$ and solve for k. Begin by isolating the exponential expression on one side.

$$
\begin{aligned}
T &= 30 + 70e^{kt} \\
80 &= 30 + 70e^{k(5)} \\
50 &= 70e^{5k} \\
\frac{50}{70} &= e^{5k} \\
\frac{5}{7} &= e^{5k}
\end{aligned}
$$

(continued on next page)

Pencil Problem #3

3. A bottle of juice initially has a temperature of 70°F. It is left to cool in a refrigerator that has a temperature of 45°F. After 10 minutes, the temperature of the juice is 55°F.

3a. Use Newton's Law of Cooling, $T = C + (T_0 - C)e^{kt}$, to find a model for the temperature of the juice, T, after t minutes.

Now reverse sides, take the natural logarithm of each side and continue to solve for k.

$$e^{5k} = \frac{5}{7}$$

$$\ln e^{5k} = \ln\left(\frac{5}{7}\right)$$

$$5k = \ln\left(\frac{5}{7}\right)$$

$$k = \frac{\ln\left(\frac{5}{7}\right)}{5} \approx -0.0673$$

Substitute -0.0673 for k in $T = 30 + 70e^{kt}$. The model is $T = 30 + 70e^{-0.0673t}$.

3b. What is the temperature of the object after 20 minutes?

Substitute 20 for t in the model found in Solved Problem #3a and evaluate with a calculator.

$T = 30 + 70e^{-0.0673(20)} \approx 48$

After 20 minutes, the temperature of the object will be approximately 48°C.

3b. What is the temperature of the juice after 15 minutes?

3c. When will the temperature of the object be 35°C?

Substitute 35 for T in the model found in Solved Problem #3a and solve for t.

$$35 = 30 + 70e^{-0.0673t}$$

$$5 = 70e^{-0.0673t}$$

$$\frac{5}{70} = e^{-0.0673t}$$

$$e^{-0.0673t} = \frac{1}{14}$$

$$\ln e^{-0.0673t} = \ln\left(\frac{1}{14}\right)$$

$$-0.0673t = \ln\left(\frac{1}{14}\right)$$

$$t = \frac{\ln\left(\frac{1}{14}\right)}{-0.0673} \approx 39$$

The object will be 35°C after approximately 39 minutes.

3c. When will the temperature of the juice be 50°F?

Objective #4: Choose an appropriate model for data.

✔ *Solved Problem #4*

4. The table shows the populations of various cities and the average walking speed of a person living in the city. Create a scatter plot for the data. Based on the scatter plot, what type of function would be a good choice for modeling the data?

Population (thousands)	Walking Speed (feet per second)
5.5	0.6
14	1.0
71	1.6
138	1.9
342	2.2

Scatter plot:

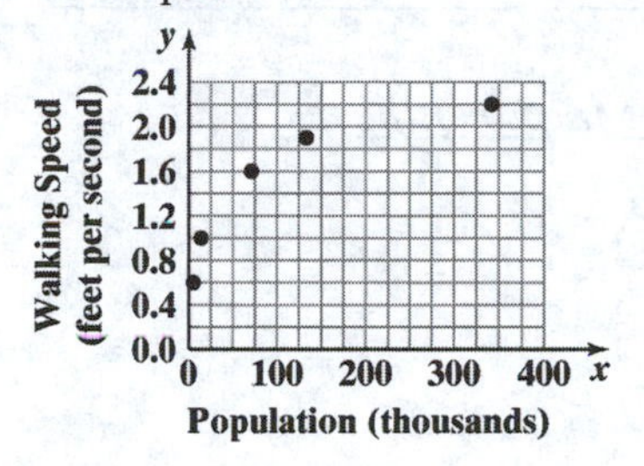

A logarithmic function would be a good choice for modeling the data.

Pencil Problem #4

4. The table shows the percentage of miscarriages by women of various ages. Create a scatter plot for the data. Determine the type of function that would be a good choice for modeling the data?

Woman's Age	Percent of Miscarriages
22	9%
27	10%
32	13%
37	20%
42	38%
47	52%

Objective #5: Express an exponential model in base *e*.

✔ *Solved Problem #5*

5. Rewrite $y = 4(7.8)^x$ in terms of base *e*. Express the answer in terms of a natural logarithm and then round to three decimal places.

$y = 4(7.8)^x$

$= 4e^{(\ln 7.8)x}$

Rounded to three decimal places:

$y = 4e^{2.054x}$

Pencil Problem #5

5. Rewrite $y = 100(4.6)^x$ in terms of base *e*. Express the answer in terms of a natural logarithm and then round to three decimal places.

Answers for Pencil Problems *(Textbook Exercise references in parentheses)*:

1a. $A = 6.04e^{0.01t}$ *(3.5 #7a)*

1b. 2040 *(3.5 #7b)*

2a. approximately 1080 people *(3.5 #37b)*

2b. 100,000 people *(3.5 #37c)*

3a. $T = 45 + 25e^{-0.0916t}$ *(3.5 #47a)* **3b.** 51° F *(3.5 #47b)* **3c.** 18 minutes *(3.5 #47c)*

4. 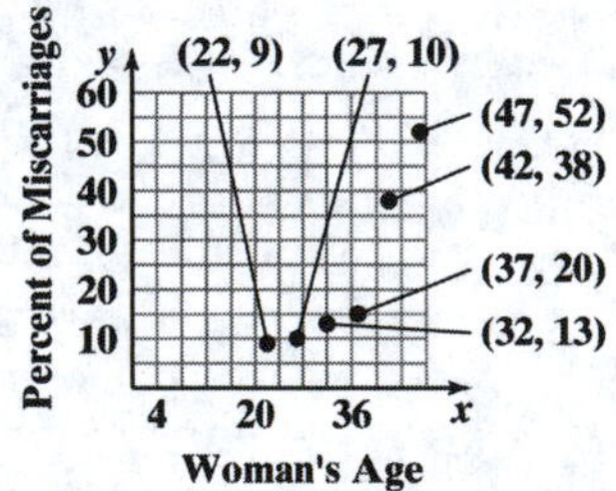

exponential function *(3.5 #51)*

5. $y = 100e^{(\ln 4.6)x} = 100e^{1.526x}$ *(3.5 #57)*

Section 4.1
Angles and Radian Measure

Ever Feel Like You're Just Going in Circles?

You're riding on a Ferris wheel and wonder how fast you are traveling. Before you got on the ride, the operator told you that the wheel completes two full revolutions every minute and that your seat is 25 feet from the center of the wheel. You just rode on the merry-go-round, which made 2.5 complete revolutions per minute. Your wooden horse was 20 feet from the center, but your friend, riding beside you was only 15 feet from the center. Were you and your friend traveling at the same rate?

In this section, we study both angular speed and linear speed and solve problems similar to those just stated.

Objective #1: Recognize and use the vocabulary of angles.

✔ *Solved Problem #1*

1a. True or false: When an angle is in standard position, its initial side is along the positive y-axis.

False; When an angle is in standard position, its initial side is along the positive x-axis.

1b. Fill in the blank to make a true statement: If the terminal side of an angle in standard position lies on the x-axis or the y-axis, the angle is called a/an ___________ angle.

Such an angle is called a quadrantal angle.

✎ *Pencil Problem #1*

1a. True or false: When an angle is in standard position, its vertex lies in quadrant I.

1b. Fill in the blank to make a true statement: A negative angle is generated by a __________ rotation.

Objective #2: Use degree measure.

✔ *Solved Problem #2*

2. Fill in the blank to make a true statement: An angle that is formed by $\frac{1}{2}$ of a complete rotation measures _________ degrees and is called a/an _________ angle.

Such an angle measures 180 degrees and is called a straight angle.

✎ *Pencil Problem #2*

2. Fill in the blank to make a true statement: An angle that is formed by $\frac{1}{4}$ of a complete rotation measures _________ degrees and is called a/an _________ angle.

Objective #3: Use radian measure.

✔ *Solved Problem #3*

3. A central angle, θ, in a circle of radius 12 feet intercepts an arc of length 42 feet. What is the radian measure of θ?

The radian measure of the central angle, θ, is the length of the intercepted arc, s, divided by the radius of the circle, r: $\theta = \frac{s}{r}$. In this case, $s = 42$ feet and $r = 12$ feet.

$$\theta = \frac{s}{r} = \frac{42 \text{ feet}}{12 \text{ feet}} = 3.5$$

The radian measure of θ is 3.5.

Pencil Problem #3

3. A central angle, θ, in a circle of radius 10 inches intercepts an arc of length 40 inches. What is the radian measure of θ?

Objective #4: Convert between degrees and radians.

✔ *Solved Problem #4*

4a. Convert 60° to radians.
To convert from degrees to radians, multiply by $\frac{\pi \text{ radians}}{180°}$. Then simplify.

$$60° \cdot \frac{\pi \text{ radians}}{180°} = \frac{60\pi \text{ radians}}{180} = \frac{\pi}{3} \text{ radians}$$

4b. Convert –300° to radians.

$$-300° \cdot \frac{\pi \text{ radians}}{180°} = -\frac{300\pi \text{ radians}}{180} = -\frac{5\pi}{3} \text{ radians}$$

Pencil Problem #4

4a. Convert 135° to radians. Express your answer as a multiple of π.

4b. Convert –225° to radians. Express your answer as a multiple of π.

4c. Convert $\frac{\pi}{4}$ radians to degrees.

To convert from radians to degrees, multiply by $\frac{180°}{\pi \text{ radians}}$. Then simplify.

$$\frac{\pi}{4} \text{ radians} \cdot \frac{180°}{\pi \text{ radians}} = \frac{180°}{4} = 45°$$

4c. Convert $\frac{\pi}{2}$ radians to degrees.

4d. Convert 6 radians to degrees.

$$6 \text{ radians} \cdot \frac{180°}{\pi \text{ radians}} = \frac{1080°}{\pi} \approx 343.8°$$

4d. Convert 2 radians to degrees. Round to two decimal places.

Objective #5: Draw angles in standard position.

✔ *Solved Problem #5*

5a. Draw the angle $\theta = -\frac{\pi}{4}$ in standard position.

Since the angle is negative, it is obtained by a clockwise rotation. Express the angle as a fractional part of 2π.

$$\left|-\frac{\pi}{4}\right| = \frac{\pi}{4} = \frac{1}{8} \cdot 2\pi$$

The angle $\theta = -\frac{\pi}{4}$ is $\frac{1}{8}$ of a full rotation in the clockwise direction.

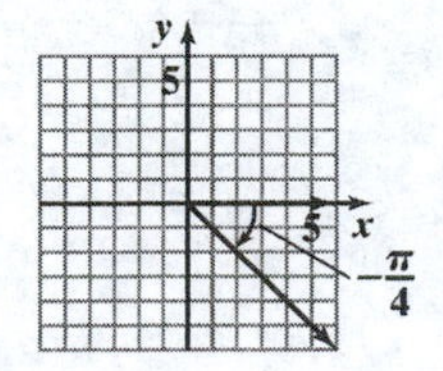

Pencil Problem #5

5a. Draw the angle $\theta = -\frac{5\pi}{4}$ in standard position.

5b. Draw the angle $\alpha = \frac{3\pi}{4}$ in standard position.

Since the angle is positive, it is obtained by a counterclockwise rotation. Express the angle as a fractional part of 2π.

$\frac{3\pi}{4} = \frac{3}{8} \cdot 2\pi$

The angle $\alpha = \frac{3\pi}{4}$ is $\frac{3}{8}$ of a full rotation in the counterclockwise direction.

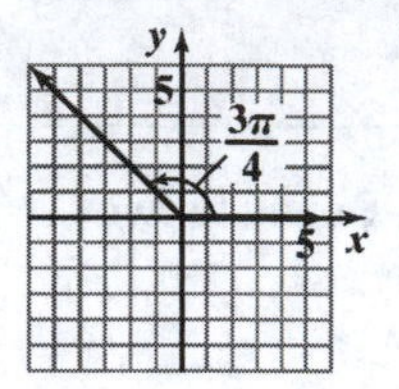

5b. Draw the angle $\alpha = \frac{7\pi}{6}$ in standard position.

5c. Draw the angle $\gamma = \frac{13\pi}{4}$ in standard position.

Since the angle is positive, it is obtained by a counterclockwise rotation. Express the angle as a fractional part of 2π.

$\frac{13\pi}{4} = \frac{13}{8} \cdot 2\pi$

The angle $\gamma = \frac{13\pi}{4}$ is $\frac{13}{8}$ or $1\frac{5}{8}$ full rotation in the counterclockwise direction. Complete one full rotation and then $\frac{5}{8}$ of a full rotation.

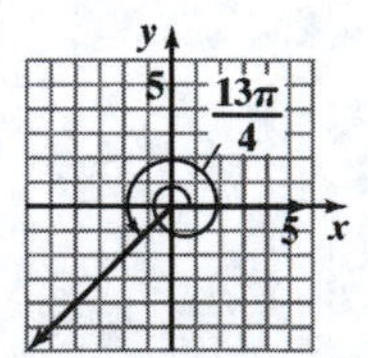

5c. Draw the angle $\gamma = \frac{16\pi}{3}$ in standard position.

Objective #6: Find coterminal angles.

✔ ***Solved Problem #6***

6a. Find a positive angle less than 360° that is coterminal with a 400° angle.

Since 400° is greater than 360°, we subtract 360°.

$400° - 360° = 40°$

A 40° angle is positive, less than 360°, and coterminal with a 400° angle.

6b. Find a positive angle less than 2π that is coterminal with a $-\frac{\pi}{15}$ angle.

Since $-\frac{\pi}{15}$ is negative, we add 2π.

$$-\frac{\pi}{15} + 2\pi = -\frac{\pi}{15} + \frac{30\pi}{15} = \frac{29\pi}{15}$$

A $\frac{29\pi}{15}$ angle is positive, less than 2π, and coterminal with a $-\frac{\pi}{15}$ angle.

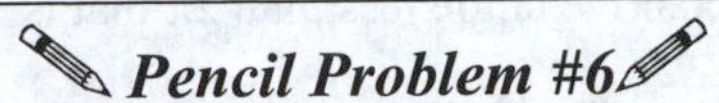

Pencil Problem #6

6a. Find a positive angle less than 360° that is coterminal with a 395° angle.

6b. Find a positive angle less than 2π that is coterminal with a $-\frac{\pi}{50}$ angle.

6c. Find a positive angle less than 2π that is coterminal with a $\frac{17\pi}{3}$ angle.

Since $\frac{17\pi}{3}$ is greater than 4π, we subtract two multiples of 2π.

$$\frac{17\pi}{3} - 2 \cdot 2\pi = \frac{17\pi}{3} - 4\pi = \frac{17\pi}{3} - \frac{12\pi}{3} = \frac{5\pi}{3}$$

A $\frac{5\pi}{3}$ angle is positive, less than 2π, and coterminal with a $\frac{17\pi}{3}$ angle.

6c. Find a positive angle less than 2π that is coterminal with a $-\frac{31\pi}{7}$ angle.

Objective #7: Find the length of a circular arc.

✔ *Solved Problem #7*

7. A circle has a radius of 6 inches. Find the length of the arc intercepted by a central angle of 45°. Express arc length in terms of π. Then round your answer to two decimal places.

We begin by converting 45° to radians.

$$45^\circ \cdot \frac{\pi \text{ radians}}{180^\circ} = \frac{45\pi \text{ radians}}{180} = \frac{\pi}{4} \text{ radians}$$

Now we use the arc length formula $s = r\theta$ with the radius $r = 6$ inches and the angle $\theta = \frac{\pi}{4}$ radians.

$$s = r\theta = (6 \text{ in.})\left(\frac{\pi}{4}\right) = \frac{6\pi}{4} \text{ in.} = \frac{3\pi}{2} \text{ in.} \approx 4.71 \text{ in.}$$

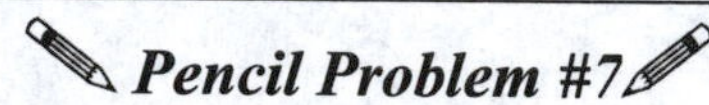

Pencil Problem #7

7. A circle has a radius of 8 feet. Find the length of the arc intercepted by a central angle of 225°. Express arc length in terms of π. Then round your answer to two decimal places.

Objective #8: Use linear and angular speed to describe motion on a circular path.

✔ Solved Problem #8

8. A 45-rpm record has an angular speed of 45 revolutions per minute. Find the linear speed, in inches per minute, at the point where the needle is 1.5 inches from the record's center.

We are given the angular speed in revolutions per minute: $\omega = 45$ revolutions per minute. We must express ω in radians per minute.

$$\omega = \frac{45 \text{ revolutions}}{1 \text{ minute}} \cdot \frac{2\pi \text{ radians}}{1 \text{ revolution}}$$

$$= \frac{90\pi \text{ radians}}{1 \text{ minute}} \text{ or } \frac{90\pi}{1 \text{ minute}}$$

Now we use the formula $v = r\omega$.

$$v = r\omega = 1.5 \text{ in.} \cdot \frac{90\pi}{1 \text{ min}} = \frac{135\pi \text{ in.}}{\text{min}} \approx 424 \text{ in./min}$$

Pencil Problem #8

8. A Ferris wheel has a radius of 25 feet. The wheel is rotating at two revolutions per minute. Find the linear speed, in feet per minute, of a seat on this Ferris wheel.

Answers for Pencil Problems *(Textbook Exercise references in parentheses)*:

1a. False **1b.** clockwise

2. 90; right

3. 4 radians *(4.1 #7)*

4a. $\frac{3\pi}{4}$ radians *(4.1 #15)* **4b.** $-\frac{5\pi}{4}$ radians *(4.1 #19)* **4c.** 90° *(4.1 #21)*

4d. 114.59° *(4.1 #35)*

5a. 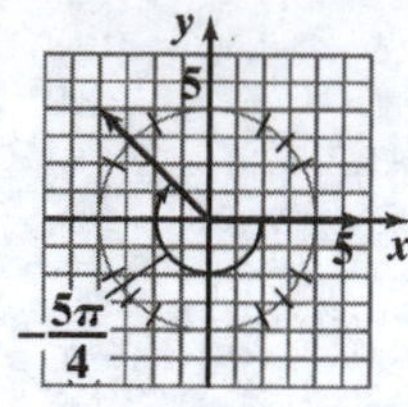

(4.1 #47)

5b. 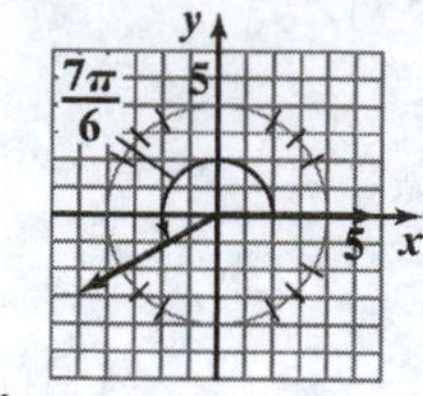

(4.1 #41)

5c. 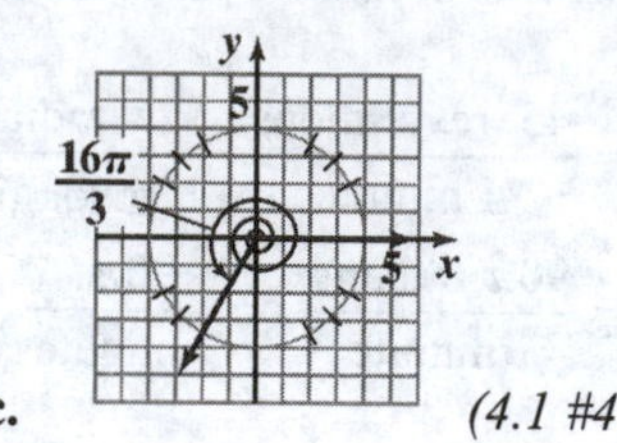

(4.1 #49)

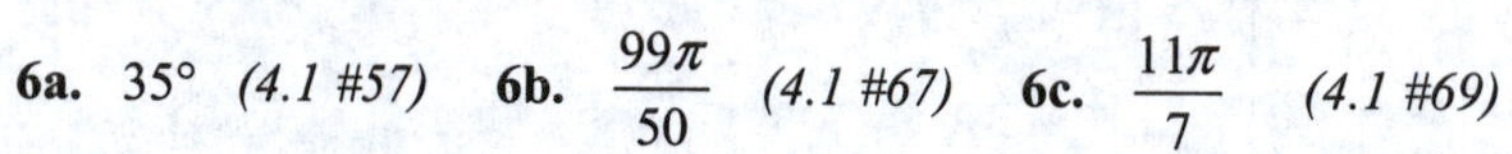

6a. 35° *(4.1 #57)* **6b.** $\frac{99\pi}{50}$ *(4.1 #67)* **6c.** $\frac{11\pi}{7}$ *(4.1 #69)*

7. 10π ft ≈ 31.42 ft *(4.1 #73)*

8. 100π ft/min ≈ 314 ft/min *(4.1 #98)*

Section 4.2
Trigonometric Functions: The Unit Circle

Could you repeat that?

There are many repetitive patterns in nature. Tides cycle through a pattern of low and high tides in a very predictable manner. The number of hours of daylight on a given day varies throughout the year, but the pattern throughout the year repeats itself year after year. In this section, we define the *trigonometric functions* using movement around a unit circle. In doing so, we can see that the values of the trigonometric functions repeat themselves each time we make a complete trip around the circle. The repetitive properties of the trigonometric functions make them useful for modeling cyclic phenomena.

Objective #1: Use a unit circle to define trigonometric functions of real numbers.

✔ Solved Problem #1

1. Use the figure to find the values of the trigonometric functions at t.

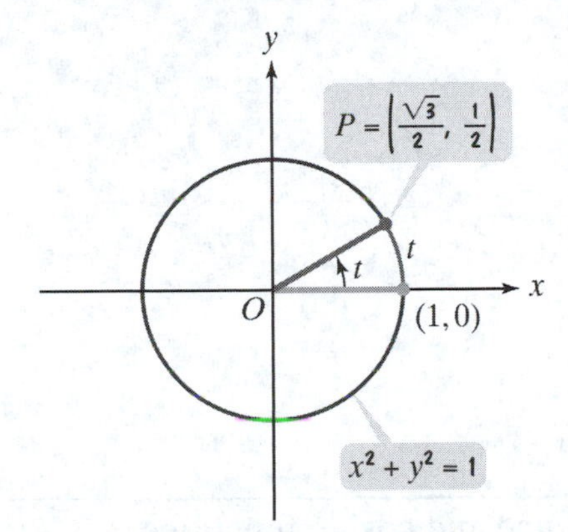

The point on the circle has coordinates $\left(\dfrac{\sqrt{3}}{2}, \dfrac{1}{2}\right)$.

(continued on next page)

Pencil Problem #1

1. Use the figure to find the values of the trigonometric functions at t.

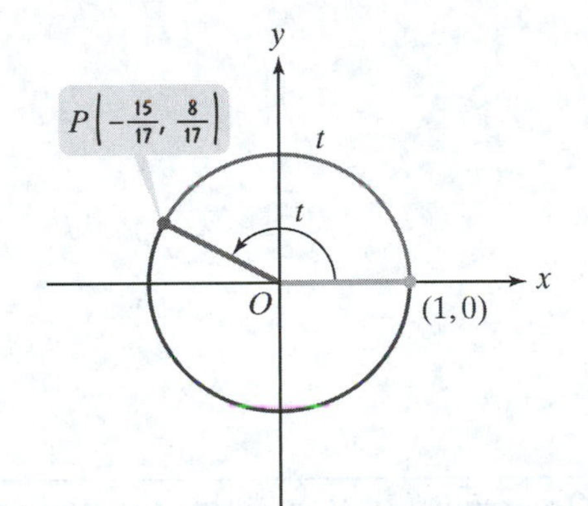

We use $x = \frac{\sqrt{3}}{2}$ and $y = \frac{1}{2}$ in the definitions.

$$\sin t = y = \frac{1}{2}$$

$$\cos t = x = \frac{\sqrt{3}}{2}$$

$$\tan t = \frac{y}{x} = \frac{\frac{1}{2}}{\frac{\sqrt{3}}{2}} = \frac{1}{\sqrt{3}} = \frac{1}{\sqrt{3}} \cdot \frac{\sqrt{3}}{\sqrt{3}} = \frac{\sqrt{3}}{3}$$

$$\csc t = \frac{1}{y} = \frac{1}{\frac{1}{2}} = 2$$

$$\sec t = \frac{1}{x} = \frac{1}{\frac{\sqrt{3}}{2}} = \frac{2}{\sqrt{3}} = \frac{2}{\sqrt{3}} \cdot \frac{\sqrt{3}}{\sqrt{3}} = \frac{2\sqrt{3}}{\sqrt{3}}$$

$$\cot t = \frac{x}{y} = \frac{\frac{\sqrt{3}}{2}}{\frac{1}{2}} = \sqrt{3}$$

Objective #2: Recognize the domain and range of sine and cosine functions.

✔ ***Solved Problem #2***

2. True or false: The range of the cosine function is all real numbers, $(-\infty, \infty)$, so there is a real number t in the domain of the cosine function for which $\cos t = 10$.

False; the range of cosine is $[-1, 1]$ and 10 is not in this interval, so there is no real number t in the domain of the cosine function for which $\cos t = 10$.

Pencil Problem #2

2. True or false: There is a real number t in the domain of the sine function for which $\sin t = -\frac{\sqrt{10}}{2}$.

Objective #3: Find exact values of the trigonometric functions at $\frac{\pi}{4}$.

✔ *Solved Problem #3*

3. Find $\sin\frac{\pi}{4}$ and $\sec\frac{\pi}{4}$.

The point P on the unit circle that corresponds to $t=\frac{\pi}{4}$ has coordinates $\left(\frac{\sqrt{2}}{2},\frac{\sqrt{2}}{2}\right)$. We use $x=\frac{\sqrt{2}}{2}$ and $y=\frac{\sqrt{2}}{2}$ in the definitions of sine and secant.

$$\sin\frac{\pi}{4}=y=\frac{\sqrt{2}}{2}$$

$$\sec\frac{\pi}{4}=\frac{1}{x}=\frac{1}{\frac{\sqrt{2}}{2}}=\frac{2}{\sqrt{2}}=\sqrt{2}$$

Pencil Problem #3

3. Find $\csc\frac{\pi}{4}$ and $\tan\frac{\pi}{4}$.

Objective #4: Use even and odd trigonometric functions.

✔ *Solved Problem #4*

4a. Given that $\sec\frac{\pi}{4}=\sqrt{2}$, find $\sec\left(-\frac{\pi}{4}\right)$.

Since secant is an even function, $\sec(-t)=\sec t$.

$$\sec\left(-\frac{\pi}{4}\right)=\sec\frac{\pi}{4}=\sqrt{2}$$

4b. Given that $\sin\frac{\pi}{4}=\frac{\sqrt{2}}{2}$, find $\sin\left(-\frac{\pi}{4}\right)$.

Since sine is an odd function, $\sin(-t)=-\sin t$.

$$\sin\left(-\frac{\pi}{4}\right)=-\sin\frac{\pi}{4}=-\frac{\sqrt{2}}{2}$$

Pencil Problem #4

4a. Given that $\cos\frac{\pi}{6}=\frac{\sqrt{3}}{2}$, find $\cos\left(-\frac{\pi}{6}\right)$.

4b. Given that $\tan\frac{5\pi}{3}=-\sqrt{3}$, find $\tan\left(-\frac{5\pi}{3}\right)$.

Objective #5: Recognize and use fundamental identities.

✔ *Solved Problem #5*

5a. Given $\sin t = \frac{2}{3}$ and $\cos t = \frac{\sqrt{5}}{3}$, find the value of each of the four remaining trigonometric functions.

Find $\tan t$ using a quotient identity.

$$\tan t = \frac{\sin t}{\cos t} = \frac{\frac{2}{3}}{\frac{\sqrt{5}}{3}} = \frac{2}{3} \cdot \frac{3}{\sqrt{5}} = \frac{2}{\sqrt{5}} \cdot \frac{\sqrt{5}}{\sqrt{5}} = \frac{2\sqrt{5}}{5}$$

Use reciprocal identities to find the remaining three function values.

$$\csc t = \frac{1}{\sin t} = \frac{1}{\frac{2}{3}} = \frac{3}{2}$$

$$\sec t = \frac{1}{\cos t} = \frac{1}{\frac{\sqrt{5}}{3}} = \frac{3}{\sqrt{5}} \cdot \frac{\sqrt{5}}{\sqrt{5}} = \frac{3\sqrt{5}}{5}$$

$$\cot t = \frac{1}{\tan t} = \frac{1}{\frac{2}{\sqrt{5}}} = \frac{\sqrt{5}}{2}$$

5b. Given that $\sin t = \frac{1}{2}$ and $0 \le t < \frac{\pi}{2}$, find the value of $\cos t$ using a trigonometric identity.

Use the Pythagorean identity $\sin^2 t + \cos^2 t = 1$.

Because $0 \le t < \frac{\pi}{2}$, $\cos t$ is positive.

$$\left(\frac{1}{2}\right)^2 + \cos^2 t = 1$$

$$\frac{1}{4} + \cos^2 t = 1$$

$$\cos^2 t = 1 - \frac{1}{4} = \frac{3}{4}$$

$$\cos t = \sqrt{\frac{3}{4}} = \frac{\sqrt{3}}{2}$$

✎ *Pencil Problem #5* ✎

5a. Given $\sin t = \frac{1}{3}$ and $\cos t = \frac{2\sqrt{2}}{3}$, find the value of each of the four remaining trigonometric functions.

5b. Given that $\sin t = \frac{6}{7}$ and $0 \le t < \frac{\pi}{2}$, find the value of $\cos t$ using a trigonometric identity.

Objective #6: Use periodic properties.

✔ *Solved Problem #6*

6a. Find the value of $\cot\frac{5\pi}{4}$.

The period of cotangent is π: $\cot(t+\pi)=\cot t$.

$\cot\frac{5\pi}{4}=\cot\left(\frac{\pi}{4}+\pi\right)=\cot\frac{\pi}{4}=1$

6b. Find the value of $\cos\left(-\frac{9\pi}{4}\right)$.

The cosine function is even: $\cos(-t)=\cos t$.
The period of cosine is 2π: $\cos(t+2\pi)=\cos t$.

$\cos\left(-\frac{9\pi}{4}\right)=\cos\frac{9\pi}{4}=\cos\left(\frac{\pi}{4}+2\pi\right)=\cos\frac{\pi}{4}=\frac{\sqrt{2}}{2}$

Pencil Problem #6

6a. Find the value of $\tan\frac{5\pi}{4}$.

6b. Find the value of $\sin\left(-\frac{9\pi}{4}\right)$.

Objective #7: Evaluate trigonometric functions with a calculator.

✔ *Solved Problem #7*

7a. Use a calculator to find the value of $\sin\frac{\pi}{4}$ to four decimal places.

Use radian mode.

On a scientific calculator, enter $\pi\div 4$, and then press = and the SIN key.

On a graphing calculator, press the SIN key, and then enter $\pi\div 4$, and press ENTER.

The display, rounded to four places, should be 0.7071.

Pencil Problem #7

7a. Use a calculator to find the value of $\cos\frac{\pi}{10}$ to four decimal places.

7b. Use a calculator to find the value of $\csc 1.5$ to four decimal places.

Use radian mode.

On a scientific calculator, enter 1.5, and then press the SIN key followed by the reciprocal key labeled $1/x$.

On a graphing calculator, open a set of parentheses, press the SIN key, and then enter 1.5. Close the parentheses, press the reciprocal key labeled x^{-1}, and press ENTER. The display, rounded to four places, should be 1.0025.

7b. Use a calculator to find the value of sec 1 to four decimal places.

Answers for Pencil Problems *(Textbook Exercise references in parentheses)*:

1. $\sin t = \frac{8}{17}$, $\cos t = -\frac{15}{17}$, $\tan t = -\frac{8}{15}$, $\csc t = \frac{17}{8}$, $\sec t = -\frac{17}{15}$, $\cot t = -\frac{15}{8}$ *(4.2 #1)*

2. false

3. $\csc\frac{\pi}{4} = \sqrt{2}$; $\tan\frac{\pi}{4} = 1$ *(4.2 Check Point #3)*

4a. $\frac{\sqrt{3}}{2}$ *(4.2 #19b)* **4b.** $\sqrt{3}$ *(4.2 #23b)*

5a. $\tan t = \frac{\sqrt{2}}{4}$, $\csc t = 3$, $\sec t = \frac{3\sqrt{2}}{4}$, $\cot t = 2\sqrt{2}$ *(4.2 #27)* **5b.** $\frac{\sqrt{13}}{7}$ *(4.2 #29)*

6a. 1 *(4.2 #43)* **6b.** $-\frac{\sqrt{2}}{2}$ *(4.2 #41)*

7a. 0.9511 *(4.2 #67)* **7b.** 1.8508 *(4.2 #66)*

Section 4.3
Right Triangle Trigonometry

Measuring Up, Way Up

Did you ever wonder how you could measure the height of a building or a tree? How can you find the distance across a lake or some other body of water?

In this section, we show how to model such situations using a right triangle and then using relationships among the lengths its sides and the measures of its angles to find distances that are otherwise difficult to measure. These relationships lead to a second approach to the trigonometric functions.

Objective #1: Use right triangles to evaluate trigonometric functions.

✔ *Solved Problem #1*

1a. Find the value of each of the six trigonometric functions of θ in the figure.

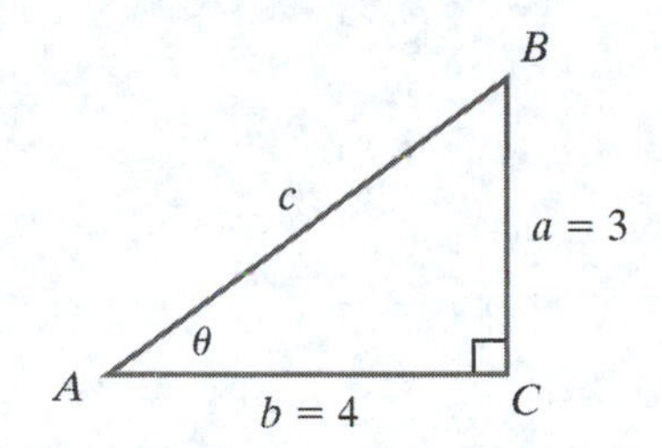

We first need to find c, the length of the hypotenuse. We use the Pythagorean Theorem.

$$c^2 = a^2 + b^2 = 3^2 + 4^2 = 9 + 16 = 25$$

$$c = \sqrt{25} = 5$$

We apply the right triangle definitions of the six trigonometric functions. Note that the side labeled $a = 3$ is opposite angle θ and the side labeled $b = 4$ is adjacent to angle θ.

$$\sin\theta = \frac{\text{opposite}}{\text{hypotenuse}} = \frac{3}{5}$$

$$\cos\theta = \frac{\text{adjacent}}{\text{hypotenuse}} = \frac{4}{5}$$

$$\tan\theta = \frac{\text{opposite}}{\text{adjacent}} = \frac{3}{4}$$

$$\csc\theta = \frac{\text{hypotenuse}}{\text{opposite}} = \frac{5}{3}$$

$$\sec\theta = \frac{\text{hypotenuse}}{\text{adjacent}} = \frac{5}{4}$$

$$\cot\theta = \frac{\text{adjacent}}{\text{opposite}} = \frac{4}{3}$$

Pencil Problem #1

1a. Find the value of each of the six trigonometric functions of θ in the figure.

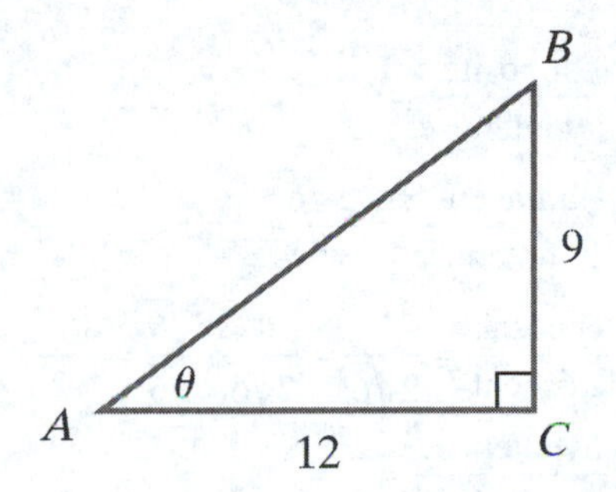

1b. Find the value of each of the six trigonometric functions of θ in the figure. Express each value in simplified form.

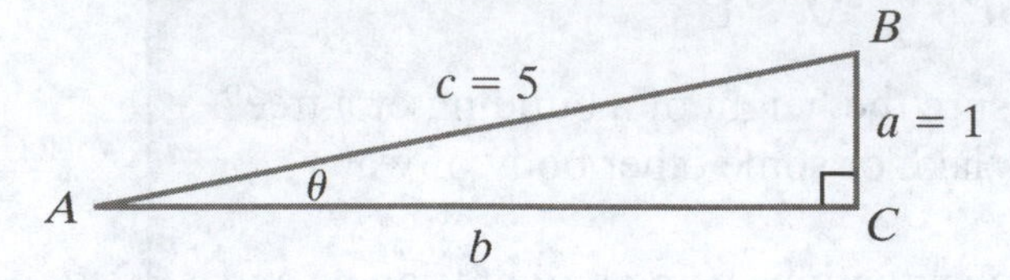

We first need to find b.

$$a^2 + b^2 = c^2$$
$$1^2 + b^2 = 5^2$$
$$1 + b^2 = 25$$
$$b^2 = 24$$
$$b = \sqrt{24} = 2\sqrt{6}$$

We apply the right triangle definitions of the six trigonometric functions.

$$\sin\theta = \frac{\text{opposite}}{\text{hypotenuse}} = \frac{1}{5}$$

$$\cos\theta = \frac{\text{adjacent}}{\text{hypotenuse}} = \frac{2\sqrt{6}}{5}$$

$$\tan\theta = \frac{\text{opposite}}{\text{adjacent}} = \frac{1}{2\sqrt{6}} = \frac{1}{2\sqrt{6}} \cdot \frac{\sqrt{6}}{\sqrt{6}} = \frac{\sqrt{6}}{12}$$

$$\csc\theta = \frac{\text{hypotenuse}}{\text{opposite}} = \frac{5}{1} = 5$$

$$\sec\theta = \frac{\text{hypotenuse}}{\text{adjacent}} = \frac{5}{2\sqrt{6}} = \frac{5}{2\sqrt{6}} \cdot \frac{\sqrt{6}}{\sqrt{6}} = \frac{5\sqrt{6}}{12}$$

$$\cot\theta = \frac{\text{adjacent}}{\text{opposite}} = \frac{2\sqrt{6}}{1} = 2\sqrt{6}$$

1b. Find the value of each of the six trigonometric functions of θ in the figure. Express each value in simplified form.

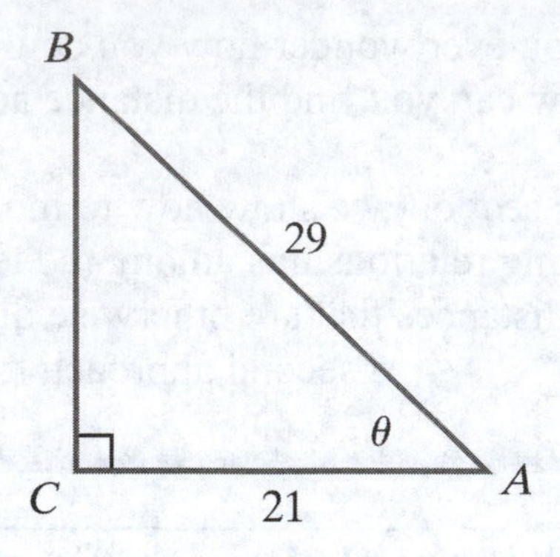

Objective #2: Find function values for $30°\left(\frac{\pi}{6}\right)$, $45°\left(\frac{\pi}{4}\right)$, and $60°\left(\frac{\pi}{3}\right)$.

✔ *Solved Problem #2*

2a. Use the right triangle to find $\csc 45°$.

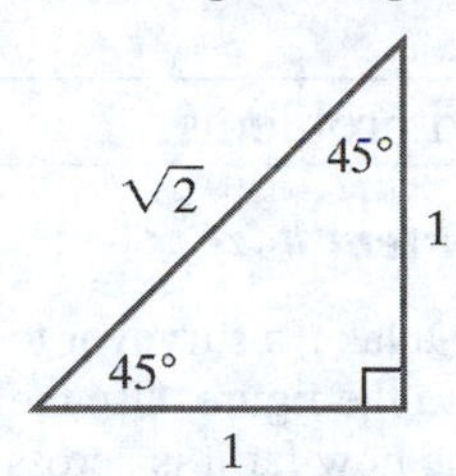

Use the definition of the cosecant function.

$$\csc 45° = \frac{\text{hypotenuse}}{\text{opposite}} = \frac{\sqrt{2}}{1} = \sqrt{2}$$

2b. Use the right triangle to find $\tan 60°$.

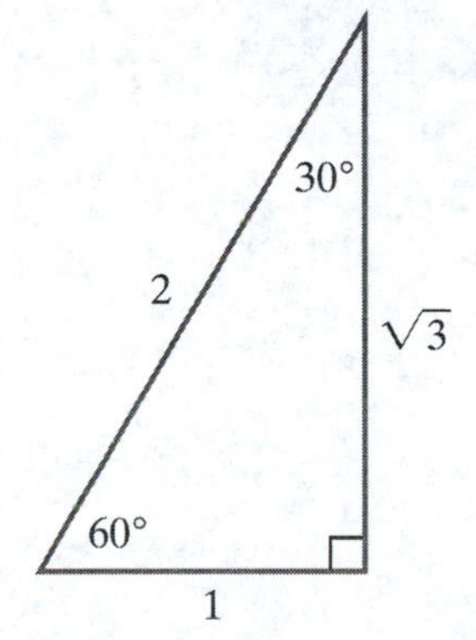

Use the definition of the tangent function and the angle marked 60° in the triangle.

$$\tan 60° = \frac{\text{opposite}}{\text{adjacent}} = \frac{\sqrt{3}}{1} = \sqrt{3}$$

Pencil Problem #2

2a. Use the right triangle in Solved Problem #2a to find $\sec 45°$.

2b. Use the right triangle in Solved Problem #2b to find $\cos 30°$.

Objective #3: Use equal cofunctions of complements.

Solved Problem #3

3a. Find a cofunction with the same value as $\sin 46°$.

$$\sin 46° = \cos(90° - 46°) = \cos 44°$$

Pencil Problem #3

3a. Find a cofunction with the same value as $\sin 7°$.

3b. Find a cofunction with the same value as $\cot\frac{\pi}{12}$.

$$\cot\frac{\pi}{12} = \tan\left(\frac{\pi}{2} - \frac{\pi}{12}\right) = \tan\left(\frac{6\pi}{12} - \frac{\pi}{12}\right) = \tan\left(\frac{5\pi}{12}\right)$$

3b. Find a cofunction with the same value as $\tan\frac{\pi}{9}$.

Objective #4: Use right triangle trigonometry to solve applied problems.

✔ *Solved Problem #4*

4. The distance across a lake, *a*, is unknown. To find this distance, a surveyor took the measurements shown in the figure. What is the distance across the lake?

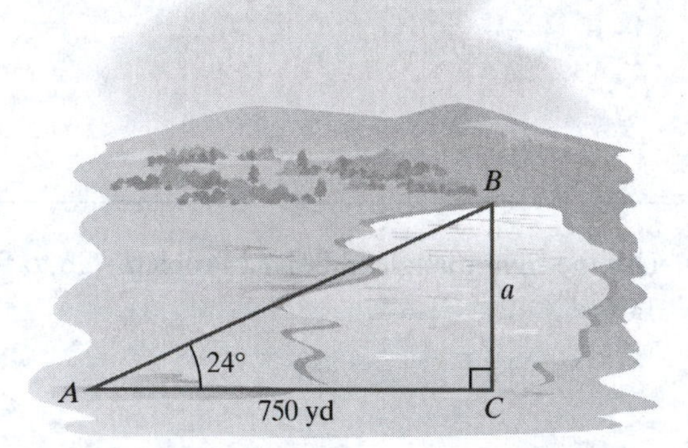

We know the measurements of one angle and the leg adjacent to the angle. We need to know the length of the side opposite the known angle. We use the tangent function.

$$\tan 24° = \frac{a}{750}$$

$$a = 750\tan 24°$$

$$a \approx 333.9$$

The distance across the lake is approximately 333.9 yards.

Pencil Problem #4

4. To find the distance across a lake, a surveyor took the measurements shown in the figure. Use the measurements to determine how far it is across the lake. Round to the nearest yard.

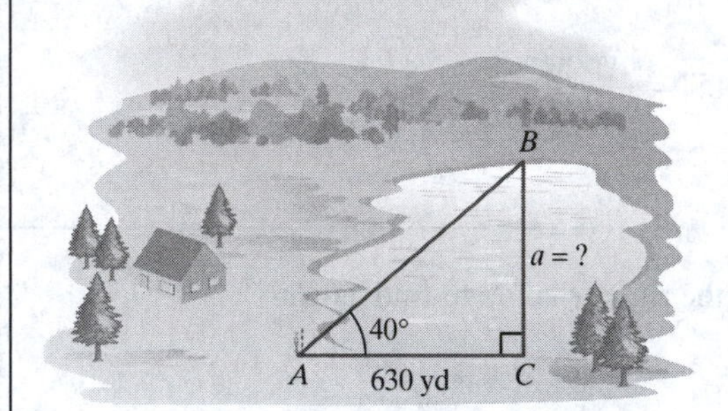

Answers for Pencil Problems *(Textbook Exercise references in parentheses)*:

1a. $\sin\theta = \frac{3}{5}$, $\cos\theta = \frac{4}{5}$, $\tan\theta = \frac{3}{4}$, $\csc\theta = \frac{5}{3}$, $\sec\theta = \frac{5}{4}$, $\cot\theta = \frac{4}{3}$ *(4.3 #1)*

1b. $\sin\theta = \frac{20}{29}$, $\cos\theta = \frac{21}{29}$, $\tan\theta = \frac{20}{21}$, $\csc\theta = \frac{29}{20}$, $\sec\theta = \frac{29}{21}$, $\cot\theta = \frac{21}{20}$ *(4.3 #3)*

2a. $\sqrt{2}$ *(4.3 #11)* **2b.** $\frac{\sqrt{3}}{2}$ *(4.3 #9)*

3a. $\cos 83°$ *(4.3 #21)* **3b.** $\cot\frac{7\pi}{18}$ *(4.3 #25)*

4. 529 yd *(4.3 #53)*

Section 4.4
Trigonometric Functions of Any Angle

Covering All the Angles

We've already considered two ways of defining the six trigonometric functions. In the previous section, we defined the functions for acute angles of a right triangle. In this section, we extend the definitions of the trigonometric functions to include all angles.

Using this approach, even more patterns in the function values become apparent. These patterns allow us to evaluate the trigonometric functions more efficiently.

Objective #1: Use the definitions of the trigonometric functions of any angle.

✔ *Solved Problem #1*

1a. Let $P = (1, -3)$ be a point on the terminal side of θ. Find each of the six trigonometric functions of θ.

We are given that $x = 1$ and $y = -3$. We need the value of r.

$$r = \sqrt{x^2 + y^2} = \sqrt{1^2 + (-3)^2} = \sqrt{1+9} = \sqrt{10}$$

Now we use the definitions of the trigonometric functions of any angle.

$$\sin\theta = \frac{y}{r} = \frac{-3}{\sqrt{10}} = \frac{-3}{\sqrt{10}} \cdot \frac{\sqrt{10}}{\sqrt{10}} = -\frac{3\sqrt{10}}{10}$$

$$\cos\theta = \frac{x}{r} = \frac{1}{\sqrt{10}} = \frac{1}{\sqrt{10}} \cdot \frac{\sqrt{10}}{\sqrt{10}} = \frac{\sqrt{10}}{10}$$

$$\tan\theta = \frac{y}{x} = \frac{-3}{1} = -3$$

$$\csc\theta = \frac{r}{y} = \frac{\sqrt{10}}{-3} = -\frac{\sqrt{10}}{3}$$

$$\sec\theta = \frac{r}{x} = \frac{\sqrt{10}}{1} = \sqrt{10}$$

$$\cot\theta = \frac{x}{y} = \frac{1}{-3} = -\frac{1}{3}$$

Pencil Problem #1

1a. Let $P = (-2, -5)$ be a point on the terminal side of θ. Find each of the six trigonometric functions of θ.

1b. Evaluate, if possible: $\csc 180°$.

The terminal side of $\theta = 180°$ is on the negative x-axis. We select the point $(-1, 0)$ on the terminal side of the angle, which is 1 unit from the origin, so $x = -1$, $y = 0$, and $r = 1$.

$\csc\theta = \frac{r}{y} = \frac{1}{0}$ (crossed out)

$\csc 180°$ is undefined.

1b. Evaluate, if possible: $\tan\frac{3\pi}{2}$.

Objective #2: Use the signs of the trigonometric functions.

✔ *Solved Problem #2*

2a. If $\sin\theta < 0$ and $\cos\theta < 0$, name the quadrant in which θ lies.

When $\sin\theta < 0$, θ lies in quadrant III or IV. When $\cos\theta < 0$, θ lies in quadrant II or III. When both conditions are met, θ must lie in quadrant III.

2b. Given that $\tan\theta = -\frac{1}{3}$ and $\cos\theta < 0$, find $\sin\theta$ and $\sec\theta$.

Because both the tangent and cosine are negative, θ lies in quadrant II, where x is negative and y is positive.

$$\tan\theta = -\frac{1}{3} = \frac{y}{x} = \frac{1}{-3}$$

So, $x = -3$ and $y = 1$. Find r.

$$r = \sqrt{x^2 + y^2} = \sqrt{(-3)^2 + 1^2} = \sqrt{9+1} = \sqrt{10}$$

Now use the definitions of the trigonometric functions of any angle.

$$\sin\theta = \frac{y}{r} = \frac{1}{\sqrt{10}} = \frac{1}{\sqrt{10}} \cdot \frac{\sqrt{10}}{\sqrt{10}} = \frac{\sqrt{10}}{10}$$

$$\sec\theta = \frac{r}{x} = \frac{\sqrt{10}}{-3} = -\frac{\sqrt{10}}{3}$$

Pencil Problem #2

2a. If $\tan\theta < 0$ and $\cos\theta < 0$, name the quadrant in which θ lies.

2b. Given that $\tan\theta = -\frac{2}{3}$ and $\sin\theta > 0$, find $\cos\theta$ and $\csc\theta$.

Objective #3: Find reference angles.

✔ *Solved Problem #3*

3a. Find the reference angle for $\theta = 210°$.

The angle lies in quadrant III. The reference angle is $\theta' = 210° - 180° = 30°$.

3b. Find the reference angle for $\theta = \dfrac{7\pi}{4}$.

The angle lies in quadrant IV. The reference angle is $\theta' = 2\pi - \dfrac{7\pi}{4} = \dfrac{8\pi}{4} - \dfrac{7\pi}{4} = \dfrac{\pi}{4}$.

3c. Find the reference angle for $\theta = -240°$.

The angle lies in quadrant II. The positive acute angle formed by the terminal side of θ and the x-axis is 60°. The reference angle is $\theta' = 60°$.

3d. Find the reference angle for $\theta = 665°$.

Subtract 360° to find a positive coterminal angle less than 360°: $665° - 360° = 305°$.

The angle $\alpha = 305°$ lies in quadrant IV. The reference angle is
$\alpha' = 360° - 305° = 55°$.

Pencil Problem #3

3a. Find the reference angle for $\theta = 160°$.

3b. Find the reference angle for $\theta = \dfrac{5\pi}{6}$.

3c. Find the reference angle for $\theta = -335°$.

3d. Find the reference angle for $\theta = 565°$.

3e. Find the reference angle for $\theta = -\frac{11\pi}{3}$.

Add 4π to find a positive coterminal angle less than 2π: $-\frac{11\pi}{3} + 4\pi = -\frac{11\pi}{3} + \frac{12\pi}{3} = \frac{\pi}{3}$.

The angle $\alpha = \frac{\pi}{3}$ lies in quadrant I. The reference angle is $\alpha' = \frac{\pi}{3}$.

3e. Find the reference angle for $\theta = -\frac{11\pi}{4}$.

Objective #4: Use reference angles to evaluate trigonometric functions.

✔ *Solved Problem #4*

4a. Use a reference angle to find the exact value of $\sin 300°$.

A 300° angle lies in quadrant IV, where the sine function is negative. The reference angle is $\theta' = 360° - 300° = 60°$.

$$\sin 300° = -\sin 60° = -\frac{\sqrt{3}}{2}$$

4b. Use a reference angle to find the exact value of $\tan\frac{5\pi}{4}$.

A $\frac{5\pi}{4}$ angle lies in quadrant III, where the tangent function is positive. The reference angle is $\theta' = \frac{5\pi}{4} - \pi = \frac{5\pi}{4} - \frac{4\pi}{4} = \frac{\pi}{4}$.

$$\tan\frac{5\pi}{4} = +\tan\frac{\pi}{4} = 1$$

Pencil Problem #4

4a. Use a reference angle to find the exact value of $\cos 225°$.

4b. Use a reference angle to find the exact value of $\sin\frac{2\pi}{3}$.

4c. Use a reference angle to find the exact value of $\sec\left(-\frac{\pi}{6}\right)$.

A $-\frac{\pi}{6}$ angle lies in quadrant IV, where the secant function is positive. Furthermore, a $-\frac{\pi}{6}$ angle forms an acute of $\frac{\pi}{6}$ with the x-axis. The reference angle is $\theta' = \frac{\pi}{6}$.

$$\sec\left(-\frac{\pi}{6}\right) = +\sec\frac{\pi}{6} = \frac{2\sqrt{3}}{3}$$

4c. Use a reference angle to find the exact value of $\tan\left(-\frac{\pi}{4}\right)$.

4d. Use a reference angle to find the exact value of $\cos\frac{17\pi}{6}$.

Subtract 2π to find a positive coterminal angle less than 2π: $\frac{17\pi}{6} - 2\pi = \frac{17\pi}{6} - \frac{12\pi}{6} = \frac{5\pi}{6}$.

A $\frac{5\pi}{6}$ angle lies in quadrant II, where the cosine function is negative. The reference angle is $\theta' = \pi - \frac{5\pi}{6} = \frac{6\pi}{6} - \frac{5\pi}{6} = \frac{\pi}{6}$.

$$\cos\frac{17\pi}{6} = -\cos\frac{\pi}{6} = -\frac{\sqrt{3}}{2}$$

4d. Use a reference angle to find the exact value of $\cot\frac{19\pi}{6}$.

Answers for Pencil Problems *(Textbook Exercise references in parentheses)*:

1a. $\sin\theta = -\frac{5\sqrt{29}}{29}$, $\cos\theta = -\frac{2\sqrt{29}}{29}$, $\tan\theta = \frac{5}{2}$, $\csc\theta = -\frac{\sqrt{29}}{5}$, $\sec\theta = -\frac{\sqrt{29}}{2}$, $\cot\theta = \frac{2}{5}$ *(4.4 #7)*

1b. undefined *(4.4 #13)*

2a. quadrant II *(4.4 #21)* **2b.** $\cos\theta = -\frac{3\sqrt{13}}{13}$, $\csc\theta = \frac{\sqrt{13}}{2}$ *(4.4 #29)*

3a. 20° *(4.4 #35)* **3b.** $\frac{\pi}{6}$ *(4.4 #43)* **3c.** 25° *(4.4 #47)* **3d.** 25° *(4.4 #51)* **3e.** $\frac{\pi}{4}$ *(4.4 #57)*

4a. $-\frac{\sqrt{2}}{2}$ *(4.4 #61)* **4b.** $\frac{\sqrt{3}}{2}$ *(4.4 #67)* **4c.** -1 *(4.4 #75)* **4d.** $\sqrt{3}$ *(4.4 #79)*

Section 4.5
Graphs of Sine and Cosine Functions

The Graph That Looks Like a Rollercoaster

In this section, we study the graphs of the sine and cosine functions. The ups and downs of the graphs may remind you of a rollercoaster. The graphs rise to peaks at maximum points then plunge to minimum points where they promptly change direction and rise again.

However, unlike a rollercoaster ride, the ups and downs of these graphs go on forever. The shapes of these graphs and their repetitive properties make it even more obvious why trigonometric functions are used to model cyclic behavior.

Objective #1: Understand the graph of $y = \sin x$.

✔ *Solved Problem #1*

1. True or false: The graph of $y = \sin x$ is symmetric with respect to the y-axis.

False; the sine function is an odd function and is symmetric with respect to the origin but not the y-axis.

Pencil Problem #1

1. True or false: The graph of $y = \sin x$ has no gaps or holes and extends indefinitely in both directions.

Objective #2: Graph variations of $y = \sin x$.

✔ *Solved Problem #2*

2a. Determine the amplitude and the period of $y = 2\sin\frac{1}{2}x$. Then graph the function for $0 \le x \le 8\pi$.

Comparing $y = 2\sin\frac{1}{2}x$ to $y = A\sin Bx$, we see that $A = 2$ and $B = \frac{1}{2}$.

Amplitude: $|A| = |2| = 2$

Period: $\frac{2\pi}{B} = \frac{2\pi}{\frac{1}{2}} = 4\pi$

The amplitude tells us that the maximum value of the function is 2 and the minimum value is –2. The period tells us that the graph completes one cycle between 0 and 4π.

(continued on next page)

Pencil Problem #2

2a. Determine the amplitude and the period of $y = 3\sin\frac{1}{2}x$. Then graph one period of the function.

Divide the period by 4: $\frac{4\pi}{4}=\pi$. The x-values of the five key points are $x_1=0$, $x_2=0+\pi=\pi$, $x_3=\pi+\pi=2\pi$, $x_4=2\pi+\pi=3\pi$, and $x_5=3\pi+\pi=4\pi$. Now find the value of y for each of these x-values.

$$y=2\sin\left(\frac{1}{2}\cdot 0\right)=2\sin 0=2\cdot 0=0:(0,\ 0)$$

$$y=2\sin\left(\frac{1}{2}\cdot\pi\right)=2\sin\frac{\pi}{2}=2\cdot 1=2:(\pi,\ 2)$$

$$y=2\sin\left(\frac{1}{2}\cdot 2\pi\right)=2\sin\pi=2\cdot 0=0:(2\pi,\ 0)$$

$$y=2\sin\left(\frac{1}{2}\cdot 3\pi\right)=2\sin\frac{3\pi}{2}=2(-1)=-2:(3\pi,\ -2)$$

$$y=2\sin\left(\frac{1}{2}\cdot 4\pi\right)=2\sin 2\pi=2\cdot 0=0:(4\pi,\ 0)$$

Notice the pattern: x-intercept, maximum, x-intercept, minimum, x-intercept. Plot these points and connect them with a smooth curve. Extend the graph one period to the right in order to graph from $0\le x\le 8\pi$.

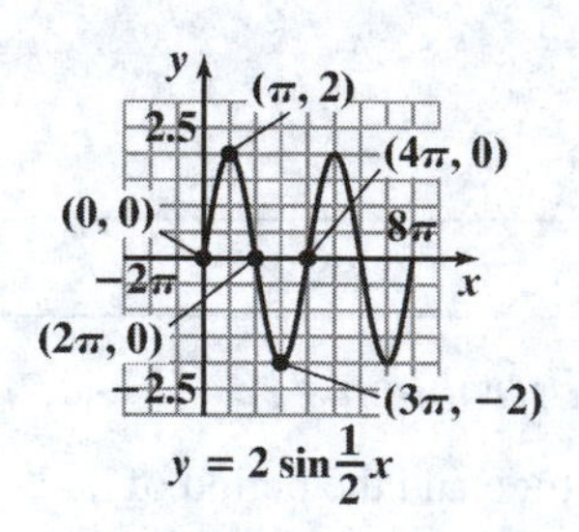

2b. Determine the amplitude, period, and phase shift of $y = 3\sin\left(2x - \frac{\pi}{3}\right)$. Then graph one period of the function.

Comparing $y = 3\sin\left(2x - \frac{\pi}{3}\right)$ to $y = A\sin(Bx - C)$, we see that $A = 3$, $B = 2$, and $C = \frac{\pi}{3}$.

Amplitude: $|A| = |3| = 3$

Period: $\frac{2\pi}{B} = \frac{2\pi}{2} = \pi$

Phase shift: $\frac{C}{B} = \frac{\frac{\pi}{3}}{2} = \frac{\pi}{3} \cdot \frac{1}{2} = \frac{\pi}{6}$

The amplitude tells us that the maximum value of the function is 3 and the minimum value is –3. The period tells us that each cycle is of length π. The phase shift tells us that a cycle starts at $\frac{\pi}{6}$.

Divide the period by 4: $\frac{\pi}{4}$. The x-values of the five key points are $x_1 = \frac{\pi}{6}$,

$x_2 = \frac{\pi}{6} + \frac{\pi}{4} = \frac{2\pi}{12} + \frac{3\pi}{12} = \frac{5\pi}{12}$,

$x_3 = \frac{5\pi}{12} + \frac{\pi}{4} = \frac{5\pi}{12} + \frac{3\pi}{12} = \frac{8\pi}{12} = \frac{2\pi}{3}$,

$x_4 = \frac{2\pi}{3} + \frac{\pi}{4} = \frac{8\pi}{12} + \frac{3\pi}{12} = \frac{11\pi}{12}$, and

$x_5 = \frac{11\pi}{12} + \frac{\pi}{4} = \frac{11\pi}{12} + \frac{3\pi}{12} = \frac{14\pi}{12} = \frac{7\pi}{6}$. Now find the value of y for each of these x-values.

$$y = 3\sin\left(2 \cdot \frac{\pi}{6} - \frac{\pi}{3}\right) = 3\sin 0 = 0 : \left(\frac{\pi}{6}, 0\right)$$

$$y = 3\sin\left(2 \cdot \frac{5\pi}{12} - \frac{\pi}{3}\right) = 3\sin\frac{\pi}{2} = 3 : \left(\frac{5\pi}{12}, 3\right)$$

$$y = 3\sin\left(2 \cdot \frac{2\pi}{3} - \frac{\pi}{3}\right) = 3\sin\pi = 0 : \left(\frac{2\pi}{3}, 0\right)$$

$$y = 3\sin\left(2 \cdot \frac{11\pi}{12} - \frac{\pi}{3}\right) = 3\sin\frac{3\pi}{2} = -3 : \left(\frac{11\pi}{12}, -3\right)$$

$$y = 3\sin\left(2 \cdot \frac{7\pi}{6} - \frac{\pi}{3}\right) = 3\sin 2\pi = 0 : \left(\frac{7\pi}{6}, 0\right)$$

(continued on next page)

2b. Determine the amplitude, period, and phase shift of $y = 3\sin(2x - \pi)$. Then graph one period of the function.

Notice the pattern: x-intercept, maximum, x-intercept, minimum, x-intercept. Plot these points and connect them with a smooth curve.

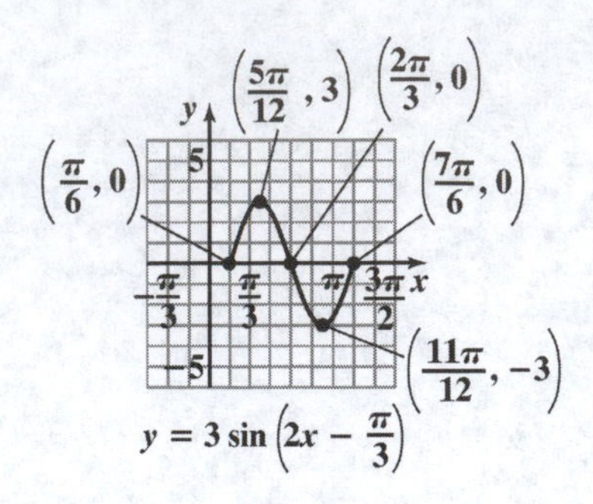

Objective #3: Understand the graph of $y = \cos x$..

✔ *Solved Problem #3*

3. True or false: The graph of $y = \cos x$ is symmetric with respect to the y-axis.

True; the cosine function is an even function, and the graphs of even functions have symmetry with respect to the y-axis.

Pencil Problem #3

3. True or false: The graph of $y = \cos x$ illustrates that the range of the cosine function is $(-\infty, \infty)$.

Objective #4: Graph variations of $y = \cos x$.

✔ *Solved Problem #4*

4. Determine the amplitude, period, and phase shift of $y = \frac{3}{2}\cos(2x + \pi)$. Then graph one period of the function.

Comparing $y = \frac{3}{2}\cos(2x + \pi)$ to $y = A\sin(Bx - C)$,

we see that $A = \frac{3}{2}$, $B = 2$, and $C = -\pi$.

Amplitude: $|A| = \left|\frac{3}{2}\right| = \frac{3}{2}$

Period: $\frac{2\pi}{B} = \frac{2\pi}{2} = \pi$

Phase shift: $\frac{C}{B} = \frac{-\pi}{2} = -\frac{\pi}{2}$

(continued on next page)

Pencil Problem #4

4. Determine the amplitude, period, and phase shift of $y = \frac{1}{2}\cos\left(3x + \frac{\pi}{2}\right)$. Then graph one period of the function.

The amplitude tells us that the maximum value of the function is $\frac{3}{2}$ and the minimum value is $-\frac{3}{2}$. The period tells us that each cycle is of length π. The phase shift tells us that a cycle starts at $-\frac{\pi}{2}$.

Divide the period by 4: $\frac{\pi}{4}$. The x-values of the five key points are $x_1 = -\frac{\pi}{2}$,

$x_2 = -\frac{\pi}{2} + \frac{\pi}{4} = -\frac{2\pi}{4} + \frac{\pi}{4} = -\frac{\pi}{4}$, $x_3 = -\frac{\pi}{4} + \frac{\pi}{4} = 0$,

$x_4 = 0 + \frac{\pi}{4} = \frac{\pi}{4}$, and $x_5 = \frac{\pi}{4} + \frac{\pi}{4} = \frac{2\pi}{4} = \frac{\pi}{2}$. Now find the value of y for each of these x-values by evaluating the function. The five key points are $\left(-\frac{\pi}{2}, \frac{3}{2}\right)$, $\left(-\frac{\pi}{4}, 0\right)$, $\left(0, -\frac{3}{2}\right)$, $\left(\frac{\pi}{4}, 0\right)$, and $\left(\frac{\pi}{2}, \frac{3}{2}\right)$. Notice the pattern: maximum, x-intercept, minimum, x-intercept, maximum. Plot these points and connect them with a smooth curve.

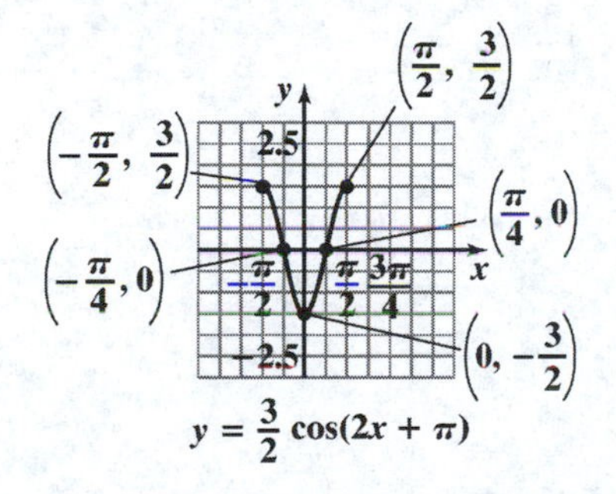

Objective #5: Use vertical shifts of sine and cosine curves.

✔ Solved Problem #5

5. Graph one period of the function $y = 2\cos x + 1$.

The graph of $y = 2\cos x + 1$ is the graph of $y = 2\cos x$ shifted one unit up. The amplitude is 2 and the period is 2π. The graph oscillates 2 units below and 2 units above the line $y = 1$.

The x-values for the five key points are 0, $\frac{\pi}{2}$, π, $\frac{3\pi}{2}$, and 2π. The five key points are (0, 3), $\left(\frac{\pi}{2}, 1\right)$, $(\pi, -1)$, $\left(\frac{3\pi}{2}, 1\right)$, and $(2\pi, 3)$. Plot these points and connect them with a smooth curve.

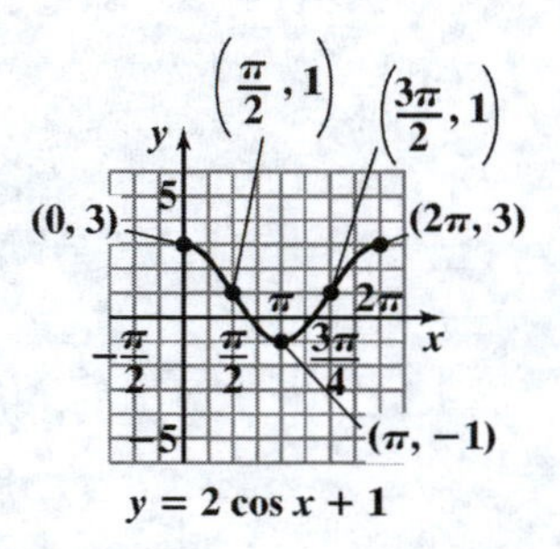

$y = 2\cos x + 1$

Pencil Problem #5

5. Graph one period of the function $y = \sin x + 2$.

Objective #6: Model periodic behavior.

✔ *Solved Problem #6*

6. A region that is 30° north of the Equator averages a minimum of 10 hours of daylight in December. Hours of daylight are at a maximum of 14 hours in June. Let x represent the month of the year, with 1 for January, 2 for February, 3 for March, and 12 for December. If y represents the number of hours of daylight in month x, use a sine function of the form $y = A\sin(Bx - C) + D$ to model the hours of daylight.

We need to determine values for A, B, C, and D in $y = A\sin(Bx - C) + D$.

To find D, notice that the values range between a minimum of 10 and a maximum of 14. The middle value is 12, so $D = 12$.

To find A, notice that the minimum, 10, and maximum, 14, are 2 units below and above the middle value, $D = 12$. The amplitude is 2, so $A = 2$.

To find B, notice that the period is 12 months (one year). Use the period formula and solve for B.

$$\frac{2\pi}{B} = 12$$

$$2\pi = 12B$$

$$B = \frac{2\pi}{12} = \frac{\pi}{6}$$

To find C, notice that a middle value occurs in March ($x = 3$), so we can begin a cycle in March. Use the phase shift formula with this value of x and the value of B just found.

$$Bx - C = 0$$

$$\frac{\pi}{6} \cdot 3 - C = 0$$

$$\frac{\pi}{2} = C$$

Substitute the values for A, B, C, and D into $y = A\sin(Bx - C) + D$. The model is

$$y = 2\sin\left(\frac{\pi}{6}x - \frac{\pi}{2}\right) + 12.$$

Pencil Problem #6

6. The figure shows the depth of water at the end of a boat dock. The depth is 6 feet at low tide and 12 feet at high tide. On a certain day, low tide occurs at 6 a.m. and high tide occurs at noon. If y represents the depth of the water x hours after midnight, use a cosine function of the form $y = A\cos Bx + D$ to model the water's depth.

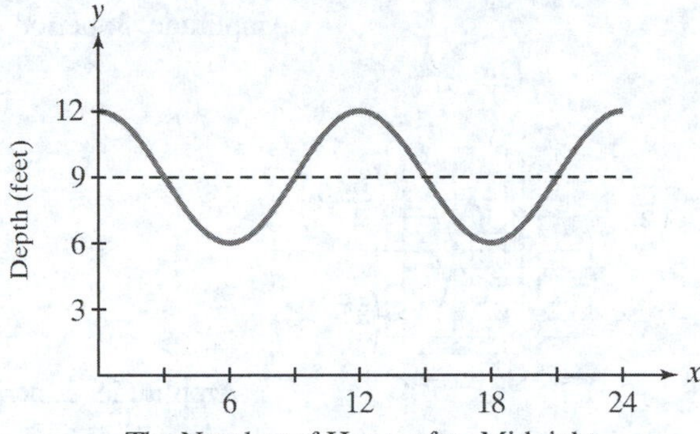

Answers for Pencil Problems *(Textbook Exercise references in parentheses)*:

1. true

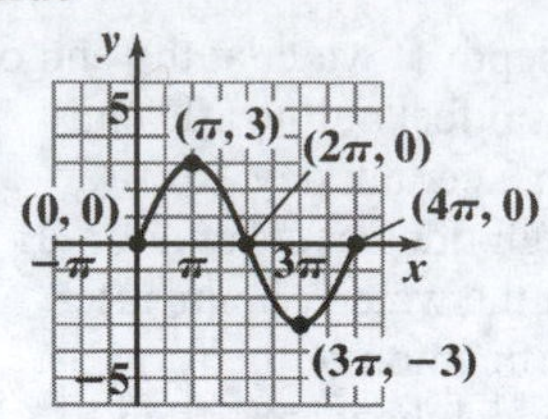

2a. Amplitude: 3; period: 4π *(4.5 #9)*

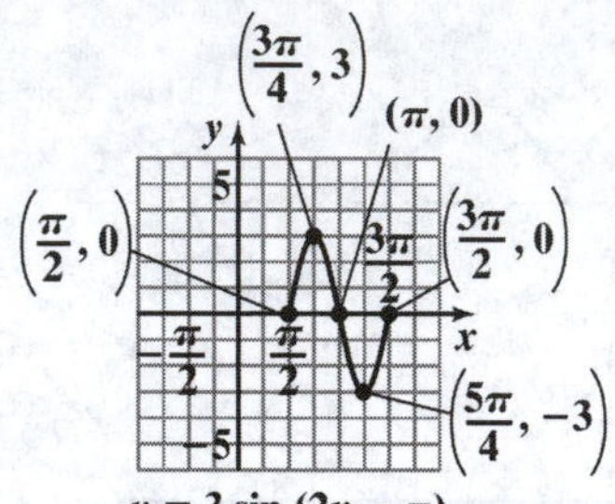

2b. Amplitude: 3; period: π; phase shift: $\frac{\pi}{2}$ *(4.5 #21)*

3. false

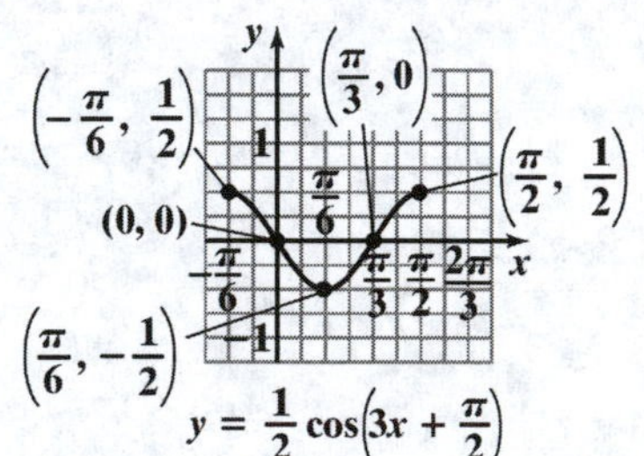

4a. Amplitude: $\frac{1}{2}$; period: $\frac{2\pi}{3}$; phase shift: $-\frac{\pi}{6}$ *(4.5 #47)*

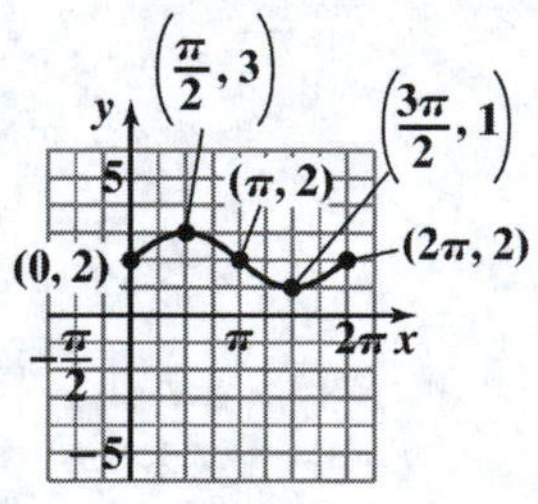

5. *(4.5 #53)*

6. $y = 3\cos\frac{\pi x}{6} + 9$ *(4.5 #87)*

Section 4.6
Graphs of Other Trigonometric Functions

Trig Functions Without Bounds

We have seen that sine and cosine functions can be used to model phenomena that are cyclic in nature, such as the number of daylight hours in a day at a specific location over a period of years. But it's not just the periodic properties of sine and cosine that are important in these situations; it's also their limited ranges. The fact that the ranges of sine and cosine functions are bounded makes them suitable for modeling phenomena that never exceed certain minimum and maximum values.

In this section, we will see that the remaining trigonometric functions, while periodic like sine and cosine, do not have bounded ranges like sine and cosine. These functions are more suitable for describing cyclic phenomena that do not have natural limits, such as the location where a beam of light from a rotating source strikes a flat surface.

Objective #1: Understand the graph of $y = \tan x$.

✔ *Solved Problem #1*

1. True or false: The graph of $y = \tan x$ is symmetric with respect to the y-axis.

False; the tangent function is an odd function and is symmetric with respect to the origin but not the y-axis.

Pencil Problem #1

1. True or false: The graph of $y = \tan x$ has no gaps or holes and extends indefinitely in both directions.

Objective #2: Graph variations of $y = \tan x$.

✔ *Solved Problem #2*

2a. Graph $y = 3\tan 2x$ for $-\frac{\pi}{4} < x < \frac{3\pi}{4}$.

Find two consecutive asymptotes by solving $-\frac{\pi}{2} < Bx - C < \frac{\pi}{2}$. In this case, $Bx - C = 2x$.

$$-\frac{\pi}{2} < 2x < \frac{\pi}{2}$$
$$-\frac{\pi}{4} < x < \frac{\pi}{4}$$

(continued on next page)

Pencil Problem #2

2a. Graph two full periods of $y = \frac{1}{2}\tan 2x$.

The graph completes one cycle on the interval $\left(-\frac{\pi}{4}, \frac{\pi}{4}\right)$ and has consecutive vertical asymptotes at $x = -\frac{\pi}{4}$ and $x = \frac{\pi}{4}$. The x-intercept is midway between the asymptotes at (0, 0). The x-values $-\frac{\pi}{8}$ and $\frac{\pi}{8}$ are $\frac{1}{4}$ and $\frac{3}{4}$ of the way between the asymptotes; the points $\left(-\frac{\pi}{8}, -3\right)$ and $\left(\frac{\pi}{8}, 3\right)$ are on the graph. Plot these three points and the asymptotes. Graph one period of the function by drawing a smooth curve through the points and approaching the asymptotes. Complete one more period to the right in order to show the graph for $-\frac{\pi}{4} < x < \frac{3\pi}{4}$.

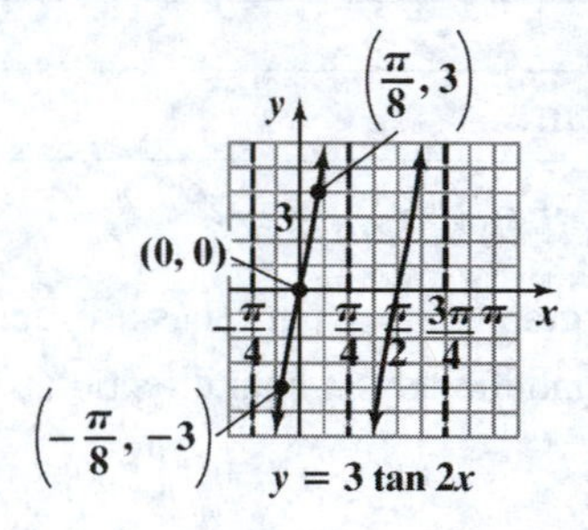

2b. Graph two full periods of $y = \tan\left(x - \frac{\pi}{2}\right)$.

Find two consecutive asymptotes by solving $-\frac{\pi}{2} < Bx - C < \frac{\pi}{2}$. In this case, $Bx - C = x - \frac{\pi}{2}$.

$$-\frac{\pi}{2} < x - \frac{\pi}{2} < \frac{\pi}{2}$$
$$-\frac{\pi}{2} + \frac{\pi}{2} < x < \frac{\pi}{2} + \frac{\pi}{2}$$
$$0 < x < \pi$$

(continued on next page)

2b. Graph two full periods of $y = \tan(x - \pi)$.

The graph completes one cycle on the interval $(0, \pi)$ and has consecutive vertical asymptotes at $x = 0$ and $x = \pi$. The x-intercept is midway between the asymptotes at $\left(\frac{\pi}{2}, 0\right)$. The x-values $\frac{\pi}{4}$ and $\frac{3\pi}{4}$ are $\frac{1}{4}$ and $\frac{3}{4}$ of the way between the asymptotes; the points $\left(\frac{\pi}{4}, -1\right)$ and $\left(\frac{3\pi}{4}, 1\right)$ are on the graph.

Plot these three points and the asymptotes. Graph one period of the function by drawing a smooth curve through the points and approaching the asymptotes. Complete one more period to the left or to the right in order to show two full periods.

The graph of $y = \tan\left(x - \frac{\pi}{2}\right)$ is the graph of $y = \tan x$ shifted $\frac{\pi}{2}$ units to the right.

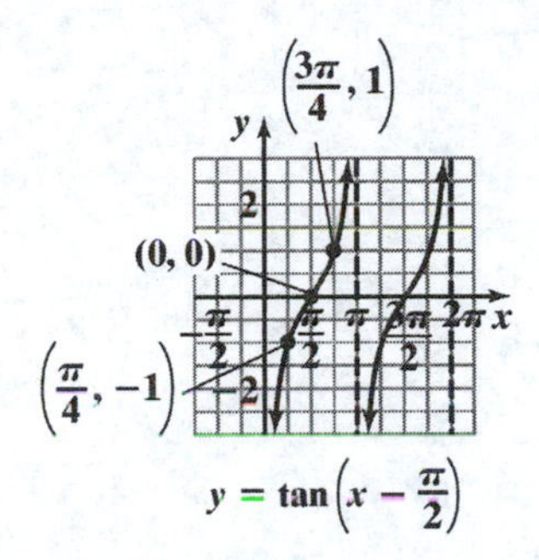

Objective #3: Understand the graph of $y = \cot x$.

✔ *Solved Problem #3*

3. True or false: The graph of $y = \cot x$ has asymptotes at values of x where the graph of $y = \tan x$ has x-intercepts.

True; the cotangent function is undefined at values of x for which tangent is 0.

Pencil Problem #3

3. True or false: The graph of $y = \cot x$ illustrates that the range of the cotangent function is $(-\infty, \infty)$.

Objective #4: Graph variations of $y = \cot x$.

✔ *Solved Problem #4*

4. Graph $y = \frac{1}{2}\cot\frac{\pi}{2}x.$

Find two consecutive asymptotes by solving $0 < Bx - C < \pi.$ In this case, $Bx - C = \frac{\pi}{2}x.$

$$0 < \frac{\pi}{2}x < \pi$$

$$\frac{2}{\pi}\cdot 0 < x < \frac{2}{\pi}\cdot \pi$$

$$0 < x < 2$$

The graph completes one cycle on the interval (0, 2) and has consecutive vertical asymptotes at $x = 0$ and $x = 2.$ The x-intercept is midway between the asymptotes at (1, 0). The x-values $\frac{1}{2}$ and $\frac{3}{2}$ are $\frac{1}{4}$ and $\frac{3}{4}$ of the way between the asymptotes; the points $\left(\frac{1}{2}, \frac{1}{2}\right)$ and $\left(\frac{3}{2}, -\frac{1}{2}\right)$ are on the graph.

Plot these three points and the asymptotes. Graph one period of the function by drawing a smooth curve through the points and approaching the asymptotes.

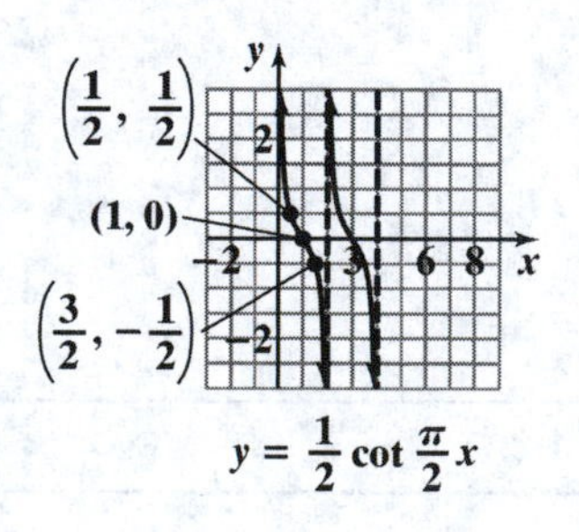

$y = \frac{1}{2}\cot\frac{\pi}{2}x$

Pencil Problem #4

4. Graph two periods of $y = \frac{1}{2}\cot 2x.$

Objective #5: Understand the graphs of $y = \csc x$ and $y = \sec x$.

✔ *Solved Problem #5*

5. True or false: The graph of $y = \csc x$ illustrates that the range of the cosecant function is $(-\infty, \infty)$.

False; the range of the cosecant function is $(-\infty, -1] \cup [1, \infty)$. It does not include values of y between -1 and 1.

5. True or false: The graph of $y = \sec x$ has asymptotes at values of x where the graph of $y = \sin x$ has x-intercepts.

Objective #6: Graph variations of $y = \csc x$ and $y = \sec x$.

✔ *Solved Problem #6*

6. Graph $y = 2\sec 2x$ for $-\frac{3\pi}{4} < x < \frac{3\pi}{4}$.

Begin by graphing $y = 2\cos 2x$ where secant has been replaced by cosine, its reciprocal function. The amplitude of $y = 2\cos 2x$ is 2 and the period is $\frac{2\pi}{2} = \pi$. Note that for $-\frac{3\pi}{4} < x < \frac{3\pi}{4}$ the graph of $y = 2\cos 2x$ has x-intercepts at $x = -\frac{3\pi}{4}$, $x = -\frac{\pi}{4}$, $x = \frac{\pi}{4}$, and $x = \frac{3\pi}{4}$. The graph of $y = 2\cos 2x$ attains its minimum value of -2 at $x = -\frac{\pi}{2}$ and $x = \frac{\pi}{2}$ and its maximum value of 2 at $x = 0$.

At each x-intercept of the graph of $y = 2\cos 2x$ draw a vertical asymptote. Graph $y = 2\sec 2x$ by starting at each minimum or maximum point and approaching the asymptotes in each direction.

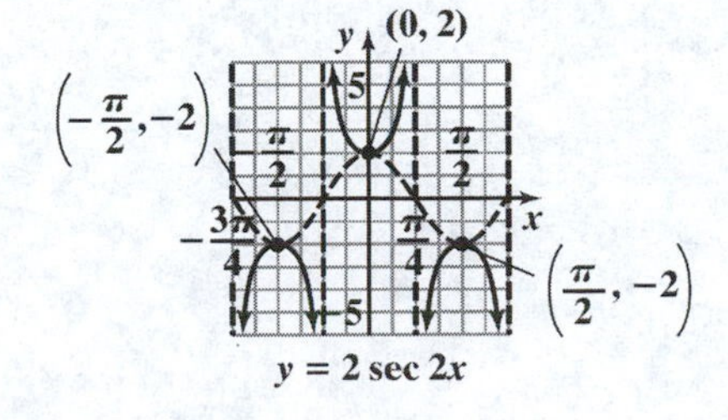

6. Graph two periods of $y = \frac{1}{2}\csc\frac{x}{2}$.

Answers for Pencil Problems *(Textbook Exercise references in parentheses)*:

1. false

2a. 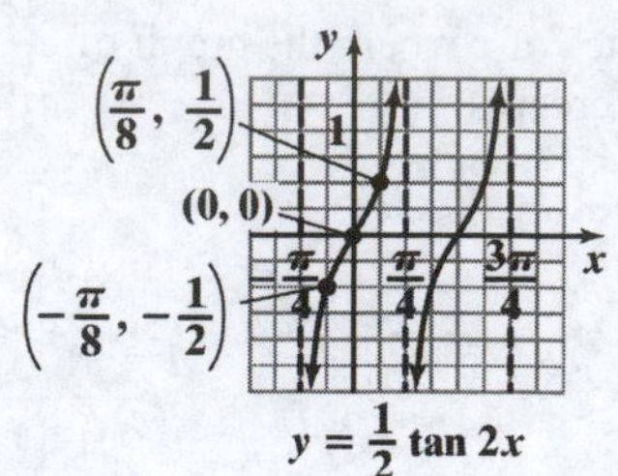

(4.6 #7)

2b. 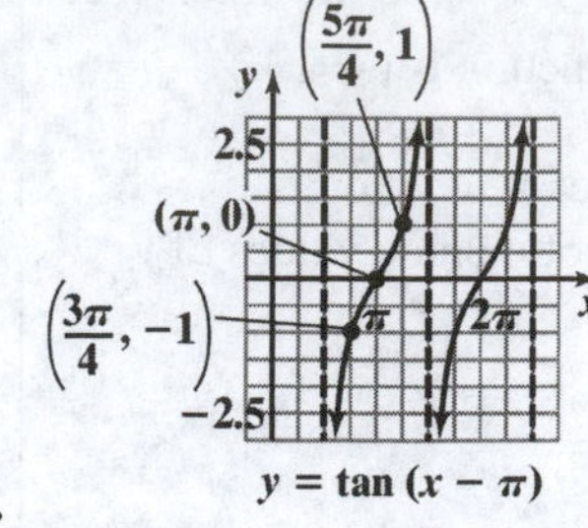

(4.6 #11)

3. true

4. 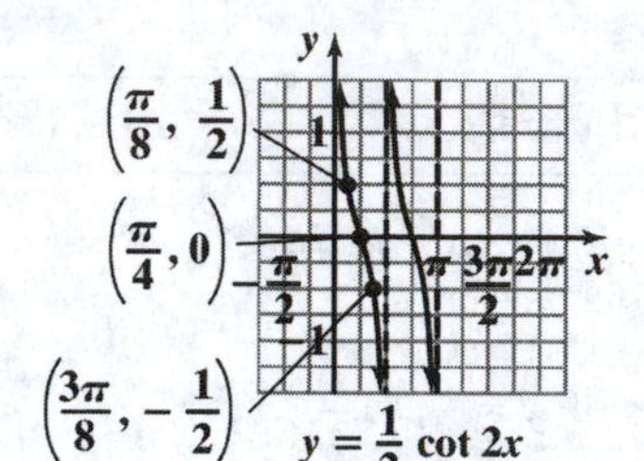

(4.6 #19)

5. false

6. 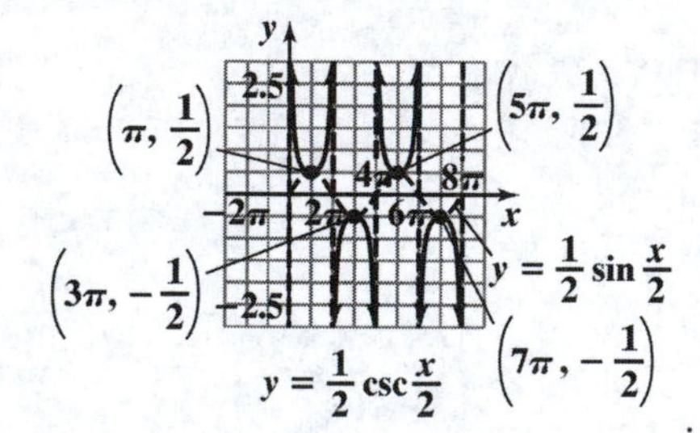

. *(4.6 #31)*

Section 4.7
Inverse Trigonometric Functions

For Your Viewing Pleasure

Your total movie experience at your local cinema is affected by many variables, including your distance from the screen. If you sit too close, your viewing angle is too small. If you sit too far back, the image is too small.

In this section's Exercise Set, you will see how inverse trigonometric functions can be used to model your viewing angle in terms of your distance from the screen and help you find the seat that will optimize your viewing pleasure.

Objective #1: Understand and use the inverse sine function.

✔ *Solved Problem #1*

1a. Find the exact value of $\sin^{-1}\frac{\sqrt{3}}{2}$.

Let $\theta = \sin^{-1}\frac{\sqrt{3}}{2}$. Then $\sin\theta = \frac{\sqrt{3}}{2}$ where $-\frac{\pi}{2} \le \theta \le \frac{\pi}{2}$.

We must find the angle θ, $-\frac{\pi}{2} \le \theta \le \frac{\pi}{2}$, whose sine equals $\frac{\sqrt{3}}{2}$. Using a table of values for the sine function for $-\frac{\pi}{2} \le \theta \le \frac{\pi}{2}$, we see that $\sin\frac{\pi}{3} = \frac{\sqrt{3}}{2}$.

Thus, $\sin^{-1}\frac{\sqrt{3}}{2} = \frac{\pi}{3}$.

Pencil Problem #1

1a. Find the exact value of $\sin^{-1}\frac{1}{2}$.

1b. Find the exact value of $\sin^{-1}\left(-\frac{\sqrt{2}}{2}\right)$.

Let $\theta = \sin^{-1}\left(-\frac{\sqrt{2}}{2}\right)$.

Then $\sin\theta = -\frac{\sqrt{2}}{2}$ where $-\frac{\pi}{2} \le \theta \le \frac{\pi}{2}$.

We must find the angle θ, $-\frac{\pi}{2} \le \theta \le \frac{\pi}{2}$, whose sine equals $-\frac{\sqrt{2}}{2}$. Using a table of values for the sine function for $-\frac{\pi}{2} \le \theta \le \frac{\pi}{2}$, we see that $\sin\left(-\frac{\pi}{4}\right) = -\frac{\sqrt{2}}{2}$.

Thus, $\sin^{-1}\left(-\frac{\sqrt{2}}{2}\right) = -\frac{\pi}{4}$.

1b. Find the exact value of $\sin^{-1}\left(-\frac{1}{2}\right)$.

Objective #2: Understand and use the inverse cosine function

✔ ***Solved Problem #2***

2. Find the exact value of $\cos^{-1}\left(-\frac{1}{2}\right)$.

Let $\theta = \cos^{-1}\left(-\frac{1}{2}\right)$.

Then $\cos\theta = -\frac{1}{2}$ where $0 \le \theta \le \pi$.

We must find the angle θ, $0 \le \theta \le \pi$, whose cosine equals $-\frac{1}{2}$. Using a table of values for the cosine function for $0 \le \theta \le \pi$, we see that $\cos\frac{2\pi}{3} = -\frac{1}{2}$.

Thus, $\cos^{-1}\left(-\frac{1}{2}\right) = \frac{2\pi}{3}$.

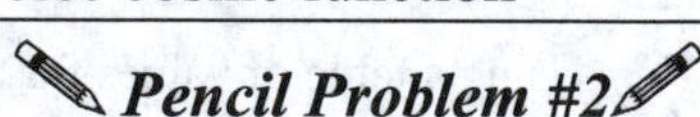

Pencil Problem #2

2. Find the exact value of $\cos^{-1}\left(-\frac{\sqrt{2}}{2}\right)$.

Objective #3: Understand and use the inverse tangent function.

✔ *Solved Problem #3*

3. Find the exact value of $\tan^{-1}(-1)$.

Let $\theta = \tan^{-1}(-1)$.

Then $\tan\theta = -1$ where $-\frac{\pi}{2} < \theta < \frac{\pi}{2}$.

We must find the angle θ, $-\frac{\pi}{2} < \theta < \frac{\pi}{2}$, whose tangent equals -1. Using a table of values for the tangent function for $-\frac{\pi}{2} < \theta < \frac{\pi}{2}$, we see that $\tan\left(-\frac{\pi}{4}\right) = -1$.

Thus, $\tan^{-1}(-1) = -\frac{\pi}{4}$.

Pencil Problem #3

3. Find the exact value of $\tan^{-1} 0$.

Objective #4: Use a calculator to evaluate inverse trigonometric functions.

✔ *Solved Problem #4*

4a. Use a calculator to find the value of $\cos^{-1}\frac{1}{3}$ to four decimal places.

To access the inverse cosine function, you will need to press the 2nd function key and then the COS key. Use radian mode.

Scientific calculator: $1 \div 3 =$ 2nd COS
Graphing calculator: 2nd COS $(1 \div 3)$ ENTER

The display should read 1.2310, rounded to four places.

Pencil Problem #4

4a. Use a calculator to find the value of $\cos^{-1}\frac{3}{8}$ to two decimal places.

4b. Use a calculator to find the value of $\tan^{-1}(-35.85)$ to four decimal places.

To access the inverse tangent function, you will need to press the 2nd function key and then the TAN key.

Use radian mode.

Scientific calculator: 35.85 +/– 2nd TAN
Graphing calculator: 2nd TAN (–) 35.85 ENTER

The display should read –1.5429, rounded to four places.

4b. Use a calculator to find the value of $\sin^{-1}(-0.32)$ to two decimal places.

Objective #5: Find exact values of composite functions with inverse trigonometric functions.

✔ Solved Problem #5

5a. Find the exact value, if possible: $\cos(\cos^{-1} 0.7)$.

Since 0.7 is in the interval $[-1, 1]$, we can use the inverse property $\cos(\cos^{-1} x) = x$.

$\cos(\cos^{-1} 0.7) = 0.7$

5b. Find the exact value, if possible: $\sin^{-1}(\sin \pi)$.

Since π is not in the $\left[-\frac{\pi}{2}, \frac{\pi}{2}\right]$, we cannot use the inverse property $\sin^{-1}(\sin x) = x$. We first evaluate $\sin \pi = 0$, and then evaluate $\sin^{-1} 0$.

$\sin^{-1}(\sin \pi) = \sin^{-1} 0 = 0$

Pencil Problem #5

5a. Find the exact value, if possible: $\sin(\sin^{-1} 0.9)$.

5b. Find the exact value, if possible: $\sin^{-1}\left(\sin \frac{5\pi}{6}\right)$.

5c. Find the exact value, if possible: $\cos[\cos^{-1}(-1.2)]$.

Since -1.2 is not in the domain of the inverse cosine function, $[-1, 1]$, $\cos[\cos^{-1}(-1.2)]$ is not defined.

5c. Find the exact value, if possible: $\sin(\sin^{-1}\pi)$.

5d. Find the exact value of $\sin\left(\tan^{-1}\frac{3}{4}\right)$.

Let θ represent the angle in $\left(-\frac{\pi}{2}, \frac{\pi}{2}\right)$ whose tangent is $\frac{3}{4}$. Thus, $\theta = \tan^{-1}\frac{3}{4}$.

So, $\tan\theta = \frac{3}{4}$, where $-\frac{\pi}{2} < \theta < \frac{\pi}{2}$. Since $\tan\theta$ is positive, θ must be in $\left(0, \frac{\pi}{2}\right)$. Thus, θ lies in quadrant I, where both x and y are positive.

$\tan\theta = \frac{3}{4} = \frac{y}{x}$, so $x = 4$ and $y = 3$

Find r and then use the value of r to find $\sin\theta$.

$$r = \sqrt{4^2 + 3^2} = \sqrt{16+9} = \sqrt{25} = 5$$

$$\sin\theta = \frac{y}{r} = \frac{3}{5}$$

Thus, $\sin\left(\tan^{-1}\frac{3}{4}\right) = \sin\theta = \frac{3}{5}$.

5d. Find the exact value of $\cos\left(\sin^{-1}\frac{4}{5}\right)$.

5e. Find the exact value of $\cos\left[\sin^{-1}\left(-\frac{1}{2}\right)\right]$.

Let θ represent the angle in $\left[-\frac{\pi}{2}, \frac{\pi}{2}\right]$ whose sine is $-\frac{1}{2}$. Thus, $\theta = \sin^{-1}\left(-\frac{1}{2}\right)$.

So, $\sin\theta = -\frac{1}{2}$, where $-\frac{\pi}{2} \le \theta \le \frac{\pi}{2}$. Since $\sin\theta$ is negative, θ must be in $\left[-\frac{\pi}{2}, 0\right)$. Thus, θ lies in quadrant IV, where both x is positive and y is negative.

$\sin\theta = -\frac{1}{2} = \frac{y}{r} = \frac{-1}{r}$, so $y = -1$ and $r = 2$

Find x and then use the value of x to find $\cos\theta$.

$$r = \sqrt{x^2 + y^2}$$
$$2 = \sqrt{x^2 + (-1)^2}$$
$$4 = x^2 + 1$$
$$3 = x^2$$
$$x = \sqrt{3}, \text{ since } x > 0$$
$$\cos\theta = \frac{x}{r} = \frac{\sqrt{3}}{2}$$

Thus, $\cos\left[\sin^{-1}\left(-\frac{1}{2}\right)\right] = \cos\theta = \frac{\sqrt{3}}{2}$.

In this problem, we know how to find the exact value of $\sin^{-1}\left(-\frac{1}{2}\right)$, so we could have also proceeded as follows:

$$\cos\left[\sin^{-1}\left(-\frac{1}{2}\right)\right] = \cos\left(-\frac{\pi}{6}\right) = \frac{\sqrt{3}}{2}.$$

5e. Find the exact value of $\csc\left[\cos^{-1}\left(-\frac{\sqrt{3}}{2}\right)\right]$.

5f. If $x > 0$, write $\sec(\tan^{-1} x)$ as an algebraic expression in x.

Let θ represent the angle in $\left(-\frac{\pi}{2}, \frac{\pi}{2}\right)$ whose tangent is x.

Thus, $\theta = \tan^{-1} x$ and $\tan\theta = x$, where $\left(-\frac{\pi}{2}, \frac{\pi}{2}\right)$.

Because $x > 0$, $\tan\theta$ is positive. Thus, θ is a first-quadrant angle.

Draw a right triangle and label one of the acute angles θ. Since

$$\tan\theta = x = \frac{x}{1} = \frac{\text{opposite}}{\text{adjacent}},$$

the length of the side opposite θ is x and the length of the adjacent side is 1.

Use the Pythagorean theorem to find the hypotenuse.

$$c^2 = x^2 + 1^2$$
$$c = \sqrt{x^2 + 1}$$

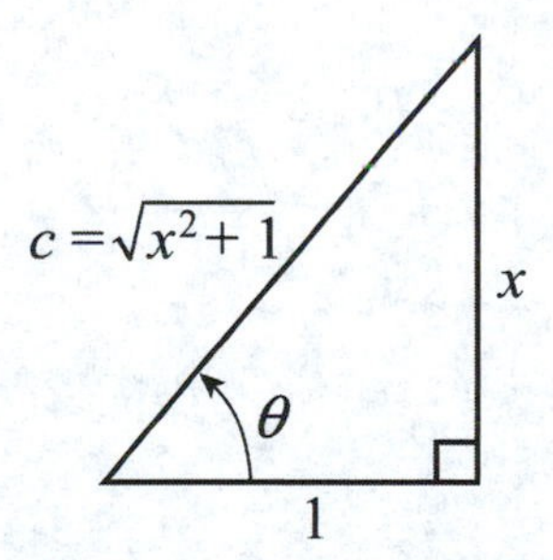

Thus, $\sec\theta = \frac{\text{hypotenuse}}{\text{adjacent}} = \frac{\sqrt{x^2+1}}{1} = \sqrt{x^2+1}$ and consequently,

$$\sec(\tan^{-1} x) = \sec\theta = \sqrt{x^2+1}.$$

5f. If $x > 0$, write $\tan(\cos^{-1} x)$ as an algebraic expression in x.

Answers for Pencil Problems *(Textbook Exercise references in parentheses)*:

1a. $\frac{\pi}{6}$ *(4.7 #1)* **1b.** $-\frac{\pi}{6}$ *(4.7 #5)*

2. $\frac{3\pi}{4}$ *(4.7 #9)*

3. 0 *(4.7 #15)*

4a. 1.19 *(4.7 #23)* **4b.** –0.33 *(4.7 #21)*

5a. 0.9 *(4.7 #31)* **5b.** $\frac{\pi}{6}$ *(4.7 #35)* **5c.** undefined *(4.7 #45)* **5d.** $\frac{3}{5}$ *(4.7 #47)*

5e. 2 *(4.7 #59)* **5f.** $\frac{\sqrt{1-x^2}}{x}$ *(4.7 #63)*

Section 4.8
Applications of Trigonometric Functions

Up, Up, & Away, Around the Corner, and Back & Forth

From finding the heights of tall buildings to modeling cyclic behavior, trigonometry is very useful. In this section, you will see how finding missing parts of right triangles has many practical applications, how ships at sea can be located using bearings and trigonometry, and even how the motion of a ball attached to a spring can be described using a sinusoidal function.

Objective #1: Solve a right triangle.

✔ *Solved Problem #1*

1a. Let $A = 62.7°$ and $a = 8.4$ in the triangle shown below. Solve the right triangle, rounding lengths to two decimal places.

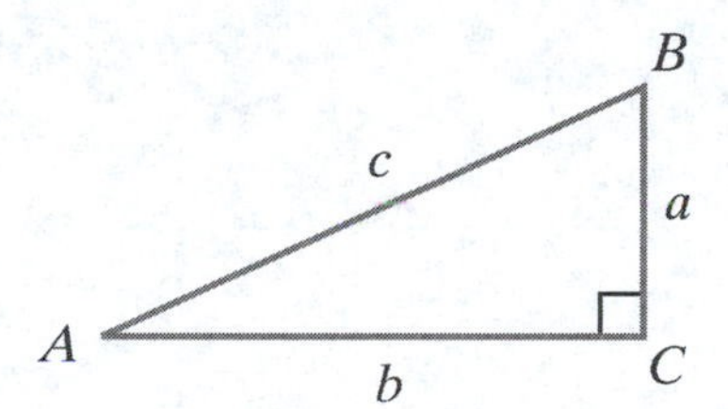

To solve the triangle, we need to find B, b, and c.

We begin with B. We know that $A + B = 90°$.

$$62.7° + B = 90°$$
$$B = 90° - 62.7° = 27.3°$$

Next we find b. Note that b is opposite $B = 27.3°$ and $a = 8.4$ is adjacent to $B = 27.3°$. We set up an equation relating a, b, and B using the tangent function. Then we solve for b.

$$\tan B = \frac{\text{side opposite } B}{\text{side adjacent to } B} = \frac{b}{a}$$

$$\tan 27.3° = \frac{b}{8.4}$$

$$b = 8.4 \tan 27.3° \approx 4.34$$

(continued on next page)

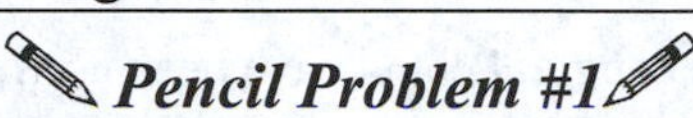

Pencil Problem #1

1a. Let $A = 23.5°$ and $b = 10$ in the triangle shown in Solved Problem #1a. Solve the right triangle, rounding lengths to two decimal places.

When finding c, we choose not to use the value of b just found because it was rounded. We will again use $B = 27.3°$ and $a = 8.4$. Since a is adjacent to B and we are finding the hypotenuse, we use the cosine function.

$$\cos B = \frac{\text{side adjacent to } B}{\text{hypotenuse}} = \frac{a}{c}$$

$$\cos 27.3° = \frac{8.4}{c}$$

$$c \cos 27.3° = 8.4$$

$$c = \frac{8.4}{\cos 27.3°} \approx 9.45$$

Thus, $B = 27.3°$, $b \approx 4.34$, and $c \approx 9.45$.

Note that angle A could have been used in the calculations to find b and c, $\tan A = \frac{a}{b}$ and $\sin A = \frac{a}{c}$, without using rounded values.

1b. From a point on level ground 80 feet from the base of a tower, the angle of elevation is 85.4°. Approximate the height of the tower to the nearest foot.

We draw a right triangle to illustrate the situation. The tower and the ground form the right angle. One of the acute angles measures 85.4°, and the side adjacent to it is 80 feet. The tower is opposite the 85.4° angle; we let h represent the height of the tower.

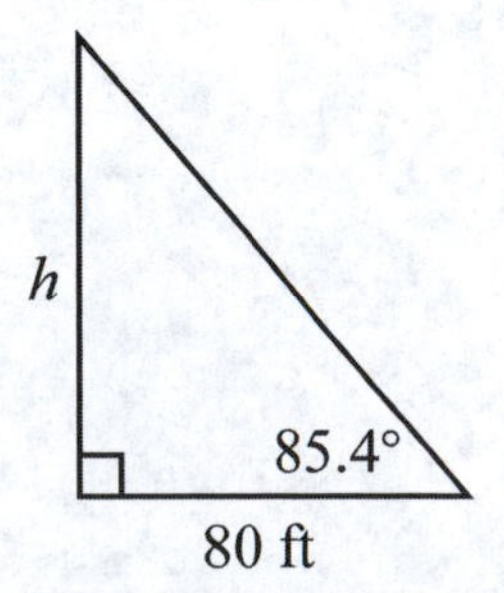

(continued on next page)

1b. From a point on level ground 5280 feet (1 mile) from the base of a television transmitting tower, the angle of elevation is 21.3°. Approximate the height of the tower to the nearest foot.

Since we are looking for the side opposite the known acute angle and we know the side adjacent to that angle, we will use tangent.

$$\tan 85.4° = \frac{\text{opposite side}}{\text{adjacent side}} = \frac{h}{80}$$

$$h = 80 \tan 85.4° \approx 994$$

The tower is approximately 994 feet tall.

1c. A guy wire is 13.8 yards long and is attached from the ground to a pole at a point 6.7 yards above the ground. Find the angle, to the nearest tenth, that the wire makes with the ground.

We draw a right triangle to illustrate the situation. The pole and the ground form the right angle. We are looking for the angle opposite the pole; call it A. The side opposite a is 6.7 yards, and the hypotenuse is 13.8 yards.

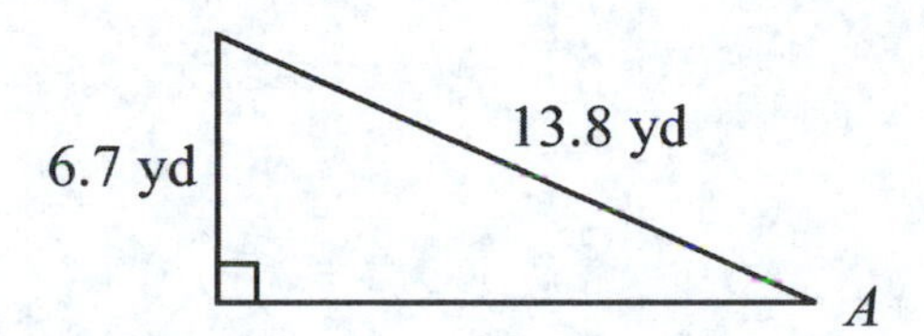

Since we know the side opposite A and the hypotenuse, we will use the sine function.

$$\sin A = \frac{\text{side opposite } A}{\text{hypotenuse}} = \frac{6.7}{13.8}$$

$$A = \sin^{-1} \frac{6.7}{13.8} \approx 29.0°$$

The wire makes an angle of approximately 29.0° with the ground.

1c. A wheelchair ramp is built beside the steps to the campus library. Find the angle of elevation of the 23-foot ramp, to the nearest tenth of a degree, if its final height is 6 feet.

1d. You are standing on level ground 800 feet from Mt. Rushmore, looking at the sculpture of Abraham Lincoln's face. The angle of elevation to the bottom of the sculpture is 32°, and the angle of elevation to the top is 35°. Find the height of the sculpture of Lincoln's face to the nearest tenth of a foot.

Refer to the figure below, where a represents the distance to the bottom of the sculpture and b represents the distance to the top of the sculpture. The height of the sculpture is then $b-a$. We need to find a and b and then subtract.

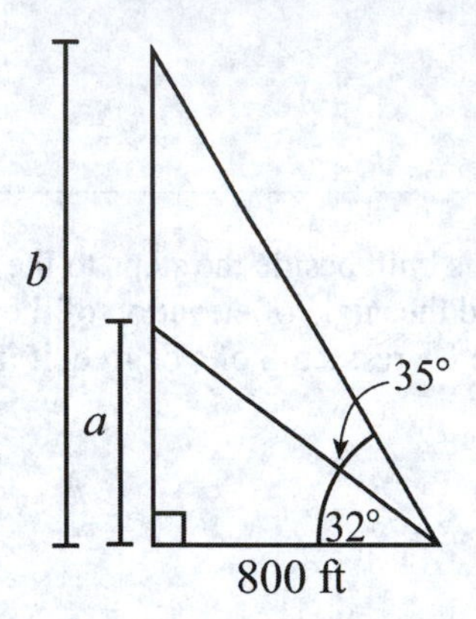

Since in each case we are finding the length of the leg opposite a known angle and we also know the length of the leg adjacent to the known angle, we will use the tangent function twice.

$$\tan 32° = \frac{\text{opposite side}}{\text{adjacent side}} = \frac{a}{800}$$

$$a = 800 \tan 32° \approx 499.9$$

$$\tan 35° = \frac{\text{opposite side}}{\text{adjacent side}} = \frac{b}{800}$$

$$b = 800 \tan 35° \approx 560.2$$

$$b - a = 560.2 - 499.9 = 60.3$$

Lincoln's face is approximately 60.3 feet tall.

1d. A hot-air balloon is rising vertically. From a point on level ground 125 feet from the point directly under the passenger compartment, the angle of elevation changes from 19.2° to 31.7°. How far, to the nearest tenth of a foot, does the balloon rise during this period?

Objective #2: Solve problems involving bearings.

✔ Solved Problem #2

2. You hike 2.3 miles on a bearing of S 31° W. Then you turn 90° clockwise and hike 3.5 miles on a bearing of N 59° W. At that time, what is your bearing, to the nearest tenth of a degree, from your starting point?

The figure below illustrates the situation. Notice that we have formed a triangle by drawing a segment from the starting point to the ending point. The triangle is a right triangle because of the 90° change in direction. We know the lengths of the legs of the right triangle.

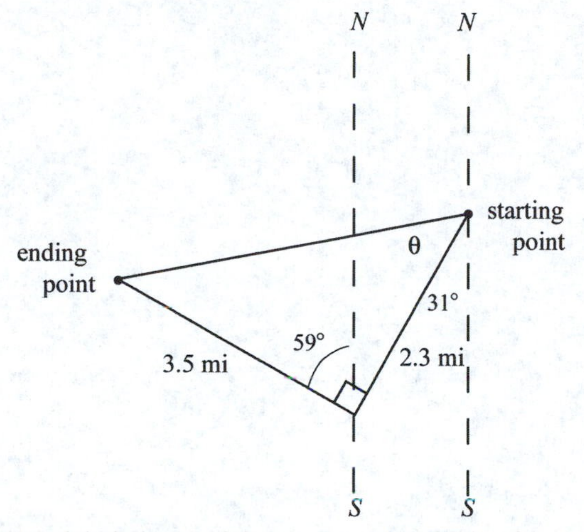

We labeled one of the acute angles θ. The side opposite θ is 3.5 miles and the side adjacent to θ is 2.3 miles. We can use the tangent function to find θ.

$$\tan\theta = \frac{\text{opposite side}}{\text{adjacent side}} = \frac{3.5}{2.3}$$

$$\theta = \tan^{-1}\frac{3.5}{2.3} \approx 56.7^\circ$$

Now the angle formed by the north-south line through the starting point and the segment between the starting and ending points is

$\theta + 31^\circ \approx 56.7^\circ + 31^\circ = 87.7^\circ$.

The bearing of the ending point from the starting point is S 87.7° W.

Pencil Problem #2

2. After takeoff, a jet flies 5 miles on a bearing of N 35° E. Then it turns 90° and flies 7 miles on a bearing of S 55° E. At that time, what is the bearing of the jet, to the nearest tenth of a degree, from the point where it took off?

Objective #3: Model simple harmonic motion.

✔ *Solved Problem #3*

3a. A ball on a string is pulled 6 inches below its rest position and then released. The period for the motion is 4 seconds. Write the equation for the ball's simple harmonic motion.

At $t = 0$ seconds, $d = -6$ inches. We use a negative value for d because the motion begins with the ball below its rest position. Also, because the motion begins when the ball is at its greatest distance from rest, we use the model containing cosine rather than the one containing sine.

We need to find values for a and ω in $d = a\cos\omega t$.

Since the maximum displacement is 6 inches and the ball is initially below its rest position, $a = -6$.

The period is given as 4 seconds. Use the period formula to find ω.

$$\frac{2\pi}{\omega} = 4$$

$$2\pi = 4\omega$$

$$\omega = \frac{2\pi}{4} = \frac{\pi}{2}$$

The equation is $d = -6\cos\frac{\pi}{2}t$.

Pencil Problem #3

3a. A ball on a string is pulled 8 inches below its rest position and then released. The period for the motion is 2 seconds. Write the equation for the ball's simple harmonic motion.

3b. An object moves in simple harmonic motion described by $d = 12\cos\frac{\pi}{4}t,$ where t is measured in seconds and d in centimeters. Find the maximum displacement, the frequency, and the time required for one cycle.

The maximum displacement is the amplitude. Because $a = 12$, the maximum displacement is 12 centimeters.

The frequency, f, is

$$f = \frac{\omega}{2\pi} = \frac{\frac{\pi}{4}}{2\pi} = \frac{\cancel{\pi}}{4} \cdot \frac{1}{2\cancel{\pi}} = \frac{1}{8}.$$

The frequency is $\frac{1}{8}$ oscillation per second.
The time required for one cycle is the period.

$$\text{period} = \frac{2\pi}{\omega} = \frac{2\pi}{\frac{\pi}{4}} = 2\cancel{\pi} \cdot \frac{4}{\cancel{\pi}} = 8$$

The time required for one cycle is 8 seconds.

3b. An object moves in simple harmonic motion described by $d = 5\cos\frac{\pi}{2}t,$ where t is measured in seconds and d in inches. Find the maximum displacement, the frequency, and the time required for one cycle.

Answers for Pencil Problems *(Textbook Exercise references in parentheses)*:

1a. $B = 66.5°$, $a \approx 4.35$, $c \approx 10.90$ *(4.8 #1)* **1b.** 2059 feet *(4.8 #41)* **1c.** 15.1° *(4.8 #47)*
1d. 33.7 feet *(4.8 #49)*

2. N 89.5° E *(4.8 #57)*

3a. $d = -8\cos \pi t$ *(4.8 #18)*

3b. maximum displacement: 5 inches; frequency: $\frac{1}{4}$ oscillation per second; time for one cycle: 4 seconds *(4.8 #21)*

Section 5.1
Verifying Trigonometric Identities

Do you Enjoy Solving Puzzles?

We have already established some basic relationships among the trigonometric functions. The reciprocal, quotient, and Pythagorean identities follow easily from the definitions of the trigonometric functions. The even-odd identities are established using properties of the Cartesian coordinate system.
In this section, we see how new identities can be verified using identities that we already know. The process will involve some trial and error and may remind you of solving a puzzle.

Objective #1: Use fundamental trigonometric identities to verify identities.

1a. Verify the identity: $\csc x \tan x = \sec x$.

In many cases, it is helpful to rewrite all the trigonometric functions on one side in terms of sines and cosines using reciprocal and quotient identities. In general, we start by working with the more complicated side, so we begin with the left side and rewrite it in terms of sines and cosines using the reciprocal identity $\csc x = \dfrac{1}{\sin x}$ and the quotient identity $\tan x = \dfrac{\sin x}{\cos x}$.

$$\csc x \tan x = \frac{1}{\sin x} \cdot \frac{\sin x}{\cos x} = \frac{1}{\cancel{\sin x}} \cdot \frac{\cancel{\sin x}}{\cos x} = \frac{1}{\cos x} = \sec x$$

Note the use of the reciprocal identity $\sec x = \dfrac{1}{\cos x}$ in the last step. Also note how we started with the left side of the given identity, $\csc x \tan x = \sec x$, and ended with the right side.

1a. Verify the identity: $\sin x \sec x = \tan x$.

1b. Verify the identity: $\cos x \cot x + \sin x = \csc x$.

Once again we will rewrite all the trigonometric functions on one side in terms of sines and cosines. We start by working with the more complicated side, the left side, and rewrite it using the quotient identity $\cot x = \dfrac{\cos x}{\sin x}$.

$$\begin{aligned}\cos x \cot x + \sin x &= \frac{\cos x}{1} \cdot \frac{\cos x}{\sin x} + \sin x \\ &= \frac{\cos^2 x}{\sin x} + \frac{\sin x}{1} \cdot \frac{\sin x}{\sin x} \\ &= \frac{\cos^2 x}{\sin x} + \frac{\sin^2 x}{\sin x} \\ &= \frac{\cos^2 x + \sin^2 x}{\sin x} \\ &= \frac{1}{\sin x} \\ &= \csc x\end{aligned}$$

Note how we expressed both terms on the left side as fractions with a common denominator and then added; this technique is often helpful when one side of the given identity has two terms and the other side has only one term. Also note the use of the Pythagorean identity $\cos^2 x + \sin^2 x = 1$.

1b. Verify the identity: $\csc \theta - \sin \theta = \cot \theta \cos \theta$. [Hints: The Pythagorean identity $\cos^2 \theta + \sin^2 \theta = 1$ can also be used in the form $1 - \sin^2 \theta = \cos^2 \theta$ and a fraction of the form $\dfrac{a^2}{b}$ can be rewritten as $\dfrac{a}{b} \cdot \dfrac{a}{1}$.]

1c. Verify the identity: $\sin x - \sin x \cos^2 x = \sin^3 x$.

Sometimes factoring out a common factor is helpful.

We start by working with the more complicated side, the left side, and factor out the common factor $\sin x$.

Then we will use the Pythagorean identity $\cos^2 x + \sin^2 x = 1$ in the form $1 - \cos^2 x = \sin^2 x$.

$$\begin{aligned}\sin x - \sin x \cos^2 x &= \sin x(1 - \cos^2 x) \\ &= \sin x \cdot \sin^2 x = \sin^3 x\end{aligned}$$

1c. Verify the identity by using factoring first: $\sec x - \sec x \sin^2 x = \cos x$. [Hint: After factoring out the common factor on the left side and using a Pythagorean identity as we did in Solved Problem #1c, rewrite secant in terms of cosine.]

1d. Verify the identity: $\dfrac{1+\cos\theta}{\sin\theta} = \csc\theta + \cot\theta.$

There is usually more than one way to verify an identity. Here we will verify the identity in two ways. In the first method, we will start with the left side and work toward the right. In the second method, we will start with the right side and work toward the left.

Method 1: Start with the left side and write the single fraction as the sum of two fractions. Then use appropriate reciprocal and quotient identities.

$$\frac{1+\cos\theta}{\sin\theta} = \frac{1}{\sin\theta} + \frac{\cos\theta}{\sin\theta} = \csc\theta + \cot\theta$$

Method 2: Start with the right side and rewrite the given functions in terms of sines and cosines. Then add the resulting fractions.

$$\csc\theta + \cot\theta = \frac{1}{\sin\theta} + \frac{\cos\theta}{\sin\theta} = \frac{1+\cos\theta}{\sin\theta}$$

Although the equation in Method 2 could be obtained by reversing the equation in Method 1, the thought processes are different. Perhaps one direction is more obvious to you than the other.

1d. Verify the identity in two different ways as in Solved Problem #1d: $\dfrac{1-\sin\theta}{\cos\theta} = \sec\theta - \tan\theta.$

1e. Verify the identity: $\dfrac{\sin x}{1+\cos x} + \dfrac{1+\cos x}{\sin x} = 2\csc x.$

We will work with the left side and begin by getting a common denominator so that we can combine the two fractions into one fraction. The LCD is $(1+\cos x)\sin x$. Watch for opportunities to factor and simplify within the fraction and to apply basic identities. The Pythagorean identity $\cos^2 x + \sin^2 x = 1$ will be used to replace $\cos^2 x + \sin^2 x$ with 1 in a key step along the way.

(continued on next page)

1e. Verify the identity: $\dfrac{\cos x}{1-\sin x} + \dfrac{1-\sin x}{\cos x} = 2\sec x.$

$$\frac{\sin x}{1+\cos x}+\frac{1+\cos x}{\sin x}=\frac{\sin x}{1+\cos x}\cdot\frac{\sin x}{\sin x}+\frac{1+\cos x}{\sin x}\cdot\frac{1+\cos x}{1+\cos x}$$
$$=\frac{\sin^2 x}{(1+\cos x)\sin x}+\frac{(1+\cos x)^2}{(1+\cos x)\sin x}$$
$$=\frac{\sin^2 x+(1+\cos x)^2}{(1+\cos x)\sin x}$$
$$=\frac{\sin^2 x+1+2\cos x+\cos^2 x}{(1+\cos x)\sin x}$$
$$=\frac{(\sin^2 x+\cos^2 x)+1+2\cos x}{(1+\cos x)\sin x}$$
$$=\frac{1+1+2\cos x}{(1+\cos x)\sin x}$$
$$=\frac{2+2\cos x}{(1+\cos x)\sin x}$$
$$=\frac{2(1+\cos x)}{(1+\cos x)\sin x}$$
$$=\frac{2\cancel{(1+\cos x)}}{\cancel{(1+\cos x)}\sin x}$$
$$=\frac{2}{\sin x}$$
$$=2\csc x$$

Answers for Pencil Problems *(Textbook Exercise references in parentheses)*:

1a. $\sin x\sec x=\frac{\sin x}{1}\cdot\frac{1}{\cos x}=\frac{\sin x}{\cos x}=\tan x$ *(5.1 #1)*

1b. $\csc\theta-\sin\theta=\frac{1}{\sin\theta}-\frac{\sin\theta}{1}\cdot\frac{\sin\theta}{\sin\theta}=\frac{1}{\sin\theta}-\frac{\sin^2\theta}{\sin\theta}=\frac{1-\sin^2\theta}{\sin\theta}=\frac{\cos^2\theta}{\sin\theta}=\frac{\cos\theta}{\sin\theta}\cdot\frac{\cos\theta}{1}=\cot\theta\cos\theta.$ *(5.1 #11)*

1c. $\sec x-\sec x\sin^2 x=\sec x(1-\sin^2 x)=\sec x\cos^2 x=\frac{1}{\cos x}\cdot\cos^2 x=\cos x$ *(5.1 #7)*

1d. $\frac{1-\sin\theta}{\cos\theta}=\frac{1}{\cos\theta}-\frac{\sin\theta}{\cos\theta}=\sec\theta-\tan\theta;\ \sec\theta-\tan\theta=\frac{1}{\cos\theta}-\frac{\sin\theta}{\cos\theta}=\frac{1-\sin\theta}{\cos\theta}$ *(5.1 #24)*

1e. $\frac{\cos x}{1-\sin x}+\frac{1-\sin x}{\cos x}=\frac{\cos x\cdot\cos x}{(1-\sin x)\cos x}+\frac{(1-\sin x)(1-\sin x)}{(1-\sin x)\cos x}=\frac{\cos^2 x+(1-\sin x)^2}{(1-\sin x)\cos x}=\frac{\cos^2 x+1-2\sin x+\sin^2 x}{(1-\sin x)\cos x}$

$=\frac{(\cos^2 x+\sin^2 x)+1-2\sin x}{(1-\sin x)\cos x}=\frac{1+1-2\sin x}{(1-\sin x)\cos x}=\frac{2-2\sin x}{(1-\sin x)\cos x}=\frac{2(1-\sin x)}{(1-\sin x)\cos x}=\frac{2}{\cos x}=2\sec x$ *(5.1 #31)*

Section 5.2
Sum and Difference Formulas

More Identities?

In this section, you will learn to use another set of identities, known as the sum and difference formulas. The first of these is established using the definitions of the trigonometric functions and the properties of the Cartesian plane. Once the first formula is established, the other three can be verified using it and other basic identities.
Once verified, these formulas can be used to find function values as well as verify other identities. In the Exercise Set, you will see how these formulas can be used to simplify functions that model sound vibrations.

Objective #1: Use the formula for the cosine of the difference of two angles.

✔ *Solved Problem #1*

1a. We know that $\cos 30° = \frac{\sqrt{3}}{2}$. Use the formula for the cosine of the difference of two angles to find the exact value of $\cos 30° = \cos(90° - 60°)$.

The formula is
$\cos(\alpha - \beta) = \cos\alpha\cos\beta + \sin\alpha\sin\beta.$

In $\cos(90° - 60°)$, $\alpha = 90°$ and $\beta = 60°$. Apply the formula above, substitute exact values for the four resulting trigonometric expressions, and simplify.

$$\cos(90° - 60°) = \cos 90° \cos 60° + \sin 90° \sin 60°$$
$$= 0 \cdot \frac{1}{2} + 1 \cdot \frac{\sqrt{3}}{2} = 0 + \frac{\sqrt{3}}{2} = \frac{\sqrt{3}}{2}$$

Note that this result is the same as the known value of $\cos 30° = \frac{\sqrt{3}}{2}$, as it should be.

Pencil Problem #1

1a. Use the formula for the cosine of the difference of two angles to find the exact value of $\cos 15° = \cos(45° - 30°)$.

1b. Use the formula for the cosine of the difference of two angles to find the exact value of $\cos 70° \cos 40° + \sin 70° \sin 40°$.

The formula is
$\cos(\alpha - \beta) = \cos\alpha\cos\beta + \sin\alpha\sin\beta.$

Notice that $\cos 70° \cos 40° + \sin 70° \sin 40°$ looks like the right side of the formula with $\alpha = 70°$ and $\beta = 40°$.

$$\begin{aligned}\cos 70° \cos 40° + \sin 70° \sin 40° &= \sin(70° - 40°)\\ &= \sin 30°\\ &= \frac{1}{2}\end{aligned}$$

1b. Use the formula for the cosine of the difference of two angles to find the exact value of $\cos 50° \cos 20° + \sin 50° \sin 20°$.

1c. Verify the identity: $\dfrac{\cos(\alpha - \beta)}{\cos\alpha\cos\beta} = 1 + \tan\alpha\tan\beta.$

We begin with the left side and apply the formula for the cosine of the difference of two angles in the numerator. We then rewrite the fraction as the sum of two fractions and simplify.

$$\begin{aligned}\frac{\cos(\alpha - \beta)}{\cos\alpha\cos\beta} &= \frac{\cos\alpha\cos\beta + \sin\alpha\sin\beta}{\cos\alpha\cos\beta}\\ &= \frac{\cos\alpha\cos\beta}{\cos\alpha\cos\beta} + \frac{\sin\alpha\sin\beta}{\cos\alpha\cos\beta}\\ &= 1 + \frac{\sin\alpha}{\cos\alpha}\cdot\frac{\sin\beta}{\cos\beta}\\ &= 1 + \tan\alpha\tan\beta\end{aligned}$$

1c. Verify the identity: $\dfrac{\cos(\alpha - \beta)}{\cos\alpha\sin\beta} = \tan\alpha + \cot\beta.$

Objective #2: Use sum and difference formulas for cosines and sines.

✔ Solved Problem #2

2. We are given that $\sin\alpha = \frac{4}{5}$ for a quadrant II angle α and $\sin\beta = \frac{1}{2}$ for a quadrant I angle β. Use this information in Solved Problems #2a–d.

2a. Find the exact value of $\cos\alpha$.

The value of $\cos\alpha$ is negative, since α is a quadrant II angle. Use the Pythagorean identity $\sin^2\alpha + \cos^2\alpha = 1$ to find $\cos\alpha$, choosing the negative value.

$$\left(\frac{4}{5}\right)^2 + \cos^2\alpha = 1$$

$$\cos^2\alpha = 1 - \left(\frac{4}{5}\right)^2 = \frac{25}{25} - \frac{16}{25} = \frac{9}{25}$$

$$\cos\alpha = -\sqrt{\frac{9}{25}} = -\frac{3}{5}$$

2b. Find the exact value of $\cos\beta$.

The value of $\cos\beta$ is positive, since β is a quadrant I angle. Use the Pythagorean identity $\sin^2\beta + \cos^2\beta = 1$ to find $\cos\beta$, choosing the positive value.

$$\left(\frac{1}{2}\right)^2 + \cos^2\beta = 1$$

$$\cos^2\beta = 1 - \left(\frac{1}{2}\right)^2 = \frac{4}{4} - \frac{1}{4} = \frac{3}{4}$$

$$\cos\beta = \sqrt{\frac{3}{4}} = \frac{\sqrt{3}}{2}$$

✎ Pencil Problem #2 ✎

2. We are given that $\sin\alpha = \frac{3}{5}$ for a quadrant I angle α and $\sin\beta = \frac{5}{13}$ for a quadrant II angle β. Use this information in Pencil Problems #2a–d.

2a. Find the exact value of $\cos\alpha$.

2b. Find the exact value of $\cos\beta$.

2c. Find the exact value of $\cos(\alpha+\beta)$.

We use the sum formula for cosine and substitute the given values, $\sin\alpha = \frac{4}{5}$ and $\sin\beta = \frac{1}{2}$, as well as the values we just found, $\cos\alpha = -\frac{3}{5}$ and $\cos\beta = \frac{\sqrt{3}}{2}$.

$$\cos(\alpha+\beta) = \cos\alpha\cos\beta - \sin\alpha\sin\beta$$
$$= -\frac{3}{5}\cdot\frac{\sqrt{3}}{2} - \frac{4}{5}\cdot\frac{1}{2}$$
$$= -\frac{3\sqrt{3}}{10} - \frac{4}{10} = \frac{-3\sqrt{3}-4}{10}$$

2c. Find the exact value of $\cos(\alpha+\beta)$.

2d. Find the exact value of $\sin(\alpha+\beta)$.

We use the sum formula for sine and substitute the values $\sin\alpha = \frac{4}{5}$, $\sin\beta = \frac{1}{2}$, $\cos\alpha = -\frac{3}{5}$, and $\cos\beta = \frac{\sqrt{3}}{2}$.

$$\sin(\alpha+\beta) = \sin\alpha\cos\beta + \cos\alpha\sin\beta$$
$$= \frac{4}{5}\cdot\frac{\sqrt{3}}{2} + \left(-\frac{3}{5}\right)\cdot\frac{1}{2}$$
$$= \frac{4\sqrt{3}}{10} - \frac{3}{10} = \frac{4\sqrt{3}-3}{10}$$

2d. Find the exact value of $\sin(\alpha+\beta)$.

Objective #3: Use sum and difference formulas for tangents.

✔ *Solved Problem #3*

3. Verify the identity: $\tan(x+\pi) = \tan x$.

Start with the left side and apply the sum formula for tangent. Use the fact that $\tan \pi = 0$.

$$\begin{aligned} \tan(x+\pi) &= \frac{\tan x + \tan \pi}{1 - \tan x \tan \pi} \\ &= \frac{\tan x + 0}{1 - \tan x \cdot 0} \\ &= \frac{\tan x}{1} \\ &= \tan x \end{aligned}$$

Pencil Problem #3

3. Verify the identity: $\tan(2\pi - x) = -\tan x$.

Answers for Pencil Problems *(Textbook Exercise references in parentheses)*:

1a. $\dfrac{\sqrt{6}+\sqrt{2}}{4}$ *(5.2 #1)*

1b. $\dfrac{\sqrt{3}}{2}$ *(5.2 #5)*

1c. $\dfrac{\cos(\alpha-\beta)}{\cos\alpha\sin\beta}=\dfrac{\cos\alpha\cos\beta+\sin\alpha\sin\beta}{\cos\alpha\sin\beta}=\dfrac{\cancel{\cos\alpha}\cos\beta}{\cancel{\cos\alpha}\sin\beta}+\dfrac{\sin\alpha\cancel{\sin\beta}}{\cos\alpha\cancel{\sin\beta}}=\cot\beta+\tan\alpha=\tan\alpha+\cot\beta$ *(5.2 #9)*

2a. $\dfrac{4}{5}$ **2b.** $-\dfrac{12}{13}$ **2c.** $-\dfrac{63}{65}$ *(5.2 #57a)* **2d.** $-\dfrac{16}{65}$ *(5.2 #57b)*

3. $\tan(2\pi-x)=\dfrac{\tan 2\pi-\tan x}{1+\tan 2\pi\tan x}=\dfrac{0-\tan x}{1+0\cdot\tan x}=\dfrac{-\tan x}{1}=-\tan x$ *(5.2 #37)*

Section 5.3
Double Angle, Power-Reducing, and Half-Angle Formulas

How Far Can You Throw That?

When you throw an object, the distance that the object will travel before hitting the ground depends on the initial speed of the object as well as the angle at which the object leaves your hand. This distance can be modeled by a formula that involves both sines and cosines.
In the Exercise Set, you will see how a trigonometric identity introduced in this section can be used to rewrite a formula involving two trigonometric functions as a formula involving only one such function. The simpler form can be used to determine an angle that maximizes throwing distance.

Objective #1: Use the double-angle formulas.

✔ *Solved Problem #1*

1a. If $\sin\theta = \frac{4}{5}$ and θ lies in quadrant II, find the exact value of $\sin 2\theta$, $\cos 2\theta$, and $\tan 2\theta$.

The formulas $\sin 2\theta = 2\sin\theta\cos\theta$ and $\cos 2\theta = \cos^2\theta - \sin^2\theta$ require that we know both $\sin\theta$ and $\cos\theta$. We are given that $\sin\theta = \frac{4}{5}$. We need to find $\cos\theta$. Since θ lies in quadrant II, the value of $\cos\theta$ is negative. Use the Pythagorean identity $\sin^2\theta + \cos^2\theta = 1$ to find $\cos\theta$, choosing the negative value.

$$\left(\frac{4}{5}\right)^2 + \cos^2\theta = 1$$

$$\cos^2\theta = 1 - \left(\frac{4}{5}\right)^2 = \frac{25}{25} - \frac{16}{25} = \frac{9}{25}$$

$$\cos\theta = -\sqrt{\frac{9}{25}} = -\frac{3}{5}$$

1a. If $\sin\theta = \frac{15}{17}$ and θ lies in quadrant II, find the exact value of $\sin 2\theta$, $\cos 2\theta$, and $\tan 2\theta$.

Now substitute the values into the formulas.

$$\sin 2\theta = 2\sin\theta\cos\theta = 2\left(\frac{4}{5}\right)\left(-\frac{3}{5}\right) = -\frac{24}{25}$$

$$\cos 2\theta = \cos^2\theta - \sin^2\theta$$
$$= \left(-\frac{3}{5}\right)^2 - \left(\frac{4}{5}\right)^2 = \frac{9}{25} - \frac{16}{25} = -\frac{7}{25}$$

The formula $\tan 2\theta = \dfrac{2\tan\theta}{1-\tan^2\theta}$ requires that we know $\tan\theta$. Since we know $\sin\theta = \dfrac{4}{5}$ and $\cos\theta = -\dfrac{3}{5}$, we can use a quotient identity to find $\tan\theta$.

$$\tan\theta = \frac{\sin\theta}{\cos\theta} = \frac{\frac{4}{5}}{-\frac{3}{5}} = \frac{4}{5}\cdot\left(-\frac{5}{3}\right) = -\frac{4}{3}$$

Now we can use this value in the formula.

$$\tan 2\theta = \frac{2\tan\theta}{1-\tan^2\theta} = \frac{2\left(-\frac{4}{3}\right)}{1-\left(-\frac{4}{3}\right)^2}$$
$$= \frac{-\frac{8}{3}}{1-\frac{16}{9}} = \frac{-\frac{8}{3}}{-\frac{7}{9}} = -\frac{8}{3}\cdot\left(-\frac{9}{7}\right) = \frac{24}{7}$$

Note that we can also find $\tan 2\theta$ using the values of $\sin 2\theta$ and $\cos 2\theta$ and a quotient identity.

$$\tan 2\theta = \frac{\sin 2\theta}{\cos 2\theta} = \frac{-\frac{24}{25}}{-\frac{7}{25}} = \frac{24}{7}$$

1b. Find the exact value of $\cos^2 15° - \sin^2 15°$.

The given expression looks like the right side of the double-angle formula $\cos 2\theta = \cos^2\theta - \sin^2\theta$ with $\theta = 15°$. We use the formula to rewrite the expression and then evaluate.

$$\cos^2 15° - \sin^2 15° = \cos(2 \cdot 15°) = \cos 30° = \frac{\sqrt{3}}{2}$$

1b. Find the exact value of $2\sin 15° \cos 15°$.

1c. Verify the identity: $\sin 3\theta = 3\sin\theta - 4\sin^3\theta$.

Notice that the sine on the left side has 3θ as its argument while the ones on the right have just θ, so we need to use one or more identities that allow us to change the argument. The right side may look more complicated, but we will start with the left side because it contains the more complicated argument and we see more opportunities to apply identities.

We first rewrite $\sin 3\theta$ as $\sin(2\theta + \theta)$ and use the sum formula for sine. Then we use double-angle formulas to simplify the occurrences of $\sin 2\theta$ and $\cos 2\theta$ that result from the first step. Since the right side contains only sine, we choose to use the form of the double-angle formula for cosine that expresses $\cos 2\theta$ in terms of sine only. Still keeping in mind that the right side contains only sine, we replace an occurrence of $\cos^2\theta$ with $1 - \sin^2\theta$, using a form of the Pythagorean identity $\sin^2\theta + \cos^2\theta = 1$.

$$\begin{aligned}
\sin 3\theta &= \sin(2\theta + \theta) \\
&= \sin 2\theta \cos\theta + \cos 2\theta \sin\theta \\
&= (2\sin\theta\cos\theta)\cos\theta + (1 - 2\sin^2\theta)\sin\theta \\
&= 2\sin\theta\cos^2\theta + \sin\theta - 2\sin^3\theta \\
&= 2\sin\theta(1 - \sin^2\theta) + \sin\theta - 2\sin^3\theta \\
&= 2\sin\theta - 2\sin^3\theta + \sin\theta - 2\sin^3\theta \\
&= 3\sin\theta - 4\sin^3\theta
\end{aligned}$$

1c. Verify the identity:
$\sin 4t = 4\sin t\cos^3 t - 4\sin^3 t\cos t$.

Objective #2: Use the power-reducing formulas.

✔ *Solved Problem #2*

2. Write an equivalent expression for $\sin^4 x$ that does not contain powers of trigonometric functions greater than 1.

We begin by writing $\sin^4 x$ as $(\sin^2 x)^2$ and applying the power-reducing formula for sine. After some simplification, we will use the power-reducing formula for cosine.

$$\begin{aligned}\sin^4 x &= (\sin^2 x)^2 \\ &= \left(\frac{1-\cos 2x}{2}\right)^2 \\ &= \frac{1-2\cos 2x+\cos^2 2x}{4} \\ &= \frac{1}{4}-\frac{1}{2}\cos 2x+\frac{1}{4}\cos^2 2x \\ &= \frac{1}{4}-\frac{1}{2}\cos 2x+\frac{1}{4}\left[\frac{1+\cos 2(2x)}{2}\right] \\ &= \frac{1}{4}-\frac{1}{2}\cos 2x+\frac{1}{8}+\frac{1}{8}\cos 4x \\ &= \frac{3}{8}-\frac{1}{2}\cos 2x+\frac{1}{8}\cos 4x\end{aligned}$$

Pencil Problem #2

2. Write an equivalent expression for $\sin^2 x\cos^2 x$ that does not contain powers of trigonometric functions greater than 1.

Objective #3: Use the half-angle formulas.

✔ *Solved Problem #3*

3. Use $\cos 210° = -\frac{\sqrt{3}}{2}$ and a half-angle formula to find the exact value of $\cos 105°$.

Since an angle that measures 105° lies in quadrant II and cosine is negative in quadrant II, we choose the negative sign in the half-angle formula for cosine.

$$\begin{aligned}\cos 105° &= \cos\frac{210°}{2}\\ &= -\sqrt{\frac{1+\cos 210°}{2}}\\ &= -\sqrt{\frac{1+\left(-\frac{\sqrt{3}}{2}\right)}{2}}\\ &= -\sqrt{\frac{2-\sqrt{3}}{4}}\\ &= -\frac{\sqrt{2-\sqrt{3}}}{2}\end{aligned}$$

Note that to simplify the fraction $\frac{1+\left(-\frac{\sqrt{3}}{2}\right)}{2}$ we multiplied the numerator and denominator both by 2: $\frac{\left[1+\left(-\frac{\sqrt{3}}{2}\right)\right]\cdot 2}{2\cdot 2} = \frac{1\cdot 2+\left(-\frac{\sqrt{3}}{2}\right)\cdot 2}{4} = \frac{2-\sqrt{3}}{4}$.

Pencil Problem #3

3. Use $\cos 315° = \frac{\sqrt{2}}{2}$ and a half-angle formula to find the exact value of $\cos 157.5°$.

Answers for Pencil Problems *(Textbook Exercise references in parentheses)*:

1a. $\sin 2\theta = -\dfrac{240}{289}$; $\cos 2\theta = -\dfrac{161}{289}$; $\tan 2\theta = \dfrac{240}{161}$ *(5.3 #7)*

1b. $\dfrac{1}{2}$ *(5.3 #15)*

1c. $\sin 4t = \sin(2 \cdot 2t) = 2\sin 2t \cos 2t = 2(2\sin t \cos t)(\cos^2 t - \sin^2 t) = 4\sin t \cos^3 t - 4\sin^3 t \cos t$ *(5.3 #33)*

2. $\dfrac{1}{8} - \dfrac{1}{8}\cos 4x$ *(5.3 #37)*

3. $-\dfrac{\sqrt{2+\sqrt{2}}}{2}$ *(5.3 #41)*

Section 5.4
Product-to-Sum and Sum-to-Product Formulas

Music to Your Ears

Each time you push a button on a touch-tone phone a sound is produced. This sound can be modeled by the sum of two sine functions. In the Exercise Set, you will learn how to play *Mary Had a Little Lamb* on your phone and you will write the sum of sines producing a note as a product.

Objective #1: Use the product-to-sum formulas.

✔ *Solved Problem #1*

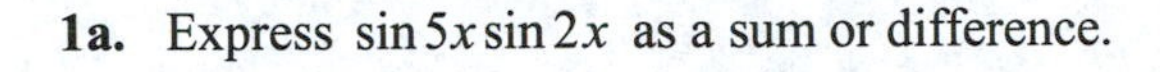

1a. Express $\sin 5x \sin 2x$ as a sum or difference.

We use the formula for the product of two sines:

$$\sin\alpha\sin\beta = \frac{1}{2}[\cos(\alpha-\beta)-\cos(\alpha+\beta)].$$

$$\sin 5x\sin 2x = \frac{1}{2}[\cos(5x-2x)-\cos(5x+2x)]$$
$$= \frac{1}{2}[\cos 3x - \cos 7x]$$

1b. Express $\cos 7x \cos x$ as a sum or difference.

We use the formula for the product of two cosines:

$$\cos\alpha\cos\beta = \frac{1}{2}[\cos(\alpha-\beta)+\cos(\alpha+\beta)].$$

$$\cos 7x\cos x = \frac{1}{2}[\cos(7x-x)+\cos(7x+x)]$$
$$= \frac{1}{2}[\cos 6x + \cos 8x]$$

Pencil Problem #1

1a. Express $\sin 6x \sin 2x$ as a sum or difference.

1b. Express $\sin x \cos 2x$ as a sum or difference.

Objective #2: Use the sum-to-product formulas.

Solved Problem #2

2a. Express $\sin 7x + \sin 3x$ as a product.

We use the formula for the sum of two sines:

$$\sin\alpha + \sin\beta = 2\sin\frac{\alpha+\beta}{2}\cos\frac{\alpha-\beta}{2}.$$

$$\begin{aligned}\sin 7x + \sin 3x &= 2\sin\frac{7x+3x}{2}\cos\frac{7x-3x}{2}\\ &= 2\sin\frac{10x}{2}\cos\frac{4x}{2}\\ &= 2\sin 5x\cos 2x\end{aligned}$$

2b. Express $\cos 3x + \cos 2x$ as a product.

We use the formula for the sum of two cosines:

$$\cos\alpha + \cos\beta = 2\cos\frac{\alpha+\beta}{2}\cos\frac{\alpha-\beta}{2}.$$

$$\begin{aligned}\cos 3x + \cos 2x &= 2\cos\frac{3x+2x}{2}\cos\frac{3x-2x}{2}\\ &= 2\cos\frac{5x}{2}\cos\frac{x}{2}\end{aligned}$$

Pencil Problem #2

2a. Express $\sin 6x + \sin 2x$ as a product.

2b. Express $\cos 4x + \cos 2x$ as a product.

<u>**Answers** for Pencil Problems *(Textbook Exercise references in parentheses)*</u>:

1a. $\frac{1}{2}[\cos 4x - \cos 8x]$ *(5.4 #1)* **1b.** $\frac{1}{2}[\sin 3x - \sin x]$ *(5.4 #5)*

2a. $2\sin 4x\cos 2x$ *(5.4 #9)* **2b.** $2\cos 3x\cos x$ *(5.4 #13)*

Section 5.5
Trigonometric Equations

How Many Solutions are There?

We have seen that trigonometric functions can be used to model cyclic phenomena such as tides, temperatures, and hours of daylight. For example, we have worked with models where we could estimate the number of hours of daylight at a location on a particular day of the year by evaluating the function for a value of the independent variable representing the day.

In this section's Exercise Set, you will solve trigonometric equations to find the dates on which a certain city has 10.5 hours of daylight. In general, the equation has infinitely many solutions, but when we restrict ourselves to one cycle, one year, there are exactly two solutions. In this section, pay attention to whether you are looking for all solutions or just the solutions that are in a given interval.

Objective #1: Find all solutions of a trigonometric equation.

✔ *Solved Problem #1*

1. Find all solutions of $5\sin x = 3\sin x + \sqrt{3}$.

First isolate $\sin x$ on the left side by first subtracting $3\sin x$ from both sides and then dividing both sides by the resulting coefficient of $\sin x$.

$$5\sin x = 3\sin x + \sqrt{3}$$
$$5\sin x - 3\sin x = 3\sin x - 3\sin x + \sqrt{3}$$
$$2\sin x = \sqrt{3}$$
$$\frac{\not{2}\sin x}{\not{2}} = \frac{\sqrt{3}}{2}$$
$$\sin x = \frac{\sqrt{3}}{2}$$

The sine function is positive in quadrants I and II and sine equals $\frac{\sqrt{3}}{2}$ at $\frac{\pi}{3}$ in quadrant I. In quadrant II, sine equals $\frac{\sqrt{3}}{2}$ at $\pi - \frac{\pi}{3} = \frac{2\pi}{3}$. In one period of the sine function, there are two solutions of the given equation, $x = \frac{\pi}{3}$ and $x = \frac{2\pi}{3}$.

(continued on next page)

✎ *Pencil Problem #1* ✎

1. Find all solutions of $4\sin\theta - 1 = 2\sin\theta$.

We are asked to find all solutions. Because of the periodic property of the sine function, if we add or subtract any integer multiple of 2π to either of the solutions on the previous page, we will get another solution. Thus, the solutions are

$$x = \frac{\pi}{3} + 2n\pi \text{ and } x = \frac{2\pi}{3} + 2n\pi,$$

where n is any integer.

Objective #2: Solve equations with multiple angles.

Solved Problem #2

2a. Solve the equation: $\tan 2x = \sqrt{3},\ 0 \le x < 2\pi.$

The equation already has the tangent expression isolated on the left. Notice that the argument, $2x$, is a double-angle. We solve for $2x$ first and then for x. We also note that we are only looking for solutions x that satisfy $0 \le x < 2\pi.$ However, we first identify all solutions and then select those in the given interval.

In Solved Problem #1, we worked with sine which has period 2π. In this problem, we are working with tangent which has period π. In one period, tangent equals $\sqrt{3}$ only once, at $\frac{\pi}{3}$. Thus,

$$2x = \frac{\pi}{3} + n\pi, \text{ where } n \text{ is any integer.}$$

Dividing by 2, we find all solutions of the given equation:

$$x = \frac{\pi}{6} + \frac{n\pi}{2}, \text{ where } n \text{ is any integer.}$$

(continued on next page)

Pencil Problem #2

2a. Solve the equation: $\tan 3x = \frac{\sqrt{3}}{3},\ 0 \le x < 2\pi.$

We want to find the solutions x that satisfy $0 \le x < 2\pi$. For any negative integer n, x is negative and does not satisfy $0 \le x < 2\pi$. For any integer $n \ge 4$, $x \ge 2\pi$ and does not satisfy $0 \le x < 2\pi$. Here are the values of x for $n = 0, 1, 2,$ and 3:

$$x = \frac{\pi}{6} + \frac{0 \cdot \pi}{2} = \frac{\pi}{6}$$

$$x = \frac{\pi}{6} + \frac{1 \cdot \pi}{2} = \frac{\pi}{6} + \frac{3\pi}{6} = \frac{4\pi}{6} = \frac{2\pi}{3}$$

$$x = \frac{\pi}{6} + \frac{2 \cdot \pi}{2} = \frac{\pi}{6} + \frac{6\pi}{6} = \frac{7\pi}{6}$$

$$x = \frac{\pi}{6} + \frac{3 \cdot \pi}{2} = \frac{\pi}{6} + \frac{9\pi}{6} = \frac{10\pi}{6} = \frac{5\pi}{3}$$

The solutions that satisfy $0 \le x < 2\pi$ are $\frac{\pi}{6}, \frac{2\pi}{3}, \frac{7\pi}{6}$, and $\frac{5\pi}{3}$.

2b. Solve the equation: $\sin\frac{x}{3} = \frac{1}{2},\ 0 \le x < 2\pi.$

The expression involving sine is already isolated on the left. In one period, the value of sine is $\frac{1}{2}$ at $\frac{\pi}{6}$ in quadrant I and at $\pi - \frac{\pi}{6} = \frac{5\pi}{6}$ in quadrant II.

Thus,

$$\frac{x}{3} = \frac{\pi}{6} + 2n\pi \text{ or } \frac{x}{3} = \frac{5\pi}{6} + 2n\pi,$$

where n is any integer. Multiplying by 3 in each, we obtain

$$x = \frac{\pi}{2} + 6n\pi \text{ or } x = \frac{5\pi}{2} + 6n\pi,$$

where n is any integer. Letting $n = 0$ in $x = \frac{\pi}{2} + 6n\pi$, we obtain $x = \frac{\pi}{2}$. No other integer n results in a value of $x = \frac{\pi}{2} + 6n\pi$ that satisfies $0 \le x < 2\pi$. For $x = \frac{5\pi}{2} + 6n\pi$, there is no integer n that results in a value of x that satisfies $0 \le x < 2\pi$.

The only solution is $\frac{\pi}{2}$.

2b. Solve the equation: $\sin\frac{2\theta}{3} = -1,\ 0 \le \theta < 2\pi.$

Objective #3: Solve trigonometric equations quadratic in form.

✔ Solved Problem #3

3a. Solve the equation:
$2\sin^2 x - 3\sin x + 1 = 0,\ 0 \le x < 2\pi.$

Notice that if we replace $\sin x$ with u on the left side of the equation we obtain $2u^2 - 3u + 1$, which factors as $(2u-1)(u-1)$. Thus, the left side of the equation factors as $(2\sin x - 1)(\sin x - 1)$. Thus, we can solve this equation by factoring and setting each factor equal to 0.

$$2\sin^2 x - 3\sin x + 1 = 0$$
$$(2\sin x - 1)(\sin x - 1) = 0$$
$$2\sin x - 1 = 0 \quad \text{or} \quad \sin x - 1 = 0$$
$$2\sin x = 1 \qquad \sin x = 1$$
$$\sin x = \frac{1}{2}$$

The equation $\sin x = \frac{1}{2}$ has two solutions satisfying $0 \le x < 2\pi$: $\frac{\pi}{6}$ and $\frac{5\pi}{6}$. The equation $\sin x = 1$ has one solution satisfying $0 \le x < 2\pi$: $\frac{\pi}{2}$. The solutions are $\frac{\pi}{6}$, $\frac{\pi}{2}$, and $\frac{5\pi}{6}$.

3b. Solve the equation: $4\cos^2 x - 3 = 0,\ 0 \le x < 2\pi.$

We solve the equation by isolating the squared expression and then using the square root property.

$$4\cos^2 x - 3 = 0$$
$$4\cos^2 x = 3$$
$$\cos^2 x = \frac{3}{4}$$
$$\cos x = \sqrt{\frac{3}{4}} = \frac{\sqrt{3}}{2} \quad \text{or} \quad \cos x = -\sqrt{\frac{3}{4}} = -\frac{\sqrt{3}}{2}$$

✎ Pencil Problem #3 ✎

3a. Solve the equation:
$2\sin^2 x - \sin x - 1 = 0,\ 0 \le x < 2\pi.$

3b. Solve the equation: $4\cos^2 x - 1 = 0,\ 0 \le x < 2\pi.$

The equation $\cos x = \frac{\sqrt{3}}{2}$ has two solutions satisfying $0 \le x < 2\pi$: $\frac{\pi}{6}$ and $\frac{11\pi}{6}$. The equation $\cos x = -\frac{\sqrt{3}}{2}$ has two solutions satisfying $0 \le x < 2\pi$: $\frac{5\pi}{6}$ and $\frac{7\pi}{6}$. The solutions are $\frac{\pi}{6}, \frac{5\pi}{6}, \frac{7\pi}{6}$, and $\frac{11\pi}{6}$.

Objective #4: Use factoring to separate different functions in trigonometric equations.

✔ *Solved Problem #4*

4. Solve the equation: $\sin x \tan x = \sin x,\ 0 \le x < 2\pi$.

We begin by subtracting $\sin x$ from both sides, obtaining 0 on the right. Then we factor out the common factor of $\sin x$ from the terms on the left and set each factor equal to 0.

$$\sin x \tan x = \sin x$$
$$\sin x \tan x - \sin x = 0$$
$$\sin x(\tan x - 1) = 0$$
$$\sin x = 0 \text{ or } \tan x - 1 = 0$$
$$\tan x = 1$$

The equation $\sin x = 0$ has solutions 0 and π satisfying $0 \le x < 2\pi$. The equation $\tan x = 1$ has solutions $\frac{\pi}{4}$ and $\frac{5\pi}{4}$ satisfying $0 \le x < 2\pi$. The solutions are 0, $\frac{\pi}{4}$, π, and $\frac{5\pi}{4}$.

4. Solve the equation:
$\sin x + 2\sin x \cos x = 0,\ 0 \le x < 2\pi$.

Objective #5: Use identities to solve trigonometric equations.

✔ Solved Problem #5

5a. Solve the equation: $\cos 2x + \sin x = 0,\ 0 \le x < 2\pi.$

We use the double-angle formula $\cos 2x = 1 - 2\sin^2 x$ to rewrite the left side in terms of sine only so the equation becomes quadratic in form.

$$\begin{aligned} \cos 2x + \sin x &= 0 \\ 1 - 2\sin^2 x + \sin x &= 0 \\ -2\sin^2 x + \sin x + 1 &= 0 \\ 2\sin^2 x - \sin x - 1 &= 0 \\ (2\sin x + 1)(\sin x - 1) &= 0 \end{aligned}$$

$$2\sin x + 1 = 0 \quad \text{or} \quad \sin x - 1 = 0$$
$$2\sin x = -1 \qquad \sin x = 1$$
$$\sin x = -\frac{1}{2}$$

The equation $\sin x = -\frac{1}{2}$ has solutions $\frac{7\pi}{6}$ and $\frac{11\pi}{6}$ satisfying $0 \le x < 2\pi.$ The equation $\sin x = 1$ has only the solution $\frac{\pi}{2}$ satisfying $0 \le x < 2\pi.$ The solutions are $\frac{\pi}{2}$, $\frac{7\pi}{6}$, and $\frac{11\pi}{6}$.

Pencil Problem #5

5a. Solve the equation: $\sin 2x = \cos x,\ 0 \le x < 2\pi.$

5b. Solve the equation: $\cos x - \sin x = -1,\ 0 \le x < 2\pi.$

We begin by squaring each side of the equation and then applying an identity. Remember that squaring may introduce extraneous solutions, so it is important to check all proposed solutions.

$$\cos x - \sin x = -1$$
$$(\cos x - \sin x)^2 = (-1)^2$$
$$\cos^2 x - 2\cos x \sin x + \sin^2 x = 1$$
$$(\cos^2 x + \sin^2 x) - 2\cos x \sin x = 1$$

Next we apply the identity $\sin^2 x + \cos^2 x = 1.$

$$1 - 2\cos x \sin x = 1$$
$$-2\cos x \sin x = 0$$
$$\cos x = 0 \text{ or } \sin x = 0$$

The first equation has solutions $\frac{\pi}{2}$ and $\frac{3\pi}{2}$ satisfying $0 \le x < 2\pi.$ The second equation has solutions 0 and π satisfying $0 \le x < 2\pi.$

Check $\frac{\pi}{2}$: $\cos\frac{\pi}{2} - \sin\frac{\pi}{2} = -1?$
$0 - 1 = -1?$
$-1 = -1$, true

Check $\frac{3\pi}{2}$: $\cos\frac{3\pi}{2} - \sin\frac{3\pi}{2} = -1?$
$0 - (-1) = -1?$
$1 = -1$, false

Check 0: $\cos 0 - \sin 0 = -1?$
$1 - 0 = -1?$
$1 = -1$, false

Check π: $\cos\pi - \sin\pi = -1?$
$-1 - 0 = -1?$
$-1 = -1$, true

The solutions are $\frac{\pi}{2}$ and $\pi.$

5b. Solve the equation: $\sin x + \cos x = 1,\ 0 \le x < 2\pi.$

Objective #6: Use a calculator to solve trigonometric equations.

✔ Solved Problem #6

6a. Solve $\tan x = 3.1044$, correct to four decimal places, for $0 \le x < 2\pi$.

Note that tangent is positive in quadrants I and III and that $\tan^{-1} 3.1044$ will be a value in quadrant I.

The solution in quadrant I is
$x = \tan^{-1} 3.1044 \approx 1.2592.$

The solution in quadrant III is
$x \approx \pi + 1.2592 \approx 4.4008.$

The solutions are 1.2592 and 4.4008.

6b. Solve $\sin x = -0.2315$, correct to four decimal places, for $0 \le x < 2\pi$.

Sine is negative in quadrants III and IV and $\sin^{-1} 0.2315 \approx 0.2336$ is in quadrant I.

The solution in quadrant III is
$x \approx \pi + 0.2336 \approx 3.3752.$

The solution in quadrant IV is
$x \approx 2\pi - 0.2336 \approx 6.0496.$

The solutions are 3.3752 and 6.0496.

✎ Pencil Problem #6 ✎

6a. Solve $\sin x = 0.8246$, correct to four decimal places, for $0 \le x < 2\pi$.

6b. Solve $\tan x = -3$, correct to four decimal places, for $0 \le x < 2\pi$.

Answers for Pencil Problems *(Textbook Exercise references in parentheses)*:

1. $\theta = \frac{\pi}{6} + 2n\pi,\ \theta = \frac{5\pi}{6} + 2n\pi,$ where n is any integer *(5.5 #21)*

2a. $\frac{\pi}{18}, \frac{7\pi}{18}, \frac{13\pi}{18}, \frac{19\pi}{18}, \frac{25\pi}{18}, \frac{31\pi}{18}$ *(5.5 #29)* **2b.** no solution *(5.5 #33)*

3a. $\frac{\pi}{2}, \frac{7\pi}{6}, \frac{11\pi}{6}$ *(5.5 #39)* **3b.** $\frac{\pi}{3}, \frac{2\pi}{3}, \frac{4\pi}{3}, \frac{5\pi}{3}$ *(5.5 #47)*

4. $0, \frac{2\pi}{3}, \pi, \frac{4\pi}{3}$ *(5.5 #59)*

5a. $\frac{\pi}{6}, \frac{\pi}{2}, \frac{5\pi}{6}, \frac{3\pi}{2}$ *(5.5 #69)* **5b.** $0, \frac{\pi}{2}$ *(5.5 #77)*

6a. 0.9695, 2.1721 *(5.5 #85)* **6b.** 1.8926, 5.0342 *(5.5 #89)*

Section 6.1
The Law of Sines

I SEE A FIRE, BUT WHERE IS IT?

We have solved many applied problems using the properties of right triangles and the relationships among the sides and angles. However, not every situation involving a triangle involves a right triangle. In this section and the next, we look at two laws that allow us to solve problems involving oblique triangles. In this section's Exercise Set, you will see how we can locate a fire, measure the height of a tree on a hillside, or find the distance a shot put is tossed using the Law of Sines.

Objective #1: Use the Law of Sines to solve oblique triangles.

✔ *Solved Problem #1*

1a. Solve the triangle shown in the figure. Round lengths of sides to the nearest tenth.

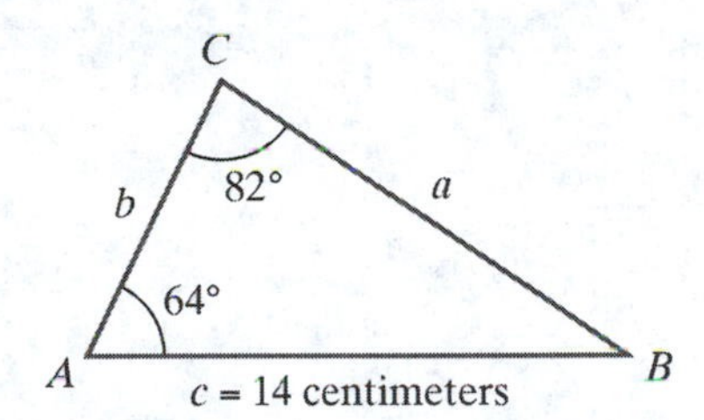

We know two angles and the side adjacent to one of the angles. This is an SAA triangle and the Law of Sines applies.

First, find the third angle.

$$\begin{aligned} A + B + C &= 180° \\ 64° + B + 82° &= 180° \\ B + 146° &= 180° \\ B &= 34° \end{aligned}$$

We know the length of one side, $c = 14$ centimeters, and the measure of the angle opposite this side, $C = 82°$. We use the ratio $\dfrac{14}{\sin 82°}$ and the Law of Sines to find the lengths of the other two sides.

(continued on next page)

1a. Solve the triangle shown in the figure. Round lengths of sides to the nearest tenth.

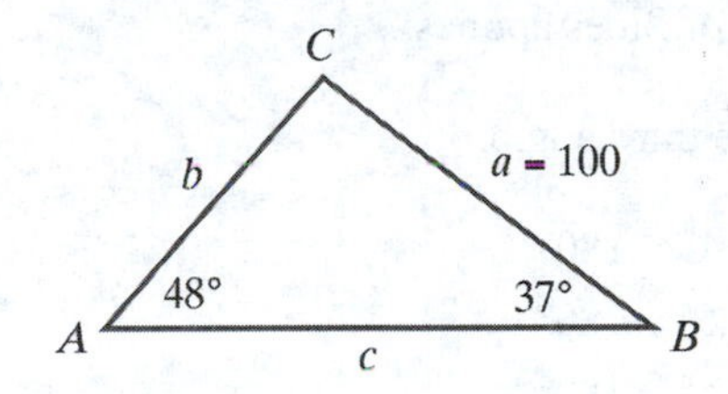

$$\frac{a}{\sin A} = \frac{c}{\sin C} \qquad \frac{b}{\sin B} = \frac{c}{\sin C}$$

$$\frac{a}{\sin 64°} = \frac{14}{\sin 82°} \qquad \frac{b}{\sin 34°} = \frac{14}{\sin 82°}$$

$$a = \frac{14 \sin 64°}{\sin 82°} \qquad b = \frac{14 \sin 34°}{\sin 82°}$$

$$a \approx 12.7 \text{ cm} \qquad b \approx 7.9 \text{ cm}$$

Thus, $B = 34°$, $a \approx 12.7$ cm, and $b \approx 7.9$ cm.

1b. Solve triangle ABC if $A = 40°$, $C = 22.5°$, and $b = 12$. Round lengths of sides to the nearest tenth.

You may want to draw a triangle and label it with the given information. We know two angles and the side between the angles. This is an ASA triangle and the Law of Sines applies.

First, find the third angle.

$$A + B + C = 180°$$
$$40° + B + 22.5° = 180°$$
$$B + 62.5° = 180°$$
$$B = 117.5°$$

We know the length of one side, $b = 12$, and the measure of the angle opposite this side, $B = 117.5°$. We use the ratio $\frac{12}{\sin 117.5°}$ and the Law of Sines to find the lengths of the other two sides.

$$\frac{a}{\sin A} = \frac{b}{\sin B} \qquad \frac{c}{\sin C} = \frac{b}{\sin B}$$

$$\frac{a}{\sin 40°} = \frac{12}{\sin 117.5°} \qquad \frac{c}{\sin 22.5°} = \frac{12}{\sin 117.5°}$$

$$a = \frac{12 \sin 40°}{\sin 117.5°} \qquad c = \frac{12 \sin 22.5°}{\sin 117.5°}$$

$$a \approx 8.7 \qquad c \approx 5.2$$

Thus, $B = 117.5°$, $a \approx 8.7$, and $c \approx 5.2$ cm.

1b. Solve triangle ABC if $A = 65°$, $B = 65°$, and $c = 6$. Round lengths of sides to the nearest tenth.

Objective #2: Use the Law of Sines to solve, if possible, the triangle or triangles in the ambiguous case.

✔ Solved Problem #2

2a. Solve triangle ABC if $A = 57°$, $a = 33$, and $b = 26$. Round lengths of sides to the nearest tenth and angle measures to the nearest degree.

We know two sides and the angle opposite one of the sides. This is an SSA case, where we first need to determine whether no triangle, exactly one triangle, or two triangles that satisfy the conditions exist.

We use the Law of Sines to find $\sin B$.

$$\frac{a}{\sin A} = \frac{b}{\sin B}$$

$$\frac{33}{\sin 57°} = \frac{26}{\sin B}$$

$$33 \sin B = 26 \sin 57°$$

$$\sin B = \frac{26 \sin 57°}{33} \approx 0.6608$$

There are two angles between 0° and 180° that satisfy $\sin B \approx 0.6608$:

$B_1 \approx \sin^{-1} 0.6608 \approx 41°$ and

$B_2 \approx 180° - 41° = 139°$.

Note that the angle sum of the given angle, $A = 57°$, and $B_2 \approx 139°$ is 196°, which exceeds 180°, so no triangle is possible with these two angles. However, this problem does not occur for angle B_1, so we have exactly one triangle.

We let $B = B_1 \approx 41°$. Now we find C and c.

$$A + B + C = 180°$$

$$57° + 41° + C \approx 180°$$

$$C + 98° \approx 180°$$

$$C \approx 82°$$

(continued on next page)

Pencil Problem #2

2a. Solve triangle ABC if $A = 40°$, $a = 20$, and $b = 15$. Round lengths of sides to the nearest tenth and angle measures to the nearest degree.

Use the Law of Sines to find c. We try to use exact values where possible, so we choose to use A in our second ratio instead of B, since B is rounded. We have no choice but to use the value for C that we calculated in the previous step.

$$\frac{c}{\sin C} = \frac{a}{\sin A}$$
$$\frac{c}{\sin 82°} = \frac{33}{\sin 57°}$$
$$c = \frac{33 \sin 82°}{\sin 57°} \approx 39.0$$

Thus, there is one triangle with $B \approx 41°$, $C \approx 82°$, and $c \approx 39.0$.

2b. Solve triangle ABC if $A = 50°$, $a = 10$, and $b = 20$.

We know two sides and the angle opposite one of the sides. This is an SSA case, where we first need to determine whether no triangle, exactly one triangle, or two triangles that satisfy the conditions exist.

We use the Law of Sines to find $\sin B$.

$$\frac{a}{\sin A} = \frac{b}{\sin B}$$
$$\frac{10}{\sin 50°} = \frac{20}{\sin B}$$
$$10 \sin B = 20 \sin 50°$$
$$\sin B = \frac{20 \sin 50°}{10} \approx 1.5321$$

Because the value of sine cannot exceed 1, there is no value of B for which $\sin B \approx 1.5321$. There is no triangle with the given measurements.

2b. Solve triangle ABC if $A = 30°$, $a = 10$, and $b = 40$.

2c. Solve triangle ABC if $A = 35°$, $a = 12$, and $b = 16$. Round lengths of sides to the nearest tenth and angle measures to the nearest degree.

We know two sides and the angle opposite one of the sides. This is an SSA case, where we first need to determine whether no triangle, exactly one triangle, or two triangles that satisfy the conditions exist.

We use the Law of Sines to find $\sin B$.

$$\frac{a}{\sin A} = \frac{b}{\sin B}$$
$$\frac{12}{\sin 35°} = \frac{16}{\sin B}$$
$$12\sin B = 16\sin 35°$$
$$\sin B = \frac{16\sin 35°}{12} \approx 0.7648$$

There are two angles between 0° and 180° that satisfy $\sin B \approx 0.7648$:

$B_1 \approx \sin^{-1} 0.7648 \approx 50°$ and
$B_2 \approx 180° - 50° = 130°$.

Note that the given angle, $A = 35°$, can be added to either of these angles without exceeding 180°, so we have two triangles. Find C_1 and C_2.

$$A + B_1 + C_1 = 180° \qquad A + B_2 + C_2 = 180°$$
$$35° + 50° + C_1 \approx 180° \qquad 35° + 130° + C_2 \approx 180°$$
$$C_1 + 85° \approx 180° \qquad C_2 + 165° \approx 180°$$
$$C_1 \approx 95° \qquad C_2 \approx 15°$$

Now find c_1 and c_2. As in Solved Problem #2a, we use exact values where possible.

$$\frac{c_1}{\sin C_1} = \frac{a}{\sin A} \qquad \frac{c_2}{\sin C_2} = \frac{a}{\sin A}$$
$$\frac{c_1}{\sin 95°} = \frac{12}{\sin 35°} \qquad \frac{c_2}{\sin 15°} = \frac{12}{\sin 35°}$$
$$c_1 = \frac{12\sin 95°}{\sin 35°} \qquad c_2 = \frac{12\sin 15°}{\sin 35°}$$
$$c_1 \approx 20.8 \qquad c_2 \approx 5.4$$

In summary, there are two possible triangles with the given measurements. One has $B_1 \approx 50°$, $C_1 \approx 95°$, and $c_1 \approx 20.8$. The other has $B_2 \approx 130°$, $C_2 \approx 15°$, and $c_2 \approx 5.4$.

2c. Solve triangle ABC if $A = 60°$, $a = 16$, and $b = 18$. Round lengths of sides to the nearest tenth and angle measures to the nearest degree.

Objective #3: Find the area of an oblique triangle using the sine function.

✔ Solved Problem #3

3. Find the area of a triangle having two sides of lengths 8 meters and 12 meters and an included angle of 135°. Round to the nearest square meter.

$$\text{Area } = \frac{1}{2}(8)(12)\sin 135° \approx 34 \text{ m}^2$$

Pencil Problem #3

3. Find the area of a triangle having two sides of lengths 20 feet and 40 feet and an included angle of 48°. Round to the nearest square foot.

Objective #4: Solve applied problems using the Law of Sines.

✔ Solved Problem #4

4. Two fire-lookout stations are 13 miles apart, with station B directly east of station A. Both stations spot a fire. The bearing of the fire from station A is N35°E and the bearing of the fire from station B is N49°W. How far, to the nearest tenth of a mile, is the fire from station B?

Refer to the figure. Note that the angle measurements given in the problem are not interior angles of the triangle shown. To find the respective interior angles, we find the complements of the given angles: $A = 90° - 35° = 55°$ and $B = 90° - 49° = 41°$.

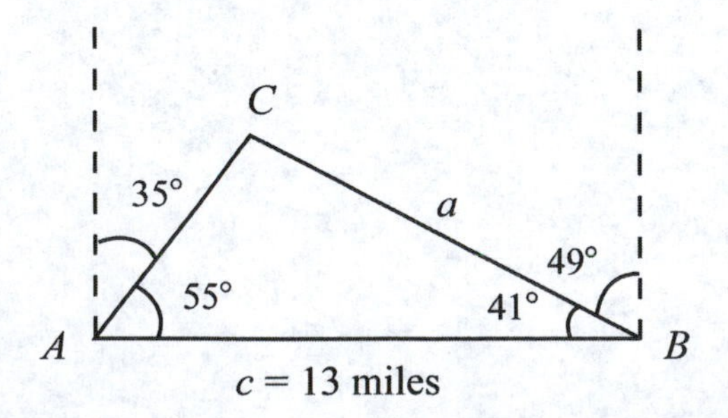

Pencil Problem #4

4. Two fire-lookout stations are 10 miles apart, with station B directly east of station A. Both stations spot a fire. The bearing of the fire from station A is N25°E and the bearing of the fire from station B is N56°W. How far, to the nearest tenth of a mile, is the fire from station B?

The fire is at C. We need to find the distance of the fire from B. This distance is labeled a in the figure, since it is opposite angle A. The only known side is c, so we need to find angle C before we can use the Law of Sines.

$$C = 180° - 55° - 41° = 84°$$

Now we use the Law of Sines to find a.

$$\frac{a}{\sin A} = \frac{c}{\sin C}$$
$$\frac{a}{\sin 55°} = \frac{13}{\sin 84°}$$
$$a = \frac{13 \sin 55°}{\sin 84°}$$
$$a \approx 10.7$$

The fire is approximately 10.7 miles from station B.

Answers for Pencil Problems *(Textbook Exercise references in parentheses)*:

Note: It is possible to have slightly different answers due to rounding in intermediate steps.

1a. $C = 95°$, $b \approx 81.0$, $c \approx 134.1$ *(6.1 #5)* **1b.** $C = 50°$, $a \approx 7.1$, $b \approx 7.1$ *(6.1 #15)*

2a. one triangle: $B \approx 29°$, $C \approx 111°$, $c \approx 29.0$ *(6.1 #17)* **2b.** no triangle *(6.1 #23)*
2c. two triangles: $B_1 \approx 77°$, $C_1 \approx 43°$, $c_1 \approx 12.6$ and $B_2 \approx 103°$, $C_2 \approx 17°$, $c_2 \approx 5.4$ *(6.1 #25)*

3. 297 square feet *(6.1 #33)*

4. approximately 9.2 miles *(6.1 #47)*

Section 6.2
The Law of Cosines

JURASSIC MATH

Fossil footprints allow scientists to study the movement of dinosaurs. In this section, you will learn the Law of Cosines which will help you determine whether a dinosaur was an efficient walker from its footprints.
Like the Law of Sines, the Law of Cosines is used to solve oblique triangles. The information given in a problem will determine which one you use. Be sure to pay attention not only to how to use the formulas but also to which formula to use.

Objective #1: Use the Law of Cosines to solve oblique triangles.

✔ *Solved Problem #1*

1a. Solve the triangle shown in the figure. Round lengths of sides to the nearest tenth and angles to the nearest degree.

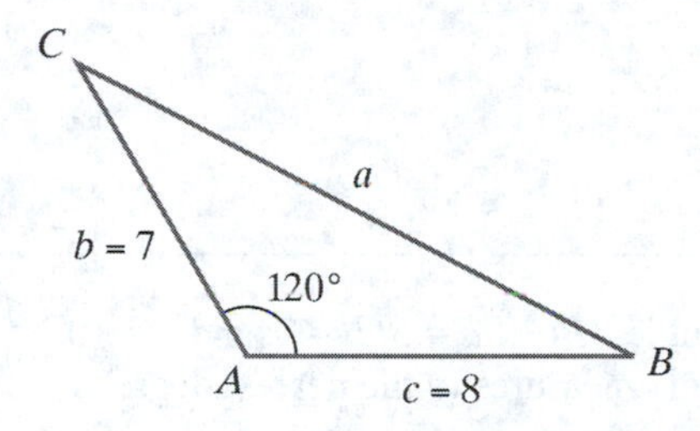

We know two sides and the included angle. This is an SAS triangle. In this case, we use the Law of Cosines to find the missing side, the side opposite the known angle, first. Since this is side a, we use the Law of Cosines in the form $a^2 = b^2 + c^2 - 2ab\cos A$, with $b = 7$, $c = 8$, and $A = 120°$. Use a calculator in degree mode to simplify the right side.

$$a^2 = 7^2 + 8^2 - 2(7)(8)\cos 120° = 169$$
$$a = \sqrt{169} = 13$$

Now we use the Law of Sines to find the angle opposite the shorter of the two given sides, $b = 7$ and $c = 8$. Since b is shorter, we find angle B.

(continued on next page)

Pencil Problem #1

1a. Solve the triangle shown in the figure. Round lengths of sides to the nearest tenth and angles to the nearest degree.

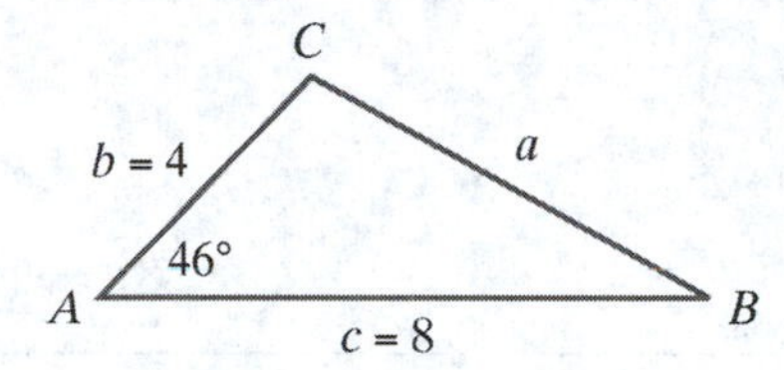

$$\frac{b}{\sin B} = \frac{a}{\sin A}$$
$$\frac{7}{\sin B} = \frac{13}{\sin 120°}$$
$$13\sin B = 7\sin 120°$$
$$\sin B = \frac{7\sin 120°}{13} \approx 0.4663$$
$$B \approx \sin^{-1} 0.4663 \approx 28°$$

(Since the angle opposite the shorter side must be acute, we do not need to consider a value for B in quadrant II.)

Now we find angle C.

$$C \approx 180° - 120° - 28° = 32°$$

In summary, $a = 13$, $B \approx 28°$, and $C \approx 32°$.

1b. Solve triangle ABC if $a = 8$, $b = 10$, and $c = 5$. Round angle measures to the nearest degree.

In this case, we know the lengths of all three sides. This is an SSS situation. We use the Law of Cosines first to find the angle opposite the longest side. (If the triangle has an obtuse angle, we want to find it first.) The longest side is $b = 10$, so we find angle B first. We use the form of the Law of Cosines that contains B. Begin by solving the formula for $\cos B$.

$$b^2 = a^2 + c^2 - 2ac\cos B$$
$$2ac\cos B = a^2 + c^2 - b^2$$
$$\cos B = \frac{a^2 + c^2 - b^2}{2ac}$$

Now substitute the given values, $a = 8$, $b = 10$, and $c = 5$, on the right side and simplify.

$$\cos B = \frac{8^2 + 5^2 - 10^2}{2(8)(5)} = -\frac{11}{80}$$

Now solve for B.

$$B = \cos^{-1}\left(-\frac{11}{80}\right) \approx 98°$$

(continued on next page)

1b. Solve triangle ABC if $a = 5$, $b = 7$, and $c = 10$. Round angle measures to the nearest degree.

The two remaining angles are both acute. We will use the Law of Sines to find one of them. We will find angle A.

$$\frac{a}{\sin A}=\frac{b}{\sin B}$$

$$\frac{8}{\sin A}=\frac{10}{\sin 98°}$$

$$10\sin A = 8\sin 98°$$

$$\sin A=\frac{8\sin 98°}{10}\approx 0.7922$$

$$B\approx \sin^{-1}0.7922\approx 52°$$

Finally, we find C.

$$C\approx 180°-52°-98°=30°$$

In summary, $A\approx 52°$, $B\approx 98°$, and $C\approx 30°$.

Objective #2: Solve applied problems using the Law of Cosines.

✔ *Solved Problem #2*

2. Two airplanes leave an airport at the same time on different runways. One flies directly north at 400 miles per hour. The other airplane flies on a bearing of N75°E at 350 miles per hour. How far apart will the airplanes be after two hours?

Refer to the figure. The first airplane has flown 800 miles in two hours, and the second has flown 700 miles. The angle between their paths is 75°. We are asked to find the distance between the planes, which is the length of the side opposite the 75° angle.

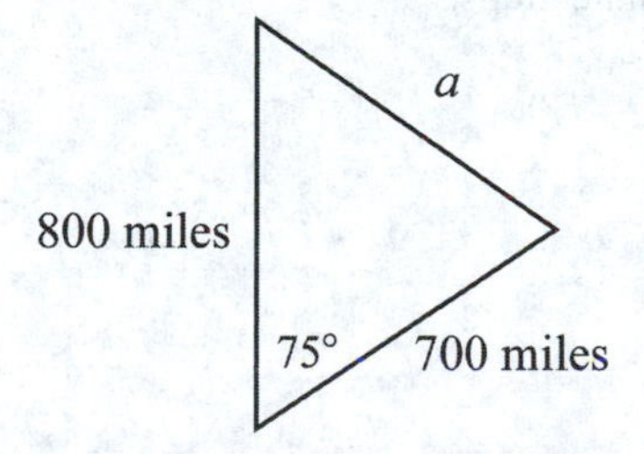

Since we know two sides and the included angle (SAS), we use the Law of Cosines. We substitute $b=800$, $c=700$, and $A=75°$ into the right side of the formula and simplify with a calculator.

$$a^2=b^2+c^2-2ab\cos A$$

$$a^2=800^2+700^2-2(800)(700)\cos 75°\approx 840{,}123$$

$$a\approx\sqrt{840{,}123}\approx 917$$

The planes are approximately 917 miles apart after two hours.

Pencil Problem #2

2. Find the distance across the lake from A to C, to the nearest yard, using the measurements shown in the figure.

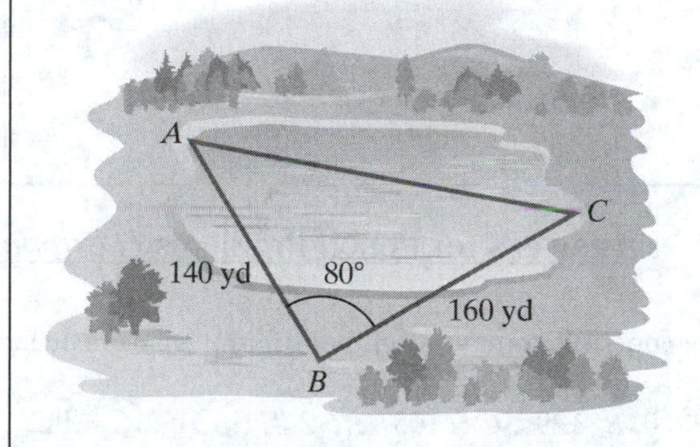

Objective #3: Use Heron's formula to find the area of a triangle.

✔ ***Solved Problem #3***

3. Find the area of a triangle with $a = 6$ meters, $b = 16$ meters, and $c = 18$ meters. Round to the nearest meter.

To use Heron's formula, we first find s, one-half the perimeter.

$$s = \frac{1}{2}(a+b+c) = \frac{1}{2}(6+16+18) = 20$$

Now we use Heron's formula to find the area.

$$\begin{aligned} \text{Area} &= \sqrt{s(s-a)(s-b)(s-c)} \\ &= \sqrt{20(20-6)(20-16)(20-18)} \\ &= \sqrt{2240} \approx 47 \text{ m}^2 \end{aligned}$$

Pencil Problem #3

3. Find the area of a triangle with $a = 11$ yards, $b = 9$ yards, and $c = 7$ yards. Round to the nearest square yard.

Answers for Pencil Problems *(Textbook Exercise references in parentheses)*:

Note: It is possible to have slightly different answers due to rounding in intermediate steps.

1a. $a \approx 6.0, B \approx 29°, C \approx 105°$ *(6.2 #1)* **1b.** $C \approx 112°, A \approx 28°, B \approx 40°$ *(6.2 #17)*

2. approximately 193 yards *(6.2 #41)*

3. 31 square yards *(6.2 #29)*

Section 6.3
Polar Coordinates

Getting a New Perspective

We have worked extensively in the Cartesian coordinate system, plotting points, graphing equations, and using the properties of the Cartesian plane to investigate functions and solve problems. In this section, we introduce a different coordinate system that will give you a new perspective of the plane. In this system, points do not have unique ordered pairs associated with them and some complicated-looking graphs have very simple equations.

Objective #1: Plot points in the polar coordinate system.

✔ *Solved Problem #1*

1a. Plot the point with polar coordinates (3, 315°).

Draw an angle measuring 315° in standard position. Then plot a point 3 units from the origin (pole) along the terminal side of the angle.

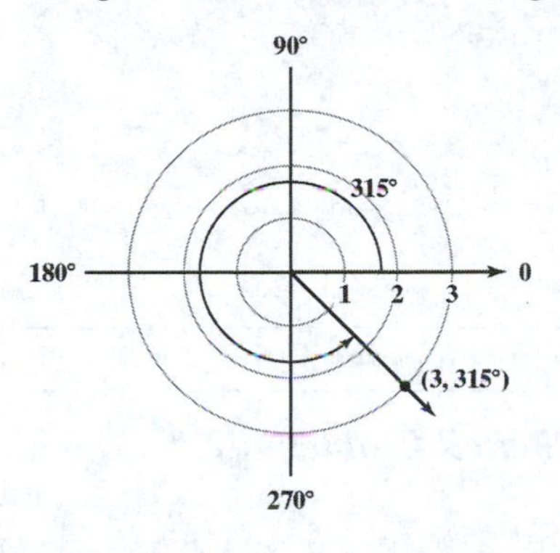

1b. Plot the point with polar coordinates $(-2, \pi)$.

Draw an angle measuring π radians in standard position. Then plot a point 2 units from the origin (pole) along a line in the opposite direction from the terminal side of the angle. Since the terminal side of the angle points to the left, this means that we move 2 units to the right.

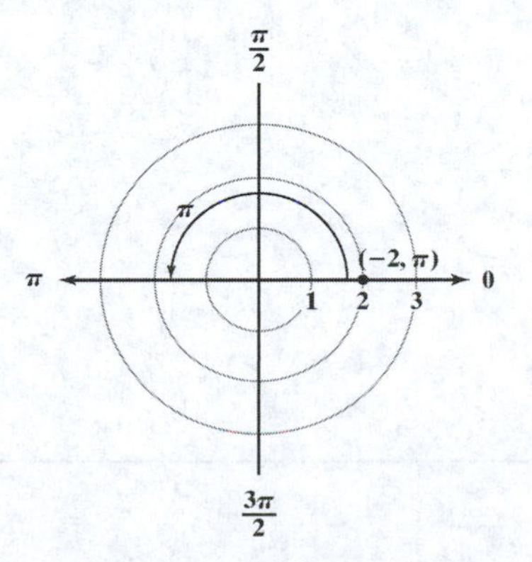

✎ *Pencil Problem #1*

1a. Plot the point with polar coordinates (2, 45°).

1b. Plot the point with polar coordinates $(-1, \pi)$.

1c. Plot the point with polar coordinates $\left(-1, -\frac{\pi}{2}\right)$.

Draw an angle measuring $-\frac{\pi}{2}$ radians in standard position. Then plot a point 1 unit from the origin (pole) along a line in the opposite direction from the terminal side of the angle. Since the terminal side of the angle points down, this means that we move 2 units up.

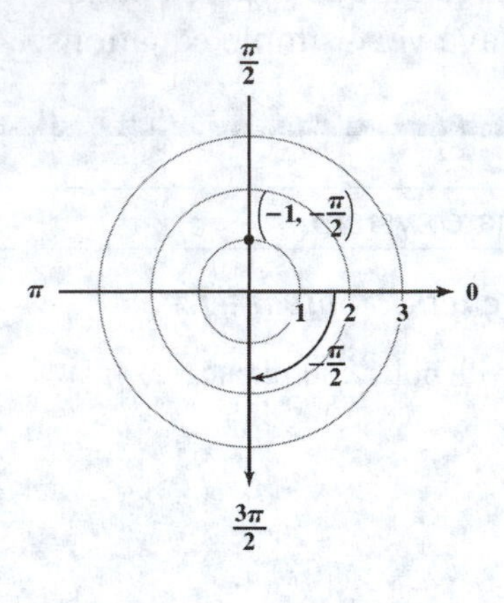

1c. Plot the point with polar coordinates $\left(-2, -\frac{\pi}{2}\right)$.

Objective #2: Find multiple sets of polar coordinates for a given point.

✔ *Solved Problem #2*

2a. Find a representation of $\left(5, \frac{\pi}{4}\right)$ in which r is positive and $2\pi < \theta < 4\pi$.

Add 2π to the angle and do not change r.

$$\left(5, \frac{\pi}{4}\right) = \left(5, \frac{\pi}{4} + 2\pi\right) = \left(5, \frac{\pi}{4} + \frac{8\pi}{4}\right) = \left(5, \frac{9\pi}{4}\right)$$

Pencil Problem #2

2a. Find a representation of $\left(10, \frac{3\pi}{4}\right)$ in which r is positive and $2\pi < \theta < 4\pi$.

2b. Find a representation of $\left(5, \frac{\pi}{4}\right)$ in which r is negative and $0 < \theta < 2\pi$.

Add π to the angle and replace r with $-r$.

$$\left(5, \frac{\pi}{4}\right) = \left(-5, \frac{\pi}{4} + \pi\right) = \left(-5, \frac{\pi}{4} + \frac{4\pi}{4}\right) = \left(-5, \frac{5\pi}{4}\right)$$

2b. Find a representation of $\left(10, \frac{3\pi}{4}\right)$ in which r is negative and $0 < \theta < 2\pi$.

2c. Find a representation of $\left(5, \frac{\pi}{4}\right)$ in which r is positive and $-2\pi < \theta < 0$.

Subtract 2π from the angle and do not change r.

$$\left(5, \frac{\pi}{4}\right) = \left(5, \frac{\pi}{4} - 2\pi\right) = \left(5, \frac{\pi}{4} - \frac{8\pi}{4}\right) = \left(5, -\frac{7\pi}{4}\right)$$

2c. Find a representation of $\left(10, \frac{3\pi}{4}\right)$ in which r is positive and $-2\pi < \theta < 0$.

Objective #3: Convert a point from polar to rectangular coordinates.

3a. Find the rectangular coordinates of the point with polar coordinates $(3, \pi)$.

$x = r\cos\theta = 3\cos\pi = 3(-1) = -3$

$y = r\sin\theta = 3\sin\pi = 3(0) = 0$

The rectangular coordinates are $(-3, 0)$.

3a. Find the rectangular coordinates of the point with polar coordinates $\left(2, \frac{\pi}{3}\right)$.

3b. Find the rectangular coordinates of the point with polar coordinates $\left(-10, \frac{\pi}{6}\right)$.

$$x = r\cos\theta = -10\cos\frac{\pi}{6} = -10\left(\frac{\sqrt{3}}{2}\right) = -5\sqrt{3}$$

$$y = r\sin\theta = -10\sin\frac{\pi}{6} = -10\left(\frac{1}{2}\right) = -5$$

The rectangular coordinates are $(-5\sqrt{3},\ -5)$.

3b. Find the rectangular coordinates of the point with polar coordinates $\left(-4, \frac{\pi}{2}\right)$.

Objective #4: Convert a point from rectangular to polar coordinates.

✔ *Solved Problem #4*

4a. Find polar coordinates of the point with rectangular coordinates $(1,\ -\sqrt{3})$.

The point $(1,\ -\sqrt{3})$ is in quadrant IV. Find r.

$$r = \sqrt{x^2 + y^2} = \sqrt{1^2 + (-\sqrt{3})^2} = \sqrt{1+3} = \sqrt{4} = 2$$

Find θ. Since $\tan^{-1}\sqrt{3} = \frac{\pi}{3}$ and θ lies in quadrant IV,

$$\theta = 2\pi - \frac{\pi}{3} = \frac{6\pi}{3} - \frac{\pi}{3} = \frac{5\pi}{3}.$$

The polar coordinates are $\left(2, \frac{5\pi}{3}\right)$.

4a. Find polar coordinates of the point with rectangular coordinates $(-2, 2)$.

4b. Find polar coordinates of the point with rectangular coordinates (0, –4).

The point (0, –4) lies on the negative y-axis 4 units below the origin. Thus, $r = 4$ and $\theta = \frac{3\pi}{2}$.

The polar coordinates are $\left(4, \frac{3\pi}{2}\right)$.

4b. Find polar coordinates of the point with rectangular coordinates (5, 0).

Objective #5: Convert an equation from rectangular to polar coordinates.

✔ Solved Problem #5

5a. Covert $3x - y = 6$ to a polar equation.

Replace x with $r\cos\theta$ and y with $r\sin\theta$ and solve for r.

$$3x - y = 6$$
$$3r\cos\theta - r\sin\theta = 6$$
$$r(3\cos\theta - \sin\theta) = 6$$
$$r = \frac{6}{3\cos\theta - \sin\theta}$$

✎ Pencil Problem #5 ✎

5a. Covert $3x + y = 7$ to a polar equation.

5b. Covert $x^2+(y+1)^2=1$ to a polar equation.

Replace x with $r\cos\theta$ and y with $r\sin\theta$ and solve for r.

$$x^2+(y+1)^2=1$$
$$(r\cos\theta)^2+(r\sin\theta+1)^2=1$$
$$r^2\cos^2\theta+r^2\sin^2\theta+2r\sin\theta+1=1$$
$$r^2(\cos^2\theta+\sin^2\theta)+2r\sin\theta=0$$
$$r^2+2r\sin\theta=0$$
$$r(r+2\sin\theta)=0$$
$$r=0 \quad \text{or} \quad r+2\sin\theta=0$$
$$r=-2\sin\theta$$

The graph of $r=0$ is the pole. The graph of $r=-2\sin\theta$ also includes the pole, so it is not necessary to include the equation $r=0$.

The polar equation is $r=-2\sin\theta$.

5b. Covert $(x-2)^2+y^2=4$ to a polar equation.

Objective #6: Convert an equation from polar to rectangular coordinates.

✔ *Solved Problem #6*

6a. Convert $r=4$ to a rectangular equation.

Square each side in anticipation of using $r^2=x^2+y^2$.

$$r=4$$
$$r^2=4^2$$
$$x^2+y^2=16$$

The rectangular equation is $x^2+y^2=16$, which we recognize as the equation of a circle centered at the origin with radius 4.

Pencil Problem #6

6a. Convert $r=8$ to a rectangular equation.

6b. Convert $\theta = \frac{3\pi}{4}$ to a rectangular equation.

Take the tangent of each side and use $\tan\theta = \frac{y}{x}$.

Then solve for y.

$$\tan\theta = \tan\frac{3\pi}{4}$$

$$\frac{y}{x} = -1$$

$$y = -x$$

The rectangular equation is $y = -x$, which we recognize as the equation of a line passing through the origin with slope -1.

6b. Convert $\theta = \frac{\pi}{2}$ to a rectangular equation.

6c. Convert $r = -2\sec\theta$ to a rectangular equation.

Use a reciprocal identity to rewrite the secant in terms of cosine. Multiply each side by $\cos\theta$ and replace $r\cos\theta$ with x.

$$r = -2\sec\theta$$

$$r = \frac{-2}{\cos\theta}$$

$$r\cos\theta = -2$$

$$x = -2$$

The rectangular equation is $x = -2$, which we recognize as the equation of a vertical line with x-intercept -2.

6c. Convert $r = 4\csc\theta$ to a rectangular equation.

6d. Convert $r = 10\sin\theta$ to a rectangular equation.

Multiply both sides by r and then replace r^2 with $x^2 + y^2$ and $r\sin\theta$ with y.

$$r = 10\sin\theta$$

$$r^2 = 10r\sin\theta$$

$$x^2 + y^2 = 10y$$

$$x^2 + y^2 - 10y = 0$$

$$x^2 + (y^2 - 10y + 25) = 25$$

$$x^2 + (y-5)^2 = 25$$

The rectangular equation is $x^2 + (y-5)^2 = 25$, which we recognize as the equation of a circle centered at (0, 5) with radius 5.

6d. Convert $r = 12\cos\theta$ to a rectangular equation.

Answers for Pencil Problems *(Textbook Exercise references in parentheses)*:

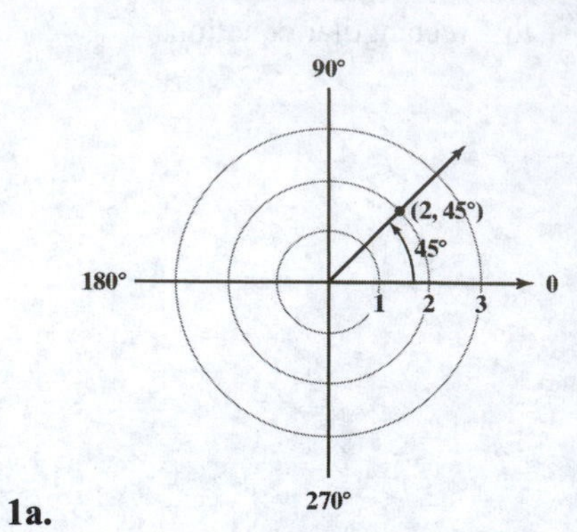

1a. *(6.3 #11)*

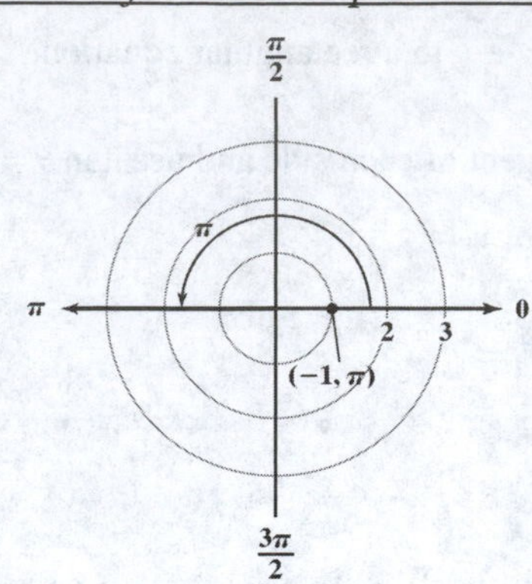

1b. *(6.3 #17)*

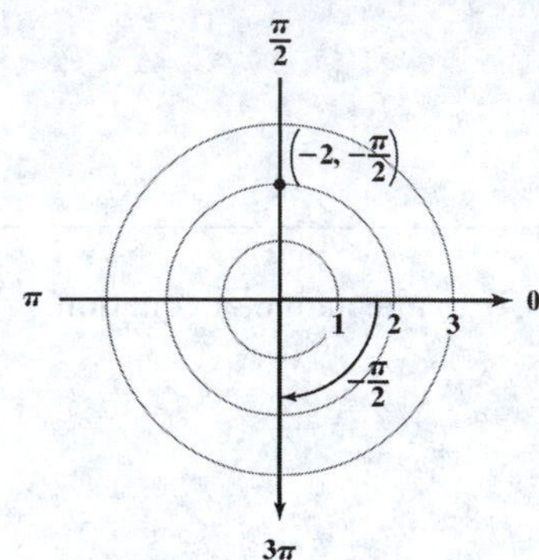

1c. *(6.3 #19)*

2a. $\left(10, \dfrac{11\pi}{4}\right)$ **2b.** $\left(-10, \dfrac{7\pi}{4}\right)$ **2c.** $\left(10, -\dfrac{5\pi}{4}\right)$ *(6.3 #23)*

3a. $(1, \sqrt{3})$ *(6.3 #35)* **3b.** $(0, -4)$ *(6.3 #37)*

4a. $\left(2\sqrt{2}, \dfrac{3\pi}{4}\right)$ *(6.3 #41)* **4b.** $(5, 0)$ *(6.3 #47)*

5a. $r = \dfrac{7}{3\cos\theta + \sin\theta}$ *(6.3 #49)* **5b.** $r = 4\cos\theta$ *(6.3 #55)*

6a. $x^2 + y^2 = 64$ *(6.3 #59)* **6b.** $x = 0$ *(6.3 #61)*

6c. $y = 4$ *(6.3 #65)* **6d.** $(x-6)^2 + y^2 = 36$ *(6.3 #69)*

Section 6.4
Graphs of Polar Equations

Snails, Roses, and Propellers?

In the previous section, we looked at polar equations representing lines and circles that could be easily converted to rectangular form. In this section, we will graph polar equations without converting them. These polar equations will be relatively simple but in some cases extremely difficult to represent or graph in rectangular coordinates. The graphs will have interesting shapes that may remind you of snails, roses, and propellers.

Objective #1: Use point plotting to graph polar equations.

✔ *Solved Problem #1*

1. Graph $r = 4\sin\theta$ with θ in radians. Use multiples of $\frac{\pi}{6}$ from 0 to π to generate coordinates for points.

We begin by creating a table of coordinates for some points on the graph.

θ	$r = 4\sin\theta$	(r,θ)
0	$r = 4\sin 0 = 4\cdot 0 = 0$	$(0,0)$
$\frac{\pi}{6}$	$r = 4\sin\frac{\pi}{6} = 4\cdot\frac{1}{2} = 2$	$\left(2,\frac{\pi}{6}\right)$
$\frac{\pi}{3}$	$r = 4\sin\frac{\pi}{3} = 4\cdot\frac{\sqrt{3}}{2} = 2\sqrt{3}$	$\left(2\sqrt{3},\frac{\pi}{3}\right)$
$\frac{\pi}{2}$	$r = 4\sin\frac{\pi}{2} = 4\cdot 1 = 4$	$\left(4,\frac{\pi}{2}\right)$
$\frac{2\pi}{3}$	$r = 4\sin\frac{2\pi}{3} = 4\cdot\frac{\sqrt{3}}{2} = 2\sqrt{3}$	$\left(2\sqrt{3},\frac{2\pi}{3}\right)$
$\frac{5\pi}{6}$	$r = 4\sin\frac{5\pi}{6} = 4\cdot\frac{1}{2} = 2$	$\left(2,\frac{5\pi}{6}\right)$
π	$r = 4\sin\pi = 4\cdot 0 = 0$	$(0,\pi)$

The third column of the table lists the polar coordinates of several points on the graph. Note that the first and last points both represent the pole.

(continued on next page)

✎ *Pencil Problem #1* ✎

1. Graph $r = 2\cos\theta$ with θ in radians. Use multiples of $\frac{\pi}{6}$ from 0 to π to generate coordinates for points.

Plot the points and connect them with a smooth curve. The graph is a circle of radius 2 centered at (0, 2) in rectangular coordinates.

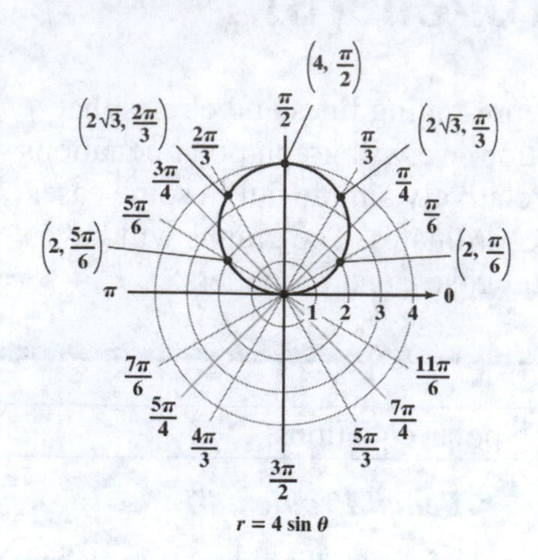

Objective #2: Use symmetry to graph polar equations.

✔ *Solved Problem #2*

2a. Check for symmetry and then graph the polar equation: $r = 1 + \cos\theta$.

Polar Axis: Replace θ with $-\theta$ and note that $\cos(-\theta) = \cos\theta$ since cosine is an even function.

$r = 1 + \cos(-\theta)$

$r = 1 + \cos\theta$

The result is equivalent to the original equation. The graph is symmetric with respect to the polar axis.

The Line $\theta = \frac{\pi}{2}$: Replace θ with $-\theta$ and r with $-r$.

$-r = 1 + \cos(-\theta)$

$-r = 1 + \cos\theta$

$r = -1 - \cos\theta$

The result is not equivalent to the original equation. The graph may or may not be symmetric with respect to the line $\theta = \frac{\pi}{2}$.

(continued on next page)

Pencil Problem #2

2a. Check for symmetry and then graph the polar equation: $r = 1 - \sin\theta$.

The Pole: Replace r with $-r$.
$-r = 1 + \cos\theta$

$r = -1 - \cos\theta$

The result is not equivalent to the original equation. The graph may or may not be symmetric with respect to the pole.

Because the period of cosine is 2π, we only need to consider values of θ between 0 and 2π. Since we have symmetry with respect to the polar axis, we can further restrict ourselves to values of θ between 0 and π. We will plot a few points, connect them, and then complete the graph by reflecting this portion about the polar axis.

θ	0	$\frac{\pi}{6}$	$\frac{\pi}{3}$	$\frac{\pi}{2}$	$\frac{2\pi}{3}$	$\frac{5\pi}{6}$	π
r	2	1.87	$\frac{3}{2}$	1	$\frac{1}{2}$	0.13	0

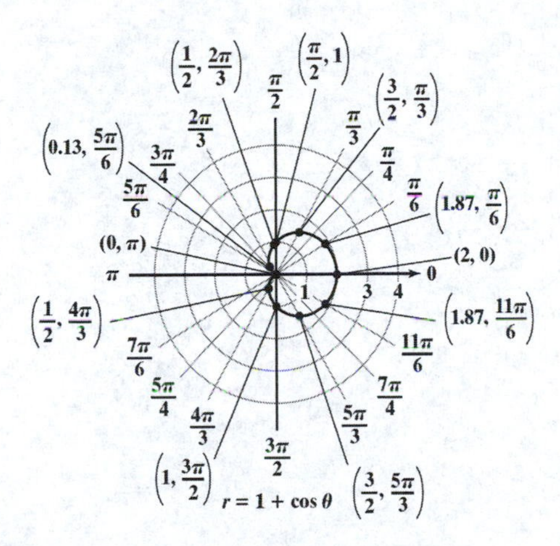

The shape of this graph is called a cardiod.

2b. Check for symmetry and then graph the polar equation: $r = 1 - 2\sin\theta$.

In the tests below, remember that the sine function is odd, so that $\sin(-\theta) = -\sin\theta$.

Polar Axis: $r = 1 - 2\sin(-\theta)$

$r = 1 + 2\sin\theta$

The Line $\theta = \frac{\pi}{2}$: $-r = 1 - 2\sin(-\theta)$

$-r = 1 + 2\sin\theta$

$r = -1 - 2\sin\theta$

The Pole: $-r = 1 + 2\sin\theta$

$r = -1 - 2\sin\theta$

None of the tests results in an equivalent equation, so the graph may or may not have each of these types of symmetry.

Because we do not have any known symmetry, we must consider values of θ from 0 to 2π.

θ	0	$\frac{\pi}{6}$	$\frac{\pi}{2}$	$\frac{5\pi}{6}$	π	$\frac{7\pi}{6}$	$\frac{3\pi}{2}$	$\frac{11\pi}{6}$	2π
r	1	0	−1	0	1	2	3	2	1

Plot the points and connect them with a smooth curve. Be sure to connect them in the order in which they appear in the table from left to right.

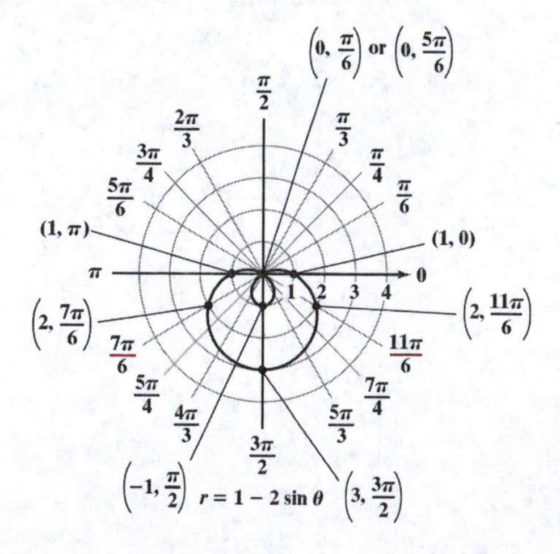

Notice that the graph does have symmetry with respect to the line $\theta = \frac{\pi}{2}$, even though the equation failed the test for this kind of symmetry. This graph is called a limaçon with an inner loop.

2b. Check for symmetry and then graph the polar equation: $r = 1 + 2\cos\theta$.

2c. Check for symmetry and then graph the polar equation: $r = 3\cos 2\theta$.

Polar Axis: $r = 3\cos 2(-\theta)$

$r = 3\cos(-2\theta)$

$r = 3\cos 2\theta$

The Line $\theta = \dfrac{\pi}{2}$: $-r = 3\cos 2(-\theta)$

$-r = 3\cos(-2\theta)$

$-r = 3\cos 2\theta$

$r = -3\cos 2\theta$

The Pole: $-r = 3\cos 2\theta$

$r = -3\cos 2\theta$

Only the test for symmetry with respect to the polar axis results in an equivalent equation, so the graph has this type of symmetry. It may or may not have the other two types.

We draw the graph for $0 \le \theta \le \pi$ first and then reflect this portion about the polar axis to complete the graph.

θ	0	$\frac{\pi}{6}$	$\frac{\pi}{4}$	$\frac{\pi}{3}$	$\frac{\pi}{2}$	$\frac{2\pi}{3}$	$\frac{3\pi}{4}$	$\frac{5\pi}{6}$	π
r	3	$\frac{3}{2}$	0	$-\frac{3}{2}$	-3	$-\frac{3}{2}$	0	$\frac{3}{2}$	3

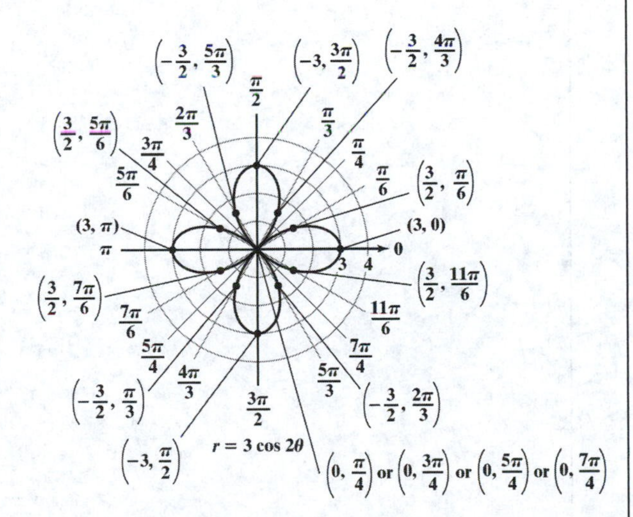

Notice that the graph has all three types of symmetry, even though it failed two of the three tests. The shape of the graph is called a rose curve with 4 petals.

2c. Check for symmetry and then graph the polar equation: $r = 4\sin 3\theta$.

2d. Check for symmetry and then graph the polar equation: $r^2 = 4\cos 2\theta$.

Polar Axis: $r^2 = 4\cos 2(-\theta)$

$r^2 = 4\cos(-2\theta)$

$r^2 = 4\cos 2\theta$

The Line $\theta = \frac{\pi}{2}$: $(-r)^2 = 4\cos 2(-\theta)$

$r^2 = 4\cos(-2\theta)$

$r^2 = 4\cos 2\theta$

The Pole: $(-r)^2 = 4\cos 2\theta$

$r^2 = 4\cos 2\theta$

All three tests result in the original equation. The graph has all three types of symmetry. We only need to plot points for $0 \le \theta \le \frac{\pi}{2}$. However, notice that for $\frac{\pi}{4} < \theta \le \frac{\pi}{2}$, $4\cos 2\theta$ is negative and the equation has no solutions. So we only need to consider $0 \le \theta \le \frac{\pi}{4}$.

Then we can reflect this portion of the graph about the polar axis and the line $\theta = \frac{\pi}{2}$, as necessary.

θ	0	$\frac{\pi}{6}$	$\frac{\pi}{4}$
r	± 2	$\pm\sqrt{2}$	0

Notice that the entries in the table represent five points, since for two of the angles we have two values for r.

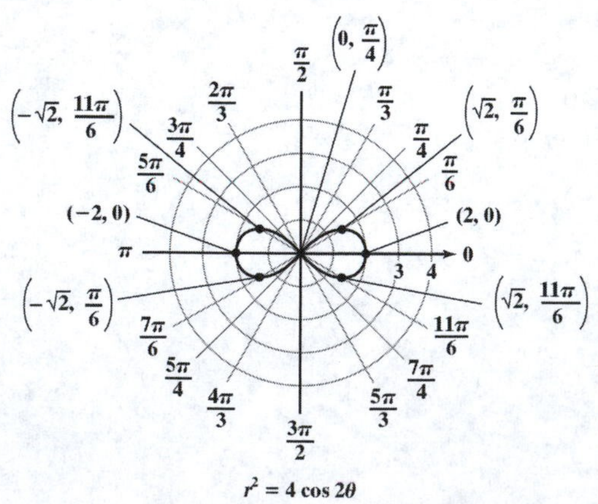

$r^2 = 4\cos 2\theta$

This propeller-shaped graph is called a lemniscate.

2d. Check for symmetry and then graph the polar equation: $r^2 = 9\cos 2\theta$.

Answers for Pencil Problems *(Textbook Exercise references in parentheses)*:

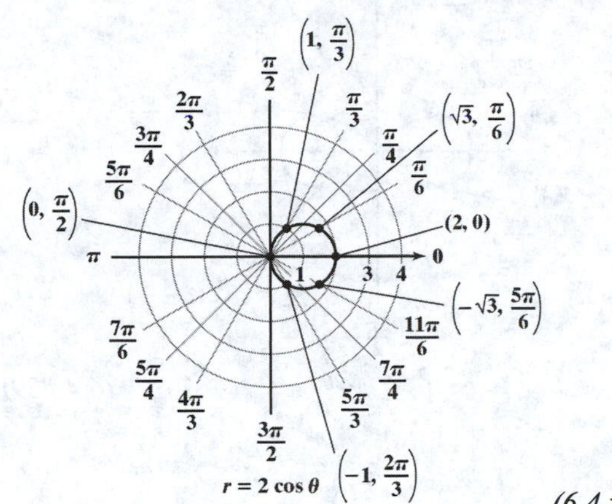

1. *(6.4 #13)*

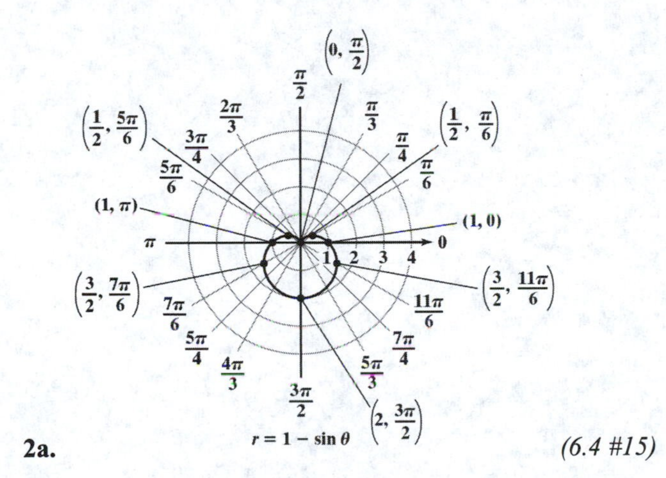

2a. *(6.4 #15)*

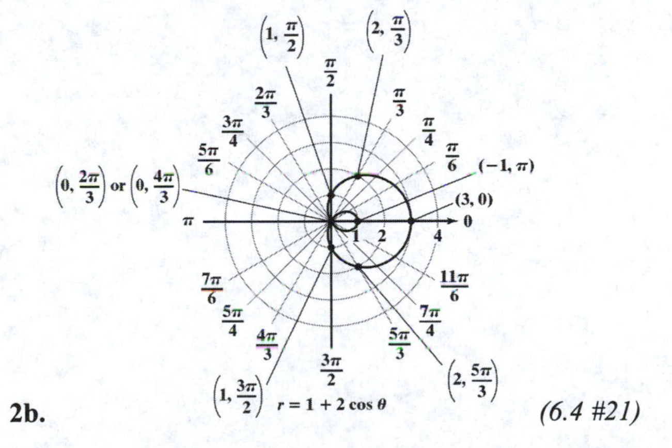

2b. *(6.4 #21)*

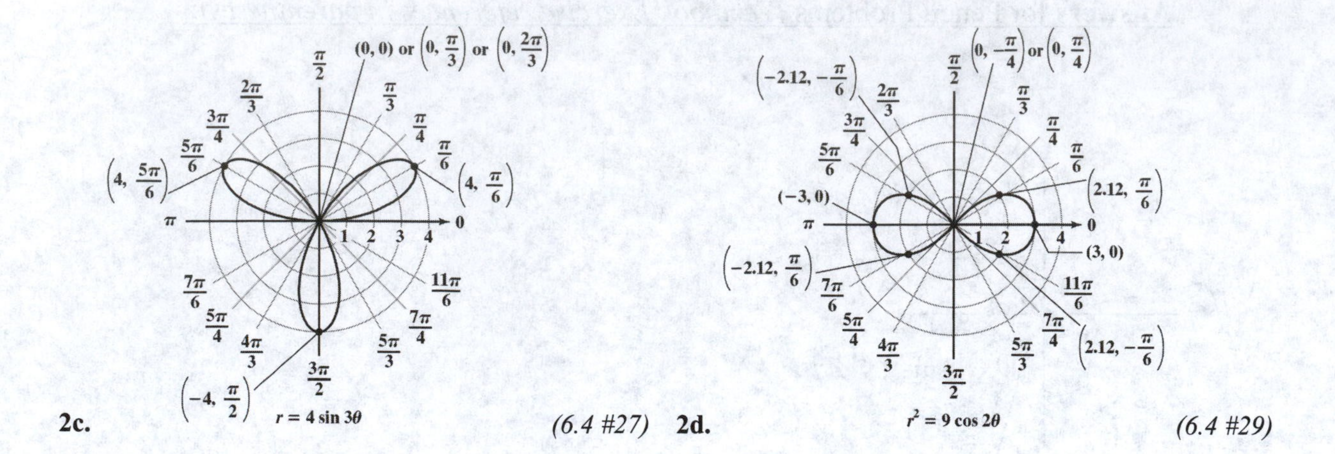

(0, 0) or (0, π/3) or (0, 2π/3)
π/2
2π/3
π/3
3π/4
π/4
5π/6
π/6
(4, 5π/6)
(4, π/6)
π
1 2 3 4
0
7π/6
11π/6
5π/4
7π/4
4π/3
5π/3
3π/2
(−4, π/2)
r = 4 sin 3θ
2c.
(6.4 #27)
(−2.12, −π/6)
(0, −π/4) or (0, π/4)
(−3, 0)
(2.12, π/6)
(3, 0)
(−2.12, π/6)
(2.12, −π/6)
r² = 9 cos 2θ
2d.
(6.4 #29)

Section 6.5
Complex Numbers in Polar Form; DeMoivre's Theorem

So We Have Finally Devolved into Chaos?

In the previous section, we saw how some complicated graphs had very simple equations in polar coordinates. In this section, we will see how writing complex numbers in polar form makes finding products, quotients, powers, and roots of complex numbers a simple task. In the Exercise Set, you will see how complex numbers are used in the study of seemingly random phenomena that are not actually random at all. Such phenomena are called "chaos" in mathematics.

Objective #1: Plot complex numbers in the complex plane.

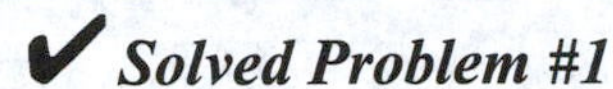

1a. Plot $z = 2 + 3i$ in the complex plane.

Move 2 units to the right along the real axis and then 3 units up parallel to the imaginary axis.

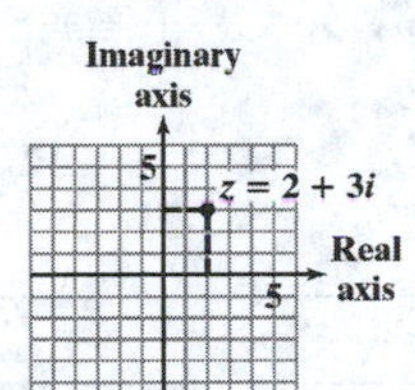

1b. Plot $z = -3 - 5i$ in the complex plane.

Move 3 units to the left along the real axis and then 5 units down parallel to the imaginary axis.

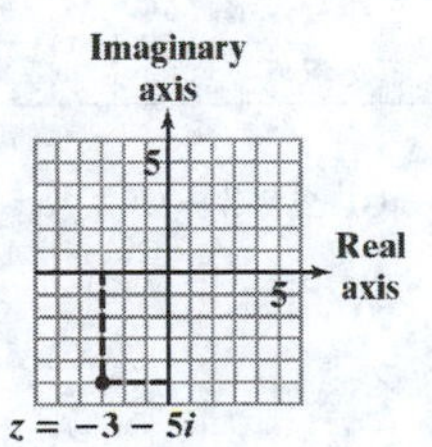

Pencil Problem #1

1a. Plot $z = 3 + 2i$ in the complex plane.

1b. Plot $z = -3 + 4i$ in the complex plane.

1c. Plot $z = -4$ in the complex plane.

Move 4 units to the left along the real axis.

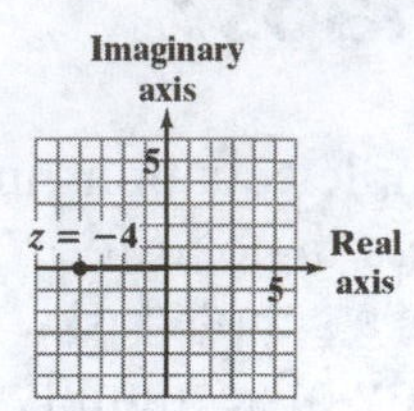

1c. Plot $z = 3$ in the complex plane.

1d. Plot $z = -i$ in the complex plane.

Move 1 unit down along the imaginary axis.

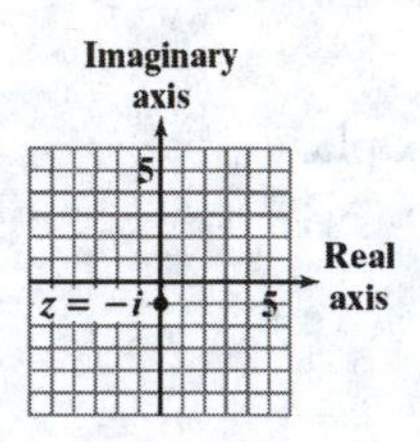

1d. Plot $z = 4i$ in the complex plane.

Objective #2: Find the absolute value of a complex number.

✔ *Solved Problem #2*

2a. Determine the absolute value of $z = 12 + 5i$.

$|z| = \sqrt{5^2 + 12^2} = \sqrt{25 + 144} = \sqrt{169} = 13$

The distance from the origin to the point $z = 12 + 5i$ is 13 units.

2b. Determine the absolute value of $z = 2 - 3i$.

$|z| = \sqrt{2^2 + (-3)^2} = \sqrt{4 + 9} = \sqrt{13}$

The distance from the origin to the point $z = 2 - 3i$ is $\sqrt{13}$ units.

Pencil Problem #2

2a. Determine the absolute value of $z = -3 + 4i$.

2b. Determine the absolute value of $z = 3 - i$.

Objective #3: Write complex numbers in polar form.

✔ *Solved Problem #3*

3. Plot $z = -1 - i\sqrt{3}$ in the complex plane. Then write z in polar form. Express the argument in radians.

To plot $z = -1 - i\sqrt{3}$, move 1 unit left and then $\sqrt{3} \approx 1.73$ units down

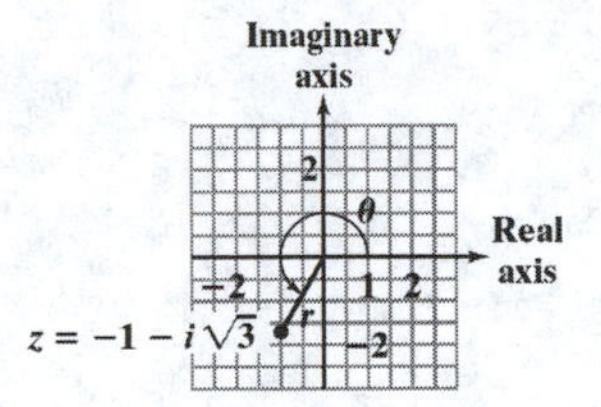

To write the number in polar form, first find the modulus, r. We use $r = \sqrt{a^2 + b^2}$ with $a = -1$ and $b = -\sqrt{3}$.

$$r = \sqrt{(-1)^2 + (-\sqrt{3})^2} = \sqrt{1+3} = \sqrt{4} = 2$$

To find θ, we use $\tan\theta = \dfrac{b}{a}$ with $a = -1$ and $b = -\sqrt{3}$.

$$\tan\theta = \frac{-\sqrt{3}}{-1} = \sqrt{3}$$

We know that $\tan\dfrac{\pi}{3} = \sqrt{3}$ and that θ lies in quadrant III, so

$$\theta = \pi + \frac{\pi}{3} = \frac{3\pi}{3} + \frac{\pi}{3} = \frac{4\pi}{3}$$

The polar form of $z = -1 - i\sqrt{3}$ is

$$z = r(\cos\theta + i\sin\theta) = 2\left(\cos\frac{4\pi}{3} + i\sin\frac{4\pi}{3}\right).$$

Pencil Problem #3

3. Plot $z = 2 + 2i$ in the complex plane. Then write z in polar form. Express the argument in radians.

Objective #4: Convert a complex number from polar to rectangular form.

✔ *Solved Problem #4*

4. Write $z = 4(\cos 30° + i\sin 30°)$ in rectangular form.

We substitute the exact values $\cos 30° = \frac{\sqrt{3}}{2}$ and $\sin 30° = \frac{1}{2}$ into $z = 4(\cos 30° + i\sin 30°)$ and simplify.

$$z = 4(\cos 30° + i\sin 30°) = 4\left(\frac{\sqrt{3}}{2} + \frac{1}{2}i\right) = 2\sqrt{3} + 2i$$

Pencil Problem #4

4. Write $z = 4(\cos 240° + i\sin 240°)$ in rectangular form.

Objective #5: Find products of complex numbers in polar form.

✔ *Solved Problem #5*

5. Find the product of the complex numbers
$z_1 = 6(\cos 40° + i\sin 40°)$ and
$z_2 = 5(\cos 20° + i\sin 20°)$.
Leave the answer in polar form.

Multiply the moduli and add the arguments.

$$\begin{aligned} z_1 z_2 &= [6(\cos 40° + i\sin 40°)][5(\cos 20° + i\sin 20°)] \\ &= (6 \cdot 5)[\cos(40° + 20°) + i\sin(40° + 20°)] \\ &= 30(\cos 60° + i\sin 60°) \end{aligned}$$

Pencil Problem #5

5. Find the product of the complex numbers
$z_1 = 3\left(\cos\frac{\pi}{5} + i\sin\frac{\pi}{5}\right)$ and
$z_2 = 4\left(\cos\frac{\pi}{10} + i\sin\frac{\pi}{10}\right)$.
Leave the answer in polar form.

Objective #6: Find quotients of complex numbers in polar form.

✔ *Solved Problem #6*

6. Find the quotient $\frac{z_1}{z_2}$ of the complex numbers

$z_1 = 50\left(\cos\frac{4\pi}{3} + i\sin\frac{4\pi}{3}\right)$ and

$z_2 = 5\left(\cos\frac{\pi}{3} + i\sin\frac{\pi}{3}\right)$. Leave the answer in polar form.

Divide the moduli and subtract the arguments.

$$\frac{z_1}{z_2} = \frac{50\left(\cos\frac{4\pi}{3} + i\sin\frac{4\pi}{3}\right)}{5\left(\cos\frac{\pi}{3} + i\sin\frac{\pi}{3}\right)}$$

$$= \frac{50}{5}\left[\cos\left(\frac{4\pi}{3} - \frac{\pi}{3}\right) + i\sin\left(\frac{4\pi}{3} - \frac{\pi}{3}\right)\right]$$

$$= 10\left(\cos\frac{3\pi}{3} + i\sin\frac{3\pi}{3}\right)$$

$$= 10(\cos\pi + i\sin\pi)$$

Pencil Problem #6

6. Find the quotient $\frac{z_1}{z_2}$ of the complex numbers

$z_1 = 20(\cos 75° + i\sin 75°)$ and

$z_2 = 4(\cos 25° + i\sin 25°)$. Leave the answer in polar form.

Objective #7: Find powers of complex numbers in polar form.

✔ *Solved Problem #7*

7a. Find $[2(\cos 30° + i\sin 30°)]^5$. Write the answer in rectangular form.

Raise the modulus to the 5th power and multiply the argument by 5. Then replace the trigonometric expressions with exact values and simplify.

$$[2(\cos 30° + i\sin 30°)]^5$$

$$= 2^5[\cos(5\cdot 30°) + i\sin(5\cdot 30°)]$$

$$= 32(\cos 150° + i\sin 150°)$$

$$= 32\left(-\frac{\sqrt{3}}{2} + \frac{1}{2}i\right)$$

$$= -16\sqrt{3} + 16i$$

Pencil Problem #7

7a. Find $[4(\cos 15° + i\sin 15°)]^3$. Write the answer in rectangular form.

7b. Find $(1+i)^4$ using DeMoivre's Theorem. Write the answer in rectangular form.

First write $1+i$ in polar form. This number is in the form $z = a+bi$ with $a = 1$ and $b = 1$. The modulus is

$$r = \sqrt{a^2+b^2} = \sqrt{1^2+1^2} = \sqrt{1+1} = \sqrt{2}.$$

The point lies in quadrant I and
$\tan\theta = \frac{b}{a} = \frac{1}{1} = 1.$

We know that $\tan\frac{\pi}{4} = 1$ and $\frac{\pi}{4}$ is a quadrant I angle, so $\theta = \frac{\pi}{4}.$

Thus, $1+i = \sqrt{2}\left(\cos\frac{\pi}{4} + i\sin\frac{\pi}{4}\right).$

Now use DeMoivre's Theorem.

$$\begin{aligned}(1+i)^4 &= \left[\sqrt{2}\left(\cos\frac{\pi}{4} + i\sin\frac{\pi}{4}\right)\right]^4 \\ &= (\sqrt{2})^4\left[\cos\left(4\cdot\frac{\pi}{4}\right) + i\sin\left(4\cdot\frac{\pi}{4}\right)\right] \\ &= 4(\cos\pi + i\sin\pi) \\ &= 4(-1+i\cdot 0) \\ &= -4\end{aligned}$$

7b. Find $(1+i)^5$ using DeMoivre's Theorem. Write the answer in rectangular form.

Objective #8: Find roots of complex numbers in polar form.

✔ Solved Problem #8

8. Find all the complex fourth roots of $16(\cos 60° + i\sin 60°)$. Write roots in polar form with θ in degrees.

We use DeMoivre's Theorem to find complex roots.

$$z_k = \sqrt[n]{r}\left[\cos\left(\frac{\theta + 360° \cdot k}{n}\right) + i\sin\left(\frac{\theta + 360° \cdot k}{n}\right)\right] \text{ for}$$

$k = 0, 1, 2, \ldots, n - 1$

The complex number $16(\cos 60° + i\sin 60°)$ is in polar form with $r = 16$ and $\theta = 60°$. Since we are looking for fourth roots, $n = 4$ and we evaluate the formula for $k = 0, 1, 2,$ and 3.

$$z_0 = \sqrt[4]{16}\left[\cos\left(\frac{60° + 360° \cdot 0}{4}\right) + i\sin\left(\frac{60° + 360° \cdot 0}{4}\right)\right]$$

$$= 2\left(\cos\frac{60°}{4} + i\sin\frac{60°}{4}\right) = 2(\cos 15° + i\sin 15°)$$

$$z_1 = \sqrt[4]{16}\left[\cos\left(\frac{60° + 360° \cdot 1}{4}\right) + i\sin\left(\frac{60° + 360° \cdot 1}{4}\right)\right]$$

$$= 2\left(\cos\frac{420°}{4} + i\sin\frac{420°}{4}\right) = 2(\cos 105° + i\sin 105°)$$

$$z_2 = \sqrt[4]{16}\left[\cos\left(\frac{60° + 360° \cdot 2}{4}\right) + i\sin\left(\frac{60° + 360° \cdot 2}{4}\right)\right]$$

$$= 2\left(\cos\frac{780°}{4} + i\sin\frac{780°}{4}\right) = 2(\cos 195° + i\sin 195°)$$

$$z_3 = \sqrt[4]{16}\left[\cos\left(\frac{60° + 360° \cdot 3}{4}\right) + i\sin\left(\frac{60° + 360° \cdot 3}{4}\right)\right]$$

$$= 2\left(\cos\frac{1140°}{4} + i\sin\frac{1140°}{4}\right) = 2(\cos 285° + i\sin 285°)$$

Pencil Problem #8

8. Find all the complex cube roots of $8(\cos 210° + i\sin 210°)$. Write roots in polar form with θ in degrees.

Answers for Pencil Problems *(Textbook Exercise references in parentheses)*:

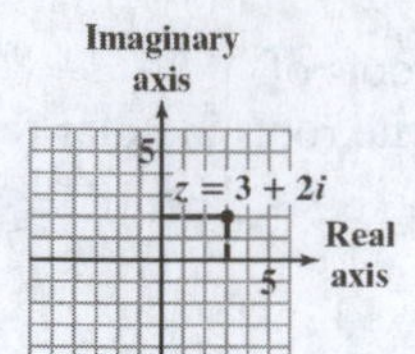

1a. *(6.5 #5)*

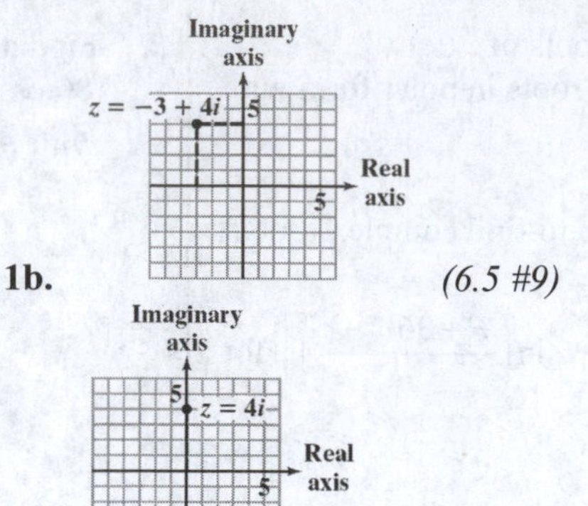

1b. *(6.5 #9)*

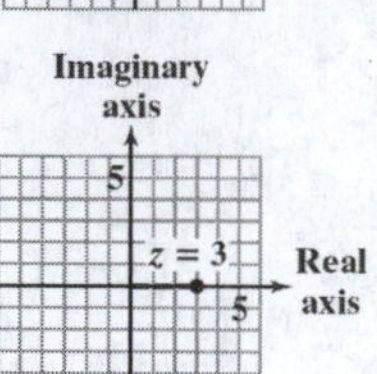

1c. *(6.5 #3)*

1d. *(6.5 #1)*

2a. 5 *(6.5 #29)* **2b.** $\sqrt{10}$ *(6.5 #29)*

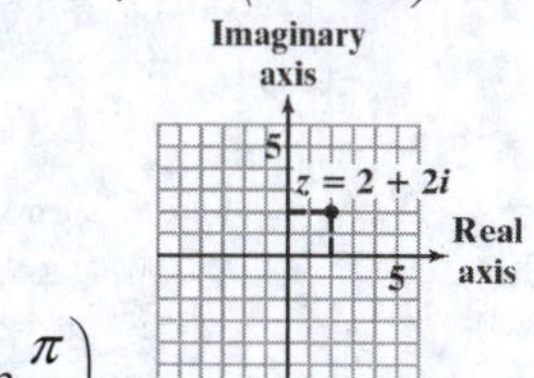

3. $2\sqrt{2}\left(\cos\dfrac{\pi}{4}+i\sin\dfrac{\pi}{4}\right)$ *(6.5 #11)*

4. $-2-2i\sqrt{3}$ *(6.5 #29)*

5. $12\left(\cos\dfrac{3\pi}{10}+i\sin\dfrac{3\pi}{10}\right)$ *(6.5 #39)*

6. $5(\cos 50°+i\sin 50°)$ *(6.5 #45)*

7a. $32\sqrt{2}+32i\sqrt{2}$ *(6.5 #53)* **7b.** $-4-4i$ *(6.5 #61)*

8. $z_0=2(\cos 70°+i\sin 70°); z_1=2(\cos 190°+i\sin 190°); z_2=2(\cos 310°+i\sin 310°)$ *(6.5 #67)*

Section 6.6
Vectors

The Force is With You

Whether you know it or not, forces are at work all around you. The force of gravity is acting on all objects pulling them downward. To keep an object from falling on the floor, you must apply an equal force in the opposite direction.

In this section, we represent such forces as vectors. We then use these vectors to solve problems involving how much force is required to pull a box up a ramp and how much force is required to keep a car from sliding down a hill.

Objective #1: Use magnitude and direction to show vectors are equal.

✔ *Solved Problem #1*

1. Refer to the figure. Show that **u** = **v**.

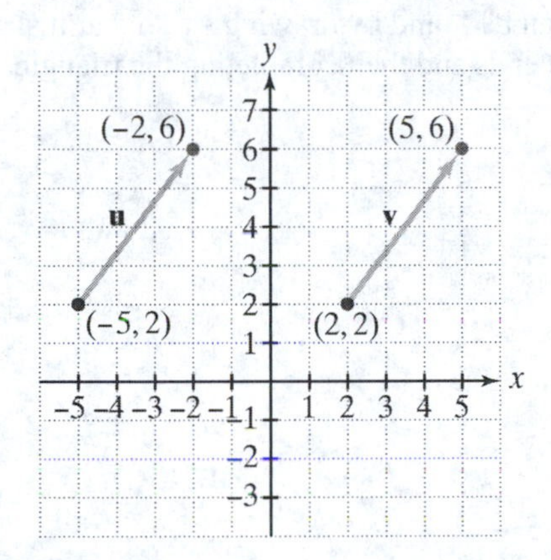

Use the distance formula to show that **u** and **v** have the same magnitude.

$$\|\mathbf{u}\| = \sqrt{[-2-(-5)]^2 + (6-2)^2}$$
$$= \sqrt{3^2 + 4^2} = \sqrt{9+16} = \sqrt{25} = 5$$
$$\|\mathbf{v}\| = \sqrt{(5-2)^2 + (6-2)^2}$$
$$= \sqrt{3^2 + 4^2} = \sqrt{9+16} = \sqrt{25} = 5$$

Both vectors have arrows that point to the upper right. Use the slope formula to show that **u** and **v** have the same direction.

Slope of **u**: $m = \dfrac{6-2}{-2-(-5)} = \dfrac{4}{3}$

Slope of **v**: $m = \dfrac{6-2}{5-2} = \dfrac{4}{3}$

The vectors **u** and **v** have the same magnitude and the same direction, so **u** = **v**.

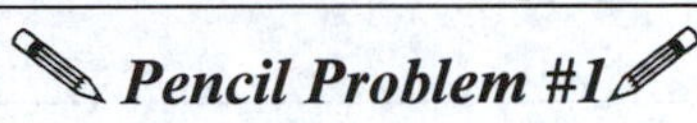

Pencil Problem #1

1. Refer to the figure. Show that **u** = **v**.

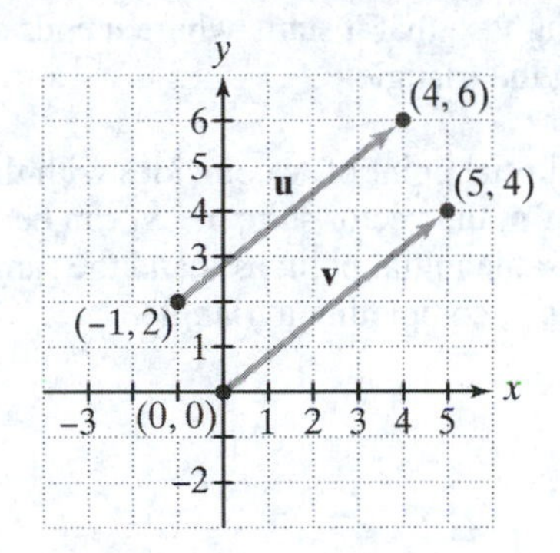

Objective #2: Visualize scalar multiplication, vector addition, and vector subtraction as geometric vectors.

✔ *Solved Problem #2*

2a. True or false: The vector 3**v** is a vector in the same direction as **v** but with 3 times the magnitude of **v**.

True; multiplying a vector by a positive scalar k does not change its direction but does change its magnitude by a factor of k. In symbols, $\|3\mathbf{v}\| = 3\|\mathbf{v}\|$.

2b. True or false: The sum of two vectors, **u** + **v**, can be found by drawing **v** so that it starts where **u** ends and then completing the triangle.

True; when the initial point of **v** coincides with the terminal point of **u**, the vector sum, **u** + **v**, can be drawn with the same initial point as **u** and the same terminal point as **v**, completing a triangle.

Pencil Problem #2

2a. True or false: The vector −**v** is a vector with the same magnitude as **v** but with opposite direction.

2b. True or false: The difference of two vectors, **u** − **v**, can be found by drawing −**v** so that it starts where **u** ends and then completing the triangle.

Objective #3: Represent vectors in the rectangular coordinate system.

✔ *Solved Problem #3*

3a. Sketch the vector **v** = 3**i** − 3**j** and find its magnitude.

The vector is in the form **v** = a**i** + b**j** with $a = 3$ and $b = -3$. Sketch the vector beginning at the origin, (0, 0), and ending at $(a, b) = (3, -3)$.

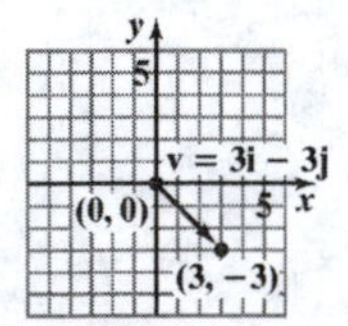

$$\|\mathbf{v}\| = \sqrt{a^2 + b^2} = \sqrt{3^2 + (-3)^2} = \sqrt{9+9} = \sqrt{18} = 3\sqrt{2}$$

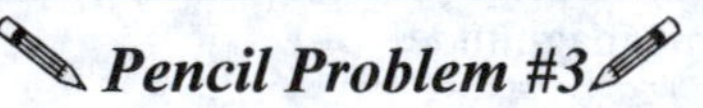

Pencil Problem #3

3a. Sketch the vector **v** = 3**i** + **j** and find its magnitude.

3b. Let **v** be the vector from initial point $P_1 = (-1, 3)$ to terminal point $P_2 = (2, 7)$. Write **v** in terms of **i** and **j**.

Let $P_1 = (x_1, y_1) = (-1, 3)$ and $P_2 = (x_2, y_2) = (2, 7)$.

$$\mathbf{v} = (x_2 - x_1)\mathbf{i} + (y_2 - y_1)\mathbf{j}$$
$$= [2-(-1)]\mathbf{i} + (7-3)\mathbf{j} = 3\mathbf{i} + 4\mathbf{j}$$

3b. Let **v** be the vector from initial point $P_1 = (-4, -4)$ to terminal point $P_2 = (6, 2)$. Write **v** in terms of **i** and **j**.

Objective #4: Perform operations with vectors in terms of i and j.

✔ *Solved Problem #4*

4a. If $\mathbf{v} = 7\mathbf{i} + 3\mathbf{j}$ and $\mathbf{w} = 4\mathbf{i} - 5\mathbf{j}$, find $\mathbf{v} + \mathbf{w}$.

$\mathbf{v} + \mathbf{w}$
$= (7\mathbf{i} + 3\mathbf{j}) + (4\mathbf{i} - 5\mathbf{j})$
$= (7 + 4)\mathbf{i} + [3 + (-5)]\mathbf{j}$
$= 11\mathbf{i} - 2\mathbf{j}$

4b. If $\mathbf{v} = 7\mathbf{i} + 10\mathbf{j}$, find $-5\mathbf{v}$.

$-5\mathbf{v} = -5(7\mathbf{i} + 10\mathbf{j}) = -35\mathbf{i} - 50\mathbf{j}$

4c. If $\mathbf{v} = 7\mathbf{i} + 3\mathbf{j}$ and $\mathbf{w} = 4\mathbf{i} - 5\mathbf{j}$, find $6\mathbf{v} - 3\mathbf{w}$.

$6\mathbf{v} - 3\mathbf{w}$
$= 6(7\mathbf{i} + 3\mathbf{j}) - 3(4\mathbf{i} - 5\mathbf{j})$
$= 42\mathbf{i} + 18\mathbf{j} - 12\mathbf{i} + 15\mathbf{j}$
$= (42 - 12)\mathbf{i} + (18 + 15)\mathbf{j}$
$= 30\mathbf{i} + 33\mathbf{j}$

✎ *Pencil Problem #4* ✎

4a. If $\mathbf{u} = 2\mathbf{i} - 5\mathbf{j}$ and $\mathbf{v} = -3\mathbf{i} + 7\mathbf{j}$, find $\mathbf{u} + \mathbf{v}$.

4b. If $\mathbf{v} = -3\mathbf{i} + 7\mathbf{j}$, find $5\mathbf{v}$.

4c. If $\mathbf{v} = -3\mathbf{i} + 7\mathbf{j}$ and $\mathbf{w} = -\mathbf{i} - 6\mathbf{j}$, find $3\mathbf{v} - 4\mathbf{w}$.

Objective #5: Find the unit vector in the direction of **v**.

✔ ***Solved Problem #5***

5. Write a unit vector in the same direction as $\mathbf{v} = 4\mathbf{i} - 3\mathbf{j}$.

First find the magnitude of **v**.

$$\|\mathbf{v}\| = \sqrt{a^2 + b^2} = \sqrt{4^2 + (-3)^2} = \sqrt{16+9} = \sqrt{25} = 5$$

Now divide **v** by its magnitude.

$$\frac{\mathbf{v}}{\|\mathbf{v}\|} = \frac{4\mathbf{i} - 3\mathbf{j}}{5} = \frac{4}{5}\mathbf{i} - \frac{3}{5}\mathbf{j}$$

Pencil Problem #5

5. Write a unit vector in the same direction as $\mathbf{v} = 3\mathbf{i} - 4\mathbf{j}$.

Objective #6: Write a vector in terms of its magnitude and direction.

✔ ***Solved Problem #6***

6. The jet stream is blowing at 60 miles per hour in the direction of N45°E. Express its velocity as a vector **v** in terms of **i** and **j**.

Since the direction of the jet stream forms a 45° angle with a north-south line on the east side of the north-south line, the vector's direction angle is $\theta = 90° - 45° = 45°$.

Since the jet stream is blowing at 60 miles per hour, the magnitude of **v** is 60: $\|\mathbf{v}\| = 60$.

Use the formula for a vector in terms of magnitude and direction.

$$\begin{aligned}\mathbf{v} &= \|\mathbf{v}\|\cos\theta\mathbf{i} + \|\mathbf{v}\|\sin\theta\mathbf{j}\\ &= 60\cos 45°\mathbf{i} + 60\sin 45°\mathbf{j}\\ &= 60\left(\frac{\sqrt{2}}{2}\right)\mathbf{i} + 60\left(\frac{\sqrt{2}}{2}\right)\mathbf{j}\\ &= 30\sqrt{2}\mathbf{i} + 30\sqrt{2}\mathbf{j}\end{aligned}$$

Pencil Problem #6

6. A quarterback releases a football with a sped of 44 feet per second at an angle of 30° with the horizontal. Express its velocity as a vector **v** in terms of **i** and **j**.

Objective #7: Solve applied problems involving vectors.

✔ *Solved Problem #7*

7. Two forces, $\mathbf{F}_1$ and $\mathbf{F}_2$, of magnitude 30 and 60 pounds, respectively, act on an object. The direction of $\mathbf{F}_1$ is N10°E and the direction of $\mathbf{F}_2$ is N60°E. Find the magnitude, to the nearest hundredth of a pound, and the direction angle, to the nearest tenth of a degree, of the resultant force.

The direction angle for $\mathbf{F}_1$ is $90° - 10° = 80°$, and the direction angle for $\mathbf{F}_2$ is $90° - 60° = 30°$.

$$\mathbf{F}_1 = 30\cos 80°\mathbf{i} + 30\sin 80°\mathbf{j} \approx 5.21\mathbf{i} + 29.54\mathbf{j}$$
$$\mathbf{F}_2 = 60\cos 30°\mathbf{i} + 60\sin 30°\mathbf{j} \approx 51.96\mathbf{i} + 30\mathbf{j}$$

The resultant force is

$$\begin{aligned}\mathbf{F} &= \mathbf{F}_1 + \mathbf{F}_2 \\ &\approx (5.21\mathbf{i} + 29.54\mathbf{j}) + (51.96\mathbf{i} + 30\mathbf{j}) \\ &= (5.21 + 51.96)\mathbf{i} + (29.54 + 30)\mathbf{j} \\ &= 57.17\mathbf{i} + 59.54\mathbf{j}.\end{aligned}$$

Its magnitude is

$$\|\mathbf{F}\| = \sqrt{57.17^2 + 59.54^2} \approx 82.54 \text{ pounds.}$$

To find the direction angle, use $\cos\theta = \frac{a}{\|\mathbf{F}\|}$, where $\mathbf{F} = a\mathbf{i} + b\mathbf{j} = 57.17\mathbf{i} + 59.54\mathbf{j}$, so $a = 57.17$.

$$\cos\theta = \frac{a}{\|\mathbf{F}\|} = \frac{57.17}{82.54}$$

So, the direction angle is $\theta = \cos^{-1}\frac{57.17}{82.54} \approx 46.2°$.

✎ *Pencil Problem #7* ✎

7. Two forces, $\mathbf{F}_1$ and $\mathbf{F}_2$, of magnitude 70 and 50 pounds, respectively, act on an object. The direction of $\mathbf{F}_1$ is S56°E and the direction of $\mathbf{F}_2$ is N72°E. Find the magnitude, to the nearest hundredth of a pound, and the direction angle, to the nearest tenth of a degree, of the resultant force.

Answers for Pencil Problems *(Textbook Exercise references in parentheses)*:

1. Magnitude: $\|\mathbf{u}\| = \|\mathbf{v}\| = \sqrt{41}$; Direction: Both vectors have arrows that point to the upper right and slopes of $\frac{4}{5}$.

The vectors **u** and **v** have the same magnitude and the same direction, so **u** = **v**. *(6.6 #1)*

2a. true **2b.** true

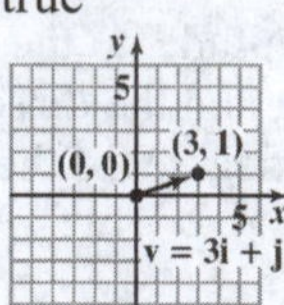

3a. $\|\mathbf{v}\| = \sqrt{10}$ *(6.6 #5)*

3b. $10\mathbf{i} + 6\mathbf{j}$ *(6.6 #13)*

4a. $-\mathbf{i} + 2\mathbf{j}$ *(6.6 #21)* **4b.** $-15\mathbf{i} + 35\mathbf{j}$ *(6.6 #27)* **4c.** $-5\mathbf{i} + 45\mathbf{j}$ *(6.6 #33)*

5. $\frac{\mathbf{v}}{\|\mathbf{v}\|} = \frac{3}{5}\mathbf{i} - \frac{4}{5}\mathbf{j}$ *(6.6 #41)*

6. $\mathbf{v} = 22\sqrt{3}\mathbf{i} + 22\mathbf{j}$ *(6.6 #65)*

7a. magnitude: ≈108.21 pounds; direction angle: ≈347.4° *(6.6 #71)*

Section 6.7
The Dot Product

Let's Get to Work!

In this section, you will learn how to find the dot product of two vectors. The dot product is very easy to compute and has many important applications, including computing work. When the force moving an object is applied in the direction of the movement, the work done is a simple product of two real numbers, the magnitude of the force and the distance the object is moved. However, when the force is applied at angle to the direction of the movement, we will use the dot product of two vectors to compute work. So, let's get ready to work!

Objective #1: Find the dot product of two vectors.

✔ ***Solved Problem #1***	✎ ***Pencil Problem #1***
1a. If $\mathbf{v} = 7\mathbf{i} - 4\mathbf{j}$ and $\mathbf{w} = 2\mathbf{i} - \mathbf{j}$, find $\mathbf{v} \cdot \mathbf{w}$. $\mathbf{v} \cdot \mathbf{w} = 7(2) + (-4)(-1) = 14 + 4 = 18$	**1a.** If $\mathbf{v} = 5\mathbf{i} - 4\mathbf{j}$ and $\mathbf{w} = -2\mathbf{i} - \mathbf{j}$, find $\mathbf{v} \cdot \mathbf{w}$.
1b. If $\mathbf{v} = 7\mathbf{i} - 4\mathbf{j}$ and $\mathbf{w} = 2\mathbf{i} - \mathbf{j}$, find $\mathbf{w} \cdot \mathbf{v}$. $\mathbf{w} \cdot \mathbf{v} = 2(7) + (-1)(-4) = 14 + 4 = 18$	**1b.** If $\mathbf{v} = 5\mathbf{i} - 4\mathbf{j}$ and $\mathbf{w} = -2\mathbf{i} - \mathbf{j}$, find $\mathbf{w} \cdot \mathbf{v}$.
1c. If $\mathbf{w} = 2\mathbf{i} - \mathbf{j}$, find $\mathbf{w} \cdot \mathbf{w}$. $\mathbf{w} \cdot \mathbf{w} = 2(2) + (-1)(-1) = 4 + 1 = 5$	**1c.** If $\mathbf{v} = 5\mathbf{i} - 4\mathbf{j}$, find $\mathbf{v} \cdot \mathbf{v}$.

Objective #2: Find the angle between two vectors.

✔ *Solved Problem #2*

2. Find the angle between the vectors $\mathbf{v} = 4\mathbf{i} - 3\mathbf{j}$ and $\mathbf{w} = \mathbf{i} + 2\mathbf{j}$. Round to the nearest tenth of a degree.

To use the formula for the angle between two vectors, $\cos\theta = \frac{\mathbf{v}\cdot\mathbf{w}}{\|\mathbf{v}\|\|\mathbf{w}\|}$, we need to know $\mathbf{v}\cdot\mathbf{w}$, $\|\mathbf{v}\|$, and $\|\mathbf{w}\|$.

$$\mathbf{v}\cdot\mathbf{w} = 4(1) + (-3)(2) = 4 - 6 = -2$$

$$\|\mathbf{v}\| = \sqrt{4^2 + (-3)^2} = \sqrt{16+9} = \sqrt{25} = 5$$

$$\|\mathbf{w}\| = \sqrt{1^2 + 2^2} = \sqrt{1+4} = \sqrt{5}$$

Now, use the formula.

$$\cos\theta = \frac{\mathbf{v}\cdot\mathbf{w}}{\|\mathbf{v}\|\|\mathbf{w}\|} = \frac{-2}{5\cdot\sqrt{5}} = -\frac{2}{5\sqrt{5}}$$

So, the angle between the vectors is

$$\theta = \cos^{-1}\left(-\frac{2}{5\sqrt{5}}\right) \approx 100.3°.$$

Pencil Problem #2

2. Find the angle between the vectors $\mathbf{v} = -3\mathbf{i} + 2\mathbf{j}$ and $\mathbf{w} = 4\mathbf{i} - \mathbf{j}$. Round to the nearest tenth of a degree.

Objective #3: Use the dot product to determine if two vectors are orthogonal.

✔ *Solved Problem #3*

3. Are the vectors $\mathbf{v} = 2\mathbf{i} + 3\mathbf{j}$ and $\mathbf{w} = 6\mathbf{i} - 4\mathbf{j}$ orthogonal?

Find the dot product.

$$\mathbf{v}\cdot\mathbf{w} = 2(6) + 3(-4) = 12 - 12 = 0$$

Since the dot product is 0, the vectors are orthogonal.

3. Are the vectors $\mathbf{v} = 2\mathbf{i} + 8\mathbf{j}$ and $\mathbf{w} = 4\mathbf{i} - \mathbf{j}$ orthogonal?

Objective #4: Find the projection of a vector onto another vector.

✔ *Solved Problem #4*

4. If $\mathbf{v} = 2\mathbf{i} - 5\mathbf{j}$ and $\mathbf{w} = \mathbf{i} - \mathbf{j}$, find the vector projection of $\mathbf{v}$ onto $\mathbf{w}$.

To use the formula for vector projection,

$\text{proj}_{\mathbf{w}}\mathbf{v} = \dfrac{\mathbf{v}\cdot\mathbf{w}}{\|\mathbf{w}\|^2}\mathbf{w}$, we need to know $\mathbf{v}\cdot\mathbf{w}$ and $\|\mathbf{w}\|$.

$$\mathbf{v}\cdot\mathbf{w} = 2(1) + (-5)(-1) = 2 + 5 = 7$$

$$\|\mathbf{w}\| = \sqrt{1^2 + (-1)^2} = \sqrt{1+1} = \sqrt{2}$$

Now, use the formula.

$$\text{proj}_{\mathbf{w}}\mathbf{v} = \frac{\mathbf{v}\cdot\mathbf{w}}{\|\mathbf{w}\|^2}\mathbf{w} = \frac{7}{(\sqrt{2})^2}\mathbf{w} = \frac{7}{2}(\mathbf{i} - \mathbf{j}) = \frac{7}{2}\mathbf{i} - \frac{7}{2}\mathbf{j}$$

Pencil Problem #4

4. If $\mathbf{v} = \mathbf{i} + 3\mathbf{j}$ and $\mathbf{w} = -2\mathbf{i} + 5\mathbf{j}$, find the vector projection of $\mathbf{v}$ onto $\mathbf{w}$.

Objective #5: Express a vector as the sum of two orthogonal vectors.

✔ *Solved Problem #5*

5. Let $\mathbf{v} = 2\mathbf{i} - 5\mathbf{j}$ and $\mathbf{w} = \mathbf{i} - \mathbf{j}$. (These are the vectors from Solved Problem #4.) Decompose $\mathbf{v}$ into two vectors, $\mathbf{v}_1$ and $\mathbf{v}_2$, where $\mathbf{v}_1$ is parallel to $\mathbf{w}$ and $\mathbf{v}_2$ is orthogonal to $\mathbf{w}$.

The vector $\mathbf{v}_1$ is the vector projection of $\mathbf{v}$ onto $\mathbf{w}$, which we found for these two vectors in Solved Problem #4. Subtract $\mathbf{v}_1$ from $\mathbf{v}$ to find $\mathbf{v}_2$.

$$\mathbf{v}_1 = \text{proj}_{\mathbf{w}}\mathbf{v} = \frac{7}{2}\mathbf{i} - \frac{7}{2}\mathbf{j}$$

$$\mathbf{v}_2 = \mathbf{v} - \mathbf{v}_1 = (2\mathbf{i} - 5\mathbf{j}) - \left(\frac{7}{2}\mathbf{i} - \frac{7}{2}\mathbf{j}\right) = -\frac{3}{2}\mathbf{i} - \frac{3}{2}\mathbf{j}$$

Pencil Problem #5

5. Let $\mathbf{v} = \mathbf{i} + 3\mathbf{j}$ and $\mathbf{w} = -2\mathbf{i} + 5\mathbf{j}$. (These are the vectors from Pencil Problem #4.) Decompose $\mathbf{v}$ into two vectors, $\mathbf{v}_1$ and $\mathbf{v}_2$, where $\mathbf{v}_1$ is parallel to $\mathbf{w}$ and $\mathbf{v}_2$ is orthogonal to $\mathbf{w}$.

Objective #6: Compute work.

✔ *Solved Problem #6*

6. A child pulls a wagon along level ground by exerting a force of 20 pounds on a handle that makes an angle of 30° with the horizontal. How much work is done pulling the wagon 150 feet? Round to the nearest foot-pound.

Use the formula $W = \|\mathbf{F}\|\|\overrightarrow{AB}\|\cos\theta$, where $\|\mathbf{F}\| = 20$ is the magnitude of the force, $\|\overrightarrow{AB}\| = 150$ is the distance the wagon is pulled, and $\theta = 30°$ is the angle between the force and the direction of the motion.

$$W = \|\mathbf{F}\|\|\overrightarrow{AB}\|\cos\theta = (20)(150)\cos 30° \approx 2598$$

The work done is approximately 2598 foot-pounds.

Pencil Problem #6

6. A wagon is pulled along level ground by exerting a force of 40 pounds on a handle that makes an angle of 32° with the horizontal. How much work is done pulling the wagon 100 feet? Round to the nearest foot-pound.

Answers for Pencil Problems *(Textbook Exercise references in parentheses)*:

1a. −6 **1b.** −6 **1c.** 41 *(6.7 #3)* **2.** ≈ 160.3° *(6.7 #19)*

3. Yes, they are orthogonal, since their dot product is 0. *(6.7 #25)*

4. $\text{proj}_{\mathbf{w}}\mathbf{v} = -\frac{26}{29}\mathbf{i} + \frac{65}{29}\mathbf{j}$ *(6.7 #35)* **5.** $\mathbf{v}_1 = \text{proj}_{\mathbf{w}}\mathbf{v} = -\frac{26}{29}\mathbf{i} + \frac{65}{29}\mathbf{j}$; $\mathbf{v}_2 = \frac{55}{29}\mathbf{i} + \frac{22}{29}\mathbf{j}$ *(6.7 #35)*

6. ≈3392 foot-pounds *(6.7 #55)*

Section 7.1
Systems of Linear Equations in Two Variables

Procrastination makes you sick!

Researchers compared college students who were procrastinators and nonprocrastinators. Early in the semester, procrastinators reported fewer symptoms of illness, but late in the semester, they reported more symptoms than their nonprocrastinating peers.

In this section of the textbook, you will identify when both groups have the same number of symptoms as the point of intersection of two lines.

Objective #1: Decide whether an ordered pair is a solution of a linear system.

✔ Solved Problem #1

1. Determine if the ordered pair $(7,6)$ is a solution of the system: $\begin{cases} 2x-3y=-4 \\ 2x+y=4 \end{cases}$

To determine if $(7,6)$ is a solution to the system, replace x with 7 and y with 6 in both equations.

$$\begin{aligned} 2x-3y&=-4 \\ 2(7)-3(6)&=-4 \\ 14-18&=-4 \\ -4&=-4, \text{ true} \end{aligned} \qquad \begin{aligned} 2x+y&=4 \\ 2(7)+6&=4 \\ 14+6&=4 \\ 20&=4, \text{ false} \end{aligned}$$

The ordered pair does not satisfy both equations, so it is not a solution to the system.

Pencil Problem #1

1. Determine if the ordered pair $(2,3)$ is a solution of the system: $\begin{cases} x+3y=11 \\ x-5y=-13 \end{cases}$

Objective #2: Solve linear systems by substitution.

✔ Solved Problem #2

2. Solve by the substitution method: $\begin{cases} 3x+2y=4 \\ 2x+y=1 \end{cases}$

Solve $2x+y=1$ for y.

$$\begin{aligned} 2x+y&=1 \\ y&=-2x+1 \end{aligned}$$

Pencil Problem #2

2. Solve by the substitution method: $\begin{cases} x+y=4 \\ y=3x \end{cases}$

Substitute: $3x + 2y = 4$

$$3x+2(\overbrace{-2x+1}^{y})=4$$
$$3x-4x+2=4$$
$$-x+2=4$$
$$-x=2$$
$$x=-2$$

Find y.

$y=-2x+1$

$y=-2(-2)+1$

$y=5$

The solution is $(-2,5)$.

The solution set is $\{(-2,5)\}$.

Objective #3: Solve linear systems by addition.

✔ *Solved Problem #3*

3. Solve the system: $\begin{cases} 4x+5y=3 \\ 2x-3y=7 \end{cases}$

Multiply each term of the second equation by –2 and add the equations to eliminate x.

$$\begin{aligned} 4x+5y &= 3 \\ -4x+6y &= -14 \\ \hline 11y &= -11 \\ y &= -1 \end{aligned}$$

Back-substitute into either of the original equations to solve for x.

$$2x-3y=7$$
$$2x-3(-1)=7$$
$$2x+3=7$$
$$2x=4$$
$$x=2$$

The solution set is $\{(2,-1)\}$.

Pencil Problem #3

3. Solve the system: $\begin{cases} 3x-4y=11 \\ 2x+3y=-4 \end{cases}$

Objective #4: Identify systems that do not have exactly one ordered-pair solution.

✔ *Solved Problem #4*

4a. Solve the system: $\begin{cases} 5x-2y=4 \\ -10x+4y=7 \end{cases}$

Multiply the first equation by 2, and then add the equations.

$$\begin{array}{r} 10x-4y=8 \\ \underline{-10x+4y=7} \\ 0=15 \end{array}$$

Since there are no pairs (x, y) for which 0 will equal 15, the system is inconsistent and has no solution.
The solution set is $\varnothing$ or $\{\ \}$.

4b. Solve the system: $\begin{cases} x=4y-8 \\ 5x-20y=-40 \end{cases}$

Substitute $4y-8$ for x in the second equation.

$$\begin{aligned} 5x-20y&=-40 \\ 5(\overbrace{4y-8}^{x})-20y&=-40 \\ 20y-40-20y&=-40 \\ -40&=-40 \end{aligned}$$

Since $-40=-40$ for all values of x and y, the system is dependent.

The solution set is $\{(x,y)|x=4y-8\}$ or $\{(x,y)|5x-20y=-40\}$.

Pencil Problem #4

4a. Solve the system: $\begin{cases} x=9-2y \\ x+2y=13 \end{cases}$

4b. Solve the system: $\begin{cases} y=3x-5 \\ 21x-35=7y \end{cases}$

Objective #5: Solve problems using systems of linear equations.

✔ *Solved Problem #5*

5. A company that manufactures running shoes has a fixed cost of \$300,000. Additionally, it costs \$30 to produce each pair of shoes. The shoes are sold at \$80 per pair.

5a. Write the cost function, C, of producing x pairs of running shoes.

$$C(x)=\overbrace{300{,}000}^{\text{fixed costs}}+\overbrace{30x}^{\text{\$30 per pair}}$$

5. A company that manufactures small canoes has a fixed cost of \$18,000. Additionally, it costs \$20 to produce each canoe. The selling price is \$80 per canoe.

5a. Write the cost function, C, of producing x canoes.

5b. Write the revenue function, R, from the sale of x pairs of running shoes.

$$R(x) = \overbrace{80x}^{\$80 \text{ per pair}}$$

5c. Determine the break-even point. Describe what this means.

The system is $\begin{cases} y = 300,000 + 30x \\ y = 80x \end{cases}$

The break-even point is where $R(x) = C(x)$.

$$\begin{aligned} R(x) &= C(x) \\ 80x &= 300,000 + 30x \\ 50x &= 300,000 \\ x &= 6000 \end{aligned}$$

Back-substitute to find y: $y = 80x$

$$\begin{aligned} y &= 80(6000) \\ y &= 480,000 \end{aligned}$$

The break-even point is (6000, 480,000).

This means the company will break even when it produces and sells 6000 pairs of shoes. At this level, both revenue and costs are $480,000.

5b. Write the revenue function, R, from the sale of x canoes.

5c. Determine the break-even point. Describe what this means.

Answers for Pencil Problems *(Textbook Exercise references in parentheses)*:

1. The ordered pair is a solution to the system. *(7.1 #1)*

2. $\{(1,3)\}$ *(7.1 #5)*

3. $\{(1,-2)\}$ *(7.1 #27)*

4a. $\varnothing$ or $\{\ \}$ *(7.1 #31)*

4b. $\{(x,y) \mid y = 3x - 5\}$ or $\{(x,y) \mid 21x - 35 = 7y\}$ *(7.1 #33)*

5a. $C(x) = 18,000 + 20x$ *(7.1 #61a)* **5b.** $R(x) = 80x$ *(7.1 #61b)*

5c. Break-even point: (300, 24,000). Which means when 300 canoes are produced the company will break-even with cost and revenue at $24,000. *(7.1 #61c)*

Section 7.2
Systems of Linear Equations in Three Variables

Hit the BRAKES!

Did you know that a mathematical model can be used to describe the relationship between the number of feet a car travels once the brakes are applied and the number of seconds the car is in motion after the brakes are applied?

In the Exercise Set of this section, using data collected by a research firm, you will be asked to write the mathematical model that describes this situation.

Objective #1: Verify the solution of a system of linear equations in three variables.

✔ Solved Problem #1

1. Show that the ordered triple $(-1,-4,5)$ is a solution of the system:

$$\begin{cases} x-2y+3z=22 \\ 2x-3y-z=5 \\ 3x+y-5z=-32 \end{cases}$$

Test the ordered triple in each equation.

$$x-2y+3z=22$$
$$(-1)-2(-4)+3(5)=22$$
$$22=22,\ \text{true}$$

$$2x-3y-z=5$$
$$2(-1)-3(-4)-(5)=5$$
$$5=5,\ \text{true}$$

$$3x+y-5z=-32$$
$$3(-1)+(-4)-5(5)=-32$$
$$-32=-32,\ \text{true}$$

The ordered triple (–1, –4, 5) makes all three equations true, so it is a solution of the system.

Pencil Problem #1

1. Determine if the ordered triple $(2,-1,3)$ is a solution of the system:

$$\begin{cases} x+y+z=4 \\ x-2y-z=1 \\ 2x-y-z=-1 \end{cases}$$

Objective #2: Solve systems of linear equations in three variables.

✔ Solved Problem #2

2. Solve the system: $\begin{cases} x+4y-z=20 \\ 3x+2y+z=8 \\ 2x-3y+2z=-16 \end{cases}$

Add the first two equations to eliminate z.

$$\begin{array}{l} x+4y-z=20 \\ \underline{3x+2y+z=\ 8} \\ 4x+6y \quad\ =28 \end{array}$$

Multiply the first equation by 2 and add it to the third equation to eliminate z again.

$$\begin{array}{l} 2x+8y-2z=40 \\ \underline{2x-3y+2z=-16} \\ 4x+5y \quad\ =24 \end{array}$$

Solve the system of two equations in two variables.

$$\begin{array}{l} 4x+6y=28 \\ 4x+5y=24 \end{array}$$

Multiply the second equation by –1 and add the equations.

$$\begin{array}{r} 4x+6y=28 \\ \underline{-4x-5y=-24} \\ y=4 \end{array}$$

Back-substitute 4 for y to find x.

$$\begin{array}{r} 4x+6y=28 \\ 4x+6(4)=28 \\ 4x+24=28 \\ 4x=4 \\ x=1 \end{array}$$

Back-substitute into an original equation.

$$\begin{array}{r} 3x+2y+z=8 \\ 3(1)+2(4)+z=8 \\ 11+z=8 \\ z=-3 \end{array}$$

The solution is $(1,4,-3)$ and the solution set is $\{(1,4,-3)\}$.

Pencil Problem #2

2. Solve the system: $\begin{cases} 4x-y+2z=11 \\ x+2y-z=-1 \\ 2x+2y-3z=-1 \end{cases}$

Objective #3: Solve problems using systems in three variables.

✔ *Solved Problem #3*

3. Find the quadratic function $y = ax^2 + bx + c$ whose graph passes through the points $(1,4)$, $(2,1)$, and $(3,4)$.

Use each ordered pair to write an equation.

$(1,4)$: $y = ax^2 + bx + c$

$$4 = a(1)^2 + b(1) + c$$
$$4 = a + b + c$$

$(2,1)$: $y = ax^2 + bx + c$

$$1 = a(2)^2 + b(2) + c$$
$$1 = 4a + 2b + c$$

$(3,4)$: $y = ax^2 + bx + c$

$$4 = a(3)^2 + b(3) + c$$
$$4 = 9a + 3b + c$$

The system of three equations in three variables is:

$$\begin{cases} a + b + c = 4 \\ 4a + 2b + c = 1 \\ 9a + 3b + c = 4 \end{cases}$$

Solve the system: $\begin{cases} a + b + c = 4 \\ 4a + 2b + c = 1 \\ 9a + 3b + c = 4 \end{cases}$

Multiply the first equation by –1 and add it to the second equation:

$$\begin{array}{r} -a - b - c = -4 \\ 4a + 2b + c = 1 \\ \hline 3a + b = -3 \end{array}$$

(continued on next page)

✎ *Pencil Problem #3* ✎

3. Find the quadratic function $y = ax^2 + bx + c$ whose graph passes through the points $(-1,6)$, $(1,4)$, and $(2,9)$.

Multiply the first equation by –1 and add it to the third equation:

$$\begin{array}{l} -a-b-c=-4 \\ \underline{9a+3b+c=\ \ 4} \\ 8a+2b\qquad =\ \ 0 \end{array}$$

Solve this system of two equations in two variables.

$$3a+b=-3$$
$$8a+2b=0$$

Multiply the first equation by –2 and add to the second equation:

$$\begin{array}{r} -6a-2b=6 \\ \underline{8a+2b=0} \\ 2a=6 \\ a=3 \end{array}$$

Back-substitute to find b:

$$\begin{aligned} 3a+b&=-3 \\ 3(3)+b&=-3 \\ 9+b&=-3 \\ b&=-12 \end{aligned}$$

Back-substitute into an original equation to find c:

$$\begin{aligned} a+b+c&=4 \\ (3)+(-12)+c&=4 \\ -9+c&=4 \\ c&=13 \end{aligned}$$

The quadratic function is $y=3x^2-12x+13$.

Answers for Pencil Problems *(Textbook Exercise references in parentheses)*:

1. not a solution *(7.2 #1)*

2. $\{(2,-1,1)\}$ *(7.2 #7)*

3. $y=2x^2-x+3$ *(7.2 #19)*

Section 7.3
Partial Fractions

Where's the "UNDO" Button?

We have learned how to write a sum or difference of rational expressions as a single rational expression and saw how this skill is necessary when solving rational inequalities. However, in calculus, it is sometimes necessary to write a single rational expression as a sum or difference of simpler rational expressions, undoing the process of adding or subtracting.
In this section, you will learn how to break up a rational expression into sums or differences of rational expressions with simpler denominators.

Objective #1: Decompose $\frac{P}{Q}$, where Q has only distinct linear factors.

✔ *Solved Problem #1*

1. Find the partial fraction decomposition of $\frac{5x-1}{(x-3)(x+4)}$.

Write a constant over each distinct linear factor in the denominator.

$$\frac{5x-1}{(x-3)(x+4)}=\frac{A}{x-3}+\frac{B}{x+4}$$

Multiply by the LCD, $(x-3)(x+4)$, to eliminate fractions. Then simplify and rearrange terms.

$$(x-3)(x+4)\frac{5x-1}{(x-3)(x+4)}=(x-3)(x+4)\frac{A}{x-3}+(x-3)(x+4)\frac{B}{x+4}$$
$$5x-1=A(x+4)+B(x-3)$$
$$5x-1=Ax+4A+Bx-3B$$
$$5x-1=(A+B)x+(4A-3B)$$

Equating the coefficients of x and equating the constant terms, we obtain a system of equations.

$$\begin{cases} A+B=5 \\ 4A-3B=-1 \end{cases}$$

Multiplying the first equation by 3 and adding it to the second equation, we obtain $7A = 14$, so $A = 2$. Substituting 2 for A in either equation, we obtain $B = 3$. The partial fraction decomposition is

$$\frac{5x-1}{(x-3)(x+4)}=\frac{2}{x-3}+\frac{3}{x+4}.$$

Pencil Problem #1

1. Find the partial fraction decomposition of $\dfrac{3x+50}{(x-9)(x+2)}$.

Objective #2: Decompose $\dfrac{P}{Q}$, where Q has repeated linear factors.

✔ *Solved Problem #2*

2. Find the partial fraction decomposition of $\dfrac{x+2}{x(x-1)^2}$.

Include one fraction for each power of $x-1$.

$$\frac{x+2}{x(x-1)^2}=\frac{A}{x}+\frac{B}{x-1}+\frac{C}{(x-1)^2}$$

Multiply by the LCD, $x(x-1)^2$, to eliminate fractions. Then simplify and rearrange terms.

$$x(x-1)^2\frac{x+2}{x(x-1)^2}=x(x-1)^2\frac{A}{x}+x(x-1)^2\frac{B}{x-1}+x(x-1)^2\frac{C}{(x-1)^2}$$

$$x+2=A(x-1)^2+Bx(x-1)+Cx$$

$$x+2=A(x^2-2x+1)+Bx(x-1)+Cx$$

$$x+2=Ax^2-2Ax+A+Bx^2-Bx+Cx$$

$$0x^2+x+2=(A+B)x^2+(-2A-B+C)x+A$$

Equating the coefficients of like terms, we obtain a system of equations.

$$\begin{cases} A+B=0 \\ -2A-B+C=1 \\ A=2 \end{cases}$$

We see immediately that $A = 2$. Substituting 2 for A in the first equation, we obtain $B = -2$. Substituting these values into the second equation, we obtain $C = 3$. The partial fraction decomposition is

$$\frac{x+2}{x(x-1)^2}=\frac{2}{x}+\frac{-2}{x-1}+\frac{3}{(x-1)^2} \text{ or } \frac{2}{x}-\frac{2}{x-1}+\frac{3}{(x-1)^2}.$$

2. Find the partial fraction decomposition of $\frac{x^2}{(x-1)^2(x+1)}$.

Objective #3: Decompose $\frac{P}{Q}$, where Q has a nonrepeated prime quadratic factor.

✔ *Solved Problem #3*

3. Find the partial fraction decomposition of $\frac{8x^2+12x-20}{(x+3)(x^2+x+2)}$.

Use a constant over the linear factor and a linear expression over the prime quadratic factor.

$$\frac{8x^2+12x-20}{(x+3)(x^2+x+2)}=\frac{A}{x+3}+\frac{Bx+C}{x^2+x+2}$$

Multiply by the LCD, $(x+3)(x^2+x+2)$, to eliminate fractions. Then simplify and rearrange terms.

$$(x+3)(x^2+x+2)\frac{8x^2+12x-20}{(x+3)(x^2+x+2)}=(x+3)(x^2+x+2)\frac{A}{x+3}+(x+3)(x^2+x+2)\frac{Bx+C}{x^2+x+2}$$

$$8x^2+12x-20=A(x^2+x+2)+(Bx+C)(x+3)$$

$$8x^2+12x-20=Ax^2+Ax+2A+Bx^2+3Bx+Cx+3C$$

$$8x^2+12x-20=(A+B)x^2+(A+3B+C)x+(2A+3C)$$

Equating the coefficients of like terms, we obtain a system of equations.

$$\begin{cases} A+B=8 \\ A+3B+C=12 \\ 2A+3C=-20 \end{cases}$$

Multiply the second equation by -3 and add to the third equation to obtain $-A-9B=-56$. Add this result to the first equation in the system above to obtain $-8B=-48$, so $B=6$. Substituting this value into the first equation, we obtain $A=2$. Substituting the value of A into the third equation, we obtain $C=-8$.

The partial fraction decomposition is

$$\frac{8x^2+12x-20}{(x+3)(x^2+x+2)}=\frac{2}{x+3}+\frac{6x-8}{x^2+x+2}.$$

✎ *Pencil Problem #3* ✎

3. Find the partial fraction decomposition of $\frac{5x^2+6x+3}{(x+1)(x^2+2x+2)}$.

Objective #4: $\frac{P}{Q}$, where Q has a prime, repeated quadratic factor

✔ *Solved Problem #4*

4. Find the partial fraction decomposition of $\frac{2x^3+x+3}{(x^2+1)^2}$.

Include one fraction with a linear numerator for each power of x^2+1.

$$\frac{2x^3+x+3}{(x^2+1)^2}=\frac{Ax+B}{x^2+1}+\frac{Cx+D}{(x^2+1)^2}$$

Multiply by the LCD, $(x^2+1)^2$, to eliminate fractions. Then simplify and rearrange terms.

$$(x^2+1)^2\frac{2x^3+x+3}{(x^2+1)^2}=(x^2+1)^2\frac{Ax+B}{x^2+1}+(x^2+1)^2\frac{Cx+D}{(x^2+1)^2}$$

$$2x^3+x+3=(Ax+B)(x^2+1)+Cx+D$$

$$2x^3+x+3=Ax^3+Ax+Bx^2+B+Cx+D$$

$$2x^3+x+3=Ax^3+Bx^2+(A+C)x+(B+D)$$

Equating the coefficients of like terms, we obtain a system of equations.

$$\begin{cases} A=2 \\ B=0 \\ A+C=1 \\ B+D=3 \end{cases}$$

We immediately see that $A=2$ and $B=0$. By performing appropriate substitutions, we obtain $C=-1$ and $D=3$. The partial fraction decomposition is

$$\frac{2x^3+x+3}{(x^2+1)^2}=\frac{2x}{x^2+1}+\frac{-x+3}{(x^2+1)^2}.$$

4. Find the partial fraction decomposition of $\frac{x^3+x^2+2}{(x^2+2)^2}$.

Answers for Pencil Problems *(Textbook Exercise references in parentheses)*:

1. $\frac{7}{x-9}-\frac{4}{x+2}$ *(7.3 #11)*
2. $\frac{1}{4(x+1)}+\frac{3}{4(x-1)}+\frac{1}{2(x-1)^2}$ *(7.3 #27)*
3. $\frac{2}{x+1}+\frac{3x-1}{x^2+2x+2}$ *(7.3 #31)*
4. $\frac{x+1}{x^2+2}-\frac{2x}{(x^2+2)^2}$ *(7.3 #37)*

Section 7.4
Systems of Nonlinear Equations in Two Variables

DO YOU FEEL SAFE????

Scientists debate the probability that a "doomsday rock" will collide with Earth. It has been estimated that an asteroid crashes into Earth about once every 250,000 years, and that such a collision would have disastrous results.

Understanding the path of Earth and the path of a comet is essential to detecting threatening space debris. Orbits about the sun are not described by linear equations. The ability to solve systems that do not contain linear equations provides NASA scientists watching for troublesome asteroids with a way to locate possible collision points with Earth's orbit.

Objective #1: Recognize systems of nonlinear equations in two variables.

✔ *Solved Problem #1*

1a. True or false: A solution of a nonlinear system in two variables is an ordered pair of real numbers that satisfies at least one equation in the system.

False; a solution must satisfy *all* equations in the system.

1b. True or false: The solution of a system of nonlinear equations corresponds to the intersection points of the graphs in the system.

True; each solution will correspond to an intersection point of the graphs.

Pencil Problem #1

1a. True or false: A system of nonlinear equations cannot contain a linear equation.

1b. True or false: The graphs of the equations in a nonlinear system could be a parabola and a circle.

Objective #2: Solve nonlinear systems by substitution.

✔ *Solved Problem #2*

2. Solve by the substitution method:

$$\begin{cases} x+2y=0 \\ (x-1)^2+(y-1)^2=5 \end{cases}$$

Solve the first equation for x: $x+2y=0$

$$x=-2y$$

Substitute the expression $-2y$ for x in the second equation and solve for y.

$$(x-1)^2+(y-1)^2=5$$
$$(\overbrace{-2y}^{x}-1)^2+(y-1)^2=5$$
$$4y^2+4y+1+y^2-2y+1=5$$
$$5y^2+2y-3=0$$
$$(5y-3)(y+1)=0$$
$$5y-3=0 \quad \text{or} \quad y+1=0$$
$$y=\tfrac{3}{5} \quad \text{or} \quad y=-1$$

If $y=\frac{3}{5}$, $x=-2\left(\frac{3}{5}\right)=-\frac{6}{5}$.
If $y=-1$, $x=-2(-1)=2$.

Check $(2,-1)$ in both original equations.

$$x+2y=0 \qquad (x-1)^2+(y-1)^2=5$$
$$2+2(-1)=0 \qquad (2-1)^2+(-1-1)^2=5$$
$$0=0,\ \text{true} \qquad 1+4=5$$
$$5=5,\ \text{true}$$

Check $\left(-\frac{6}{5},\frac{3}{5}\right)$ in both original equations.

$$x+2y=0 \qquad (x-1)^2+(y-1)^2=5$$
$$-\tfrac{6}{5}+2\left(\tfrac{3}{5}\right)=0 \qquad \left(-\tfrac{6}{5}-1\right)^2+\left(\tfrac{3}{5}-1\right)^2=5$$
$$-\tfrac{6}{5}+\tfrac{6}{5}=0 \qquad \tfrac{121}{25}+\tfrac{4}{25}=5$$
$$0=0,\ \text{true} \qquad \tfrac{125}{25}=5$$
$$5=5,\ \text{true}$$

The solution set is $\left\{\left(-\frac{6}{5},\frac{3}{5}\right),(2,-1)\right\}$.

✎ *Pencil Problem #2* ✎

2. Solve by the substitution method:

$$\begin{cases} x+y=2 \\ y=x^2-4x+4 \end{cases}$$

Objective #3: Solve nonlinear systems by addition.

✔ *Solved Problem #3*

3. Solve by the addition method:

$$\begin{cases} y = x^2 + 5 \\ x^2 + y^2 = 25 \end{cases}$$

Arrange the first equation so that variable terms appear on the left, and constants appear on the right.
Add the resulting equations to eliminate the x^2-terms and solve for y.

$$\begin{array}{r} -x^2 + y = 5 \\ \underline{x^2 + y^2 = 25} \\ y^2 + y = 30 \end{array}$$

Solve the resulting quadratic equation.

$$y^2 + y - 30 = 0$$
$$(y+6)(y-5) = 0$$
$$y + 6 = 0 \quad \text{or} \quad y - 5 = 0$$
$$y = -6 \quad \text{or} \quad y = 5$$

If $y = -6$,

$$x^2 + (-6)^2 = 25$$
$$x^2 + 36 = 25$$
$$x^2 = -11$$

When $y = -6$ there is no real solution.

If $y = 5$,

$$x^2 + (5)^2 = 25$$
$$x^2 + 25 = 25$$
$$x^2 = 0$$
$$x = 0$$

Check $(0,5)$ in both original equations.

$y = x^2 + 5$	$x^2 + y^2 = 25$
$5 = (0)^2 + 5$	$0^2 + 5^2 = 25$
$5 = 5$, true	$25 = 25$, true

The solution set is $\{(0,5)\}$.

Pencil Problem #3

3. Solve by the addition method:

$$\begin{cases} x^2 + y^2 = 13 \\ x^2 - y^2 = 5 \end{cases}$$

Objective #4: Solve problems using systems of nonlinear equations.

✔ *Solved Problem #4*

4. Find the length and width of a rectangle whose perimeter is 20 feet and whose area is 21 square feet.

The system is $\begin{cases} 2x+2y=20 \\ xy=21. \end{cases}$

Solve the second equation for x: $xy=21$

$$x=\frac{21}{y}$$

Substitute the expression $\frac{21}{y}$ for x in the first equation and solve for y.

$$2x+2y=20$$
$$2\left(\frac{21}{y}\right)+2y=20$$
$$\frac{42}{y}+2y=20$$
$$42+2y^2=20y$$
$$2y^2-20y+42=0$$
$$y^2-10y+21=0$$
$$(y-7)(y-3)=0$$
$$y-7=0 \quad \text{or} \quad y-3=0$$
$$y=7 \quad \text{or} \quad y=3$$

If $y=7$, $x=\frac{21}{7}=3.$

If $y=3$, $x=\frac{21}{3}=7.$

The dimensions are 7 feet by 3 feet.

Pencil Problem #4

4. The sum of two numbers is 10 and their product is 24. Find the numbers.

Answers for Pencil Problems *(Textbook Exercise references in parentheses)*:

1a. false *(7.4 #1)* **1b.** true *(7.4 #27)* **2.** $\{(1,1),(2,0)\}$ *(7.4 #3)*

3. $\{(-3,-2),(-3,2),(3,-2),(3,2)\}$ *(7.4 #19)* **4.** 6 and 4 *(7.4 #43)*

Section 7.5
Systems of Inequalities

Does Your Weight Fit You?

This chapter opened by noting that the modern emphasis on thinness as the ideal body shape has been suggested as a major cause of eating disorders. In this section, the textbook will demonstrate how systems of linear inequalities in two variables can enable you to establish a healthy weight range for your height and age.

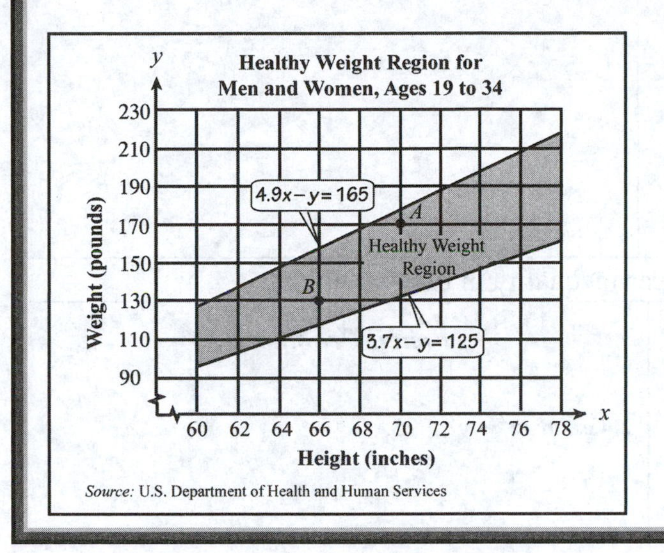

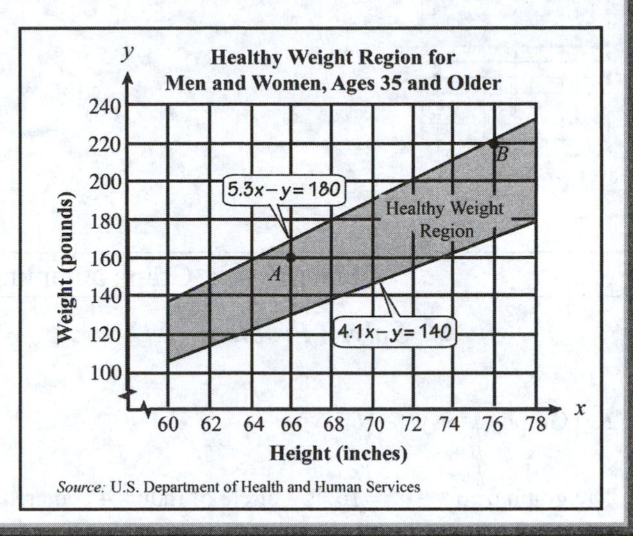

Objective #1: Graph a linear inequality in two variables.

✔ Solved Problem #1

1a. Graph: $4x-2y\geq 8$.

First, graph the equation $4x-2y=8$ with a solid line.

Find the x–intercept:
$$4x-2y=8$$
$$4x-2(0)=8$$
$$4x=8$$
$$x=2$$

Find the y–intercept:
$$4x-2y=8$$
$$4(0)-2y=8$$
$$-2y=8$$
$$y=-4$$

Next, use the origin as a test point.
$$4x-2y\geq 8$$
$$4(0)-2(0)\geq 8$$
$$0\geq 8, \text{ false}$$

Since the statement is false, shade the half-plane that does not contain the test point.

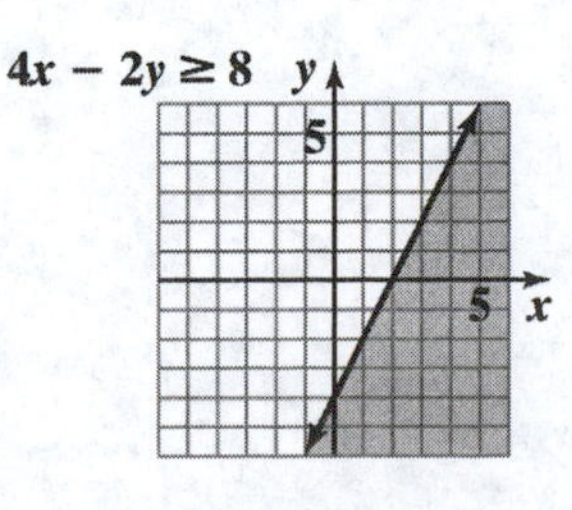

✎ Pencil Problem #1 ✎

1a. Graph: $x-2y>10$.

1b. Graph: $y > 1$.

Graph the line $y = 1$ with a dashed line.

Since the inequality is of the form $y > a$, shade the half-plane above the line.

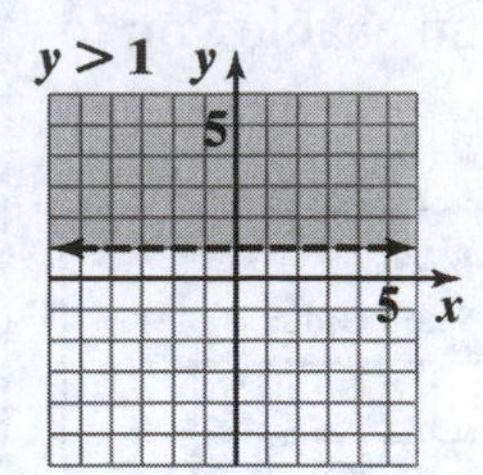

1b. Graph: $x \le 1$.

Objective #2: Graph a nonlinear inequality in two variables..

✔ *Solved Problem #2*

2. Graph: $x^2 + y^2 \ge 16$.

The graph of $x^2 + y^2 = 16$ is a circle of radius 4 centered at the origin. Use a solid circle because equality is included in ≥.

The point (0, 0) is not on the circle, so we use it as a test point. The result is $0 \ge 16$, which is false. Since the point (0, 0) is inside the circle, the region outside the circle belongs to the solution set. Shade the region outside the circle.

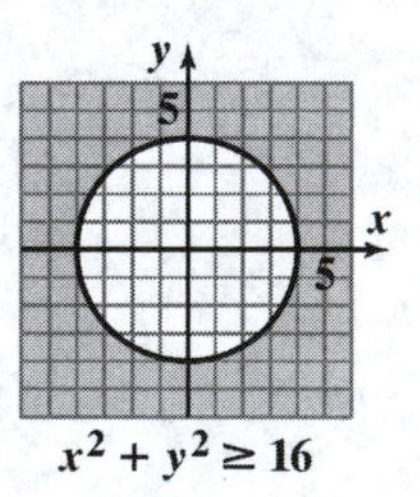

✎ *Pencil Problem #2* ✎

2. Graph: $x^2 + y^2 > 25$.

Objective #3: Use mathematical models involving systems of linear inequalities.

✔ *Solved Problem #3*

3. The healthy weight region for men and women ages 19 to 34 can be modeled by the following system of linear inequalities:

$$\begin{cases} 4.9x - y \geq 165 \\ 3.7x - y \leq 125 \end{cases}$$

Show that (66, 130) is a solution of the system of inequalities that describes healthy weight for this age group.

Substitute the coordinates of (66, 130) into both inequalities of the system.

$$\begin{cases} 4.9x - y \geq 165 \\ 3.7x - y \leq 125 \end{cases}$$

$$4.9x - y \geq 165$$
$$4.9(66) - 130 \geq 165$$
$$193.4 \geq 165, \text{ true}$$

$$3.7x - y \leq 125$$
$$3.7(66) - 130 \leq 125$$
$$114.2 \leq 125, \text{ true}$$

(66, 130) is a solution of the system.

Pencil Problem #3

3. The healthy weight region for men and women ages 35 and older can be modeled by the following system of linear inequalities:

$$\begin{cases} 5.3x - y \geq 180 \\ 4.1x - y \leq 14 \end{cases}$$

Show that (66, 160) is a solution of the system of inequalities that describes healthy weight for this age group.

Objective #4: Graph a system of linear inequalities.

✔ *Solved Problem #4*

4. Graph the solution set of the system:

$$\begin{cases} x - 3y < 6 \\ 2x + 3y \geq -6 \end{cases}$$

Graph the line $x - 3y = 6$ with a dashed line.
Graph the line $2x + 3y = -6$ with a solid line.

For $x - 3y < 6$ use a test point such as (0, 0).

$$x - 3y < 6$$
$$0 - 3(0) < 6$$
$$0 < 6, \text{ true}$$

Since the statement is true, shade the half-plane that contains the test point.

Pencil Problem #4

4. Graph the solution set of the system:

$$\begin{cases} y > 2x - 3 \\ y < -x + 6 \end{cases}$$

For $2x+3y \geq -6$ use a test point such as (0, 0).

$$2x+3y \geq -6$$
$$2(0)+3(0) \geq -6$$
$$0 \geq -6, \text{ true}$$

Since the statement is true, shade the half-plane that contains the test point.

The solution set of the system is the intersection (the overlap) of the two half-planes.

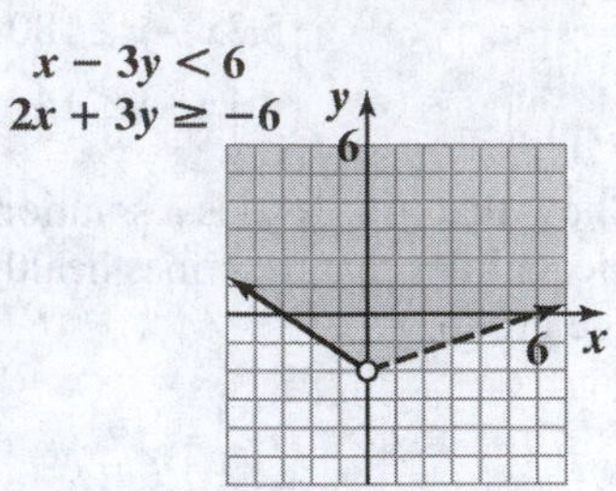

Answers for Pencil Problems *(Textbook Exercise references in parentheses)*:

1a. 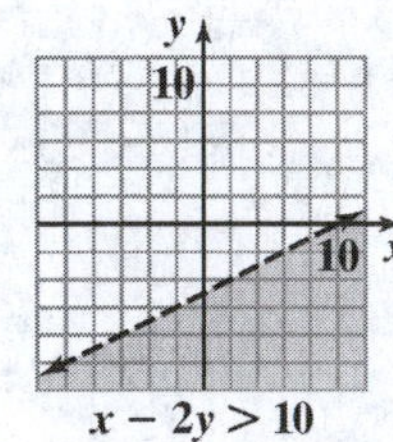

(7.5 #3)

1b. 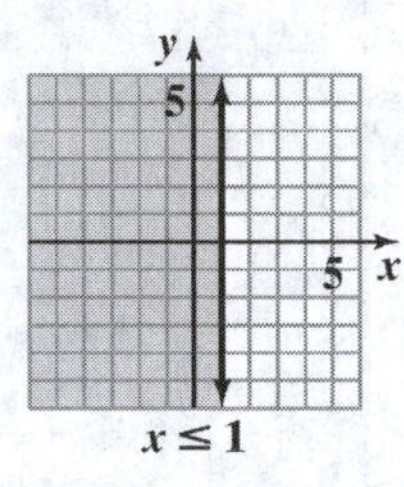

(7.5 #9)

2. 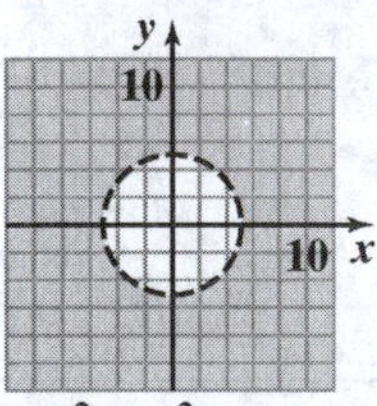

(7.5 #15)

3.

$$5.3x - y \geq 180 \qquad 4.1x - y \leq 14$$
$$5.3(66)-160 \geq 180 \qquad 4.1(66)-160 \leq 140$$
$$189.8 \geq 180, \text{ true} \qquad 110.6 \leq 140, \text{ true}$$

(7.5 #77)

4. 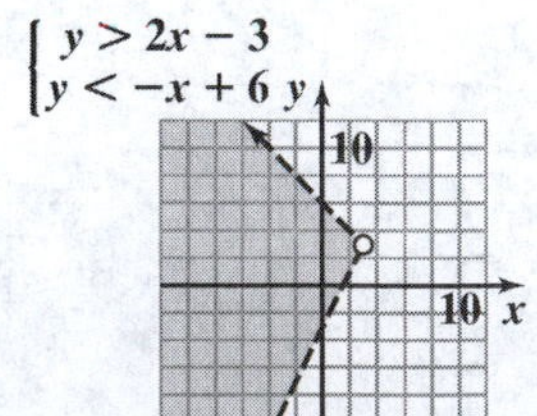

(7.5 #31)

Section 7.6
Linear Programming

***MAXIMUM* Output with *MINIMUM* Effort!**

Many situations in life involve quantities that must be maximized or minimized. Businesses are interested in maximizing profit and minimizing costs.

In the Exercise Set for this section of the textbook, you will encounter a manufacturer looking to maximize profits from selling two models of mountain bikes. But there is a limited amount of time available to assemble and paint these bikes. We will learn how to balance these constraints and determine the proper number of each bike that should be produced.

Objective #1:

Write an objective function describing a quantity that must be maximized or minimized.

✔ *Solved Problem #1*

1. A company manufactures bookshelves and desks for computers. Let x represent the number of bookshelves manufactured daily and y the number of desks manufactured daily. The company's profits are \$25 per bookshelf and \$55 per desk.

Write the objective function that models the company's total daily profit, z, from x bookshelves and y desks.

The total profit is 25 times the number of bookshelves, x, plus 55 times the number of desks, y.

$$\underbrace{z}_{\text{profit}} = \underbrace{25x}_{\text{\$25 for each bookshelf}} + \underbrace{55y}_{\text{\$55 for each desk}}$$

The objective function is $z = 25x + 55y$.

Pencil Problem #1

1. A television manufacturer makes rear-projection and plasma televisions. The profit per unit is \$125 for the rear-projection televisions and \$200 for the plasma televisions. Let x = the number of rear-projection televisions manufactured in a month and y = the number of plasma televisions manufactured in a month.

Write the objective function that models the total monthly profit.

Objective #2: Use inequalities to describe limitations in a situation.

✔ *Solved Problem #2*

2. Recall that the company in *Solved Problem #1* manufactures bookshelves and desks for computers. x represents the number of bookshelves manufactured daily and y the number of desks manufactured daily.

2a. Write an inequality that models the following constraint: To maintain high quality, the company should not manufacture more than a total of 80 bookshelves and desks per day.

$x + y \leq 80$

2b. Write an inequality that models the following constraint: To meet customer demand, the company must manufacture between 30 and 80 bookshelves per day, inclusive.

$30 \leq x \leq 80$

2c. Write an inequality that models the following constraint: The company must manufacture at least 10 and no more than 30 desks per day.

$10 \leq y \leq 30$

2d. Summarize what you have described about this company by writing the objective function for its profits (from *Solved Problem #1*) and the three constraints.

Objective function: $z = 25x + 55y$.

Constraints: $\begin{cases} x + y \leq 80 \\ 30 \leq x \leq 80 \\ 10 \leq y \leq 30 \end{cases}$

Pencil Problem #2

2. Recall that the manufacturer in *Pencil Problem #1* makes rear-projection and plasma televisions. x represents the number of rear-projection televisions manufactured monthly and y the number of plasma televisions manufactured monthly.

2a. Write an inequality that models the following constraint: Equipment in the factory allows for making at most 450 rear-projection televisions in one month.

2b. Write an inequality that models the following constraint: Equipment in the factory allows for making at most 200 plasma televisions in one month.

2c. Write an inequality that models the following constraint: The cost to the manufacturer per unit is \$600 for the rear-projection televisions and \$900 for the plasma televisions. Total monthly costs cannot exceed \$360,000.

2d. Summarize what you have described about this company by writing the objective function for its profits (from *Pencil Problem #1*) and the three constraints.

Objective #3: Use linear programming to solve problems.

✔ *Solved Problem #3*

3a. For the company in *Solved Problems #1 and 2*, how many bookshelves and how many desks should be manufactured per day to obtain maximum profit? What is the maximum daily profit?

Graph the constraints and find the corners, or vertices, of the region of intersection.

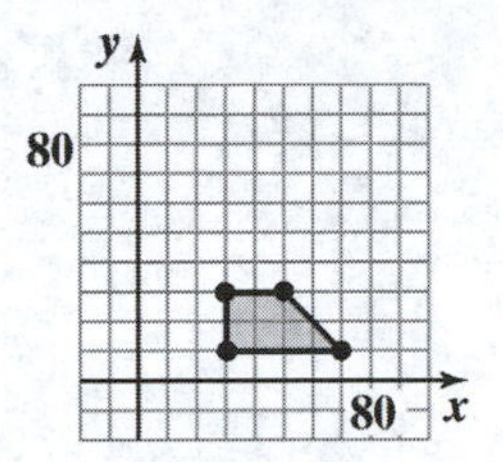

Find the value of the objective function at each corner of the graphed region.

Corner (x, y)	Objective Function $z = 25x + 55y$
$(30,10)$	$z = 25(30)+55(10)$ $= 750+550$ $= 1300$
$(30,30)$	$z = 25(30)+55(30)$ $= 750+1650$ $= 2400$
$(50,30)$	$z = 25(50)+55(30)$ $= 1250+1650$ $= 2900$ (Maximum)
$(70,10)$	$z = 25(70)+55(10)$ $= 1750+550$ $= 2300$

The maximum value of z is 2900 and it occurs at the point (50, 30).

In order to maximize profit, 50 bookshelves and 30 desks must be produced each day for a profit of $2900.

Pencil Problem #3

3a. For the company in *Pencil Problems #1 and 2*, how many rear-projection and plasma televisions should be manufactured per month to obtain maximum profit? What is the maximum monthly profit?

3b. Find the maximum value of the objective function $z = 3x + 5y$ subject to the constraints:

$$\begin{cases} x \geq 0, \quad y \geq 0 \\ x + y \geq 1 \\ x + y \leq 6 \end{cases}$$

Graph the region that represents the intersection of the constraints:

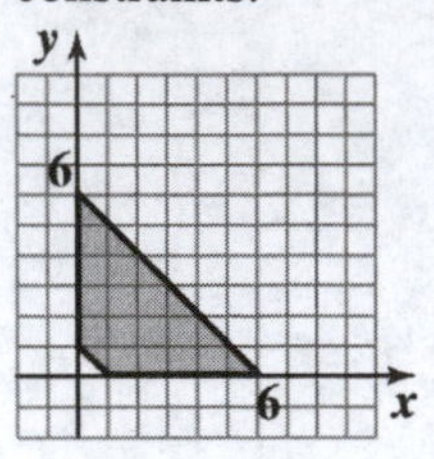

Find the value of the objective function at each corner of the graphed region.

Corner (x, y)	Objective Function $z = 3x + 5y$
(0,1)	$z = 3(0) + 5(1) = 5$
(1,0)	$z = 3(1) + 5(0) = 3$
(0,6)	$z = 3(0) + 5(6) = 30$ (Maximum)
(6,0)	$z = 3(6) + 5(0) = 18$

The maximum value is 30.

3b. Find the maximum value of the objective function $z = 4x + y$ subject to the constraints:

$$\begin{cases} x \geq 0, \quad y \geq 0 \\ 2x + 3y \leq 12 \\ x + y \geq 3 \end{cases}$$

Answers for Pencil Problems *(Textbook Exercise references in parentheses)*:

1. $z = 125x + 200y$ *(7.6 #15a)* **2a.** $x \leq 450$ *(7.6 #15b)* **2b.** $y \leq 200$ *(7.6 #15b)*

2c. $600x + 900y \leq 360{,}000$ *(7.6 #15b)* **2d.** $z = 125x + 200y$; $\begin{cases} x \leq 450 \\ y \leq 200 \\ 600x + 900y \leq 360{,}000 \end{cases}$ *(7.6 #15b)*

3a. 300 rear-projection and 200 plasma televisions; Maximum profit: \$77,500 *(7.6 #15e)* **3b.** 24 *(7.6 #7)*

Section 8.1
Matrix Solutions to Linear Systems

Gender Imbalance in the Workplace

The chart below shows the gender breakdown for various careers in the United States.

In this section of the textbook, we will look at how such data can be placed in something called a matrix and how matrices can be used to solve linear systems.

	Computer Occupations	Lawyers	CEOs of *Fortune 500* Companies	Teachers	Physicians and Surgeons	Registered Nurses
Men	74%	67%	95.4%	25%	65%	11%
Women	26%	33%	4.6%	75%	35%	89%

Source: U.S. Bureau of Labor Statistics

Objective #1: Write the augmented matrix for a linear system.

✔ Solved Problem #1

1. Write the augmented matrix for the system:

$$\begin{cases} x - 2y = 2 \\ 2x + 3y + z = 11 \\ y - 4z = -7 \end{cases}$$

The augmented matrix has a row for each equation and a vertical bar separating the coefficients of the variables on the left from the constants on the right. Coefficients of the same variable are lined up vertically in the same column. If a variable is missing from an equation, its coefficient is 0.

It may be helpful to view the system as

$$\begin{cases} 1x - 2y + 0 = 2 \\ 2x + 3y + 1z = 11 \\ 0x + 1y - 4z = -7 \end{cases}$$

The augmented matrix is

$$\left[\begin{array}{ccc|c} 1 & -2 & 0 & 2 \\ 2 & 3 & 1 & 11 \\ 0 & 1 & -4 & -7 \end{array}\right]$$

Pencil Problem #1

1. Write the augmented matrix for the system:

$$\begin{cases} 5x - 2y - 3z = 0 \\ x + y = 5 \\ 2x - 3z = 4 \end{cases}$$

Objective #2: Perform matrix row operations.

✔ Solved Problem #2

2a. Perform the row operation and write the new matrix:

$$\left[\begin{array}{ccc|c} 4 & 12 & -20 & 8 \\ 1 & 6 & -3 & 7 \\ -3 & -2 & 1 & -9 \end{array}\right];\ \frac{1}{4}R_1$$

Multiply each element in row 1 by $\frac{1}{4}$. The elements in row 2 and row 3 do not change.

$$\left[\begin{array}{ccc|c} \frac{1}{4}(4) & \frac{1}{4}(12) & \frac{1}{4}(-20) & \frac{1}{4}(8) \\ 1 & 6 & -3 & 7 \\ -3 & -2 & 1 & -9 \end{array}\right] = \left[\begin{array}{ccc|c} 1 & 3 & -5 & 2 \\ 1 & 6 & -3 & 7 \\ -3 & -2 & 1 & -9 \end{array}\right]$$

2b. Perform the row operation and write the new matrix:

$$\left[\begin{array}{ccc|c} 4 & 12 & -20 & 8 \\ 1 & 6 & -3 & 7 \\ -3 & -2 & 1 & -9 \end{array}\right];\ 3R_2 + R_3$$

Multiply each element in row 2 by 3 and add to the corresponding element in row 3. Replace the elements in row 3. Row 1 and row 2 do not change.

$$\left[\begin{array}{ccc|c} 4 & 12 & -20 & 8 \\ 1 & 6 & -3 & 7 \\ 3(1)+(-3) & 3(6)+(-2) & 3(-3)+1 & 3(7)+(-9) \end{array}\right]$$

$$= \left[\begin{array}{ccc|c} 4 & 12 & -20 & 8 \\ 1 & 6 & -3 & 7 \\ 0 & 16 & -8 & 12 \end{array}\right]$$

✎ Pencil Problem #2 ✎

2a. Perform the row operation and write the new matrix:

$$\left[\begin{array}{ccc|c} 2 & -6 & 4 & 10 \\ 1 & 5 & -5 & 0 \\ 3 & 0 & 4 & 7 \end{array}\right];\ \frac{1}{2}R_1$$

2b. Perform the row operation and write the new matrix:

$$\left[\begin{array}{ccc|c} 1 & -3 & 2 & 0 \\ 3 & 1 & -1 & 7 \\ 2 & -2 & 1 & 3 \end{array}\right];\ -3R_1 + R_2$$

Objective #3: Use matrices and Gaussian elimination to solve systems.

✔ *Solved Problem #3*

3. Use matrices to solve the system: $\begin{cases} 2x+y+2z=18 \\ x-y+2z=9 \\ x+2y-z=6 \end{cases}$

Write the augmented matrix for the system.

$$\left[\begin{array}{ccc|c} 2 & 1 & 2 & 18 \\ 1 & -1 & 2 & 9 \\ 1 & 2 & -1 & 6 \end{array}\right]$$

We want a 1 in the upper left position. One way to do this is to interchange row 1 and row 2.

$$\left[\begin{array}{ccc|c} 2 & 1 & 2 & 18 \\ 1 & -1 & 2 & 9 \\ 1 & 2 & -1 & 6 \end{array}\right] R_1 \leftrightarrow R_2 = \left[\begin{array}{ccc|c} 1 & -1 & 2 & 9 \\ 2 & 1 & 2 & 18 \\ 1 & 2 & -1 & 6 \end{array}\right]$$

Now we want zeros below the 1 in the first column.

$$\left[\begin{array}{ccc|c} 1 & -1 & 2 & 9 \\ 2 & 1 & 2 & 18 \\ 1 & 2 & -1 & 6 \end{array}\right] -2R_1+R_2 = \left[\begin{array}{ccc|c} 1 & -1 & 2 & 9 \\ 0 & 3 & -2 & 0 \\ 1 & 2 & -1 & 6 \end{array}\right]$$

$$\left[\begin{array}{ccc|c} 1 & -1 & 2 & 9 \\ 0 & 3 & -2 & 0 \\ 1 & 2 & -1 & 6 \end{array}\right] -R_1+R_3 = \left[\begin{array}{ccc|c} 1 & -1 & 2 & 9 \\ 0 & 3 & -2 & 0 \\ 0 & 3 & -3 & -3 \end{array}\right]$$

Next we want a 1 in the second row, second column.

$$\left[\begin{array}{ccc|c} 1 & -1 & 2 & 9 \\ 0 & 3 & -2 & 0 \\ 0 & 3 & -3 & -3 \end{array}\right] \frac{1}{3}R_2 = \left[\begin{array}{ccc|c} 1 & -1 & 2 & 9 \\ 0 & 1 & -\frac{2}{3} & 0 \\ 0 & 3 & -3 & -3 \end{array}\right]$$

Now we want a zero below the 1 in the second row, second column.

$$\left[\begin{array}{ccc|c} 1 & -1 & 2 & 9 \\ 0 & 1 & -\frac{2}{3} & 0 \\ 0 & 3 & -3 & -3 \end{array}\right] -3R_2+R_3 = \left[\begin{array}{ccc|c} 1 & -1 & 2 & 9 \\ 0 & 1 & -\frac{2}{3} & 0 \\ 0 & 0 & -1 & -3 \end{array}\right]$$

✎ *Pencil Problem #3* ✎

3. Use matrices to solve: $\begin{cases} x+y-z=-2 \\ 2x-y+z=5 \\ -x+2y+2z=1 \end{cases}$

Next we want a 1 in the third row, third column.

$$\left[\begin{array}{ccc|c} 1 & -1 & 2 & 9 \\ 0 & 1 & -\frac{2}{3} & 0 \\ 0 & 0 & -1 & -3 \end{array}\right] \quad -R_3 = \left[\begin{array}{ccc|c} 1 & -1 & 2 & 9 \\ 0 & 1 & -\frac{2}{3} & 0 \\ 0 & 0 & 1 & 3 \end{array}\right]$$

The resulting system is:

$$\begin{aligned} x - y + 2z &= 9 \\ y - \frac{2}{3}z &= 0 \\ z &= 3 \end{aligned}$$

Back-substitute 3 for z in the second equation.

$$\begin{aligned} y - \frac{2}{3}(3) &= 0 \\ y - 2 &= 0 \\ y &= 2 \end{aligned}$$

Back-substitute 2 for y and 3 for z in the first equation.

$$\begin{aligned} x - y + 2z &= 9 \\ x - (2) + 2(3) &= 9 \\ x - 2 + 6 &= 9 \\ x + 4 &= 9 \\ x &= 5 \end{aligned}$$

$(5,2,3)$ satisfies both equations.

The solution set is $\{(5,2,3)\}$.

Objective #4: Use matrices and Gauss-Jordan elimination to solve systems.

✔ Solved Problem #4

Solve the system by Gauss-Jordan elimination. Begin with the matrix obtained in Solved Problem #3.

$$\begin{cases} 2x + y + 2z = 18 \\ x - y + 2z = 9 \\ x + 2y - z = 6 \end{cases}$$

The final matrix from Solved Problem #3 is

$$\left[\begin{array}{ccc|c} 1 & -1 & 2 & 9 \\ 0 & 1 & -\frac{2}{3} & 0 \\ 0 & 0 & 1 & 3 \end{array}\right]$$

First we want a zero above the 1 in the second column.

$$\left[\begin{array}{ccc|c} 1 & -1 & 2 & 9 \\ 0 & 1 & -\frac{2}{3} & 0 \\ 0 & 0 & 1 & 3 \end{array}\right] \quad R_2 + R_1 = \left[\begin{array}{ccc|c} 1 & 0 & \frac{4}{3} & 9 \\ 0 & 1 & -\frac{2}{3} & 0 \\ 0 & 0 & 1 & 3 \end{array}\right]$$

Now we want zeros above the 1 in the third column.

$$\left[\begin{array}{ccc|c} 1 & 0 & \frac{4}{3} & 9 \\ 0 & 1 & -\frac{2}{3} & 0 \\ 0 & 0 & 1 & 3 \end{array}\right] \quad -\frac{4}{3}R_3 + R_1 = \left[\begin{array}{ccc|c} 1 & 0 & 0 & 5 \\ 0 & 1 & -\frac{2}{3} & 0 \\ 0 & 0 & 1 & 3 \end{array}\right]$$

$$\left[\begin{array}{ccc|c} 1 & 0 & 0 & 5 \\ 0 & 1 & -\frac{2}{3} & 0 \\ 0 & 0 & 1 & 3 \end{array}\right] \quad \frac{2}{3}R_3 + R_2 = \left[\begin{array}{ccc|c} 1 & 0 & 0 & 5 \\ 0 & 1 & 0 & 2 \\ 0 & 0 & 1 & 3 \end{array}\right]$$

This last matrix corresponds to

$$x = 5, y = 2, z = 3$$

The solution set is $\{(5, 2, 3)\}$.

✎ Pencil Problem #4 ✎

4a. Solve the system by Gauss-Jordan elimination. Begin with the matrix obtained in Pencil Problem #3.

$$\begin{cases} x + y - z = -2 \\ 2x - y + z = 5 \\ -x + 2y + 2z = 1 \end{cases}$$

Answers for Pencil Problems *(Textbook Exercise references in parentheses)*:

1. $\left[\begin{array}{ccc|c} 5 & -2 & -3 & 0 \\ 1 & 1 & 0 & 5 \\ 2 & 0 & -3 & 4 \end{array}\right]$ *(8.1 #5)*

2a. $\left[\begin{array}{ccc|c} 1 & -3 & 2 & 5 \\ 1 & 5 & -5 & 0 \\ 3 & 0 & 4 & 7 \end{array}\right]$ *(8.1 #13)* **2b.** $\left[\begin{array}{ccc|c} 1 & -3 & 2 & 0 \\ 0 & 10 & -7 & 7 \\ 2 & -2 & 1 & 3 \end{array}\right]$ *(8.1 #15)*

3. $\{(1, -1, 2)\}$ *(8.1 #21)*

4. $\{(1, -1, 2)\}$ *(8.1 #21)*

Section 8.2
Inconsistent and Dependent Systems and Their Applications

Lane Closed Ahead! Be Prepared to Stop!

You've allowed yourself barely enough time to get to campus, and now you see a sign for road construction along your normal route. Should you take an alternate route to make it to class on time?

In this section of the textbook, we use systems of equations to model traffic flow. Systems of equations with more than one solution can tell us how many cars should be directed toward alternate routes when flow along one street is limited by road work.

Objective #1: Apply Gaussian elimination to systems without unique solutions.

✔ *Solved Problem #1*

1a. Use Gaussian elimination to solve the system:

$$\begin{cases} x - 2y - z = -5 \\ 2x - 3y - z = 0 \\ 3x - 4y - z = 1 \end{cases}$$

Write the augmented matrix for the system.

$$\left[\begin{array}{ccc|c} 1 & -2 & -1 & -5 \\ 2 & -3 & -1 & 0 \\ 3 & -4 & -1 & 1 \end{array}\right]$$

Attempt to simplify the matrix to row-echelon form. The matrix already has a 1 in the upper left position. We want 0s below the 1. Multiply row 1 by -2 and add to row 2, and multiply row 1 by -3 and add to row 3.

$$\left[\begin{array}{ccc|c} 1 & -2 & -1 & -5 \\ 0 & 1 & 1 & 10 \\ 0 & 2 & 2 & 16 \end{array}\right]$$

Now we want a 0 below the 1 in the second column. Multiply row 2 by -2 and add to row 3.

$$\left[\begin{array}{ccc|c} 1 & -2 & -1 & -5 \\ 0 & 1 & 1 & 10 \\ 0 & 0 & 0 & -4 \end{array}\right]$$

The last row represents $0x + 0y + 0z = -4$, which is false. The system has no solution. The solution set is $\varnothing$, the empty set.

Pencil Problem #1

1a. Use Gaussian elimination to solve the system:

$$\begin{cases} 5x + 12y + z = 10 \\ 2x + 5y + 2z = -1 \\ x + 2y - 3z = 5 \end{cases}$$

1b. Use Gaussian elimination to solve the system:

$$\begin{cases} x - 2y - z = 5 \\ 2x - 5y + 3z = 6 \\ x - 3y + 4z = 1 \end{cases}$$

We begin with the augmented matrix.

$$\left[\begin{array}{ccc|c} 1 & -2 & -1 & 5 \\ 2 & -5 & 3 & 6 \\ 1 & -3 & 4 & 1 \end{array}\right]$$

$$\xrightarrow[-R_1+R_3]{-2R_1+R_2} \left[\begin{array}{ccc|c} 1 & -2 & -1 & 5 \\ 0 & -1 & 5 & -4 \\ 0 & -1 & 5 & -4 \end{array}\right]$$

$$\xrightarrow{-R_2} \left[\begin{array}{ccc|c} 1 & -2 & -1 & 5 \\ 0 & 1 & -5 & 4 \\ 0 & -1 & 5 & -4 \end{array}\right]$$

$$\xrightarrow{R_2+R_3} \left[\begin{array}{ccc|c} 1 & -2 & -1 & 5 \\ 0 & 1 & -5 & 4 \\ 0 & 0 & 0 & 0 \end{array}\right]$$

The original system is equivalent to the system

$$\begin{cases} x - 2y - z = 5 \\ y - 5z = 4 \end{cases}$$

The system is consistent and the equations are dependent. Express x and y in terms of z.

$$y - 5z = 4$$
$$y = 5z + 4$$

$$x - 2y - z = 5$$
$$x = 2y + z + 5$$
$$x = 2(5z + 4) + z + 5$$
$$x = 10z + 8 + z + 5$$
$$x = 11z + 13$$

Each ordered pair of the form $(11z + 13, 5z + 4, z)$ is a solution of the system. The solution set is $\{(11z + 13, 5z + 4, z)\}$.

1b. Use Gaussian elimination to solve the system:

$$\begin{cases} 8x + 5y + 11z = 30 \\ -x - 4y + 2z = 3 \\ 2x - y + 5z = 12 \end{cases}$$

Objective #2: Apply Gaussian elimination to systems with more variables than equations.

✔ Solved Problem #2

2. Use Gaussian elimination to solve the system:

$$\begin{cases} x + 2y + 3z = 70 \\ x + y + z = 60 \end{cases}$$

We begin with the augmented matrix.

$$\left[\begin{array}{ccc|c} 1 & 2 & 3 & 70 \\ 1 & 1 & 1 & 60 \end{array}\right]$$

$$\xrightarrow{-R_1+R_2} \left[\begin{array}{ccc|c} 1 & 2 & 3 & 70 \\ 0 & -1 & -2 & -10 \end{array}\right]$$

$$\xrightarrow{-R_2} \left[\begin{array}{ccc|c} 1 & 2 & 3 & 70 \\ 0 & 1 & 2 & 10 \end{array}\right]$$

The original system is equivalent to the system

$$\begin{cases} x + 2y + 3z = 70 \\ y + 2z = 10 \end{cases}$$

Express x and y in terms of z.

$$y + 2z = 10$$
$$y = -2z + 10$$
$$x + 2y + 3z = 70$$
$$x = -2y - 3z + 70$$
$$x = -2(-2z + 10) - 3z + 70$$
$$x = 4z - 20 - 3z + 70$$
$$x = z + 50$$

Each ordered pair of the form $(z + 50, -2z + 10, z)$ is a solution of the system. The solution set is $\{(z + 50, -2z + 10, z)\}$.

✎ Pencil Problem #2

2. Use Gaussian elimination to solve the system:

$$\begin{cases} 2x + y - z = 2 \\ 3x + 3y - 2z = 3 \end{cases}$$

Objective #3: Solve problems involving systems without unique solutions.

✔ *Solved Problem #3*

3. The figure shows a system of four one-way streets. The numbers in the figure denote the number of cars per minute that travel in the direction shown.

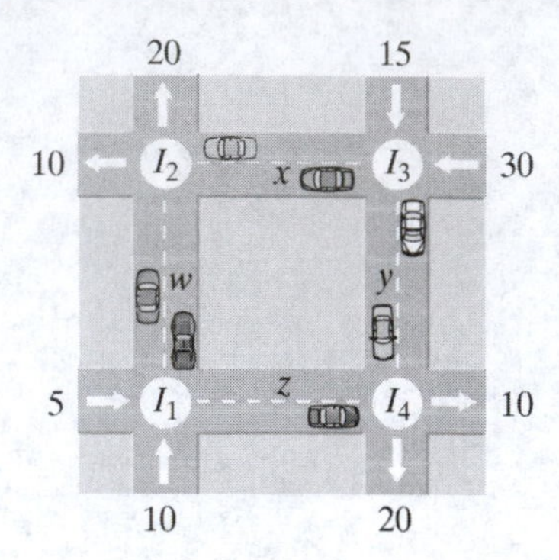

3a. Use the requirement that the number of cars entering each of the intersections per minute must equal the number of cars leaving per minute to set up a system of equations that keeps traffic moving.

Consider one intersection at a time.

I_1: $10 + 5 = 15$ cars enter and $w + z$ leave, so $w + z = 15$.

I_2: $w + x$ cars enter and $10 + 20 = 30$ leave, so $w + x = 30$.

I_3: $15 + 30 = 45$ cars enter and $x + y$ leave, so $x + y = 45$.

I_4: $y + z$ cars enter and $20 + 10 = 30$ leave, so $y + z = 30$.

The system is $\begin{cases} w + z = 15 \\ w + x = 30 \\ x + y = 45 \\ y + z = 30 \end{cases}$

Pencil Problem #3

3. The figure shows a system of four one-way streets. The numbers in the figure denote the number of cars per hour that travel in the direction shown.

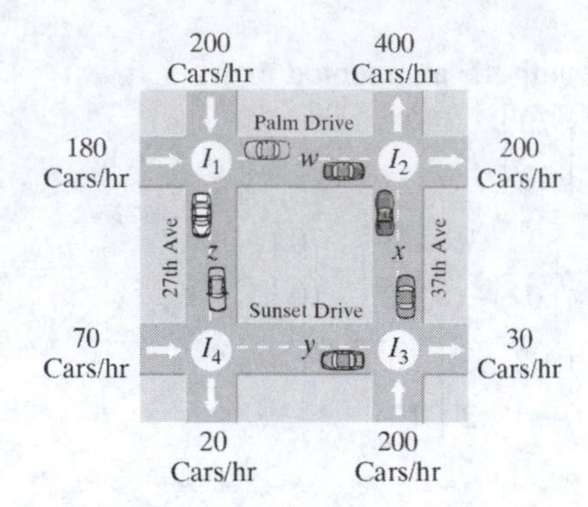

3a. Use the requirement that the number of cars entering each of the intersections per hour must equal the number of cars leaving per hour to set up a system of equations that keeps traffic moving.

3b. Use Gaussian elimination to solve the system.

We begin with the augmented matrix.

$$\left[\begin{array}{cccc|c} 1 & 0 & 0 & 1 & 15 \\ 1 & 1 & 0 & 0 & 30 \\ 0 & 1 & 1 & 0 & 45 \\ 0 & 0 & 1 & 1 & 30 \end{array}\right]$$

$$\xrightarrow{-R_1+R_2} \left[\begin{array}{cccc|c} 1 & 0 & 0 & 1 & 15 \\ 0 & 1 & 0 & -1 & 15 \\ 0 & 1 & 1 & 0 & 45 \\ 0 & 0 & 1 & 1 & 30 \end{array}\right]$$

$$\xrightarrow{-R_2+R_3} \left[\begin{array}{cccc|c} 1 & 0 & 0 & 1 & 15 \\ 0 & 1 & 0 & -1 & 15 \\ 0 & 0 & 1 & 1 & 30 \\ 0 & 0 & 1 & 1 & 30 \end{array}\right]$$

$$\xrightarrow{-R_3+R_4} \left[\begin{array}{cccc|c} 1 & 0 & 0 & 1 & 15 \\ 0 & 1 & 0 & -1 & 15 \\ 0 & 0 & 1 & 1 & 30 \\ 0 & 0 & 0 & 0 & 0 \end{array}\right]$$

From the first row, we get $w + z = 15$, so $w = 15 - z$. From the second and third rows, we get $x - z = 15$ and $y + z = 30$, respectively, so $x = 15 + z$ and $y = 30 - z$. The solution set is $\{(15 - z, 15 + z, 30 - z, z)\}$.

3b. Use Gaussian elimination to solve the system.

3c. If construction limits z to 10 cars per minute, how many cars per minute must pass between the other intersections to keep traffic flowing?

Substitute 10 for z in the system's solution.

$(15 - z, 15 + z, 30 - z, z)$
$= (15 - 10, 15 + 10, 30 - 10, 10)$
$= (5, 25, 20, 10)$

To keep traffic flowing, we must have $w = 5$, $x = 25$, and $y = 20$ cars per minute.

3c. If construction limits z to 50 cars per hour, how many cars per hour must pass between the other intersections to keep traffic flowing?

Answers for Pencil Problems *(Textbook Exercise references in parentheses)*:

1a. no solution or ∅ *(8.2 #1)* **1b.** $\{(5 - 2z, -2 + z, z)\}$ *(8.2 #7)*

2. $\left\{\left(1+\frac{1}{3}z,\ \frac{1}{3}z,\ z\right)\right\}$ *(8.2 #15)*

3a. $\begin{cases} w + z = 380 \\ w + x = 600 \\ x - y = 170 \\ y - z = 50 \end{cases}$ *(8.2 #33a)* **3b.** $\{(380 - z, 220 + z, 50 + z, z)\}$ *(8.2 #33b)*

3c. $w = 330, x = 270, y = 100$ *(8.2 #33c)*

Section 8.3
Matrix Operations and Their Applications

Making Things Clearer

Have you ever had trouble reading a document because the text didn't differ sufficiently from the background? By increasing the contrast between the text and the background, you can often make the document easier to read.

In this section of the textbook, we use matrix operations to change the contrast between a letter and its background and to transform figures through translations, stretching or shrinking, and reflections.

Objective #1: Use matrix notation.

✔ *Solved Problem #1*

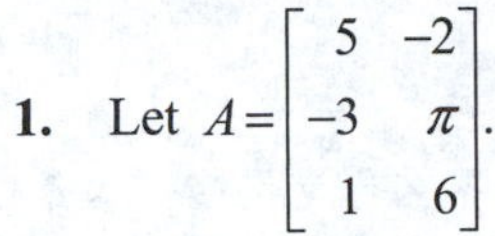

1. Let $A = \begin{bmatrix} 5 & -2 \\ -3 & \pi \\ 1 & 6 \end{bmatrix}$.

1a. What is the order of A?
The matrix has 3 rows and 2 columns, so it is of order 3×2.

1b. Identify a_{12} and a_{31}.

The element a_{12} is in the first row and second column: $a_{12} = -2$.

The element a_{31} is in the third row and first column: $a_{31} = 1$.

✎ *Pencil Problem #1* ✎

1. Let $A = \begin{bmatrix} 1 & -5 & \pi & e \\ 0 & 7 & -6 & -\pi \\ -2 & \frac{1}{2} & 11 & -\frac{1}{5} \end{bmatrix}$.

1a. What is the order of A?

1b. Identify a_{32} and a_{23}.

Objective #2: Understand what is meant by equal matrices.

✔ *Solved Problem #2*

2. Find values for the variables so that the matrices are equal.

$$\begin{bmatrix} x & y+1 \\ z & 6 \end{bmatrix} = \begin{bmatrix} 1 & 5 \\ 3 & 6 \end{bmatrix}$$

These matrices are of the same order, so they are equal if and only if corresponding elements are equal.

$x = 1$
$y + 1 = 5$, so $y = 4$
$z = 3$

✎ *Pencil Problem #2* ✎

2. Find values for the variables so that the matrices are equal.

$$\begin{bmatrix} x & 2y \\ z & 9 \end{bmatrix} = \begin{bmatrix} 4 & 12 \\ 3 & 9 \end{bmatrix}$$

Objective #3: Add and subtract matrices.

✔ *Solved Problem #3*

3. Perform the indicated matrix operations.

3a. $\begin{bmatrix} -4 & 3 \\ 7 & -6 \end{bmatrix} + \begin{bmatrix} 6 & -3 \\ 2 & -4 \end{bmatrix}$

Add corresponding elements.

$$\begin{bmatrix} -4+6 & 3+(-3) \\ 7+2 & -6+(-4) \end{bmatrix} = \begin{bmatrix} 2 & 0 \\ 9 & -10 \end{bmatrix}$$

3b. $\begin{bmatrix} 5 & 4 \\ -3 & 7 \\ 0 & 1 \end{bmatrix} - \begin{bmatrix} -4 & 8 \\ 6 & 0 \\ -5 & 3 \end{bmatrix}$

Subtract corresponding elements.

$$\begin{bmatrix} 5-(-4) & 4-8 \\ -3-6 & 7-0 \\ 0-(-5) & 1-3 \end{bmatrix} = \begin{bmatrix} 9 & -4 \\ -9 & 7 \\ 5 & -2 \end{bmatrix}$$

Pencil Problem #3

3. Perform the indicated matrix operations.

3a. $\begin{bmatrix} 1 & 3 \\ 3 & 4 \\ 5 & 6 \end{bmatrix} + \begin{bmatrix} 2 & -1 \\ 3 & -2 \\ 0 & 1 \end{bmatrix}$

3b. $\begin{bmatrix} 4 & 1 \\ 3 & 2 \end{bmatrix} - \begin{bmatrix} 5 & 9 \\ 0 & 7 \end{bmatrix}$

Objective #4: Perform scalar multiplication.

✔ *Solved Problem #4*

4. If $A = \begin{bmatrix} -4 & 1 \\ 3 & 0 \end{bmatrix}$ and $B = \begin{bmatrix} -1 & -2 \\ 8 & 5 \end{bmatrix}$, find each of the following.

4a. $-6B$

$$-6B = -6\begin{bmatrix} -1 & -2 \\ 8 & 5 \end{bmatrix} = \begin{bmatrix} -6(-1) & -6(-2) \\ -6(8) & -6(5) \end{bmatrix} = \begin{bmatrix} 6 & 12 \\ -48 & -30 \end{bmatrix}$$

Pencil Problem #4

4. If $A = \begin{bmatrix} 2 \\ -4 \\ 1 \end{bmatrix}$ and $B = \begin{bmatrix} -5 \\ 3 \\ -1 \end{bmatrix}$, find each of the following.

4a. $-4A$

4b. $3A + 2B$

$$3A+2B=3\begin{bmatrix}-4 & 1\\ 3 & 0\end{bmatrix}+2\begin{bmatrix}-1 & -2\\ 8 & 5\end{bmatrix}$$
$$=\begin{bmatrix}3(-4) & 3(1)\\ 3(3) & 3(0)\end{bmatrix}+\begin{bmatrix}2(-1) & 2(-2)\\ 2(8) & 2(5)\end{bmatrix}$$
$$=\begin{bmatrix}-12 & 3\\ 9 & 0\end{bmatrix}+\begin{bmatrix}-2 & -4\\ 16 & 10\end{bmatrix}$$
$$=\begin{bmatrix}-12+(-2) & 3+(-4)\\ 9+16 & 0+10\end{bmatrix}$$
$$=\begin{bmatrix}-14 & -1\\ 25 & 10\end{bmatrix}$$

4b. $3A + 2B$

Objective #5: Solve matrix equations.

✔ *Solved Problem #5*

5. Solve for X in the matrix equation $3X + A = B$ where $A=\begin{bmatrix}2 & -8\\ 0 & 4\end{bmatrix}$ and $B=\begin{bmatrix}-10 & 1\\ -9 & 17\end{bmatrix}$.

Begin by solving the matrix equation for X.

$$3X + A = B$$
$$3X = B - A$$
$$X = \frac{1}{3}(B - A)$$

Now use matrices A and B to find X.

$$X=\frac{1}{3}\left(\begin{bmatrix}-10 & 1\\ -9 & 17\end{bmatrix}-\begin{bmatrix}2 & -8\\ 0 & 4\end{bmatrix}\right)$$
$$=\frac{1}{3}\begin{bmatrix}-12 & 9\\ -9 & 13\end{bmatrix}$$
$$=\begin{bmatrix}-4 & 3\\ -3 & \frac{13}{3}\end{bmatrix}$$

✎ *Pencil Problem #5* ✎

5. Solve for X in the matrix equation $2X + A = B$ where $A=\begin{bmatrix}-3 & -7\\ 2 & -9\\ 5 & 0\end{bmatrix}$ and $B=\begin{bmatrix}-5 & -1\\ 0 & 0\\ 3 & -4\end{bmatrix}$.

Objective #6: Multiply matrices.

✔ *Solved Problem #6*

6a. Find AB, given $A=\begin{bmatrix}1 & 3\\ 2 & 5\end{bmatrix}$ and $B=\begin{bmatrix}4 & 6\\ 1 & 0\end{bmatrix}$.

$$AB=\begin{bmatrix}1 & 3\\ 2 & 5\end{bmatrix}\begin{bmatrix}4 & 6\\ 1 & 0\end{bmatrix}$$
$$=\begin{bmatrix}1(4)+3(1) & 1(6)+3(0)\\ 2(4)+5(1) & 2(6)+5(0)\end{bmatrix}$$
$$=\begin{bmatrix}7 & 6\\ 13 & 12\end{bmatrix}$$

6b. Find the product, if possible.

$$\begin{bmatrix}1 & 3\\ 0 & 2\end{bmatrix}\begin{bmatrix}2 & 3 & -1 & 6\\ 0 & 5 & 4 & 1\end{bmatrix}$$

The number of columns in the first matrix equals the number of rows in the second matrix, so it is possible to find the product.

$$\begin{bmatrix}1 & 3\\ 0 & 2\end{bmatrix}\begin{bmatrix}2 & 3 & -1 & 6\\ 0 & 5 & 4 & 1\end{bmatrix}$$
$$=\begin{bmatrix}1(2)+3(0) & 1(3)+3(5) & 1(-1)+3(4) & 1(6)+3(1)\\ 0(2)+2(0) & 0(3)+2(5) & 0(-1)+2(4) & 0(6)+2(1)\end{bmatrix}$$
$$=\begin{bmatrix}2 & 18 & 11 & 9\\ 0 & 10 & 8 & 2\end{bmatrix}$$

6c. Find the product, if possible.

$$\begin{bmatrix}2 & 3 & -1 & 6\\ 0 & 5 & 4 & 1\end{bmatrix}\begin{bmatrix}1 & 3\\ 0 & 2\end{bmatrix}$$

The number of columns in the first matrix does not equal the number of rows in the second matrix. The product of the matrices is undefined.

Pencil Problem #6

6a. Find AB, given $A=\begin{bmatrix}1 & 3\\ 5 & 3\end{bmatrix}$ and $B=\begin{bmatrix}3 & -2\\ -1 & 6\end{bmatrix}$.

6b. Find the product, if possible.

$$\begin{bmatrix}4 & 2\\ 6 & 1\\ 3 & 5\end{bmatrix}\begin{bmatrix}2 & 3 & 4\\ -1 & -2 & 0\end{bmatrix}$$

6c. Find the product, if possible.

$$\begin{bmatrix}2 & 3 & 4\\ -1 & -2 & 0\end{bmatrix}\begin{bmatrix}4 & 2\\ 6 & 1\\ 3 & 5\end{bmatrix}$$

Objective #7: Model applied situations with matrix operations.

✔ *Solved Problem #7*

7. The triangle with vertices (0, 0), (3, 5), and (4, 2) in a rectangular coordinate system can be represented by the matrix $\begin{bmatrix} 0 & 3 & 4 \\ 0 & 5 & 2 \end{bmatrix}$. Use matrix operations to move the triangle 3 units to the left and 1 unit down. Graph the original triangle and the transformed triangle in the same rectangular coordinate system.

We subtract 3 from each x-coordinate and subtract 1 from each y-coordinate.

$$\begin{bmatrix} 0 & 3 & 4 \\ 0 & 5 & 2 \end{bmatrix} + \begin{bmatrix} -3 & -3 & -3 \\ -1 & -1 & -1 \end{bmatrix}$$

$$= \begin{bmatrix} -3 & 0 & 1 \\ -1 & 4 & 1 \end{bmatrix}$$

The vertices of the translated triangle are (–3, –4), (0, 4), and (1, 1).

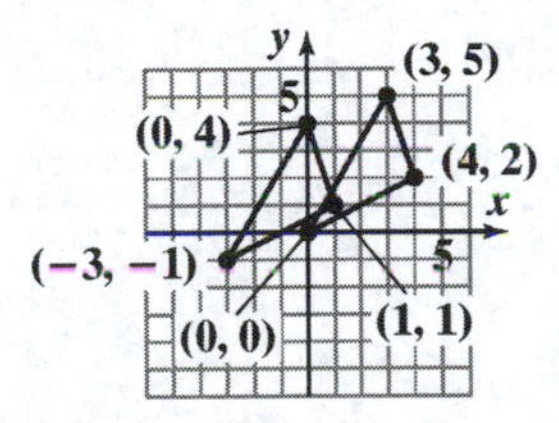

✎ *Pencil Problem #7* ✎

7. An L-shaped figure has vertices at (0, 0), (3, 0), (3, 1), (1, 1), (1, 5), and (0, 5) in a rectangular coordinate system and can be represented by the matrix $\begin{bmatrix} 0 & 3 & 3 & 1 & 1 & 0 \\ 0 & 0 & 1 & 1 & 5 & 5 \end{bmatrix}$. Use matrix operations to move the figure 2 units to the left and 3 units down.
Graph the original figure and the transformed figure in the same rectangular coordinate system.

Answers for Pencil Problems *(Textbook Exercise references in parentheses)*:

1a. 3×4 *(8.3 #3a)* **1b.** $a_{32} = \frac{1}{2}$; $a_{23} = -6$ *(8.3 #3b)*

2. $x = 4, y = 6, z = 3$ *(8.3 #7)*

3a. $\begin{bmatrix} 3 & 2 \\ 6 & 2 \\ 5 & 7 \end{bmatrix}$ *(8.3 #11a)* **3b.** $\begin{bmatrix} -1 & -8 \\ 3 & -5 \end{bmatrix}$ *(8.3 #9b)*

4a. $\begin{bmatrix} -8 \\ 16 \\ -4 \end{bmatrix}$ *(8.3 #13c)* **4b.** $\begin{bmatrix} -4 \\ -6 \\ 1 \end{bmatrix}$ *(8.3 #13d)*

5. $X = \begin{bmatrix} -1 & 3 \\ -1 & \frac{9}{2} \\ -1 & -2 \end{bmatrix}$ *(8.3 #19)*

6a. $\begin{bmatrix} 0 & 16 \\ 12 & 8 \end{bmatrix}$ *(8.3 #27a)* **6b.** $\begin{bmatrix} 6 & 8 & 16 \\ 11 & 16 & 24 \\ 1 & -1 & 12 \end{bmatrix}$ *(8.3 #33a)* **6c.** $\begin{bmatrix} 38 & 27 \\ -16 & -4 \end{bmatrix}$ *(8.3 #33b)*

7. $\begin{bmatrix} -2 & 1 & 1 & -1 & -1 & -2 \\ -3 & -3 & -2 & -2 & 2 & 2 \end{bmatrix}$; *(8.3 #53)*

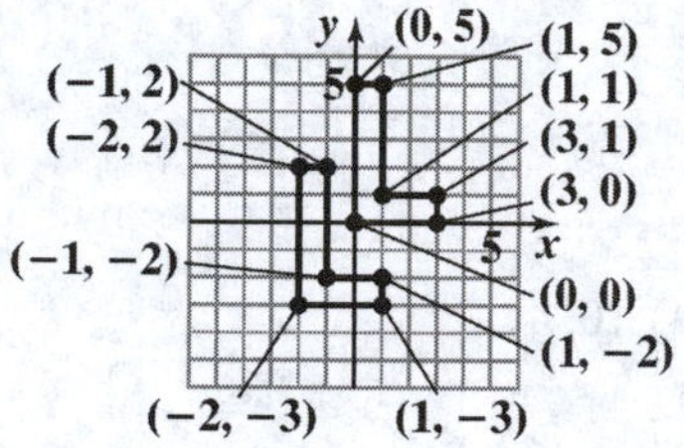

Section 8.4
Multiplicative Inverses of Matrices and Matrix Equations

THINK YOU'RE TOO OLD FOR SENDING SECRET MESSAGES?

Did you know that a secure electronic message is usually encrypted in such a way that only the receiver will be able to decrypt it to read it? The technology used is a bit more sophisticated than the cereal-box ciphers that we used to send secret messages as kids.

Matrix multiplication can be used to encrypt a message. To decrypt the message, we need to know the multiplicative inverse of the encryption matrix. Without the original matrix or its inverse, it is extremely difficult to "break the code."

Objective #1: Find the multiplicative inverse of a square matrix.

✔ *Solved Problem #1*

1a. Find the multiplicative inverse of $A=\begin{bmatrix} 3 & -2 \\ -1 & 1 \end{bmatrix}$.

$A=\begin{bmatrix} a & b \\ c & d \end{bmatrix}=\begin{bmatrix} 3 & -2 \\ -1 & 1 \end{bmatrix}$, so $a=3, b=-2, c=-1,$ and $d=1$.

$ad-bc=3(1)-(-2)(-1)=1\neq 0,$ so the matrix has an inverse.

Using the quick method,

$$A^{-1}=\frac{1}{ad-bc}\begin{bmatrix} d & -b \\ -c & a \end{bmatrix}$$
$$=\frac{1}{3(1)-(-2)(-1)}\begin{bmatrix} 1 & -(-2) \\ -(-1) & 3 \end{bmatrix}$$
$$=\frac{1}{1}\begin{bmatrix} 1 & 2 \\ 1 & 3 \end{bmatrix}=\begin{bmatrix} 1 & 2 \\ 1 & 3 \end{bmatrix}$$

You can verify the result by showing that $AA^{-1}=I_2$ and $A^{-1}A=I_2$, where $I_2=\begin{bmatrix} 1 & 0 \\ 0 & 1 \end{bmatrix}$ is the 2×2 identity matrix.

Pencil Problem #1

1a. Find the multiplicative inverse of $A=\begin{bmatrix} 2 & 3 \\ -1 & 2 \end{bmatrix}$.

1b. Find the multiplicative inverse of $A = \begin{bmatrix} 1 & 0 & 2 \\ -1 & 2 & 3 \\ 1 & -1 & 0 \end{bmatrix}$.

Form the augmented matrix $[A|I_3]$ and perform row operations to obtain a matrix of the form $[I_3|B]$.

$$\left[\begin{array}{ccc|ccc} 1 & 0 & 2 & 1 & 0 & 0 \\ -1 & 2 & 3 & 0 & 1 & 0 \\ 1 & -1 & 0 & 0 & 0 & 1 \end{array}\right]$$

$$\xrightarrow[-R_1+R_3]{R_1+R_2} \left[\begin{array}{ccc|ccc} 1 & 0 & 2 & 1 & 0 & 0 \\ 0 & 2 & 5 & 1 & 1 & 0 \\ 0 & -1 & -2 & -1 & 0 & 1 \end{array}\right]$$

$$\xrightarrow{\frac{1}{2}R_2} \left[\begin{array}{ccc|ccc} 1 & 0 & 2 & 1 & 0 & 0 \\ 0 & 1 & \frac{5}{2} & \frac{1}{2} & \frac{1}{2} & 0 \\ 0 & -1 & -2 & -1 & 0 & 1 \end{array}\right]$$

$$\xrightarrow{R_2+R_3} \left[\begin{array}{ccc|ccc} 1 & 0 & 2 & 1 & 0 & 0 \\ 0 & 1 & \frac{5}{2} & \frac{1}{2} & \frac{1}{2} & 0 \\ 0 & 0 & \frac{1}{2} & -\frac{1}{2} & \frac{1}{2} & 1 \end{array}\right]$$

$$\xrightarrow{2R_3} \left[\begin{array}{ccc|ccc} 1 & 0 & 2 & 1 & 0 & 0 \\ 0 & 1 & \frac{5}{2} & \frac{1}{2} & \frac{1}{2} & 0 \\ 0 & 0 & 1 & -1 & 1 & 2 \end{array}\right]$$

$$\xrightarrow[-\frac{5}{2}R_3+R_2]{-2R_3+R_1} \left[\begin{array}{ccc|ccc} 1 & 0 & 0 & 3 & -2 & -4 \\ 0 & 1 & 0 & 3 & -2 & -5 \\ 0 & 0 & 1 & -1 & 1 & 2 \end{array}\right]$$

The inverse matrix is

$$A^{-1} = \begin{bmatrix} 3 & -2 & -4 \\ 3 & -2 & -5 \\ -1 & 1 & 2 \end{bmatrix}.$$

You can verify the result by showing that $AA^{-1} = I_3$ and $A^{-1}A = I_3$.

1b. Find the multiplicative inverse of

$$A = \begin{bmatrix} 1 & 2 & -1 \\ -2 & 0 & 1 \\ 1 & -1 & 0 \end{bmatrix}.$$

Objective #2: Use inverses to solve matrix equations.

✔ *Solved Problem #2*

2. Solve the system by using A^{-1}, the inverse of the coefficient matrix, where $A^{-1} = \begin{bmatrix} 3 & -2 & -4 \\ 3 & -2 & -5 \\ -1 & 1 & 2 \end{bmatrix}$.

$$\begin{cases} x + 2z = 6 \\ -x + 2y + 3z = -5 \\ x - y = 6 \end{cases}$$

The system can be written as

$$\begin{bmatrix} 1 & 0 & 2 \\ -1 & 2 & 3 \\ 1 & -1 & 0 \end{bmatrix} \begin{bmatrix} x \\ y \\ z \end{bmatrix} = \begin{bmatrix} 6 \\ -5 \\ 6 \end{bmatrix},$$

which is of the form $AX = B$. The solution is $X = A^{-1}B$.

$$X = A^{-1}B = \begin{bmatrix} 3 & -2 & -4 \\ 3 & -2 & -5 \\ -1 & 1 & 2 \end{bmatrix} \begin{bmatrix} 6 \\ -5 \\ 6 \end{bmatrix}$$

$$= \begin{bmatrix} 3(6) - 2(-5) - 4(6) \\ 3(6) - 2(-5) - 5(6) \\ -1(6) + 1(-5) + 2(6) \end{bmatrix}$$

$$= \begin{bmatrix} 4 \\ -2 \\ 1 \end{bmatrix}$$

So, $x = 4$, $y = -2$, and $z = 1$.
The solution set is $\{(4, -2, 1)\}$.

Pencil Problem #2

2. Solve the system by using A^{-1}, the inverse of the coefficient matrix, where

$$A^{-1} = \begin{bmatrix} 3 & 3 & -1 \\ -2 & -2 & 1 \\ -4 & -5 & 2 \end{bmatrix}.$$

$$\begin{cases} x - y + z = 8 \\ 2y - z = -7 \\ 2x + 3y = 1 \end{cases}$$

Objective #3: Encode and decode messages.

✔ *Solved Problem #3*

3a. Use the coding matrix $\begin{bmatrix} -2 & -3 \\ 3 & 4 \end{bmatrix}$ to encode the word BASE.

The numerical equivalent of the word BASE is 2, 1, 19, 5.

The matrix for the word BASE is $\begin{bmatrix} 2 & 19 \\ 1 & 5 \end{bmatrix}$.

Multiply using the encoding matrix on the left.

$$\begin{bmatrix} -2 & -3 \\ 3 & 4 \end{bmatrix}\begin{bmatrix} 2 & 19 \\ 1 & 5 \end{bmatrix} = \begin{bmatrix} -2(2)-3(1) & -2(19)-3(5) \\ 3(2)+4(1) & 3(19)+4(5) \end{bmatrix}$$

$$= \begin{bmatrix} -7 & -53 \\ 10 & 77 \end{bmatrix}$$

The encoded message is –7, 10, –53, 77.

3b. Decode the word encoded in Solved Problem 3a.
Find the inverse of the coding matrix.

$$A^{-1} = \frac{1}{-2(4)-(-3)(3)}\begin{bmatrix} 4 & -(-3) \\ -3 & -2 \end{bmatrix}$$

$$= \frac{1}{1}\begin{bmatrix} 4 & 3 \\ -3 & -2 \end{bmatrix} = \begin{bmatrix} 4 & 3 \\ -3 & -2 \end{bmatrix}$$

Multiply A^{-1} and the coded matrix from Solved Problem 3a.

$$\begin{bmatrix} 4 & 3 \\ -3 & -2 \end{bmatrix}\begin{bmatrix} -7 & -53 \\ 10 & 77 \end{bmatrix}$$

$$= \begin{bmatrix} 4(-7)+3(10) & 4(-53)+3(77) \\ -3(-7)-2(10) & -3(-53)-2(77) \end{bmatrix}$$

$$= \begin{bmatrix} 2 & 19 \\ 1 & 5 \end{bmatrix}$$

The decoded message is 2, 1, 19, 5, or BASE.

✎ *Pencil Problem #3* ✎

3a. Use the coding matrix $\begin{bmatrix} 4 & -1 \\ -3 & 1 \end{bmatrix}$ to encode the word HELP.

3b. Decode the word encoded in Pencil Problem 3a.

Answers for Pencil Problems *(Textbook Exercise references in parentheses)*:

1a. $A^{-1} = \frac{1}{7}\begin{bmatrix} 2 & -3 \\ 1 & 2 \end{bmatrix} = \begin{bmatrix} \frac{2}{7} & -\frac{3}{7} \\ \frac{1}{7} & \frac{2}{7} \end{bmatrix}$ *(8.4 #13)* **1b.** $A^{-1} = \begin{bmatrix} 1 & 1 & 2 \\ 1 & 1 & 1 \\ 2 & 3 & 4 \end{bmatrix}$ *(8.4 #21)*

2. {(2, –1, 5)} *(8.4 #39)*

3a. 27, –19, 32, –20 *(8.4 #51)* **3b.** 8, 5, 12, 16 or HELP *(8.4 #51)*

Section 8.5
Determinants and Cramer's Rule

Look……Closer!!!

Do you see the difference between these two mathematical expressions?

$$\begin{bmatrix} 1 & 2 \\ 0 & 1 \end{bmatrix} \qquad \begin{vmatrix} 1 & 2 \\ 0 & 1 \end{vmatrix}$$

If you look carefully, you will notice that the expression on the left is surrounded by brackets, [ ], and is therefore a **matrix**. The expression on the right is surrounded by bars, | |, and represents a **determinant**.

But be careful! This section will discuss *both* determinants *and* matrices.

Objective #1: Evaluate a second-order determinant.

✔ Solved Problem #1

1. Evaluate the determinant of the matrix: $\begin{bmatrix} 10 & 9 \\ 6 & 5 \end{bmatrix}$

The determinant of the matrix $\begin{bmatrix} 10 & 9 \\ 6 & 5 \end{bmatrix}$ is $\begin{vmatrix} 10 & 9 \\ 6 & 5 \end{vmatrix}$.

$$\begin{vmatrix} 10 & 9 \\ 6 & 5 \end{vmatrix} = 10(5) - 6(9) = 50 - 54 = -4$$

Pencil Problem #1

1. Evaluate the determinant: $\begin{vmatrix} -4 & 1 \\ 5 & 6 \end{vmatrix}$

Objective #2: Solve a system of linear equations in two variables using Cramer's rule.

✔ Solved Problem #2

2. Use Cramer's rule to solve the system:

$$\begin{cases} 5x + 4y = 12 \\ 3x - 6y = 24 \end{cases}$$

$$D = \begin{vmatrix} 5 & 4 \\ 3 & -6 \end{vmatrix} = 5(-6) - 3(4) = -30 - 12 = -42$$

$$D_x = \begin{vmatrix} 12 & 4 \\ 24 & -6 \end{vmatrix} = 12(-6) - 24(4) = -72 - 96 = -168$$

$$D_y = \begin{vmatrix} 5 & 12 \\ 3 & 24 \end{vmatrix} = 5(24) - 3(12) = 120 - 36 = 84$$

$$x = \frac{D_x}{D} = \frac{-168}{-42} = 4 \quad y = \frac{D_y}{D} = \frac{84}{-42} = -2$$

The solution set is $\{(4, -2)\}$.

Pencil Problem #2

2. Use Cramer's rule to solve the system:

$$\begin{cases} 12x + 3y = 15 \\ 2x - 3y = 13 \end{cases}$$

Objective #3: Evaluate a third-order determinant.

✔ Solved Problem #3

3. Evaluate the determinant of the matrix:

$$\begin{bmatrix} 2 & 1 & 7 \\ -5 & 6 & 0 \\ -4 & 3 & 1 \end{bmatrix}$$

$$\begin{vmatrix} 2 & 1 & 7 \\ -5 & 6 & 0 \\ -4 & 3 & 1 \end{vmatrix}$$

$$= 2\begin{vmatrix} 6 & 0 \\ 3 & 1 \end{vmatrix} - (-5)\begin{vmatrix} 1 & 7 \\ 3 & 1 \end{vmatrix} - 4\begin{vmatrix} 1 & 7 \\ 6 & 0 \end{vmatrix}$$

$$= 2(6(1)-3(0))+5(1(1)-3(7))-4(1(0)-6(7))$$

$$= 2(6)+5(-20)-4(-42)$$

$$= 12-100+168$$

$$= 80$$

✎ Pencil Problem #3

3. Evaluate the determinant:

$$\begin{vmatrix} 3 & 0 & 0 \\ 2 & 1 & -5 \\ 2 & 5 & -1 \end{vmatrix}$$

Objective #4: Solve a system of linear equations in three variables using Cramer's rule.

✔ Solved Problem #4

4. Use Cramer's rule to solve the system:

$$\begin{cases} 3x-2y+z=16 \\ 2x+3y-z=-9 \\ x+4y+3z=2 \end{cases}$$

First, find D, D_x, D_y, and D_z.

$$D = \begin{vmatrix} 3 & -2 & 1 \\ 2 & 3 & -1 \\ 1 & 4 & 3 \end{vmatrix}$$

$$= 3\begin{vmatrix} 3 & -1 \\ 4 & 3 \end{vmatrix} - 2\begin{vmatrix} -2 & 1 \\ 4 & 3 \end{vmatrix} + 1\begin{vmatrix} -2 & 1 \\ 3 & -1 \end{vmatrix}$$

$$= 58$$

$$D_x = \begin{vmatrix} 16 & -2 & 1 \\ -9 & 3 & -1 \\ 2 & 4 & 3 \end{vmatrix}$$

$$= 16\begin{vmatrix} 3 & -1 \\ 4 & 3 \end{vmatrix} - (-9)\begin{vmatrix} -2 & 1 \\ 4 & 3 \end{vmatrix} + 2\begin{vmatrix} -2 & 1 \\ 3 & -1 \end{vmatrix}$$

$$= 116$$

✎ Pencil Problem #4

4. Use Cramer's rule to solve the system:

$$\begin{cases} x+y+z=0 \\ 2x-y+z=-1 \\ -x+3y-z=-8 \end{cases}$$

$$D_y = \begin{vmatrix} 3 & 16 & 1 \\ 2 & -9 & -1 \\ 1 & 2 & 3 \end{vmatrix}$$

$$= 3\begin{vmatrix} -9 & -1 \\ 2 & 3 \end{vmatrix} - 2\begin{vmatrix} 16 & 1 \\ 2 & 3 \end{vmatrix} + 1\begin{vmatrix} 16 & 1 \\ -9 & -1 \end{vmatrix}$$

$$= -174$$

$$D_z = \begin{vmatrix} 3 & -2 & 16 \\ 2 & 3 & -9 \\ 1 & 4 & 2 \end{vmatrix}$$

$$= 3\begin{vmatrix} 3 & -9 \\ 4 & 2 \end{vmatrix} - 2\begin{vmatrix} -2 & 16 \\ 4 & 2 \end{vmatrix} + 1\begin{vmatrix} -2 & 16 \\ 3 & -9 \end{vmatrix}$$

$$= 232$$

Next, use $D,\ D_x,\ D_y,$ and D_z to find $x, y,$ and z.

$D = 58$, $D_x = 116$, $D_y = -174$, and $D_z = 232$.

$$x = \frac{D_x}{D} = \frac{116}{58} = 2$$

$$y = \frac{D_y}{D} = \frac{-174}{58} = -3$$

$$z = \frac{D_z}{D} = \frac{232}{58} = 4$$

The solution set is $\{(2, -3, 4)\}$.

Objective #5: Evaluate higher-order determinants.

✔ ***Solved Problem #5***

5. Evaluate the determinant: $\begin{vmatrix} 0 & 4 & 0 & -3 \\ -1 & 1 & 5 & 2 \\ 1 & -2 & 0 & 6 \\ 3 & 0 & 0 & 1 \end{vmatrix}$.

With three 0s in the third column, we expand along the third column.

$$\begin{vmatrix} 0 & 4 & 0 & -3 \\ -1 & 1 & 5 & 2 \\ 1 & -2 & 0 & 6 \\ 3 & 0 & 0 & 1 \end{vmatrix}$$

$$= (-1)^{2+3}(5)\begin{vmatrix} 0 & 4 & -3 \\ 1 & -2 & 6 \\ 3 & 0 & 1 \end{vmatrix}$$

$$= -5\left((-1)^{2+1}(1)\begin{vmatrix} 4 & -3 \\ 0 & 1 \end{vmatrix} + (-1)^{3+1}(3)\begin{vmatrix} 4 & -3 \\ -2 & 6 \end{vmatrix}\right)$$

$$= -5\big(-1(4(1)-(-3)(0)) + 3(4(6)-(-3)(-2))\big)$$

$$= -5(-1(4-0)+3(24-6))$$

$$= -5(-4+54)$$

$$= -5(50)$$

$$= -250$$

✎ ***Pencil Problem #5*** ✎

5. Evaluate the determinant: $\begin{vmatrix} 4 & 2 & 8 & -7 \\ -2 & 0 & 4 & 1 \\ 5 & 0 & 0 & 5 \\ 4 & 0 & 0 & -1 \end{vmatrix}$.

Answers for Pencil Problems *(Textbook Exercise references in parentheses)*:

1. −29 *(8.5 #3)*

2. $\{(2,-3)\}$ *(8.5 #13)*

3. 72 *(8.5 #23)*

4. $\{(-5,-2,7)\}$ *(8.5 #29)*

5. −200 *(8.5 #37)*

Section 9.1
The Ellipse

Do You Trust Politicians?

The U.S. Capitol Building is beautiful both inside, and out. But did you know that part of its architecture includes an elliptical ceiling in Sanctuary Hall?

John Quincy Adams, while a member of the house of Representatives, discovered that he could use the reflective properties of the room, which we will study in this section, to eavesdrop on the conversations of other House members.

Objective #1: Graph ellipses centered at the origin.

✔ *Solved Problem #1*

1. Graph and locate the foci: $16x^2 + 9y^2 = 144$

First, write the equation in standard form.

$$16x^2 + 9y^2 = 144$$

$$\frac{16x^2}{144} + \frac{9y^2}{144} = \frac{144}{144}$$

$$\frac{x^2}{9} + \frac{y^2}{16} = 1$$

Because the denominator of the y^2 – term is greater than the denominator of the x^2 – term, the major axis is vertical.

Since $a^2 = 16$, $a = 4$ and the vertices are $(0,-4)$ and $(0,4)$.

Since $b^2 = 9$, $b = 3$ and endpoints of the minor axis are $(-3,0)$ and $(3,0)$.

$c^2 = a^2 - b^2 = 16 - 9 = 7$, $c = \sqrt{7}$ and the foci are $(0,-\sqrt{7})$ and $(0,\sqrt{7})$.

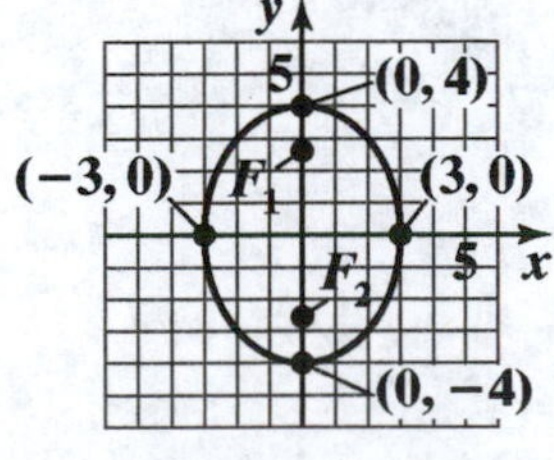

$16x^2 + 9y^2 = 144$

✎ *Pencil Problem #1* ✎

1. Graph and locate the foci: $\frac{x^2}{16} + \frac{y^2}{4} = 1$

Objective #2: Write equations of ellipses in standard form.

✔ *Solved Problem #2*

2. Find the standard form of the equation of an ellipse with foci at (–2, 0) and (2, 0) and vertices (–3, 0) and (3, 0).

Because the foci are located on the x-axis, the major axis is horizontal with the center midway between them at (0, 0). The form of the equation is $\frac{x^2}{a^2}+\frac{y^2}{b^2}=1$. We need to determine values for a^2 and b^2. The distance from the center to either vertex is 3, so $a = 3$ and $a^2 = 9$. The distance from the center to either focus is 2, so $c = 2$.

$$b^2 = a^2 - c^2 = 3^2 - 2^2 = 5$$

The equation is $\frac{x^2}{9}+\frac{y^2}{5}=1$.

Pencil Problem #2

2. Find the standard form of the equation of an ellipse with foci at (0, –4) and (0, 4) and vertices (0, –7) and (0, 7).

Objective #3: Graph ellipses not centered at the origin.

✔ *Solved Problem #3*

3. Graph: $\frac{(x+1)^2}{9}+\frac{(y-2)^2}{4}=1$. Where are the foci located?

$$\frac{(x+1)^2}{9}+\frac{(y-2)^2}{4}=1$$

The center of the ellipse is $(-1,2)$.

Because the denominator of the x^2 – term is greater than the denominator of the y^2 – term, the major axis is horizontal.

Since $a^2=9$, $a=3$ and the vertices lie 3 units to the right and left of the center.

Since $b^2=4$, $b=2$ and endpoints of the minor axis lie 2 units above and below the center.

Since $c^2=a^2-b^2=9-4=5$, $c=\sqrt{5}$ and the foci are located $\sqrt{5}$ units to the right and left of center.

The following chart summarizes these key points.

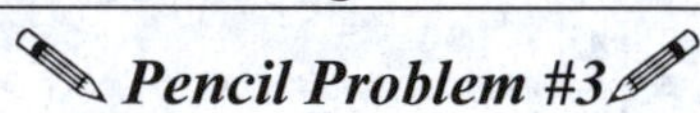

Pencil Problem #3

3. Graph: $\frac{(x-4)^2}{9}+\frac{(y+2)^2}{25}=1$. Where are the foci located?

Center	Vertices	Endpoints Minor Axis	Foci
$(-1,2)$	$(-1-3,2)$ $=(-4,2)$	$(-1,2-2)$ $=(-1,0)$	$(-1-\sqrt{5},2)$
	$(-1+3,2)$ $=(2,2)$	$(-1,2+2)$ $=(-1,4)$	$(-1+\sqrt{5},2)$

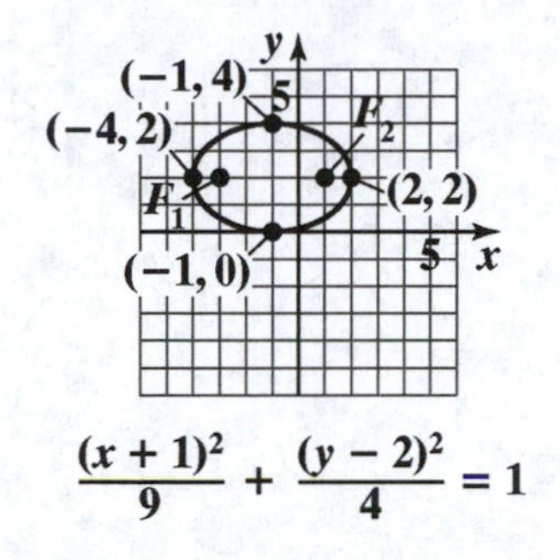

$$\frac{(x+1)^2}{9}+\frac{(y-2)^2}{4}=1$$

Objective #4: Solve applied problems involving ellipses.

✔ *Solved Problem #4*

4. A semielliptical archway over a one-way road has a height of 10 feet and a width of 40 feet. Your truck has a width of 12 feet and a height of 9 feet. Will your truck clear the opening of the archway?

Using the equation $\frac{x^2}{a^2}+\frac{y^2}{b^2}=1$ the archway can be expressed as $\frac{x^2}{20^2}+\frac{y^2}{10^2}=1$ or $\frac{x^2}{400}+\frac{y^2}{100}=1$.

Since the truck is 12 feet wide, we need to determine the height of the archway at $\frac{12}{2}=6$ feet from the center.

Substitute 6 for x to find the height y.

$$\frac{x^2}{400}+\frac{y^2}{100}=1$$

$$\frac{6^2}{400}+\frac{y^2}{100}=1$$

Pencil Problem #4

4. Will a truck that is 8 feet wide carrying a load that reaches 7 feet above the ground clear the semielliptical arch on the one-way road that passes under a bridge that has a height of 10 feet and a width of 30 feet?

Solve for y.

$$\frac{6^2}{400}+\frac{y^2}{100}=1$$

$$\frac{36}{400}+\frac{y^2}{100}=1$$

$$400\left(\frac{36}{400}+\frac{y^2}{100}\right)=400(1)$$

$$36+4y^2=400$$

$$4y^2=364$$

$$y^2=91$$

$$y=\sqrt{91}\approx 9.54$$

The height of the archway 6 feet from the center is approximately 9.54 feet.

Since the truck is 9 feet high, the truck will clear the archway.

Answers for Pencil Problems *(Textbook Exercise references in parentheses)*:

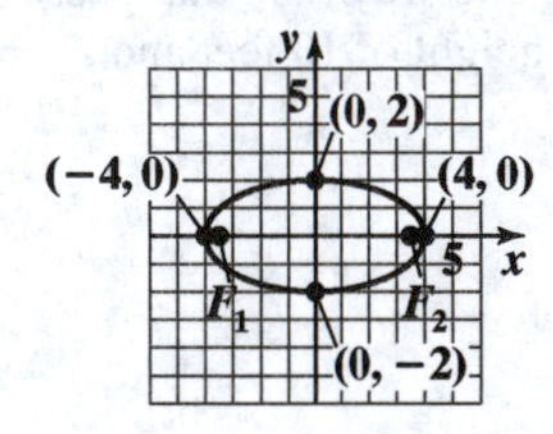

1. $\frac{x^2}{16}+\frac{y^2}{4}=1$ foci at $(-2\sqrt{3},0)$ and $(2\sqrt{3},0)$ *(9.1 #1)*

2. $\frac{x^2}{33}+\frac{y^2}{49}=1$ *(9.1 #27)*

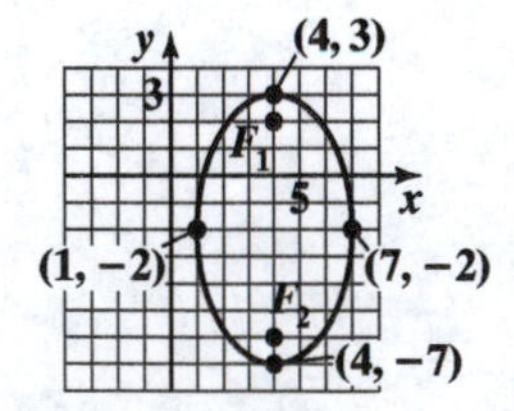

3. $\frac{(x-4)^2}{9}+\frac{(y+2)^2}{25}=1$ foci at (4, 2) and (4, −6) *(9.1 #41)*

4. Yes; the height of the archway 4 feet from the center is approximately 9.64 feet. *(9.1 #65)*

Section 9.2
The Hyperbola

Sonic Boom !

When a jet flies at a speed greater than the speed of sound, the shock wave that is created is heard as a sonic boom.

The wave has the shape of a cone.
The shape formed as the cone hits the ground is one branch of a hyperbola, the topic of this section of the textbook.

Objective #1: Locate a hyperbola's vertices and foci.

✔ _Solved Problem #1_

1a. Find the vertices and locate the foci for the hyperbola with the given equation: $\frac{x^2}{25}-\frac{y^2}{16}=1.$

The x^2 – term is positive.
Therefore, the transverse axis lies along the x–axis.

Since $a^2=25$ and $a=5,$ the vertices are $(-5,0)$ and $(5,0)$.

Since $c^2=a^2+b^2=25+16=41,$ $c=\sqrt{41}$ and the foci are $(-\sqrt{41},0)$ and $(\sqrt{41},0)$.

1b. Find the vertices and locate the foci for the hyperbola with the given equation: $\frac{y^2}{25}-\frac{x^2}{16}=1.$

The y^2 – term is positive.
Therefore, the transverse axis lies along the y–axis.

Since $a^2=25$ and $a=5,$ the vertices are $(0,-5)$ and $(0,5)$.

Since $c^2=a^2+b^2=25+16=41,$ $c=\sqrt{41}$ and the foci are $(0,-\sqrt{41})$ and $(0,\sqrt{41})$.

✎ _Pencil Problem #1_ ✎

1a. Find the vertices and locate the foci for the hyperbola with the given equation: $\frac{x^2}{4}-\frac{y^2}{1}=1.$

1b. Find the vertices and locate the foci for the hyperbola with the given equation: $\frac{y^2}{4}-\frac{x^2}{1}=1.$

Objective #2: Write equations of hyperbolas in standard form.

✔ *Solved Problem #2*

2. Find the standard form of the equation of a hyperbola with foci at (0, –5) and (0, 5) and vertices (0, –3) and (0, 3).

Because the foci are located on the y-axis, the transverse axis lies on the y-axis with the center midway between the foci at (0, 0). The form of the equation is $\frac{y^2}{a^2}-\frac{x^2}{b^2}=1$. We need to determine values for a^2 and b^2.

The distance from the center to either vertex is 3, so $a=3$ and $a^2=9$. The distance from the center to either focus is 5, so $c=5$.

$b^2=c^2-a^2=5^2-3^2=16$

The equation is $\frac{y^2}{9}-\frac{x^2}{16}=1$.

Pencil Problem #2

2. Find the standard form of the equation of a hyperbola with foci at (–4, 0) and (4, 0) and vertices (–3, 0) and (3, 0).

Objective #3: Graph hyperbolas centered at the origin.

✔ *Solved Problem #3*

3a. Graph and locate the foci: $\frac{x^2}{36}-\frac{y^2}{9}=1$. What are the equations of the asymptotes?

$$\frac{x^2}{36}-\frac{y^2}{9}=1$$

Since the x^2 – term is positive, the transverse axis lies along the x–axis.

Since $a^2=36$ and $a=6$, the vertices are $(-6,0)$ and $(6,0)$.

Construct a rectangle using –6 and 6 on the x–axis, and –3 and 3 on the y–axis.

Draw extended diagonals to obtain the asymptotes.

The equations of the asymptotes are $y=\pm\frac{3}{6}x=\pm\frac{1}{2}x$.

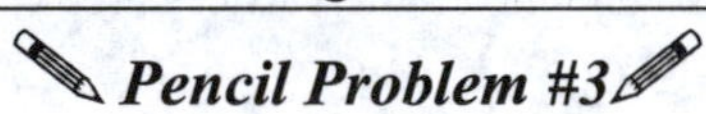

Pencil Problem #3

3a. Graph and locate the foci: $\frac{x^2}{9}-\frac{y^2}{25}=1$. What are the equations of the asymptotes?

Draw the two branches of the hyperbola by starting at each vertex and approaching the asymptotes.

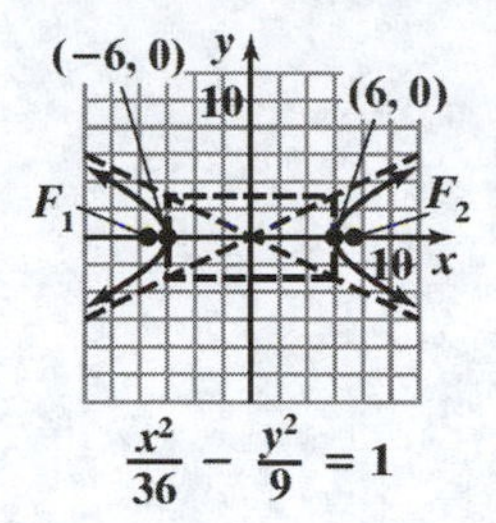

Since $c^2 = a^2 + b^2 = 36 + 9 = 45$, $c = \sqrt{45} = 3\sqrt{5}$ and the foci are located at $(-3\sqrt{5}, 0)$ and $(3\sqrt{5}, 0)$.

3b. Graph and locate the foci: $y^2 - 4x^2 = 4$. What are the equations of the asymptotes?

First write the equation in standard form.

$$y^2 - 4x^2 = 4$$

$$\frac{y^2}{4} - \frac{4x^2}{4} = \frac{4}{4}$$

$$\frac{y^2}{4} - \frac{x^2}{1} = 1$$

The equation is in the form $\frac{y^2}{a^2} - \frac{x^2}{b^2} = 1$ with

$a^2 = 4$ and $b^2 = 1$.

The transverse axis lies on the *y*-axis and the vertices are $(0, -2)$ and $(0, 2)$.

Because $a^2 = 4$ and $b^2 = 1$, $a = 2$ and $b = 1$.

Construct a rectangle using -2 and 2 on the *y*–axis, and -1 and 1 on the *x*–axis.

Draw extended diagonals to obtain the asymptotes.

The equations of the asymptotes are $y = \pm\frac{2}{1}x = \pm 2x$.

3b. Graph and locate the foci: $9y^2 - 25x^2 = 225$. What are the equations of the asymptotes?

Draw the two branches of the hyperbola by starting at each vertex and approaching the asymptotes.

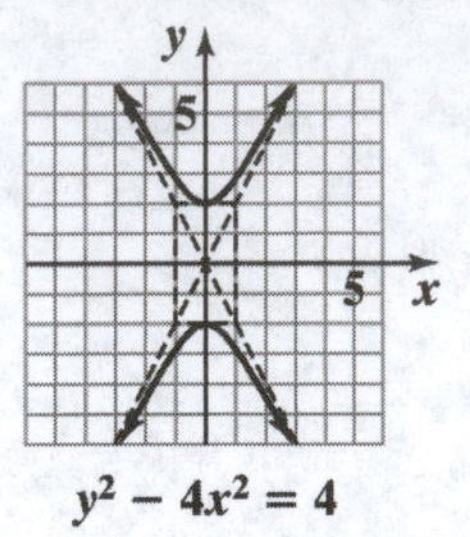

Since $c^2 = a^2 + b^2 = 4 + 1 = 5$, $c = \sqrt{5}$ and the foci are located at $(0, -\sqrt{5})$ and $(0, \sqrt{5})$.

Objective #4: Graph hyperbolas not centered at the origin.

✔ *Solved Problem #4*

4. Graph: $\dfrac{(x-3)^2}{4} - \dfrac{(y-1)^2}{1} = 1$. Where are the foci located? What are the equations of the asymptotes?

Because the term involving x^2 has the positive coefficient, the transverse axis is horizontal. Based on the standard form $\dfrac{(x-h)^2}{a^2} - \dfrac{(y-k)^2}{b^2} = 1$, we see that $h = 3$ and $k = 1$ so the center is (3, 1). We also see that $a^2 = 4$ and $b^2 = 1$, so $a = 2$ and $b = 1$.

Since $a = 2$, the vertices are 2 units to the left and right of the center at (3 – 2, 1), or (1, 1), and (3 + 2, 1), or (5, 1). Draw a rectangle using the vertices, (1, 1) and (5, 1) and the points $b = 1$ unit above and below the center. Draw the extended diagonals to obtain the asymptotes.
The asymptotes of the unshifted hyperbola are $y = \pm\dfrac{b}{a}x = \pm\dfrac{1}{2}x$. Thus, the asymptotes of the shifted hyperbola are $y - 1 = \pm\dfrac{1}{2}(x-3)$.
Draw the two branches of each hyperbola by starting at each vertex and approaching the asymptotes.

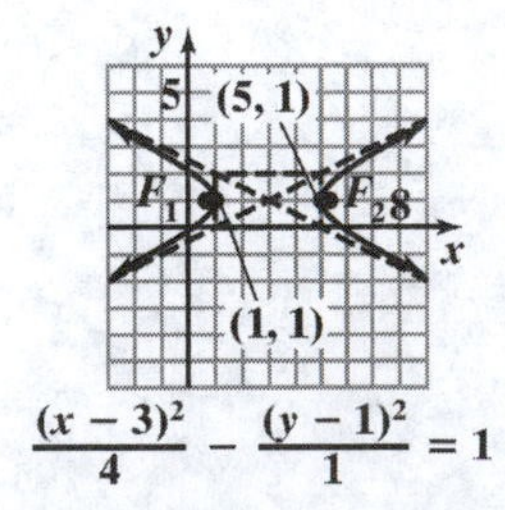

Since $c^2 = a^2 + b^2 = 4 + 1 = 5$, $c = \sqrt{5}$ and the foci are located $\sqrt{5}$ units to the left and right of center at $(3 - \sqrt{5},\ 1)$ and $(3 + \sqrt{5},\ 1)$.

✎ *Pencil Problem #4* ✎

4. Graph: $\dfrac{(y+2)^2}{4} - \dfrac{(x-1)^2}{16} = 1$. Where are the foci located? What are the equations of the asymptotes?

Objective #5: Solve applied problems involving hyperbolas.

✔ ***Solved Problem #5***

5. An explosion is recorded by two microphones that are 2 miles apart. Microphone M_1 received the sound 3 seconds before microphone M_2. Assuming sound travels at 1100 feet per second, determine the possible locations of the explosion relative to the location of the microphones.

Because 1 mile = 5280 feet, place microphone M_1 at (5280, 0) in a coordinate system. Since the microphones are two miles apart, place M_2 at (−5280, 0). Assume that the explosion is at point $P(x, y)$ in the coordinate system. The set of all possible points for the explosion is a hyperbola with the microphones at the foci.

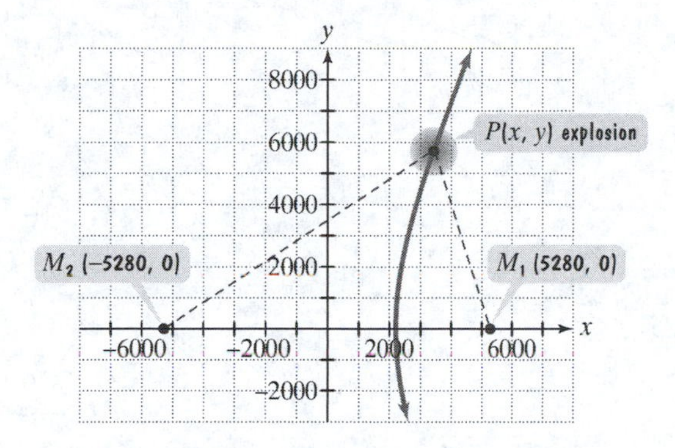

Since M_1 received the sound 3 seconds before microphone M_2 and sound travels at 1100 feet per second, the difference between the distances from P to M_1 and from P to M_2 is 3300 feet. Thus, $2a = 3300$ and $a = 1650$, so $a^2 = 2{,}722{,}500$.

The distance from the center, (0, 0), to either focus is 5280, so $c = 5280$.

$b^2 = c^2 - a^2 = 5280^2 - 1650^2 = 25{,}155{,}900$

The equation of the hyperbola is

$\frac{x^2}{2{,}722{,}500} - \frac{y^2}{25{,}155{,}900} = 1$. The explosion occurred somewhere on the right branch of this hyperbola (the branch closer to M_1).

Pencil Problem #5

5. An explosion is recorded by two microphones that are 1 mile apart. Microphone M_1 received the sound 2 seconds before microphone M_2. Assuming sound travels at 1100 feet per second, determine the possible locations of the explosion relative to the location of the microphones.

Answers for Pencil Problems *(Textbook Exercise references in parentheses)*:

1a. vertices: $(-2,0)$ and $(2,0)$; foci: $(-\sqrt{5},0)$ and $(\sqrt{5},0)$ *(9.2 #1)*

1b. vertices: $(0,-2)$ and $(0,2)$; foci: $(0,-\sqrt{5})$ and $(0,\sqrt{5})$ *(9.2 #3)*

2. $\dfrac{x^2}{9}-\dfrac{x^2}{7}=1$ *(9.2 #7)*

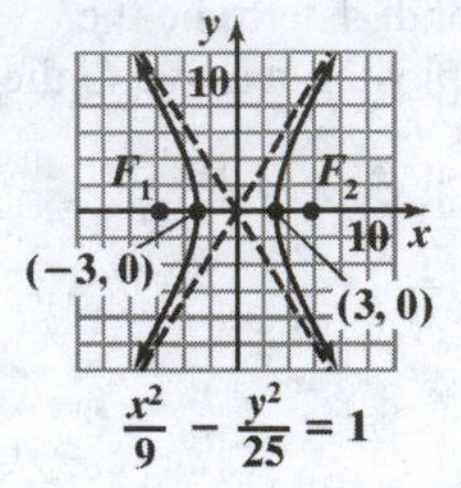

3a. asymptotes: $y=\pm\dfrac{5}{3}x$; foci: $(-\sqrt{34},0)$ and $(\sqrt{34},0)$ *(9.2 #13)*

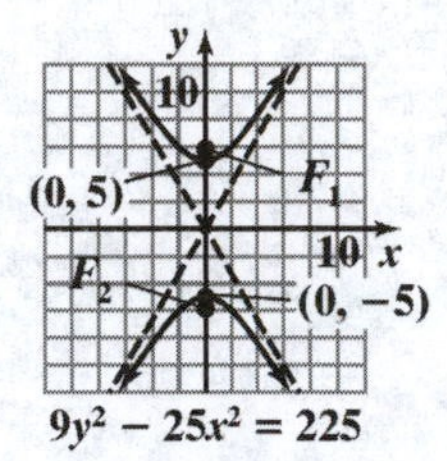

3b. asymptotes: $y=\pm\dfrac{5}{3}x$; foci: $(0,-\sqrt{34})$ and $(0,\sqrt{34})$ *(9.2 #23)*

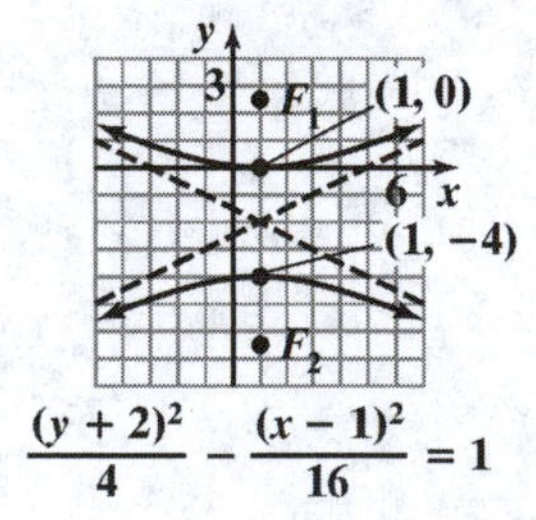

4. asymptotes: $y+2=\pm\dfrac{1}{2}(x-1)$; foci: $(1,-2-2\sqrt{5})$ and $(1,-2+2\sqrt{5})$ *(9.2 #37)*

5. If M_1 is located 2640 feet to the right of the origin on the x-axis, the explosion is located on the right branch of the hyperbola given by the equation $\dfrac{x^2}{1{,}210{,}000}-\dfrac{x^2}{5{,}759{,}600}=1$. *(9.2 #61)*

Section 9.3
The Parabola

How Good Is Your Reception?

In this section we study parabolas and their properties.
A satellite dish is in the shape of a parabolic surface.
Signals coming from a satellite strike the surface of the dish and are reflected to the focus, where the receiver is located.

The applications in the Exercise Set include concepts from each of the conic sections that we have studied.

Objective #1: Graph parabolas with vertices at the origin.

✔ *Solved Problem #1*

1a. Find the focus and directrix of the parabola given by $y^2 = 8x$. Then graph the parabola.

The equation $y^2 = 8x$ is in the standard form $y^2 = 4px$, so $4p = 8$ and $p = 2$. Because p is positive, the parabola opens to the right. The focus is 2 units to the right of the vertex, (0, 0), at $(p, 0)$ or (2, 0). The directrix is 2 units to the left of the vertex: $x = -p$ or $x = -2$.

To graph the parabola, substitute 2 for x in the equation.

$$y^2 = 8 \cdot 2$$
$$y^2 = 16$$
$$y = \pm\sqrt{16} = \pm 4$$

The points (2, 4) and (2, –4) are on the parabola above and below the focus.

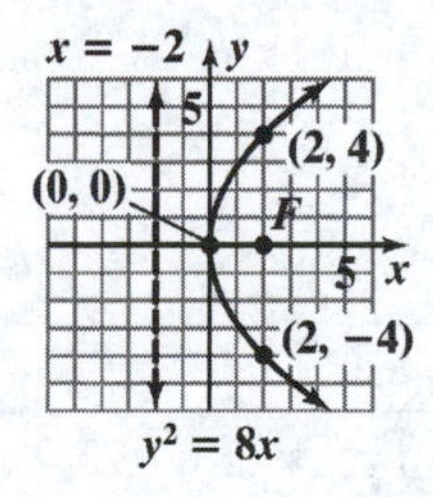

Pencil Problem #1

1a. Find the focus and directrix of the parabola given by $y^2 = 16x$. Then graph the parabola.

1b. Find the focus and directrix of the parabola given by $x^2 = -12y$. Then graph the parabola.

The equation $x^2 = -12y$ is in the standard form $x^2 = 4py$, so $4p = -12$ and $p = -3$. Because p is negative, the parabola opens downward. The focus is 3 units below the vertex, (0, 0), at $(0, p)$ or $(0, -3)$. The directrix is 3 units above the vertex: $y = -p$ or $y = 3$.

To graph the parabola, substitute -3 for y in the equation.

$$x^2 = -12(-3)$$
$$x^2 = 36$$
$$x = \pm\sqrt{36} = \pm 6$$

The points $(-6, -3)$ and $(6, -3)$ are on the parabola to the left and right of the focus.

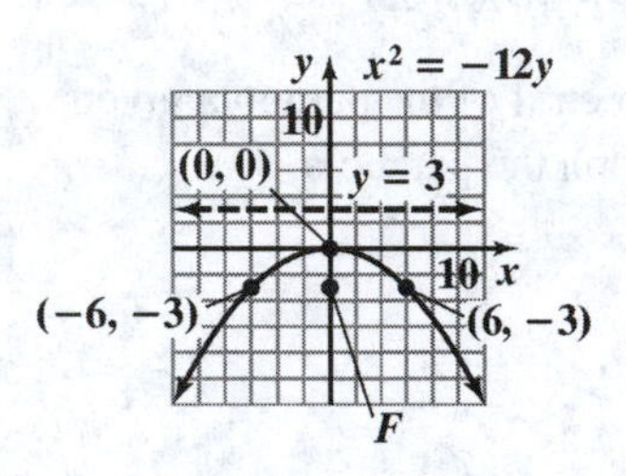

1b. Find the focus and directrix of the parabola given by $x^2 = -16y$. Then graph the parabola.

Objective #2: Write equations of parabolas in standard form.

✔ Solved Problem #2

2. Find the standard form of the equation of a parabola with focus (8, 0) and directrix $x = -8$.

The vertex of the parabola is midway between the focus and the directrix at (0, 0). Since the focus is on the x-axis, we use the standard form $y^2 = 4px$.

The focus is 8 units to the right of the vertex, so $p = 8$.

The equation is $y^2 = 4 \cdot 8x$ or $y^2 = 32x$.

Pencil Problem #2

2. Find the standard form of the equation of a parabola with focus (0, 15) and directrix $y = -15$.

Objective #3: Graph parabolas with vertices not at the origin.

✔ Solved Problem #3

3a. Find the vertex, focus, and directrix of the parabola given by $(x-2)^2 = 4(y+1)$. Then graph the parabola.

Writing the equation as $(x-2)^2 = 4(y-(-1))$, we see that $h = 2$ and $k = -1$. The vertex is $(h, k) = (2, -1)$.

Pencil Problem #3

3a. Find the vertex, focus, and directrix of the parabola given by $(x+1)^2 = -8(y+1)$. Then graph the parabola.

Because $4p = 4$, $p = 1$. The focus is 1 unit above the vertex at $(h, k + p) = (2, -1 + 1) = (2, 0)$. The directrix is 1 unit below the vertex: $y = k - p = -1 - 1$ or $y = -2$.

The length of the latus rectum is $|4p| = |4 \cdot 1| = |4| = 4$. The latus rectum extends 2 units to the left and right of the focus. The endpoints of the latus rectum are $(2 - 2, 0)$ or $(0, 0)$ and $(2 + 2, 0)$ or $(4, 0)$.

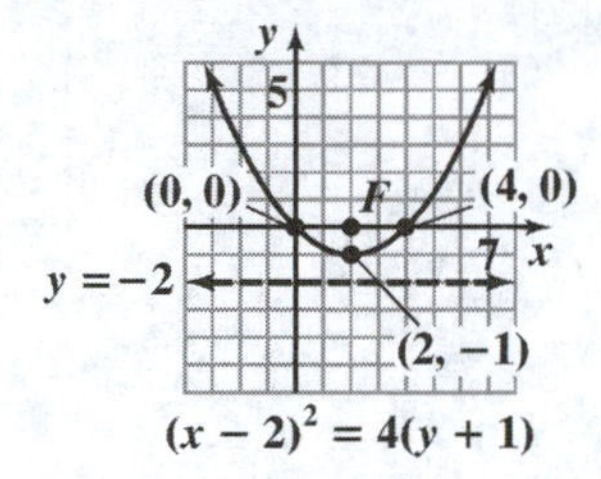

3b. Find the vertex, focus, and directrix of the parabola given by $y^2 + 2y + 4x - 7 = 0$. Then graph the parabola.

Complete the square on y.

$$y^2 + 2y + 4x - 7 = 0$$
$$y^2 + 2y = -4x + 7$$
$$y^2 + 2y + 1 = -4x + 8$$
$$(y + 1)^2 = -4(x - 2)$$

We see that $k = -1$ and $h = 2$, so the vertex is at $(h, k) = (2, -1)$. Since $4p = -4$, $p = -1$. The focus is 1 unit to the left of the vertex at $(h, k) = (2 - 1, -1) = (1, -1)$. The directrix is 1 unit to the right of the vertex: $x = h - p = 2 - (-1)$ or $x = 3$.

The length of the latus rectum is $|4p| = |4(-1)| = |-4| = 4$. The latus rectum extends 2 units above and below the focus. The endpoints of the latus rectum are $(1, -1 + 2)$ or $(1, 1)$ and $(1, -1 - 2)$ or $(1, -3)$.

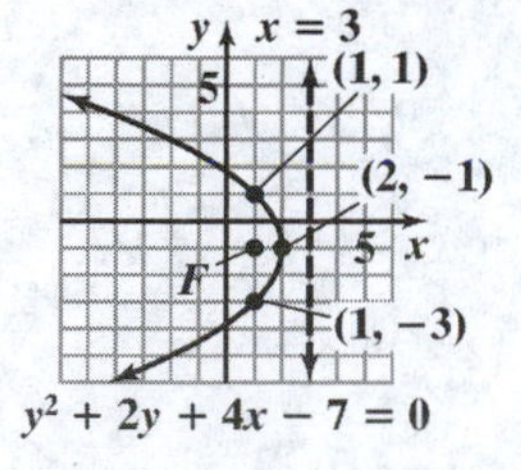

3b. Find the vertex, focus, and directrix of the parabola given by $y^2 - 2y + 12x - 35 = 0$. Then graph the parabola.

Objective #4: Solve applied problems involving parabolas.

✔ Solved Problem #2

4. An engineer is designing a flashlight using a parabolic mirror and a light source. The casting has a diameter of 6 inches and a depth of 4 inches. What is the equation of the parabola used to shape the mirror? At what point should the light source be placed relative to the mirror's vertex?

Position the parabola with its vertex at the origin and opening upward. Then the focus is at $(0, p)$ on the y-axis. Since the casting has a diameter of 6 inches, it extends 3 units to the left and right of the y-axis. Since it is 4 inches deep, the point (3, 4) is on the parabola.

We use the standard form $x^2 = 4py$. Using (3, 4), we have

$$3^2 = 4p \cdot 4$$
$$9 = 16p$$
$$\frac{9}{16} = p.$$

Thus, the equation is $x^2 = 4 \cdot \frac{9}{16} y$ or $x^2 = \frac{9}{4} y$. The light source should be placed at the focus, $(0, p) = (0, \frac{9}{16})$, or $\frac{9}{16}$ inch above the vertex along the axis of symmetry.

Pencil Problem #2

4. The reflector of a flashlight is in the shape of a parabolic surface. The casting has a diameter of 4 inches and a depth of 1 inch. What is the equation of the parabola used to shape the mirror? At what point should the light source be placed relative to the mirror's vertex?

Objective #5: Identify conics without completing the square.

✔ Solved Problem #5

5a. Identify the graph of $3x^2 + 2y^2 + 12x - 4y + 2 = 0$.

The coefficient of x^2 is 3: $A = 3$.
The coefficient of y^2 is 2: $C = 2$.
$AC = 3(2) = 6$

Since $A \neq C$ and $AC > 0$, the graph of the equation is an ellipse.

5b. Identify the graph of $x^2 + y^2 - 6x + y + 3 = 0$.

The coefficient of x^2 is 1: $A = 1$.
The coefficient of y^2 is 1: $C = 1$.

Since $A = C$, the graph of the equation is a circle.

5c. Identify the graph of $y^2 - 12x - 4y + 52 = 0$.

There is no x^2-term, so the coefficient of x^2 is 0: $A = 0$.
The coefficient of y^2 is 1: $C = 1$.
$AC = 0(1) = 0$

Since $AC = 0$, the graph of the equation is a parabola.

5d. Identify the graph of $9x^2 - 16y^2 - 90x + 64y + 17 = 0$.

The coefficient of x^2 is 9: $A = 9$.
The coefficient of y^2 is -16: $C = -16$.
$AC = 9(-16) = -144$

Since $A \neq C$ and $AC < 0$, the graph of the equation is a hyperbola.

Pencil Problem #5

5a. Identify the graph of $9x^2 + 4y^2 - 36x + 8y + 31 = 0$.

5b. Identify the graph of $4x^2 + 4y^2 + 12x + 4y + 1 = 0$.

5c. Identify the graph of $y^2 - 4x + 2y + 21 = 0$.

5d. Identify the graph of $100x^2 - 7y^2 + 90y - 368 = 0$.

Answers for Pencil Problems *(Textbook Exercise references in parentheses)*:

1a.

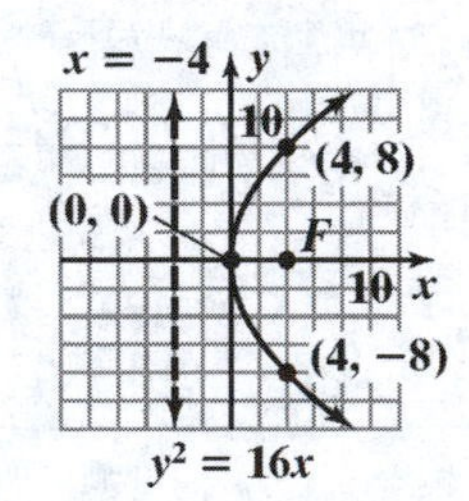

focus: (4, 0); directrix: $x = -4$ *(9.3 #5)*

1b.

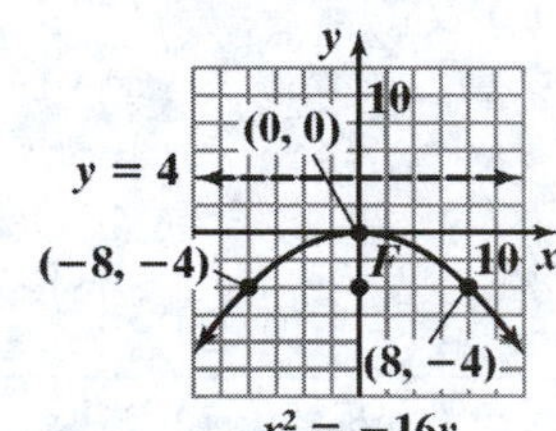

focus: (0, –4); directrix: $y = 4$ *(9.3 #11)*

2. $x^2 = 60y$ *(9.3 #21)*

3a.

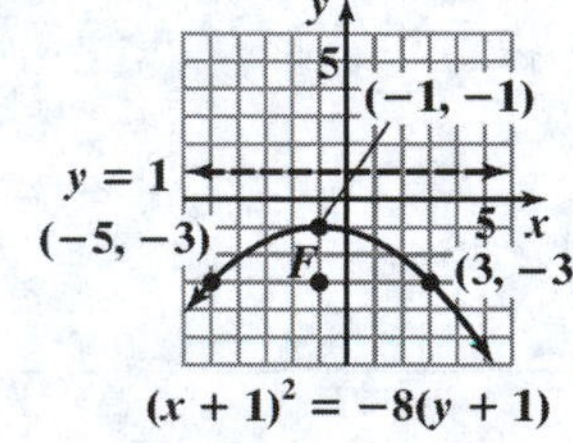

vertex: (–1, –1); focus: (–1, –3); directrix: $y = 1$ *(9.3 #37)*

3b.

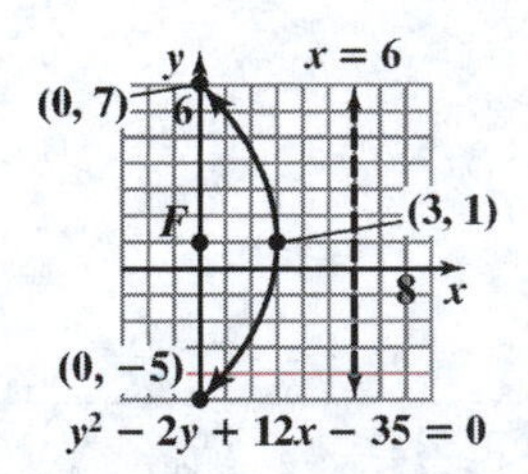

vertex: (3, 1); focus: (0, 1); directrix: $x = 6$ *(9.3 #45)*

4. $x^2 = 4y$; 1 inch above the vertex along the axis of symmetry *(9.3 #61)*

5a. ellipse *(9.3 #54)* **5b.** circle *(9.3 #53)* **5c.** parabola *(9.3 #49)* **5d.** hyperbola *(9.3 #55)*

Section 9.4
Rotation of Axes

A New Twist on Conics

Well, technically it's a rotation not a twist, but you get the idea.
Not all ellipses, hyperbolas, and parabolas have axes that are parallel to the x- or y-axis. When an axis of a conic section is not parallel to the x- or y-axis, we will introduce a new rectangular coordinate system with an x'-axis and a y'-axis that intersect at the origin of the original system. The technique is called a rotation of axes. In the rotated system, the conic will have a form that we've already studied.

Objective #1: Use rotation of axes formulas.

✔ *Solved Problem #1*

1. Write the equation $xy = 2$ in terms of a rotated $x'y'$-system if the angle of rotation from the x-axis to the x'-axis is 45°. Express the equation in standard form.

With $\theta = 45°$, the rotation formulas are

$$\begin{aligned} x &= x'\cos\theta - y'\sin\theta \\ &= x'\cos 45° - y'\sin 45° \\ &= x'\left(\frac{\sqrt{2}}{2}\right) - y'\left(\frac{\sqrt{2}}{2}\right) = \frac{\sqrt{2}}{2}(x' - y') \end{aligned}$$

$$\begin{aligned} y &= x'\sin\theta + y'\cos\theta \\ &= x'\sin 45° + y'\cos 45° \\ &= x'\left(\frac{\sqrt{2}}{2}\right) + y'\left(\frac{\sqrt{2}}{2}\right) = \frac{\sqrt{2}}{2}(x' + y') \end{aligned}$$

Now substitute into the equation $xy = 2$. Multiply and simplify.

$$xy = 2$$

$$\left[\frac{\sqrt{2}}{2}(x' - y')\right]\left[\frac{\sqrt{2}}{2}(x' + y')\right] = 2$$

$$\frac{2}{4}(x' - y')(x' + y') = 2$$

$$\frac{1}{2}(x'^2 - y'^2) = 2$$

$$\frac{1}{4}(x'^2 - y'^2) = 1$$

$$\frac{x'^2}{4} - \frac{y'^2}{4} = 1$$

This is the standard form of the equation of a hyperbola.

✎ *Pencil Problem #1* ✎

1. Write the equation $xy = -1$ in terms of a rotated $x'y'$-system if the angle of rotation from the x-axis to the x'-axis is 45°. Express the equation in standard form.

Objective #2: Write equations of rotated conics in standard form.

✔ *Solved Problem #2a*

2a. Rewrite the equation $2x^2+\sqrt{3}xy+y^2-2=0$ in a rotated $x'y'$-system without an $x'y'$-term. Express the equation in the standard form of a conic section. Graph the conic section in the rotated system.

Find $\cot 2\theta$.
The coefficient of x^2 is 2: $A = 2$.
The coefficient of xy is $\sqrt{3}$: $B = \sqrt{3}$.
The coefficient of y^2 is 1: $C = 1$.

$$\cot 2\theta = \frac{A-C}{B} = \frac{2-1}{\sqrt{3}} = \frac{1}{\sqrt{3}} \cdot \frac{\sqrt{3}}{\sqrt{3}} = \frac{\sqrt{3}}{3}$$

Find the angle of rotation θ.

Since $\cot 60° = \frac{\sqrt{3}}{3}$, we have $2\theta = 60°$. Thus, $\theta = 30°$.

Substitute $\theta = 30°$ into the rotation formulas.

$$x = x'\cos\theta - y'\sin\theta = x'\cos 30° - y'\sin 30° = x'\left(\frac{\sqrt{3}}{2}\right) - y'\left(\frac{1}{2}\right) = \frac{\sqrt{3}x'-y'}{2}$$

$$y = x'\sin\theta + y'\cos\theta = x'\sin 30° + y'\cos 30° = x'\left(\frac{1}{2}\right) + y'\left(\frac{\sqrt{3}}{2}\right) = \frac{x'+\sqrt{3}y'}{2}$$

Now substitute into the equation $2x^2+\sqrt{3}xy+y^2-2=0$.

$$2x^2+\sqrt{3}xy+y^2-2=0$$

$$2\left(\frac{\sqrt{3}x'-y'}{2}\right)^2+\sqrt{3}\left(\frac{\sqrt{3}x'-y'}{2}\right)\left(\frac{x'+\sqrt{3}y'}{2}\right)+\left(\frac{x'+\sqrt{3}y'}{2}\right)^2-2=0$$

$$2\left(\frac{3x'^2-2\sqrt{3}x'y'+y'^2}{4}\right)+\sqrt{3}\left(\frac{\sqrt{3}x'^2+3x'y'-x'y'-\sqrt{3}y'^2}{4}\right)+\left(\frac{x'^2+2\sqrt{3}x'y'+3y'^2}{4}\right)-2=0$$

$$2(3x'^2-2\sqrt{3}x'y'+y'^2)+\sqrt{3}(\sqrt{3}x'^2+2x'y'-\sqrt{3}y'^2)+(x'^2+2\sqrt{3}x'y'+3y'^2)-8=0$$

$$6x'^2-4\sqrt{3}x'y'+2y'^2+3x'^2+2\sqrt{3}x'y'-3y'^2+x'^2+2\sqrt{3}x'y'+3y'^2-8=0$$

$$10x'^2+2y'^2-8=0$$

Write the equation in standard form.

$$10x'^2+2y'^2-8=0$$

$$10x'^2+2y'^2=8$$

$$\frac{10x'^2}{8}+\frac{2y'^2}{8}=1$$

$$\frac{5x'^2}{4}+\frac{y'^2}{4}=1$$

$$\frac{x'^2}{\frac{4}{5}}+\frac{y'^2}{4}=1$$

This is the equation of an ellipse with vertices at (0, 2) and (0, –2) on the y'-axis. The endpoints of the minor axis are at $\left(\frac{2\sqrt{5}}{5}, 0\right)$ and $\left(-\frac{2\sqrt{5}}{5}, 0\right)$.

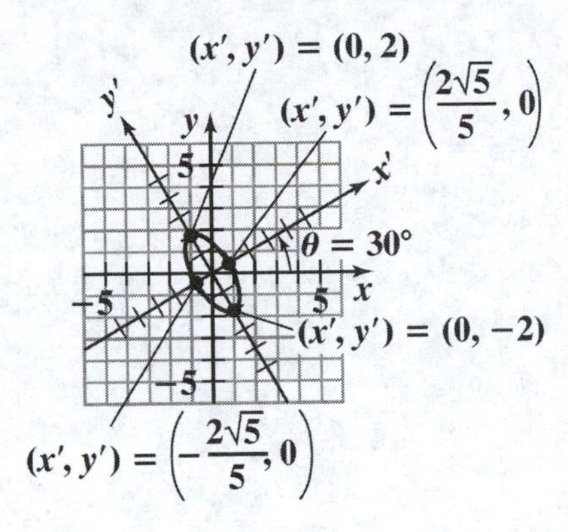

✎ Pencil Problem #2a ✎

2a. Rewrite the equation $11x^2 + 10\sqrt{3}xy + y^2 - 4 = 0$ in a rotated $x'y'$-system without an $x'y'$-term. Express the equation in the standard form of a conic section. Graph the conic section in the rotated system.

✔ Solved Problem #2b

2b. Rewrite the equation $4x^2 - 4xy + y^2 - 8\sqrt{5}x - 16\sqrt{5}y = 0$ in a rotated $x'y'$-system without an $x'y'$-term. Express the equation in the standard form of a conic section. Graph the conic section in the rotated system.

Find $\cot 2\theta$.
The coefficient of x^2 is 4: $A = 4$.
The coefficient of xy is -4: $B = -4$.
The coefficient of y^2 is 1: $C = 1$.

$$\cot 2\theta = \frac{A-C}{B} = \frac{4-1}{-4} = -\frac{3}{4}$$

Use $\cot 2\theta = -\frac{3}{4}$ to find $\sin\theta$ and $\cos\theta$. Begin by finding $\cos 2\theta$.

Since $\cot 2\theta$ is negative, 2θ is a quadrant II angle. In quadrant II, x is negative and y is positive.

$$\cot 2\theta = -\frac{3}{4} = \frac{x}{y} = \frac{-3}{4}$$

We let $x = -3$ and $y = 4$. Then $r = \sqrt{(-3)^2 + 4^2} = \sqrt{9+16} = \sqrt{25} = 5$. Use x and r to find $\cos 2\theta$.

$$\cos 2\theta = \frac{x}{r} = \frac{-3}{5} = -\frac{3}{5}$$

Now use identities to find $\sin\theta$ and $\cos\theta$. Since 2θ is a quadrant II angle, θ is a quadrant I angle. Sine and cosine are both positive in quadrant I.

$$\sin\theta = \sqrt{\frac{1-\cos 2\theta}{2}} = \sqrt{\frac{1-\left(-\frac{3}{5}\right)}{2}} = \sqrt{\frac{\frac{5}{5}+\frac{3}{5}}{2}} = \sqrt{\frac{\frac{8}{5}}{2}} = \sqrt{\frac{4}{5}} = \frac{2}{\sqrt{5}} = \frac{2\sqrt{5}}{5}$$

$$\cos\theta = \sqrt{\frac{1+\cos 2\theta}{2}} = \sqrt{\frac{1+\left(-\frac{3}{5}\right)}{2}} = \sqrt{\frac{\frac{5}{5}-\frac{3}{5}}{2}} = \sqrt{\frac{\frac{2}{5}}{2}} = \sqrt{\frac{1}{5}} = \frac{1}{\sqrt{5}} = \frac{\sqrt{5}}{5}$$

Substitute $\sin\theta = \frac{2\sqrt{5}}{5}$ and $\cos\theta = \frac{\sqrt{5}}{5}$ into the rotation formulas.

$$x = x'\cos\theta - y'\sin\theta = x'\left(\frac{\sqrt{5}}{5}\right) - y'\left(\frac{2\sqrt{5}}{5}\right) = \frac{\sqrt{5}}{5}(x' - 2y')$$

$$y = x'\sin\theta + y'\cos\theta = x'\left(\frac{2\sqrt{5}}{5}\right) + y'\left(\frac{\sqrt{5}}{5}\right) = \frac{\sqrt{5}}{5}(2x' + y')$$

Now substitute the expressions for x and y into the equation $4x^2 - 4xy + y^2 - 8\sqrt{5}x - 16\sqrt{5}y = 0$.

$$4\left[\frac{\sqrt{5}}{5}(x'-2y')\right]^2 - 4\left[\frac{\sqrt{5}}{5}(x'-2y')\right]\left[\frac{\sqrt{5}}{5}(2x'+y')\right] + \left[\frac{\sqrt{5}}{5}(2x'+y')\right]^2$$
$$- 8\sqrt{5}\left[\frac{\sqrt{5}}{5}(x'-2y')\right] - 16\sqrt{5}\left[\frac{\sqrt{5}}{5}(2x'+y')\right] = 0$$

(continued on next page)

Begin simplifying. Square and distribute as appropriate. Note that $\left(\frac{\sqrt{5}}{5}\right)^2 = \frac{(\sqrt{5})^2}{5^2} = \frac{5}{25} = \frac{1}{5}$ and

$\sqrt{5} \cdot \frac{\sqrt{5}}{5} = \frac{\sqrt{5} \cdot \sqrt{5}}{5} = \frac{5}{5} = 1.$

$$4 \cdot \frac{1}{5}(x'^2 - 4x'y' + 4y'^2) - 4 \cdot \frac{1}{5}(2x'^2 - 3x'y' - 2y'^2) + \frac{1}{5}(4x'^2 + 4x'y' + y'^2)$$
$$- 8x' + 16y' - 32x' - 16y' = 0$$

Distribute again and begin combining like terms. (Note that you could multiply by 5 to eliminate fractions.) The coefficients of the terms involving x'^2, $x'y'$, and y' each have a sum of 0.

$$\frac{4}{5}x'^2 - \frac{16}{5}x'y' + \frac{16}{5}y'^2 - \frac{8}{5}x'^2 + \frac{12}{5}x'y' + \frac{8}{5}y'^2 + \frac{4}{5}x'^2 + \frac{4}{5}x'y' + \frac{1}{5}y'^2 - 40x' = 0$$
$$\frac{16}{5}y'^2 + \frac{8}{5}y'^2 + \frac{1}{5}y'^2 - 40x' = 0$$
$$\frac{25}{5}y'^2 - 40x' = 0$$
$$5y'^2 = 40x'$$
$$y'^2 = 8x'$$

This is the standard form of the equation of parabola with vertex at (0, 0). Letting $4p = 8$, we see that $p = 2$. The focus is at (2, 0). Substituting 2 for x', we get $y'^2 = 16$, so $y' = \pm 4$. The endpoints of the latus rectum are (2, 4) and (−2, 4). By solving $\cos\theta = \frac{\sqrt{5}}{5}$ for θ, we have $\theta = \cos^{-1}\frac{\sqrt{5}}{5} \approx 63°$.

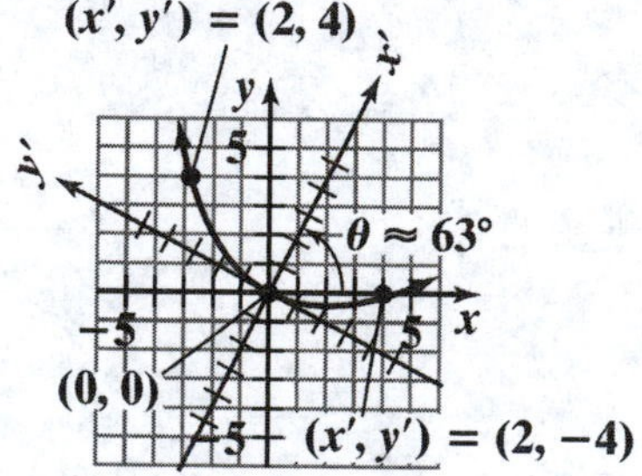

Pencil Problem #2b

2b. Rewrite the equation $34x^2 - 24xy + 41y^2 - 25 = 0$ in a rotated $x'y'$-system without an $x'y'$-term. Express the equation in the standard form of a conic section. Graph the conic section in the rotated system.

Objective #3: Identify conics without rotating axes.

✔ *Solved Problem #3*

3. Identify the graph of $3x^2 - 2\sqrt{3}xy + y^2 + 2x + 2\sqrt{3}y = 0.$

The coefficient of x^2 is 3: $A = 3$.
The coefficient of xy is $-2\sqrt{3}$: $B = -2\sqrt{3}$.
The coefficient of y^2 is 1: $C = 1$.
$B^2 - 4AC = (-2\sqrt{3})^2 - 4(3)(1) = 12 - 12 = 0$

Since $B^2 - 4AC = 0$, the graph of the equation is a parabola.

Pencil Problem #3

3. Identify the graph of $5x^2 - 2xy + 5y^2 - 12 = 0.$

Answers for Pencil Problems *(Textbook Exercise references in parentheses)*:

1. $\frac{y'^2}{2} - \frac{x'^2}{2} = 1$ *(9.4 #1)*

2a. $\frac{x'^2}{\frac{1}{4}} - \frac{y'^2}{1} = 1$

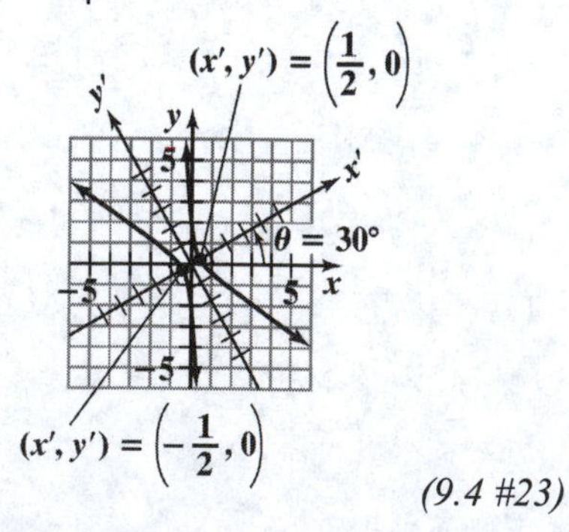

(9.4 #23)

2b. $\frac{x'^2}{1} + \frac{y'^2}{\frac{1}{2}} = 1$

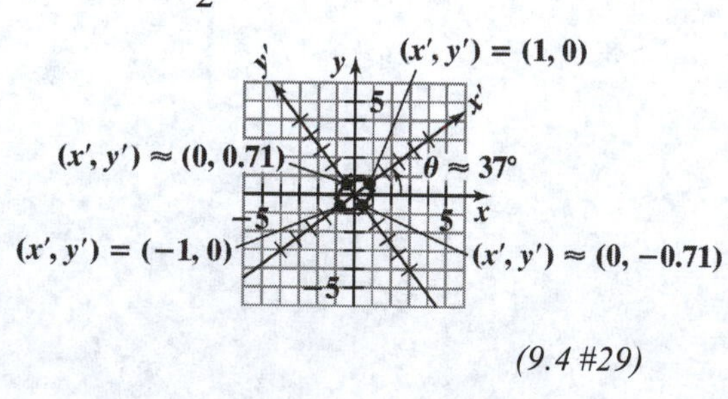

(9.4 #29)

3. ellipse or circle *(9.4 #31)*

Section 9.5
Parametric Equations

Home Run!

The ball is hit. You watch as it flies through the air. As the distance from the batter increases, the height of the ball above the ground increases until it reaches its highest point and then decreases. Will it make it? Yes, home run!

In the Exercise Set, you will see how the motion of the baseball can be described by a pair of equations, one representing the distance of the ball from the batter and the other representing it height above the ground. Such equations, called *parametric equations*, allow us to describe both the path of an object and its location on that path at a particular time.

Objective #1: Use point plotting to graph plane curves described by parametric equations.

✔ Solved Problem #1

1. Graph the plane curve defined by the parametric equations

$$x = t^2 + 1, \quad y = 3t, \quad -2 \le t \le 2.$$

We will select integer values for t on the given interval. Let $t = -2, -1, 0, 1$, and 2. Calculate values of x and y for each value of t and form the ordered pairs (x, y).

t	$x = t^2 + 1$	$y = 3t$	(x, y)
-2	$(-2)^2 + 1 = 5$	$3(-2) = -6$	$(5, -6)$
-1	$(-1)^2 + 1 = 2$	$3(-1) = -3$	$(2, -3)$
0	$0^2 + 1 = 1$	$3(0) = 0$	$(1, 0)$
1	$1^2 + 1 = 2$	$3(1) = 3$	$(2, 3)$
2	$2^2 + 1 = 5$	$3(2) = 6$	$(5, 6)$

Plot the points in order of increasing values of t and connect them with a smooth curve. Arrows along the graph indicate the direction, or orientation, of the curve as t increases from -2 to 2.

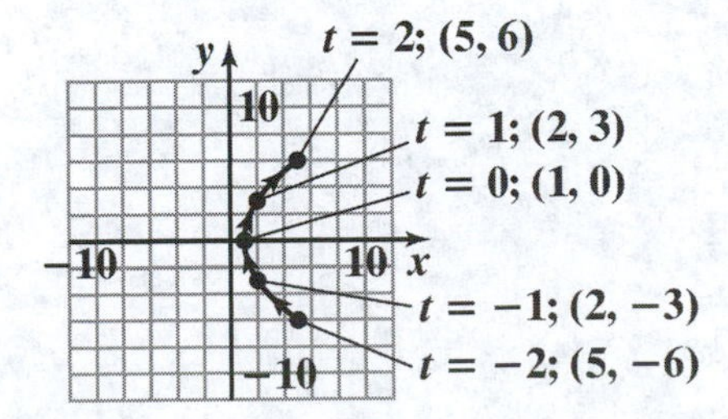

✎ Pencil Problem #1 ✎

1. Graph the plane curve defined by the parametric equations

$$x = t + 2, \quad y = t^2, \quad -2 \le t \le 2.$$

Objective #2: Eliminate the parameter.

✔ *Solved Problem #2*

2a. Sketch the plane curve represented by the parametric equations $x = \sqrt{t}$ and $y = 2t - 1$ by eliminating the parameter.

We begin by solving for t in one of the equations.
Solve $x = \sqrt{t}$ for t by squaring both sides.

$x = \sqrt{t}$

$x^2 = t,\ x \ge 0$

In the original equation, x is the principal square of t, which is never negative, so $x \ge 0$. Now substitute x^2 for t in $y = 2t - 1$.

$y = 2t - 1$

$y = 2x^2 - 1$

Graph $y = 2x^2 - 1$ on the restricted domain $x \ge 0$.

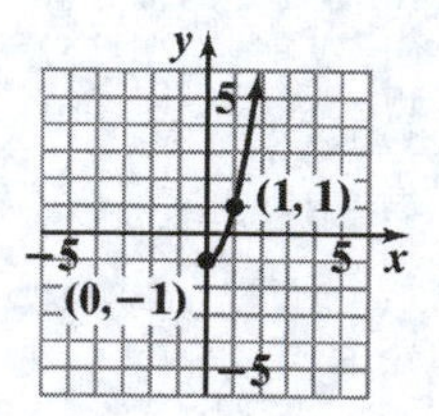

Pencil Problem #2

2a. Sketch the plane curve represented by the parametric equations $x = \sqrt{t}$ and $y = t - 1$ by eliminating the parameter.

Solved Problem #2

2b. Sketch the plane curve represented by the parametric equations $x = 6\cos t,\ \ y = 4\sin t,\ \pi \le t \le 2\pi$ by eliminating the parameter.

We will begin by rewriting both equations in anticipation of applying the identity $\sin^2 t + \cos^2 t = 1$. Solve each equation for the trigonometric function and then square each side.

$x = 6\cos t$	$y = 4\sin t$
$\dfrac{x}{6} = \cos t$	$\dfrac{y}{4} = \sin t$
$\left(\dfrac{x}{6}\right)^2 = (\cos t)^2$	$\left(\dfrac{y}{4}\right)^2 = (\sin t)^2$
$\dfrac{x^2}{36} = \cos^2 t$	$\dfrac{y^2}{16} = \sin^2 t$

Pencil Problem #2

2b. Sketch the plane curve represented by the parametric equations $x = 2\sin t$, $y = 2\cos t,\ 0 \le t \le 2\pi$ by eliminating the parameter.

Now replace $\cos^2 t$ with $\frac{x^2}{36}$ and $\sin^2 t$ with $\frac{y^2}{16}$ in the identity $\sin^2 t + \cos^2 t = 1$.

$$\sin^2 t + \cos^2 t = 1$$

$$\frac{y^2}{16} + \frac{x^2}{36} = 1 \text{ or } \frac{x^2}{36} + \frac{y^2}{16} = 1$$

This is the equation of an ellipse with vertices at (–6, 0) and (6, 0) and endpoints of the minor axis at (–4, 0) and (4, 0). However, in the original equations, we had a restriction on t: $\pi \le t \le 2\pi$. On this interval, $y = 4\sin t$ is always negative or 0, so $y \le 0$. The graph is the lower half of the ellipse just described.

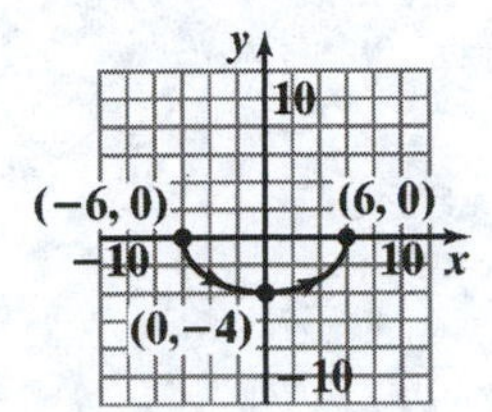

Objective #3: Find parametric equations for functions.

✔ *Solved Problem #3*

3. Find parametric equations for the parabola whose equation is $y = x^2 - 25$.

Let $x = t$. Then parametric equations for $y = x^2 - 25$ are $x = t$ and $y = t^2 - 25$.

Pencil Problem #3

3. Find parametric equations for the parabola whose equation is $y = x^2 + 4$.

Objective #4: Understand the advantages of parametric representations.

✔ Solved Problem #4

4. True or false: Parametric equations provide more information than rectangular equations because they can describe both the path of a moving object in the plane in terms of coordinates (x, y) and the time the object is at each point.

True; the parameter t in the equations that define x and y often represents time. Without, the parameter, we can describe the path of an object but not the time it is at each point along the path.

Pencil Problem #4

4. True or false: When using a graphing utility, you must eliminate the parameter before you can graph parametric equations.

Answers for Pencil Problems *(Textbook Exercise references in parentheses)*:

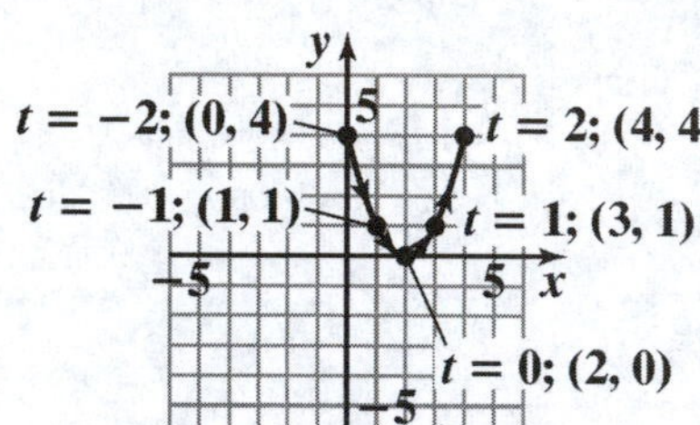

1. *(9.5 #9)*

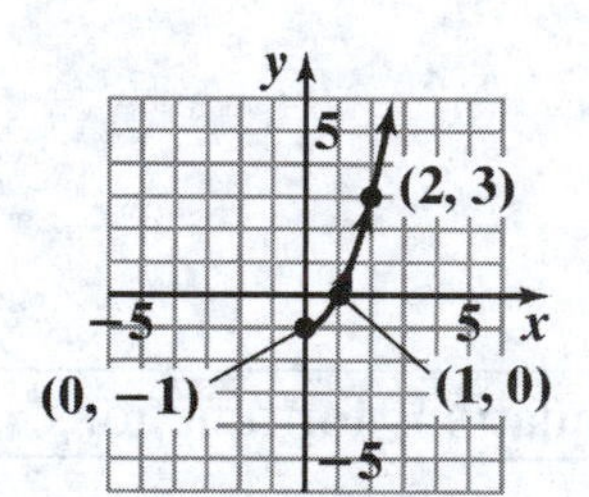

2a. $y = x^2 - 1,\ x \geq 0$ *(9.5 #25)*

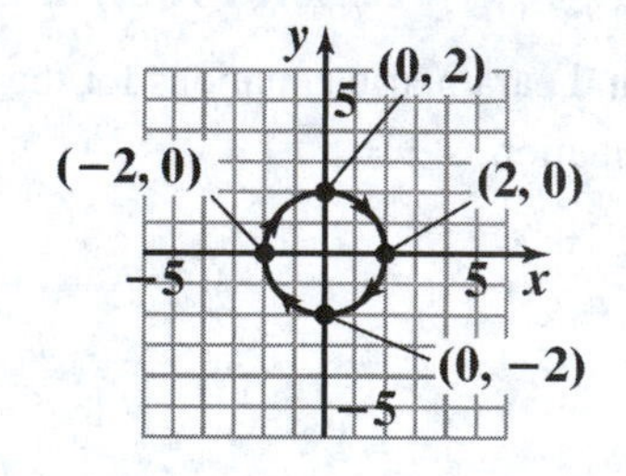

2b. $\frac{x^2}{4} + \frac{y^2}{4} = 1$ *(9.5 #27)*

3. $x = t$ and $y = t^2 + 4$ *(9.5 #55)*

4. false

Section 9.6
Conic Sections in Polar Coordinates

Back to the Drawing Board!

You know how ellipses and hyperbolas are defined in terms of distances from fixed points in the plane using rectangular coordinates. In this section, we want to study the conic sections in polar coordinates but not by converting their equations by algebraic means. Instead, we start with new, but equivalent, definitions of the conics.

Objective #1: Define conics in terms of a focus and a directrix.

1. True or false: If the eccentricity of a conic is $e = 1$, then the conic is an ellipse.

False; when the eccentricity of a conic is 1, the distance from a point on the conic to the focus is equal to its distance from the directrix: When $e = \dfrac{PF}{PD} = 1$, $PF = PD$. This describes a parabola.

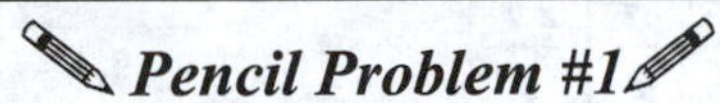

1. True or false: If the eccentricity of a conic is $e = 4$, then the conic is an ellipse.

Objective #2: Graph the polar equations of conics.

Solved Problem #2

2a. Graph the polar equation: $r = \dfrac{4}{2 - \cos\theta}$.

Notice that the constant term in the denominator is not 1, so the equation is not in standard form. Divide the numerator and the denominator by 2, so that the constant will be 1.

$$r = \frac{\frac{4}{2}}{\frac{2}{2} - \frac{\cos\theta}{2}} = \frac{2}{1 - \frac{1}{2}\cos\theta}$$

The equation is now in the standard form $r = \dfrac{ep}{1 - e\cos\theta}$, where $ep = 2$ and $e = \dfrac{1}{2}$. Using these facts, $ep = \dfrac{1}{2}p = 2$, so $p = 4$. Because $e = \dfrac{1}{2} < 1$, the conic is an ellipse.

2a. Graph the polar equation: $r = \dfrac{12}{5 + 3\cos\theta}$.

For equations in the form $r = \frac{ep}{1 - e\cos\theta}$, one focus is at the pole, the directrix is the line $x = -p$ or $x = -4$, located 4 units to the left of the pole, and the graph has polar axis symmetry.

The major axis of the ellipse is on the polar axis, so let $\theta = 0$ and $\theta = \pi$ to find the vertices. The vertices are (4, 0) and $\left(\frac{4}{3}, \pi\right)$. Pick some other values of θ between 0 and π to sketch the upper half of the ellipse.

θ	$\frac{\pi}{3}$	$\frac{\pi}{2}$	$\frac{2\pi}{3}$
r	$\frac{8}{3}$	2	$\frac{8}{5}$

Reflect the upper half about the polar axis to complete the ellipse.

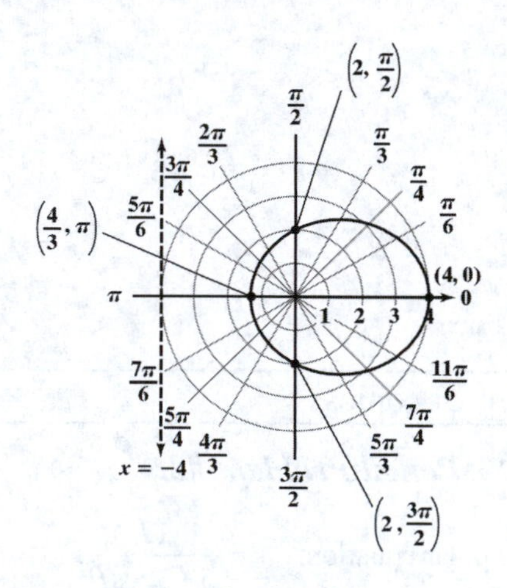

2b. Graph the polar equation: $r = \frac{8}{4 + 4\sin\theta}$.

Notice that the constant term in the denominator is not 1, so the equation is not in standard form. Divide the numerator and the denominator by 4, so that the constant will be 1.

$$r = \frac{\frac{8}{4}}{\frac{4}{4} + \frac{4\sin\theta}{4}} = \frac{2}{1 + \sin\theta}$$

2b. Graph the polar equation: $r = \frac{6}{2 - 2\sin\theta}$.

The equation is now in the standard form $r=\dfrac{ep}{1+e\sin\theta}$, where $ep=2$ and $e=1$. Using these facts, $ep=1p=2$, so $p=2$. Because $e=1$, the conic is a parabola.

For equations in the form $r=\dfrac{ep}{1+e\sin\theta}$, one focus is at the pole, the directrix is the line $y=p$ or $y=2$, located 2 units above the pole, and the graph has symmetry with respect to the line $\theta=\dfrac{\pi}{2}$.

Let $\theta=\dfrac{\pi}{2}$ to find the vertex; the vertex is $\left(1,\ \dfrac{\pi}{2}\right)$.

Pick some other values of θ between $-\dfrac{\pi}{2}$ and $\dfrac{\pi}{2}$ to sketch the right half of the parabola.

θ	$-\dfrac{\pi}{6}$	0	$\dfrac{\pi}{6}$
r	4	2	$\dfrac{4}{3}$

Reflect the right half about the line $\theta=\dfrac{\pi}{2}$ to complete the parabola.

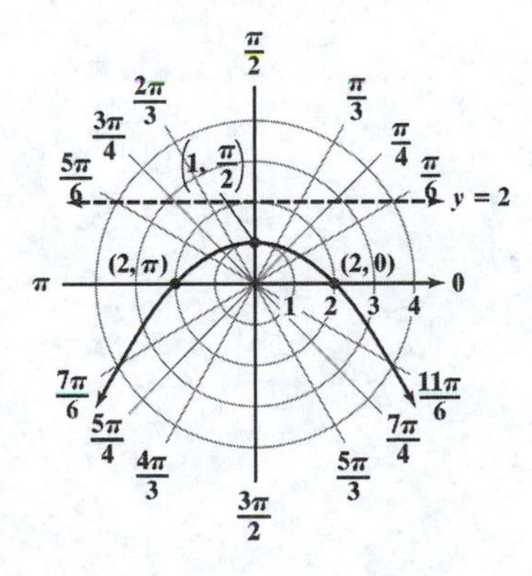

2c. Graph the polar equation: $r = \dfrac{9}{3 - 9\cos\theta}$.

Notice that the constant term in the denominator is not 1, so the equation is not in standard form. Divide the numerator and the denominator by 3, so that the constant will be 1.

$$r = \frac{\frac{9}{3}}{\frac{3}{3} - \frac{9}{3}\cos\theta} = \frac{3}{1 - 3\cos\theta}$$

The equation is now in the standard form $r = \dfrac{ep}{1 - e\cos\theta}$, where $ep = 3$ and $e = 3$. Using these facts, $ep = 3p = 3$, so $p = 1$. Because $e = 3 > 1$, the conic is a hyperbola.

For equations in the form $r = \dfrac{ep}{1 - e\cos\theta}$, one focus is at the pole, the directrix is the line $x = -p$ or $x = -1$, located 1 unit to the left of the pole, and the graph has polar axis symmetry.

The transverse axis of the hyperbola is on the polar axis, so let $\theta = 0$ and $\theta = \pi$ to find the vertices. The vertices are $\left(-\dfrac{3}{2}, 0\right)$ and $\left(\dfrac{3}{4}, \pi\right)$. Pick some other values of θ to sketch the upper right portion of the hyperbola.

θ	$\frac{\pi}{2}$	$\frac{2\pi}{3}$
r	3	$\frac{6}{5}$

Reflect the upper right portion about the polar axis to complete the right half of the hyperbola. Draw the mirror image of the right half through the other vertex to complete the graph.

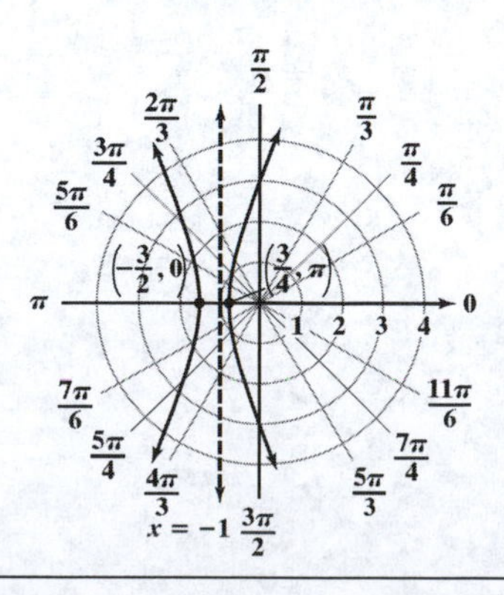

2c. Graph the polar equation: $r = \dfrac{8}{2 - 4\cos\theta}$.

Answers for Pencil Problems *(Textbook Exercise references in parentheses)*:

1. False

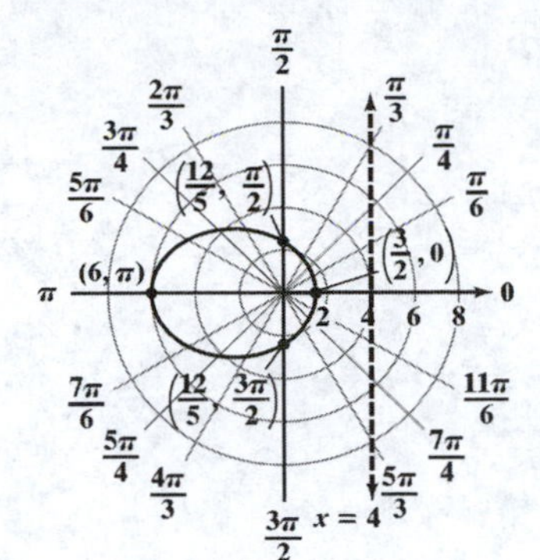

2a. *(9.6 #13)*

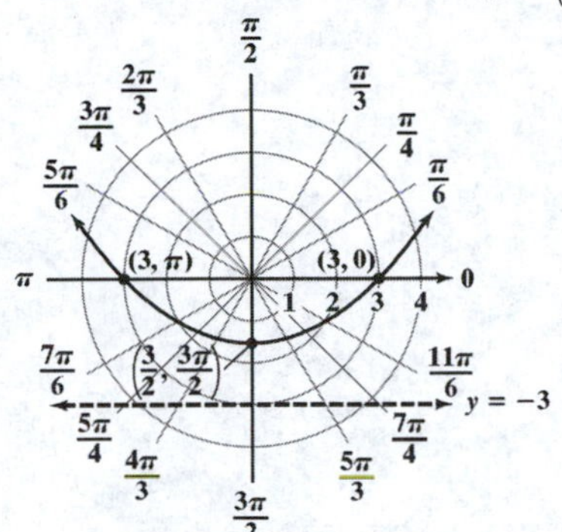

2b. *(9.6 #15)*

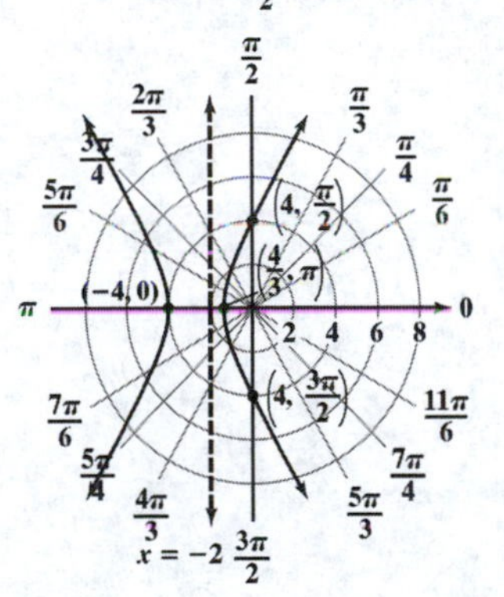

2c. *(9.6 #17)*

Section 10.1
Sequences and Summation Notation

Bees, Trees, and Piano Keys !

What can those three things possibly have in common?

In this section, we will study sequences. One amazing example is called the Fibonacci sequence, an infinite sequence of numbers investigated by Leonardo of Pisa, also known as Fibonacci, an Italian mathematician of the thirteenth century.

The sequence is generated using simple addition, and yet it shows up in some unexpected, and complex, ways.

As you read the textbook, you will find interesting areas where these concepts apply.

Objective #1: Find particular terms of a sequence from the general term.

✔ *Solved Problem #1*

1a. Write the first four terms of the sequence whose nth term, or general term, is $a_n = 2n+5$.

$$a_n = 2n+5$$
$$a_1 = 2(1)+5 = 7$$
$$a_2 = 2(2)+5 = 9$$
$$a_3 = 2(3)+5 = 11$$
$$a_4 = 2(4)+5 = 13$$

The first four terms are 7, 9, 11, and 13.

1b. Write the first four terms of the sequence whose nth term, or general term, is $a_n = \frac{(-1)^n}{2^n+1}$.

$$a_n = \frac{(-1)^n}{2^n+1}$$
$$a_1 = \frac{(-1)^1}{2^1+1} == \frac{-1}{3} - \frac{1}{3}$$
$$a_2 = \frac{(-1)^2}{2^2+1} = \frac{1}{5}$$
$$a_3 = \frac{(-1)^3}{2^3+1} = \frac{-1}{9} = -\frac{1}{9}$$
$$a_4 = \frac{(-1)^4}{2^4+1} = \frac{1}{17}$$

The first four terms are $-\frac{1}{3}, \frac{1}{5}, -\frac{1}{9},$ and $\frac{1}{17}$.

Pencil Problem #1

1a. Write the first four terms of the sequence whose nth term, or general term, is $a_n = 3n+2$.

1b. Write the first four terms of the sequence whose nth term, or general term, is $a_n = (-1)^n(n+3)$.

Objective #2: Use recursion formulas.

✔ *Solved Problem #2*

2. Find the first four terms of the sequence in which $a_1 = 3$ and $a_n = 2a_{n-1} + 5$ for $n \ge 2$.

$$a_1 = 3$$
$$a_2 = 2a_1 + 5 = 2(3) + 5 = 11$$
$$a_3 = 2a_2 + 5 = 2(11) + 5 = 27$$
$$a_4 = 2a_3 + 5 = 2(27) + 5 = 59$$

The first four terms are 3, 11, 27, and 59.

Pencil Problem #2

2. Find the first four terms of the sequence in which $a_1 = 4$ and $a_n = 2a_{n-1} + 3$ for $n \ge 2$.

Objective #3: Use factorial notation.

✔ *Solved Problem #3*

3. Write the first four terms of the sequence whose nth term is $a_n = \dfrac{20}{(n+1)!}$.

$$a_n = \frac{20}{(n+1)!}$$
$$a_1 = \frac{20}{(1+1)!} = \frac{20}{2!} = 10$$
$$a_2 = \frac{20}{(2+1)!} = \frac{20}{3!} = \frac{20}{6} = \frac{10}{3}$$
$$a_3 = \frac{20}{(3+1)!} = \frac{20}{4!} = \frac{20}{24} = \frac{5}{6}$$
$$a_4 = \frac{20}{(4+1)!} = \frac{20}{5!} = \frac{20}{120} = \frac{1}{6}$$

The first four terms are 10, $\frac{10}{3}$, $\frac{5}{6}$, and $\frac{1}{6}$.

Pencil Problem #3

3. Write the first four terms of the sequence whose nth term is $a_n = \dfrac{n^2}{n!}$.

Objective #4: Use summation notation.

✔ *Solved Problem #4*

4a. Expand and evaluate the sum: $\displaystyle\sum_{k=3}^{5}\left(2^k - 3\right)$.

$$\sum_{k=3}^{5}\left(2^k - 3\right)$$
$$= \left(2^3 - 3\right) + \left(2^4 - 3\right) + \left(2^5 - 3\right)$$
$$= (8-3) + (16-3) + (32-3)$$
$$= 5 + 13 + 29$$
$$= 47$$

Pencil Problem #4

4a. Expand and evaluate the sum: $\displaystyle\sum_{k=1}^{5} k(k+4)$.

4b. Expand and evaluate the sum: $\sum_{i=1}^{5} 4.$

$$\sum_{i=1}^{5} 4 = 4+4+4+4+4$$
$$= 20$$

4b. Expand and evaluate the sum: $\sum_{i=5}^{9} 11$

4c. Express the sum using summation notation. Use 1 as the lower limit of summation and *i* for the index of summation.

$$1^2 + 2^2 + 3^2 + \cdots + 9^2$$

The sum has nine terms, each of the form i^2, starting at $i = 1$ and ending at $i = 9$.

$$1^2 + 2^2 + 3^2 + \cdots + 9^2 = \sum_{i=1}^{9} i^2$$

4c. Express the sum using summation notation. Use 1 as the lower limit of summation and *i* for the index of summation.

$$2 + 2^2 + 2^3 + \ldots + 2^{11}$$

4d. Express the sum using summation notation. Use 1 as the lower limit of summation and *i* for the index of summation.

$$1 + \frac{1}{2} + \frac{1}{4} + \frac{1}{8} + \cdots + \frac{1}{2^{n-1}}$$

The sum has *n* terms, each of the form $\frac{1}{2^{i-1}}$, starting at $i = 1$ and ending at $i = n$.

$$1 + \frac{1}{2} + \frac{1}{4} + \frac{1}{8} + \cdots + \frac{1}{2^{n-1}} = \sum_{i=1}^{n} \frac{1}{2^{i-1}}$$

4d. Express the sum using summation notation. Use 1 as the lower limit of summation and *i* for the index of summation.

$$\frac{1}{2} + \frac{2}{3} + \frac{3}{4} + \ldots + \frac{14}{14+1}$$

Answers for Pencil Problems *(Textbook Exercise references in parentheses)*:

1a. 5, 8, 11, 14 *(10.1 #1)*

1b. –4, 5, –6, 7 *(10.1 #7)*

2. 4, 11, 25, 53 *(10.1 #17)*

3. $1, 2, \frac{3}{2}, \frac{2}{3}$ *(10.1 #19)*

4a. 115 *(10.1 #33)*

4b. 55 *(10.1 #37)*

4c. $\sum_{i=1}^{11} 2^i$ *(10.1 #45)*

4d. $\sum_{i=1}^{14} \frac{i}{i+1}$ *(10.1 #49)*

Section 11.2
Arithmetic Sequences

IT'S A FULL THEATER TONIGHT !

Some theaters have the same number of seats in each row. But other theaters are more fan-shaped.

In this section of the textbook, we will encounter such a fan-shaped theater, and we will use the techniques of this section to quickly determine the total number of seats without actually adding the number in each row.

Objective #1: Find the common difference for an arithmetic sequence.

1. True or false: An arithmetic sequence is a sequence in which each term after the first differs from the preceding term by a constant amount.

true

1. True or false: In an arithmetic sequence, each term after the first term can be obtained by adding the common difference to the preceding term.

Objective #2: Write terms of an arithmetic sequence.

✔ ***Solved Problem #2***

2. Write the first six terms of the arithmetic sequence with first term 100 and common difference −30.

$a_1 = 100$
$a_2 = 100 + (-30) = 70$
$a_3 = 70 + (-30) = 40$
$a_4 = 40 + (-30) = 10$
$a_5 = 10 + (-30) = -20$
$a_6 = -20 + (-30) = -50$

2. Write the first six terms of the arithmetic sequence with first term −7 and common difference 4.

Objective #3: Use the formula for the general term of an arithmetic sequence.

✔ Solved Problem #3

3a. Find the ninth term of the arithmetic sequence whose first term is 6 and whose common difference is –5.

$a_1 = 6,\ d = -5$

To find the ninth term, a_9, replace n in the formula with 9, replace a_1 with 6, and replace d with –5.

$$\begin{aligned} a_n &= a_1 + (n-1)d \\ a_9 &= 6 + (9-1)(-5) \\ &= 6 + 8(-5) \\ &= 6 + (-40) \\ &= -34 \end{aligned}$$

3b. In 2010, 16% of the U.S. population was Latino. On average, this is projected to increase by approximately 0.35% per year. Write a formula for the *n*th term of the arithmetic sequence that describes the percentage of the U.S. population that will be Latino *n* years after 2009.

$$\begin{aligned} a_n &= a_1 + (n-1)d \\ &= 16 + (n-1)0.35 \\ &= 0.35n + 15.65 \end{aligned}$$

3c. Use the result from the previous problem to project the percentage of the U.S. population that will be Latino in 2030.

2030 is 21 years after 2009.

$$\begin{aligned} a_n &= 0.35n + 15.65 \\ a_{20} &= 0.35(21) + 15.65 = 23 \end{aligned}$$

In 2030, 23% of the U.S. population is projected to be Latino.

✎ Pencil Problem #3 ✎

3a. Find the 50th term of the arithmetic sequence whose first term is 7 and whose common difference is 5.

3b. In 1990, 18.4% of American women ages 25 and older had graduated from college. On average, this percentage has increased by approximately 0.6 each year. Write a formula for the *n*th term of the arithmetic sequence that models the percentage of American women ages 25 and older who had or will have graduated from college *n* years after 1990.

3c. Use the result from the previous problem to project the percentage of American women ages 25 and older who will be college graduates by 2019.

Objective #4: Use the formula for the sum of the first n terms of an arithmetic sequence.

✔ Solved Problem #4

4a. Find the sum of the first 15 terms of the arithmetic sequence: 3, 6, 9, 12, ...

To find the sum of the first 15 terms, S_{15}, replace n in the formula with 15.

$$S_n = \frac{n}{2}(a_1 + a_n)$$
$$S_{15} = \frac{15}{2}(a_1 + a_{15})$$

Use the formula for the general term of a sequence to find a_{15}. The common difference, d, is 3, and the first term, a_1, is 3.

$$\begin{aligned} a_n &= a_1 + (n-1)d \\ a_{15} &= 3 + (15-1)(3) \\ &= 3 + 14(3) \\ &= 3 + 42 \\ &= 45 \end{aligned}$$

Thus, $S_{15} = \frac{15}{2}(3+45) = \frac{15}{2}(48) = 360$.

4b. Find the following sum: $\sum_{i=1}^{30}(6i-11)$.

$$\begin{aligned} &\sum_{i=1}^{30}(6i-11) \\ &= (6\cdot1-11)+(6\cdot2-11)+(6\cdot3-11)+\ldots+(6\cdot30-11) \\ &= -5+1+7+\ldots+169 \end{aligned}$$

The first term, a_1, is -5.
The common difference, d, is $1-(-5)=6$.
The last term, a_{30}, is 169.

$$\begin{aligned} S_n &= \tfrac{n}{2}(a_1 + a_n) \\ S_{30} &= \tfrac{30}{2}(-5+169) \\ &= 15(164) \\ &= 2460 \end{aligned}$$

Thus, $\sum_{i=1}^{30}(6i-11) = 2460$

✎ Pencil Problem #4 ✎

4a. Find the sum of the first 50 terms of the arithmetic sequence: $-10, -6, -2, 2, \ldots$

4b. Find the following sum: $\sum_{i=1}^{100} 4i$.

4c. The model $a_n = 1800n + 64{,}130$ describes yearly adult residential community costs *n* years after 2017. How much would it cost for the adult residential community for a ten-year period beginning in 2018?

$$a_n = 1800n + 64{,}130$$
$$a_1 = 1800(1) + 64{,}130 = 65{,}930$$
$$a_{10} = 1800(10) + 64{,}130 = 82{,}130$$

$$S_n = \frac{n}{2}(a_1 + a_n)$$
$$S_{10} = \frac{10}{2}(a_1 + a_{10})$$
$$= 5(65{,}930 + 82{,}130)$$
$$= 5(148{,}060)$$
$$= \$740{,}300$$

It would cost \$740,300 for the ten-year period beginning in 2018.

4c. A section in a stadium has 20 seats in the first row, 23 seats in the second row, increasing by 3 seats each row for a total of 38 rows. How many seats are in this section of the stadium?

Answers for Pencil Problems *(Textbook Exercise references in parentheses)*:

1. true *(10.2 #1)* **2.** $-7, -3, 1, 5, 9, 13$ *(10.2 #3)*

3a. 252 *(10.2 #17)* **3b.** $a_n = 0.6n + 17.8$ *(10.2 #61a)* **3c.** 35.8% *(10.2 #61b)*

4a. 4400 *(10.2 #37)* **4b.** 20,200 *(10.2 #49)* **4c.** 2869 seats *(10.2 #71)*

Section 10.3
Geometric Sequences and Series

How Much Will You End Up With?

Suppose you are 24 and you have just landed a job!

You decide that you can save for retirement by putting aside \$80 per month into an account which pays 5% compounded monthly.

What will the account balance be when you reach age 65?

In this section, several applications will deal with money and we will apply geometric sequences and series to find answers to a variety of financial questions.

Objective #1: Find the common ratio of a geometric sequence.

✔ ***Solved Problem #1***

1. True or False: The sequence
$$6, -12, 24, -48, 96, \ldots$$
is an example of a geometric sequence.

True. Each term after the first term is -2 times the previous term. The common ration is -2.

1. True or False: The sequence
$$2, 6, 24, 120, \ldots$$
is an example of a geometric sequence.

Objective #2: Write terms of a geometric sequence.

✔ ***Solved Problem #2***

2. Write the first six terms of the geometric sequence with first term 12 and common ratio $\frac{1}{2}$.

$a_1 = 12,\ r = \dfrac{1}{2}$

$a_2 = 12\left(\frac{1}{2}\right)^1 = 6$

$a_3 = 12\left(\frac{1}{2}\right)^2 = \frac{12}{4} = 3$

$a_4 = 12\left(\frac{1}{2}\right)^3 = \frac{12}{8} = \frac{3}{2}$

$a_5 = 12\left(\frac{1}{2}\right)^4 = \frac{12}{16} = \frac{3}{4}$

$a_6 = 12\left(\frac{1}{2}\right)^5 = \frac{12}{32} = \frac{3}{8}$

The first six terms are 12, 6, 3, $\frac{3}{2}$, $\frac{3}{4}$, and $\frac{3}{8}$.

2. Write the first five terms of the geometric sequence with first term 5 and common ratio 3.

Objective #3: Use the formula for the general term of a geometric sequence.

✔ Solved Problem #3

3a. Find the seventh term of the geometric sequence whose first term is 5 and whose common ratio is −3.

$a_1 = 5, r = -3$

$a_n = a_1 r^{n-1}$

$a_7 = 5(-3)^{7-1}$

$= 5(-3)^6$

$= 5(729)$

$= 3645$

The seventh term is 3645.

3b. Write the general term for the geometric sequence: $3, 6, 12, 24, 48, \ldots$

Then use the formula for the general term to find the eighth term.

$r = \frac{6}{3} = 2, \ a_1 = 3$

Formula for the general term:

$a_n = a_1(r)^{n-1}$

$a_n = 3(2)^{n-1}$

Find the eighth term:

$a_n = 3(2)^{n-1}$

$a_8 = 3(2)^{8-1}$

$= 3(2)^7$

$= 3(128)$

$= 384$

The eighth term is 384.

Pencil Problem #3

3a. Find the eighth term of the geometric sequence whose first term is 6 and whose common ratio is 2.

3b. Write the general term for the geometric sequence: $3, 12, 48, 192, \ldots$

Then use the formula for the general term to find the seventh term.

Objective #4: Use the formula for the sum of the first n terms of a geometric sequence.

✔ Solved Problem #4

4a. Find the sum of the first nine terms of the geometric sequence: $2, -6, 18, -54, \ldots$

$$a_1 = 2,\ r = \frac{-6}{2} = -3$$

$$S_n = \frac{a_1(1-r^n)}{1-r}$$

$$S_9 = \frac{2\left(1-(-3)^9\right)}{1-(-3)} = \frac{2(19{,}684)}{4} = 9842$$

The sum of the first nine terms is 9842.

4b. Find the following sum: $\sum_{i=1}^{8} 2\cdot 3^i$.

$$a_1 = 2\cdot(3)^1 = 6,\ r = 3$$

$$S_n = \frac{a_1(1-r^n)}{1-r}$$

$$S_8 = \frac{6\left(1-3^8\right)}{1-3} = \frac{6(-6560)}{-2} = 19{,}680$$

Thus, $\sum_{i=1}^{8} 2\cdot 3^i = 19{,}680$.

Pencil Problem #4

4a. Find the sum of the first 11 terms of the geometric sequence: $3, -6, 12, -24, \ldots$

4b. Find the following sum: $\sum_{i=1}^{10} 5\cdot 2^i$.

4c. A job pays a salary of $30,000 the first year. During the next 29 years, the salary increases by 6% each year. What is the total lifetime salary over the 30-year period? Round to the nearest dollar.

$$a_1 = 30,000,\ r = 1.06$$

$$S_n = \frac{a_1(1-r^n)}{1-r}$$

$$S_{30} = \frac{30,000\left(1-(1.06)^{30}\right)}{1-1.06} \approx 2,371,746$$

The total lifetime salary is $2,371,746.

4c. A job pays a salary of $24,000 the first year. During the next 19 years, the salary increases by 5% each year. What is the total lifetime salary over the 20-year period? Round to the nearest dollar.

Objective #5: Find the value of an annuity.

✔ *Solved Problem #5*

5. At age 30, to save for retirement, you decide to deposit $100 at the end of each month into an IRA that pays 9.5% compounded monthly. Find how much will you have in the IRA when you retire at age 65 and find how much is interest.

$$A = \frac{P\left[\left(1+\frac{r}{n}\right)^{nt} - 1\right]}{\frac{r}{n}}$$

$$P = 100,\ r = 0.095,\ n = 12,\ t = 35$$

$$A = \frac{100\left[\left(1+\frac{0.095}{12}\right)^{12\cdot 35} - 1\right]}{\frac{0.095}{12}} \approx 333,946$$

The value of the IRA will be $333,946.

Find the interest:

$$\begin{aligned}\text{Interest} &= \text{Value of IRA} - \text{Total deposits}\\ &\approx \$333,946 - \$100\cdot 12\cdot 35\\ &\approx \$333,946 - \$42,000\\ &\approx \$291,946\end{aligned}$$

Pencil Problem #5

5. At age 25, to save for retirement, you decide to deposit $50 at the end of each month into an IRA that pays 5.5% compounded monthly. Find how much will you have in the IRA when you retire at age 65 and find how much is interest.

Objective #6: Use the formula for the sum of an infinite geometric series.

✔ Solved Problem #6

6a. Find the sum of the infinite geometric series:

$$3+2+\frac{4}{3}+\frac{8}{9}+\cdots$$

$a_1 = 3,\ r=\frac{2}{3}$

$$S=\frac{a_1}{1-r}$$

$$S=\frac{3}{1-\frac{2}{3}}$$

$$=\frac{3}{\frac{1}{3}}$$

$$=9$$

The sum of this infinite geometric series is 9.

6b. Express $0.\overline{9}$ as a fraction in lowest terms.

$$0.\overline{9}=0.9999\cdots=\frac{9}{10}+\frac{9}{100}+\frac{9}{1000}+\cdots$$

$a_1=\frac{9}{10},\ r=\frac{1}{10}$

$$S=\frac{a_1}{1-r}$$

$$S=\frac{\frac{9}{10}}{1-\frac{1}{10}}$$

$$=\frac{\frac{9}{10}}{\frac{9}{10}}$$

$$=1$$

An equivalent fraction for $0.\overline{9}$ is 1.

Pencil Problem #6

6a. Find the sum of the infinite geometric series:

$$3+\frac{3}{4}+\frac{3}{4^2}+\frac{3}{4^3}+\cdots$$

6b. Express $0.\overline{5}$ as a fraction in lowest terms.

Answers for Pencil Problems *(Textbook Exercise references in parentheses)*:

1. false *(10.3 #105)*

2. 5,15,45,135,405 *(10.3 #1)*

3a. 768 *(10.3 #9)* **3b.** $a_n = 3(4)^{n-1}$; $a_7 = 12,288$ *(10.3 #17)*

4a. 2049 *(10.3 #27)* **4b.** 10,230 *(10.3 #33)* **4c.** $793,583 *(10.3 #73)*

5. $87,052; $63,052 *(10.3 #79)*

6a. 4 *(10.3 #39)* **6b.** $\frac{5}{9}$ *(10.3 #45)*

Section 10.4
Mathematical Induction

Will They ALL Fall Down?

The mathematical principle of this section can be illustrated using an unending line of dominoes. If the first domino is pushed over, it knocks down the next, which knocks down the next, and so on, in a chain reaction.

To topple all the dominoes in the infinite sequence, two conditions must be satisfied:

1. The first domino must be knocked down.
2. If the domino in position k is knocked down, then the domino in position $k+1$ must be knocked down.

If the second condition is not satisfied, it does not follow that all the dominoes will topple. For example, suppose the dominoes are spaced far enough apart so that a falling domino does not push over the next domino in the line.

Objective #1: Understand the principle of mathematical induction.

✔ *Solved Problem #1*

1a. For the given statement S_n, write the statement S_1.

$S_n: 2+4+6+\cdots+2n = n(n+1)$

If $n=1$ then the statement S_1 is obtained by writing the first term, 2, on the left, and substituting 1 for n on the right.

$S_1: 2 = 1(1+1)$

Pencil Problem #1

1a. For the given statement S_n, write the statement S_1.

$S_n: 1+3+5+(2n-1) = n^2$

✔ *Solved Problem #1b*

1b. For the given statement S_n, write the two statements S_k, and S_{k+1}.

$$S_n : 1^3 + 2^3 + 3^3 + \cdots + n^3 = \frac{n^2(n+1)^2}{4}$$

Write S_k by taking the sum of the first k terms on the left and replacing n with k on the right.

$$S_k : 1^3 + 2^3 + 3^3 + \cdots + k^3 = \frac{k^2(k+1)^2}{4}$$

Write S_{k+1} by taking the sum of the first $k+1$ terms on the left and replacing n with $k+1$ on the right.

$$S_{k+1} : 1^3 + 2^3 + 3^3 + \cdots + k^3 + (k+1)^3 = \frac{(k+1)^2(k+1+1)^2}{4}$$

$$= \frac{(k+1)^2(k+2)^2}{4}$$

✎ *Pencil Problem #1b* ✎

1b. For the given statement S_n, write the two statements S_k, and S_{k+1}.

$$S_n : 3 + 7 + 11 + \cdots + (4n-1) = n(2n+1)$$

Objective #2: Prove statements using mathematical induction.

✔ *Solved Problem #2*

2. Use mathematical induction to prove that $1^3 + 2^3 + 3^3 + \cdots + n^3 = \dfrac{n^2(n+1)^2}{4}$ for all positive integers n.

Step 1. *Show that S_1 is true:*

$$1^3 = \frac{1^2(1+1)^2}{4}$$

$$1 = \frac{1(2)^2}{4}$$

$$1 = \frac{4}{4}$$

$$1 = 1, \text{ True}$$

Step 2. *Show that if S_k is true, then S_{k+1} is true:*

Assume $1^3 + 2^3 + 3^3 + \cdots + k^3 = \dfrac{k^2(k+1)^2}{4}$ is true. Then,

$$1^3 + 2^3 + 3^3 + \cdots + k^3 + (k+1)^3 = \frac{k^2(k+1)^2}{4} + (k+1)^3$$

$$1^3 + 2^3 + 3^3 + \cdots + k^3 + (k+1)^3 = \frac{k^2(k+1)^2}{4} + \frac{4(k+1)^3}{4}$$

$$1^3 + 2^3 + 3^3 + \cdots + k^3 + (k+1)^3 = \frac{k^2(k+1)^2 + 4(k+1)^3}{4}$$

$$1^3 + 2^3 + 3^3 + \cdots + k^3 + (k+1)^3 = \frac{(k+1)^2\left(k^2 + 4(k+1)\right)}{4}$$

$$1^3 + 2^3 + 3^3 + \cdots + k^3 + (k+1)^3 = \frac{(k+1)^2\left(k^2 + 4k + 4\right)}{4}$$

$$1^3 + 2^3 + 3^3 + \cdots + k^3 + (k+1)^3 = \frac{(k+1)^2(k+2)^2}{4}$$

The final statement is S_{k+1}.

Thus, by mathematical induction, the statement $1^3 + 2^3 + 3^3 + \cdots + n^3 = \dfrac{n^2(n+1)^2}{4}$ is true for all positive integers n.

✎ *Pencil Problem #2* ✎

2. Use mathematical induction to prove that $1+2+2^2+\cdots+2^{n-1}=2^n-1$ for all positive integers *n*.

Answers for Pencil Problems *(Textbook Exercise references in parentheses)*:

1a. $S_1: 1=1^2$ *(10.4 #1)*

1b. $S_k: 3+7+11+\cdots+(4k-1)=k(2k+1)$

$S_{k+1}: 3+7+11+\cdots+[4(k+1)-1]=(k+1)[2(k+1)+1]=(k+1)(2k+3)$ *(10.4 #7)*

2. *Show that* S_1 *is true*: $1=2^1-1$

$1=2-1$

$1=1$, true

Show that if S_k *is true, then* S_{k+1} *is true*:

Assume $S_k: 1+2+2^2+\cdots+2^{k-1}=2^k-1$ is true. Then, $1+2+2^2+\cdots+2^{k-1}+2^{(k+1)-1}=2^k-1+2^{(k+1)-1}$

$$1+2+2^2+\cdots+2^{k-1}+2^{(k+1)-1}=2^k-1+2^k$$

$$1+2+2^2+\cdots+2^{k-1}+2^{(k+1)-1}=2\cdot 2^k-1$$

$$1+2+2^2+\cdots+2^{k-1}+2^{(k+1)-1}=2^{k+1}-1$$

Thus, $1+2+2^2+\cdots+2^{n-1}=2^n-1$ is true for all positive integers *n*. *(10.4 #17)*

Section 10.5
The Binomial Theorem

Who Knew That First?

Telephones, Internet, and other modern forms of communication mean that information now can spread across the globe in the blink of an eye.

In this section, we study a special array of numbers known as Pascal's triangle, named after mathematician Blaise Pascal.
However, this triangular array of numbers actually appeared centuries earlier in a Chinese document.

The same mathematics is often discovered by independent researchers separated by time, place, and culture.
But with modern communication, important discoveries are now shared much more efficiently.

Objective #1: Evaluate a binomial coefficient.

Solved Problem #1

1a. Evaluate: $\binom{6}{3}$.

$$\binom{6}{3} = \frac{6!}{3!(6-3)!}$$
$$= \frac{6!}{3!3!}$$
$$= \frac{6\cdot5\cdot4\cdot\cancel{3!}}{3\cdot2\cdot1\cdot\cancel{3!}}$$
$$= 20$$

1b. Evaluate: $\binom{6}{0}$.

$$\binom{6}{0} = \frac{6!}{0!(6-0)!}$$
$$= \frac{6!}{6!}$$
$$= 1$$

Pencil Problem #1

1a. Evaluate: $\binom{8}{3}$.

1b. Evaluate: $\binom{12}{1}$.

1c. Evaluate: $\binom{8}{2}$.

$$\binom{8}{2} = \frac{8!}{2!(8-2)!} = \frac{8!}{2!6!} = \frac{8\cdot 7}{2} = 28$$

1c. Evaluate: $\binom{100}{2}$.

1d. Evaluate: $\binom{3}{3}$.

$$\binom{3}{3} = \frac{3!}{3!(3-3)!} = \frac{3!}{3!0!} = \frac{3!}{3!} = 1$$

1d. Evaluate: $\binom{6}{6}$.

Objective #2: Expand a binomial raised to a power.

✔ *Solved Problem #2a*

2a. Expand: $(x+1)^4$

$$(x+1)^4 = \binom{4}{0}x^4 + \binom{4}{1}x^3 + \binom{4}{2}x^2 + \binom{4}{3}x + \binom{4}{4} = x^4 + 4x^3 + 6x^2 + 4x + 1$$

Pencil Problem #2a

2a. Expand: $(x+2)^3$

✔ *Solved Problem #2b*

2b. Expand: $(x-2y)^5$

$$(x-2y)^5 = \binom{5}{0}x^5(-2y)^0 + \binom{5}{1}x^4(-2y)^1 + \binom{5}{2}x^3(-2y)^2 + \binom{5}{3}x^2(-2y)^3 + \binom{5}{4}x(-2y)^4 + \binom{5}{5}x^0(-2y)^5$$

$$= x^5 \quad - 5x^4(2y) \quad + 10x^3(4y^2) \quad - 10x^2(8y^3) \quad + 5x(16y^4) \quad - 32y^5$$

$$= x^5 - 10x^4y + 40x^3y^2 - 80x^2y^3 + 80xy^4 - 32y^5$$

✎ *Pencil Problem #2b* ✎

2b. Expand: $\left(x^2+2y\right)^4$

Objective #3: Find a particular term in a binomial expansion.

Solved Problem #3

3. Find the fifth term in the expansion of $(2x+y)^9$.

Since we are looking for the 5th term, $r = 5-1 = 4$.
Thus, $r = 4$, $a = 2x$, $b = y$, and $n = 9$.

$$(r+1)\text{st term} = \binom{n}{r} a^{n-r} b^r$$

$$\text{fifth term} = \binom{9}{4}(2x)^5 y^4$$

$$= \frac{9!}{4!5!}(32x^5)y^4$$

$$= 4032x^5y^4$$

Pencil Problem #3

3. Find the sixth term in the expansion of $\left(x^2+y^3\right)^8$.

Answers for Pencil Problems *(Textbook Exercise references in parentheses)*:

1a. 56 *(10.5 #1)*

1b. 12 *(10.5 #3)*

1c. 4950 *(10.5 #7)*

1d. 1 *(10.5 #5)*

2a. $x^3+6x^2+12x+8$ *(10.5 #9)*

2b. $x^8+8x^6y+24x^4y^2+32x^2y^3+16y^4$ *(10.5 #17)*

3. $56x^6y^{15}$ *(10.5 #43)*

Section 10.6
Counting Principles, Permutations, and Combinations

I Have NOTHING to Wear!

On many mornings, we feel quite limited on the fashion statement we desire to make that day. But the truth is we usually have more clothing options than we might think. If we consider how the various components of our outfit can be mixed and matched, the number of unique outfits can be difficult to count.

Attempting to count each possibility one-by-one can be daunting in many situations.

In this section of the textbook, we will use organized mathematical methods and formulas that will allow us to count more quickly and accurately than the 1, 2, 3, 4, 5, 6, … method.

Objective #1: Use the Fundamental Counting Principle.

✔ Solved Problem #1

1a. A pizza can be ordered with three choices of size (small, medium, or large), four choices of crust (thin, thick, crispy, or regular), and six choices of toppings (ground beef, sausage, pepperoni, bacon, mushrooms, or onions). How many different one-topping pizzas can be ordered?

Multiply the number of choices for each of the three decisions:

$$\underset{\text{Size}}{3} \times \underset{\text{Crust}}{4} \times \underset{\text{Topping}}{6} = 72$$

72 different one-topping pizzas can be ordered.

1b. License plates in a particular state display two letters followed by three numbers, such as AT-887 or BB-013. How many different license plates can be manufactured?

Multiply the number of choices for each of the letters and each of the digits:

$$\overbrace{26}^{\text{Letter 1}} \times \overbrace{26}^{\text{Letter 2}} \times \overbrace{10}^{\text{Digit 1}} \times \overbrace{10}^{\text{Digit 2}} \times \overbrace{10}^{\text{Digit 3}} = 676{,}000$$

676,000 different license plates can be manufactured.

✎ Pencil Problem #1 ✎

1a. An ice cream store sells two drinks (sodas or milk shakes), in four sizes (small, medium, large, or jumbo), and five flavors (vanilla, strawberry, chocolate, coffee, or pistachio). In how many ways can a customer order a drink?

1b. You are taking a multiple-choice test that has five questions. Each of the questions has three answer choices, with one correct answer per question. If you select one of these three choices for each question and leave nothing blank, in how many ways can you answer the questions?

Objective #2: Use the permutations formula.

✔ *Solved Problem #2*

2a. A corporation has seven members on its board of directors. In how many different ways can it elect a president, vice-president, secretary, and treasurer?

The corporation is choosing 4 officers from a group of 7 people. The order in which the officers are chosen matters because the president, vice-president, secretary, and treasurer each have different responsibilities. Thus, we are looking for the number of permutations of 7 things taken 4 at a time.

$${}_7P_4 = \frac{7!}{(7-4)!} = \frac{7!}{3!} = 840$$

There are 840 ways of filling the four offices.

2b. In how many ways can 6 books be lined up along a shelf?

Because you are using all six of your books in every possible arrangement, you are arranging 6 books from a group of 6 books. Thus, we are looking for the number of permutations of 6 things taken 6 at a time.

$${}_6P_6 = \frac{6!}{(6-6)!} = \frac{6!}{0!} = \frac{6\cdot5\cdot4\cdot3\cdot2\cdot1}{1} = 720$$

There are 720 ways the 6 books can be lined up along the shelf.

Pencil Problem #2

2a. Using 15 flavors of ice cream, how many cones with three different flavors can you create if it is important to you which flavor goes on the top, middle, and bottom?

2b. What is the number of permutations of 8 things taken 0 at a time?

Objective #3: Distinguish between permutation problems and combination problems.

✔ *Solved Problem #3*

3a. Determine if the question involves combinations or permutations. (Do *not* solve the problem.)
How many ways can you select 6 free DVDs from a list of 200 DVDs?

The order in which the DVDs are selected does not matter.

Thus, this problem involves combinations.

Pencil Problem #3

3a. Determine if the question involves combinations or permutations. (Do *not* solve the problem.)
A medical researcher needs 6 people to test the effectiveness of an experimental drug. If 13 people have volunteered for the test, in how many ways can 6 people be selected?

3b. Determine if the question involves combinations or permutations. (Do *not* solve the problem.)
In a race in which there are 50 runners and no ties, in how many ways can the first three finishers come in?

The order in which the runners finish does matter.

Thus, this problem involves permutations.

3b. Determine if the question involves combinations or permutations. (Do *not* solve the problem.)
How many different four-letter passwords can be formed from the letters A, B, C, D, E, F, and G if no repetition of letters is allowed?

Objective #4: Use the combinations formula.

✔ *Solved Problem #4*

4a. From a group of 10 physicians, in how many ways can four people be selected to attend a conference on acupuncture?

The order in which the four people are selected does not matter. This is a problem of selecting 4 people from a group of 10 people. We are looking for the number of combinations of 10 things taken 4 at a time.

$$
\begin{aligned}
{}_{10}C_4 &= \frac{10!}{(10-4)!4!} \\
&= \frac{10!}{6!4!} \\
&= \frac{10 \cdot 9 \cdot 8 \cdot 7 \cdot 6!}{6! \cdot 4 \cdot 3 \cdot 2 \cdot 1} \\
&= 210
\end{aligned}
$$

The four attendees can be selected in 210 different ways.

Pencil Problem #4

4a. An election ballot asks voters to select three city commissioners from a group of six candidates. In how many ways can this be done?

4b. How many different 4-card hands can be dealt from a deck that has 16 different cards?

Because the order in which the 4 cards are dealt does not matter, this is a problem involving combinations. We are looking for the number of combinations of 16 cards drawn 4 at a time.

$$\begin{aligned} {}_{16}C_4 &= \frac{16!}{(16-4)!4!} \\ &= \frac{16!}{12!4!} \\ &= \frac{16\cdot 15\cdot 14\cdot 13\cdot 12!}{12!\cdot 4\cdot 3\cdot 2\cdot 1} \\ &= \frac{16\cdot 15\cdot 14\cdot 13\cdot \cancel{12!}}{\cancel{12!}\cdot 4\cdot 3\cdot 2\cdot 1} \\ &= \frac{16\cdot 15\cdot 14\cdot 13}{4\cdot 3\cdot 2\cdot 1} \\ &= 1820 \end{aligned}$$

There are 1820 different 4-card hands.

4b. You volunteer to help drive children at a charity event to the zoo, but you can fit only 8 of the 17 children present in your van. How many different groups of 8 children can you drive?

Answers for Pencil Problems *(Textbook Exercise references in parentheses)*:

1a. 40 *(10.6 #31)* **1b.** 243 *(10.6 #33)*

2a. 2730 *(10.6 #65)* **2b.** 1 *(10.6 #7)*

3a. combinations *(10.6 #17)* **3b.** permutations *(10.6 #19)*

4a. 20 *(10.6 #49)* **4b.** 24,310 *(10.6 #53)*

Section 10.7
Probability

The Weather Outside is Frightful!

Have you ever thought about the chances of being hit by lightning, caught in a tornado, hurricane, or some other major weather event?

In one of the application exercises in this section, mathematicians, meteorologists, and you will team up to determine such probabilities.

Objective #1: Compute empirical probability.

✔ *Solved Problem #1*

1. Use the data in the table to find the probabilities.

Mammography Screening on 100,000 U.S. Women, Ages 40 to 50	Breast Cancer	No Breast Cancer
Positive Mammogram	720	6944
Negative Mammogram	80	92,256

1a. Find the probability that a woman aged 40 to 50 has a positive mammogram.

The probability of having a positive mammogram is the number of women with a positive mammogram divided by the total number of women.

$$\begin{aligned} P(\text{positive mammogram}) &= \frac{720+6944}{100{,}000} \\ &= \frac{7664}{100{,}000} \\ &\approx 0.077 \end{aligned}$$

Pencil Problem #1

1. The table shows the distribution, by marital status and gender, of the 242 million Americans ages 18 or older. Use the table to find the probabilities.

	Never Married	Married	Widowed	Divorced
Male	40	65	3	10
Female	34	65	11	14

1a. If one person is randomly selected from the population described in the table, find the probability, to the nearest hundredth, that the person is divorced.

1b. Among women with positive mammograms, find the probability of having breast cancer.
(Use the data in the table on the previous page)

To find the probability of breast cancer among women with positive mammograms, restrict the data to women with positive mammograms:

Mammography Screening on 100,000 U.S. Women, Ages 40 to 50	Breast Cancer	No Breast Cancer
Positive Mammogram	720	6944

$$P(\text{breast cancer}) = \frac{720}{720+6944} = \frac{720}{7664} \approx 0.094$$

1b. Among those who are divorced, find the probability of selecting a woman.
(Use the data in the table on the previous page)

Objective #2: Compute theoretical probability.

✔ *Solved Problem #2*

2a. A die is rolled. Find the probability of getting a number greater than 4.

Two of the six numbers, 5 and 6, are greater than 4.

$$P(\text{greater than } 4) = \frac{2}{6} = \frac{1}{3}$$

2b. In Powerball, a minimum award of $50,000 is given to a player who correctly matches four of the five numbers drawn from 69 white balls and one number drawn from 26 red Powerballs. Find the probability of winning this consolation prize. Express the answer as a fraction.

Total number of combinations:

$$\begin{aligned} {}_{69}C_5 \times 26 &= \frac{69!}{(69-5)!5!} \times 26 \\ &= \frac{69!}{64!5!} \times 26 \\ &= \frac{69 \cdot 68 \cdot 67 \cdot 66 \cdot 65 \cdot 64!}{64!5 \cdot 4 \cdot 3 \cdot 2 \cdot 1} \times 26 \\ &= 11{,}238{,}513 \times 26 = 292{,}201{,}338 \end{aligned}$$

Number of selections that match 4 out of 5 white balls and the gold Mega Ball:

$$\underbrace{{}_5C_4}_{\text{match 4 of the 5 selected white balls}} \times \underbrace{{}_{64}C_1}_{\text{any 1 of the 64 non-selected white balls}} \times \underbrace{1}_{\text{gold Mega Ball}}$$

$$= 5 \times 64 \times 1 = 320$$

P(matching 4 of the 5 white balls and the gold Mega Ball)

$$= \frac{320}{292{,}201{,}338} = \frac{160}{146{,}100{,}669}$$

✎ *Pencil Problem #2* ✎

2a. A die is rolled. Find the probability of getting a 4.

2b. In Mega Millions, a player wins the jackpot by matching all five numbers drawn from 56 white balls and matching the number on the gold Mega Ball. There are 46 gold balls. What is the probability of winning the jackpot? Express the answer as a fraction.

Objective #3: Find the probability that an event will not occur.

✔ Solved Problem #3

3. In a 190-minute NFL TV broadcast, 63 minutes are devoted to commercials. Find the probability that a minute of an NFL broadcast is not devoted to commercials.

$$P(\text{not devoted to commercials})$$
$$= 1 - P(\text{devoted to commercials})$$
$$= 1 - \frac{63}{190}$$
$$= \frac{190}{190} - \frac{63}{190}$$
$$= \frac{127}{190}$$

Pencil Problem #3

3. If you are dealt one card from a 52-card deck, find the probability that you are *not* dealt a king.

Objective #4: Find the probability of one event or a second event occurring.

✔ Solved Problem #4

4a. If you roll a single, six-sided die, what is the probability of getting either a 4 or a 5?

These events are mutually exclusive. Thus, add their individual probabilities.

$$P(4 \text{ or } 5) = P(4) + P(5)$$
$$= \frac{1}{6} + \frac{1}{6}$$
$$= \frac{2}{6}$$
$$= \frac{1}{3}$$

Pencil Problem #4

4a. If you are dealt one card from a 52-card deck, find the probability that you are dealt a 2 or a 3.

Solved Problem #4b

4b. Each number, 1 through 8, is written on slips of paper and placed in a hat. If one number is selected at random, find the probability that the number selected will be an odd number or a number less than 5.

These events are *not* mutually exclusive. Thus, use the formula $P(A \text{ or } B) = P(A) + P(B) - P(A \text{ and } B)$.

$$P(\text{odd or less than 5})$$
$$= P(\text{odd}) + P(\text{less than 5}) - P(\text{odd and less than 5})$$
$$= \frac{4}{8} + \frac{4}{8} - \frac{2}{8}$$
$$= \frac{6}{8}$$
$$= \frac{3}{4}$$

Pencil Problem #4b

4b. Each number, 1 through 8, is written on slips of paper and placed in a hat. If one number is selected at random, find the probability that the number selected will be an odd number or a number less than 6.

Objective #5: Find the probability of one event and a second event occurring.

✔ Solved Problem #5

5a. On a roulette wheel, the ball can land with equal probability on any one of the 38 numbered slots, two of which are green. Find the probability of green occurring on two consecutive plays.

The events are independent.
Thus, use the formula $P(A \text{ and } B) = P(A) \cdot P(B)$.

$$\begin{aligned} P(\text{green and green}) &= P(\text{green}) \cdot P(\text{green}) \\ &= \frac{2}{38} \cdot \frac{2}{38} \\ &= \frac{1}{361} \\ &\approx 0.00277 \end{aligned}$$

5b. Find the probability of a family having four boys in a row.

The events are independent.
Thus, multiply their probabilities.

$$\begin{aligned} P(4 \text{ boys in a row}) &= P(\text{boy and boy and boy and boy}) \\ &= P(\text{boy}) \cdot P(\text{boy}) \cdot P(\text{boy}) \cdot P(\text{boy}) \\ &= \frac{1}{2} \cdot \frac{1}{2} \cdot \frac{1}{2} \cdot \frac{1}{2} \\ &= \frac{1}{16} \end{aligned}$$

✎ Pencil Problem #5

5a. A single die is rolled twice. Find the probability of rolling a 2 the first time and a 3 the second time.

5b. If you toss a fair coin six times, what is the probability of getting all heads?

Answers for Pencil Problems *(Textbook Exercise references in parentheses)*:

1a. 0.10 *(10.7 #1)* **1b.** 0.58 *(10.7 #7)* **2a.** $\frac{1}{6}$ *(10.7 #11)* **2b.** $\frac{1}{175{,}711{,}536}$ *(10.7 #27)*

3. $\frac{12}{13}$ *(10.7 #39)* **4a.** $\frac{2}{13}$ *(10.7 #41)* **4b.** $\frac{3}{4}$ *(10.7 #45)* **5a.** $\frac{1}{36}$ *(10.7 #49)* **5b.** $\frac{1}{64}$ *(10.7 #53)*

Section 11.1
Finding Limits Using Tables and Graphs

Knowing Your Limits

In order to succeed in calculus, you will need to understand *limits*. In this section, we take an intuitive approach to limits to help you understand what a limit is. You will learn the formal definition of a limit when you take calculus.

Many fascinating properties of the real numbers make calculus possible. In this section, as you consider *approaching a number*, remember that no matter how close two numbers may be to each other on a number line, there are always infinitely many real numbers between those two numbers.

So if you pick a number close to 0, but not equal to 0, I can pick a number even closer than yours. Then you can pick a new number closer to 0 than my number, and we could go back and forth forever, picking numbers closer to 0.

Objective #1: Understand limit notation.

✔ *Solved Problem #1*

1. Fill in the blank: The limit notation $\lim_{x\to 5} f(x) = 3$ means that the output values of the function f are approaching _____ as the input values are approaching _____.

$\lim_{x\to 5} f(x) = 3$ means that the output values of the function f are approaching 3 as the input values are approaching 5.

Pencil Problem #1

1. Fill in the blank: The limit notation $\lim_{x\to -4} g(x) = 7$ means that the output values of the function g are approaching _____ as the input values are approaching _____.

Objective #2: Find limits using tables.

✔ *Solved Problem #2*

2a. Construct a table to find $\lim_{x\to 3} 4x^2$.

As x gets closer to 3, but not equal to 3, we must determine the value that $4x^2$ is getting closer to. We choose several values of x that are less than 3, such as 2.99, 2.999, and 2.9999, and several values of x that are greater than 3, such as 3.01, 3.001, and 3.001, and evaluate $4x^2$ at each of these values.

(*continued on next page*)

2a. Construct a table to find $\lim_{x\to 2} 5x^2$.

The tables show the values we chose for x and the corresponding values of $4x^2$.

x approaches 3 from the left. →

x	2.99	2.999	2.9999	→
$4x^2$	35.7604	35.976004	35.99760004	→

← x approaches 3 from the right.

x	←	3.0001	3.001	3.01
$4x^2$	←	36.00240004	36.024004	36.2404

It appears that as x gets closer to 3, the values of $4x^2$ get closer to 36.

We infer that $\lim_{x \to 3} 4x^2 = 36.$

2b. Construct a table to find $\lim_{x \to 0} \frac{\cos x - 1}{x}$.

We choose several values of x that are less than 0 and several values of x that are greater than 0 and then evaluate $\frac{\cos x - 1}{x}$ at each of these values. The values of x are in radians. We round the values of $\frac{\cos x - 1}{x}$ to five decimal places.

x approaches 0 from the left. →

x	−0.01	−0.001	−0.0001	→
$\frac{\cos x - 1}{x}$	0.00500	0.00050	0.00005	→

← x approaches 0 from the right.

x	←	0.0001	0.001	0.01
$\frac{\cos x - 1}{x}$	←	−0.00005	−0.00050	−0.00500

It appears that as x gets closer to 0, the values of $\frac{\cos x - 1}{x}$ also get closer to 0.

We infer that $\lim_{x \to 0} \frac{\cos x - 1}{x} = 0.$

2b. Construct a table to find $\lim_{x \to 0} \frac{\tan x}{x}$.

Objective #3: Find limits using graphs.

✔ *Solved Problem #3*

3a. Use the graph to find $\lim_{x\to -2} f(x)$ and $f(-2)$.

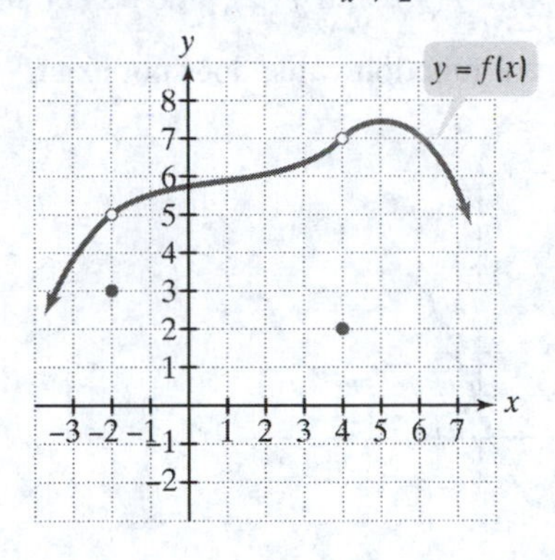

To find $\lim_{x\to -2} f(x)$, examine the graph of f near $x = -2$.

As x gets closer to -2, the values of $f(x)$ get closer to the y-coordinate of the open dot on the left. The y-coordinate of this point is 5. We conclude that $\lim_{x\to -2} f(x) = 5$.

To find $f(-2)$, examine the graph at $x = -2$. The graph of f is shown by the closed dot with coordinates $(-2, 3)$. Thus, $f(-2) = 3$.

3b. Graph the function $f(x) = \begin{cases} 3x - 2 & \text{if } x \neq 2 \\ 1 & \text{if } x = 2. \end{cases}$

Use the graph to find $\lim_{x\to 2} f(x)$.

This is a piecewise function. To graph the piece defined by $f(x) = 3x - 2$, we use the slope, 3, and the y-intercept, -2. Since this piece does not include $x = 2$, we include a hole in the graph at $(2, 4)$. To graph the other piece, we plot the point $(2, 1)$.

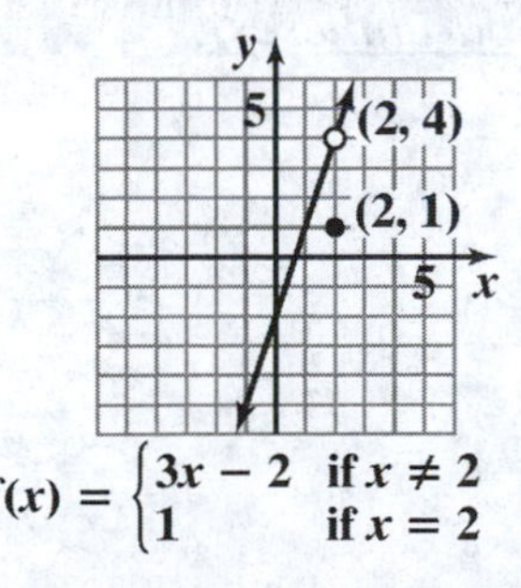

$f(x) = \begin{cases} 3x - 2 & \text{if } x \neq 2 \\ 1 & \text{if } x = 2 \end{cases}$

To find $\lim_{x\to 2} f(x)$, examine the graph of f near $x = 2$. As x gets closer to 2, the values of $f(x)$ get closer to the y-coordinate of the open dot. The y-coordinate of this point is 4. We conclude that $\lim_{x\to 2} f(x) = 4$.

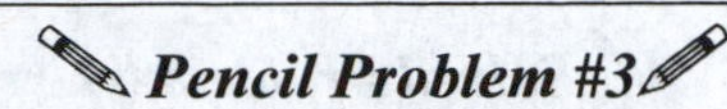

Pencil Problem #3

3a. Use the graph to find $\lim_{x\to 2} f(x)$ and $f(2)$.

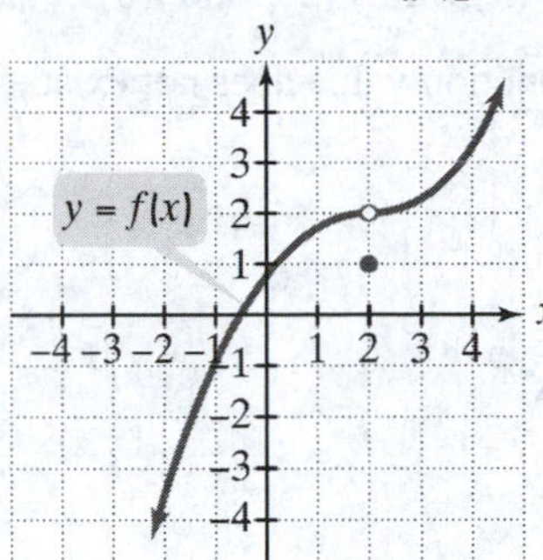

3b. Graph the function $f(x) = \begin{cases} x + 1 & \text{if } x \neq 2 \\ 5 & \text{if } x = 2. \end{cases}$ Use the graph to find $\lim_{x\to 2} f(x)$.

Objective #4: Find one-sided limits and use them to determine if a limit exists.

✔ Solved Problem #4

4. Use the graph to find $\lim_{x\to 0^-} f(x)$, $\lim_{x\to 0^+} f(x)$, $\lim_{x\to 0} f(x)$, and $f(0)$ or state that the limit or function value does not exist.

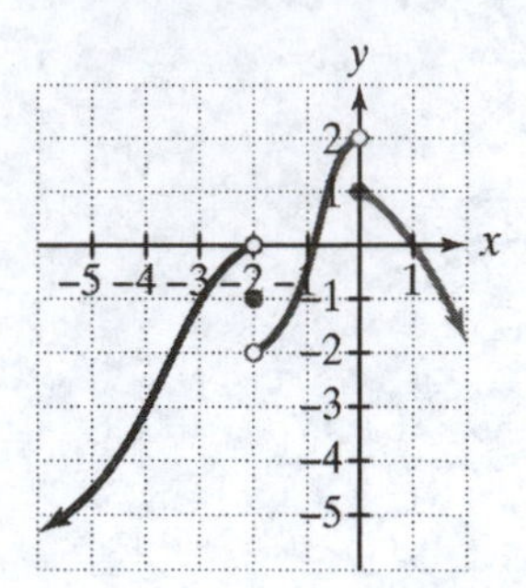

To find $\lim_{x\to 0^-} f(x)$, examine the portion of the graph near, but to the left of $x = 0$. As x approaches 0 from the left, the values of $f(x)$ get closer to the y-coordinate of the open dot at (0, 2). Thus, $\lim_{x\to 0^-} f(x) = 2$.

To find $\lim_{x\to 0^+} f(x)$, examine the portion of the graph near, but to the right of $x = 0$. As x approaches 0 from the right, the values of $f(x)$ get closer to the y-coordinate of the closed dot at (0, 1). Thus, $\lim_{x\to 0^+} f(x) = 1$.

Because $\lim_{x\to 0^-} f(x) = 2$ and $\lim_{x\to 0^+} f(x) = 1$ are not equal, $\lim_{x\to 0} f(x)$ does not exist.

To find $f(0)$, examine the graph for $x = 0$. The closed dot at (0, 1) indicates that $f(0) = 1$.

✎ Pencil Problem #4

4. Use the graph to find $\lim_{x\to 2^-} f(x)$, $\lim_{x\to 2^+} f(x)$, $\lim_{x\to 2} f(x)$, and $f(2)$ or state that the limit or function value does not exist.

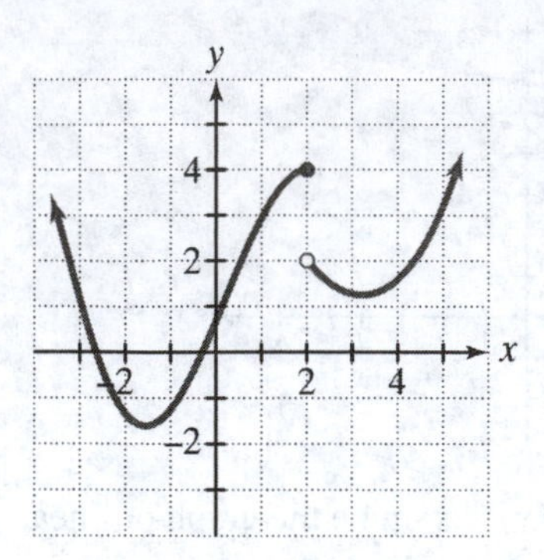

Answers for Pencil Problems *(Textbook Exercise references in parentheses)*:

1. 7; –4

2a. 20 *(11.1 #5)* **2b.** 1 *(11.1 #15)*

3a. $\lim_{x\to 2} f(x) = 2$; $f(2) = 1$ *(11.1 #21)* **3b.** 3 *(11.1 #47)*

4. $\lim_{x\to 2^-} f(x) = 4$; $\lim_{x\to 2^+} f(x) = 2$; $\lim_{x\to 2} f(x)$ does not exist; $f(2) = 4$ *(11.1 #27)*

Section 11.2
Finding Limits Using Properties of Limits

Establishing Your Limits

In the previous section, we explored limits using tables and graphs. While these methods help us to understand what limits are, they are not practical for determining the values of limits in most cases. Not only is it time consuming to construct a table or graph but it may also be difficult to find exact values for limits using these methods. If the value of a limit were $\frac{\pi}{2} \approx 1.570796...$, how would you be able to figure that out from a table or graph?

In this section, we look at more efficient methods of determining limits. In the Exercise Set, you will see some amazing results when limits are applied to traveling at velocities approaching the speed of light.

Objective #1: Find limits of constant functions and the identity function.

✔ *Solved Problem #1*

1a. Find $\lim_{x \to 8} 11$.

Since the expression following the limit is a constant, the value of the limit is that constant: $\lim_{x \to a} c = c$.

$\lim_{x \to 8} 11 = 11$

1b. Find $\lim_{x \to 0}(-9)$.

Since the expression following the limit is a constant, the value of the limit is that constant: $\lim_{x \to a} c = c$.

$\lim_{x \to 0}(-9) = -9$

1c. Find $\lim_{x \to 19} x$.

Since the expression following the limit is the identity function, x, the value of the limit is the same as the number x is approaching: $\lim_{x \to a} x = a$.

$\lim_{x \to 19} x = 19$

✎ *Pencil Problem #1* ✎

1a. Find $\lim_{x \to 2} 8$.

1b. Find $\lim_{x \to 3}(-6)$.

1c. Find $\lim_{x \to 2} x$.

Objective #2: Find limits using properties of limits.

✔ *Solved Problem #2*

2a. Find $\lim_{x \to -5} (3x - 7)$.

Use the rule for the limit of a difference: The limit of a difference is the difference of the limits. Then use the rule for the limit of a product: The limit of a product is a product of the limits.

$$\begin{aligned}\lim_{x \to -5} (3x - 7) &= \lim_{x \to -5} (3x) - \lim_{x \to -5} 7 \\ &= \lim_{x \to -5} 3 \cdot \lim_{x \to -5} x - \lim_{x \to -5} 7 \\ &= 3(-5) - 7 \\ &= -22\end{aligned}$$

2b. Find $\lim_{x \to 2} (-7x^3)$.

Use the rule for the limit of a monomial: The limit of a monomial as $x \to a$ is the monomial evaluated at a.

$$\lim_{x \to 2} (-7x^3) = -7(2)^3 = -7 \cdot 8 = -56$$

2c. Find $\lim_{x \to 2} (7x^3 + 3x^2 - 5x + 3)$.

Use the rule for the limit of a polynomial: The limit of a polynomial as $x \to a$ is the polynomial evaluated at a.

$$\begin{aligned}\lim_{x \to 2} (7x^3 + 3x^2 - 5x + 3) &= 7 \cdot 2^3 + 3 \cdot 2^2 - 5 \cdot 2 + 3 \\ &= 7 \cdot 8 + 3 \cdot 4 - 5 \cdot 2 + 3 \\ &= 56 + 12 - 10 + 3 \\ &= 61\end{aligned}$$

Pencil Problem #2

2a. Find $\lim_{x \to 6} (3x - 4)$.

2b. Find $\lim_{x \to -2} 7x^2$.

2c. Find $\lim_{x \to 5} (x^2 - 3x - 4)$.

2d. Find $\lim_{x \to 4}(3x-5)^3$.

Use the rule for the limit of a power: The limit of a power is the power of the limit. Then use the rule for the limit of a polynomial.

$$\lim_{x \to 4}(3x-5)^3 = [\lim_{x \to 4}(3x-5)]^3 = [3 \cdot 4 - 5]^3 = 7^3 = 343$$

2d. Find $\lim_{x \to 2}(5x-8)^3$.

2e. Find $\lim_{x \to -1} \sqrt{6x^2 - 4}$.

Use the rule for the limit of a root: The limit of a root is the root of the limit. Then use the rule for the limit of a polynomial.

$$\lim_{x \to -1} \sqrt{6x^2 - 4} = \sqrt{\lim_{x \to -1}(6x^2 - 4)} = \sqrt{6(-1)^2 - 4} = \sqrt{2}$$

2e. Find $\lim_{x \to -4} \sqrt{x^2 + 9}$.

2f. Find $\lim_{x \to 2} \dfrac{x^2 - 4x + 1}{3x - 5}$.

First find the limit of the denominator as $x \to 2$.
$\lim_{x \to 2}(3x - 5) = 3 \cdot 2 - 5 = 1$

Since the limit of the denominator is not 0, we can use the rule for the limit of a quotient: The limit of a quotient is the quotient of the limits. Then use the rule for the limit of a polynomial in the numerator and denominator.

$$\lim_{x \to 2} \frac{x^2 - 4x + 1}{3x - 5} = \frac{\lim_{x \to 2}(x^2 - 4x + 1)}{\lim_{x \to 2}(3x - 5)} = \frac{2^2 - 4 \cdot 2 + 1}{3 \cdot 2 - 5} = \frac{-3}{1} = -3$$

2f. Find $\lim_{x \to 2} \dfrac{x^2 - 1}{x - 1}$.

Objective #3: Find one-sided limits using properties of limits.

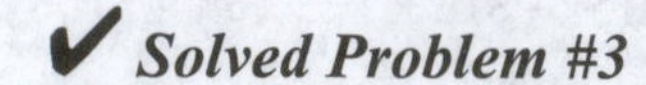

Solved Problem #3

3. Consider the piecewise function

$$f(x)=\begin{cases}-1 & \text{if } x<1\\ \sqrt[3]{2x-1} & \text{if } x\ge 1.\end{cases}$$

Find $\lim\limits_{x\to 1^-} f(x)$, $\lim\limits_{x\to 1^+} f(x)$, and $\lim\limits_{x\to 1} f(x)$, or state that the limit does not exist.

To find $\lim\limits_{x\to 1^-} f(x)$, we look at values of $f(x)$ when x is close to 1 but less than 1. Because x is less than 1, we use the first line of the function: $f(x)=-1$.

$$\lim_{x\to 1^-} f(x)=\lim_{x\to 1^-}(-1)=-1$$

To find $\lim\limits_{x\to 1^+} f(x)$, we look at values of $f(x)$ when x is close to 1 but greater than 1. Because x is greater than 1, we use the second line of the function: $f(x)=\sqrt[3]{2x-1}$.

$$\lim_{x\to 1^+} f(x)=\lim_{x\to 1^+}\sqrt[3]{2x-1}=\sqrt[3]{2\cdot 1-1}=1$$

Because $\lim\limits_{x\to 1^-} f(x)=-1$ and $\lim\limits_{x\to 1^+} f(x)=1$ are not equal, $\lim\limits_{x\to 1} f(x)$ does not exist.

Pencil Problem #3

3. Consider the piecewise function

$$f(x)=\begin{cases}x^2+5 & \text{if } x<2\\ x^3+1 & \text{if } x\ge 2.\end{cases}$$

Find $\lim\limits_{x\to 2^-} f(x)$, $\lim\limits_{x\to 2^+} f(x)$, and $\lim\limits_{x\to 2} f(x)$, or state that the limit does not exist.

Objective #4: Find limits of fractional expressions in which the limit of the denominator is zero.

✔ *Solved Problem #4*

4a. Find $\lim_{x\to 1}\frac{x^2+2x-3}{x-1}$.

First find the limit of the denominator as $x \to 1$.

$$\lim_{x\to 1}(x-1) = 1-1 = 0$$

Since the limit of the denominator is 0, we cannot use the rule for the limit of a quotient. We factor the numerator and cancel a common factor and then proceed to apply limit properties.

$$\begin{aligned}\lim_{x\to 1}\frac{x^2+2x-3}{x-1} &= \lim_{x\to 1}\frac{(x+3)(x-1)}{x-1}\\ &= \lim_{x\to 1}\frac{(x+3)\cancel{(x-1)}}{\cancel{x-1}}\\ &= \lim_{x\to 1}(x+3)\\ &= 1+3 = 4\end{aligned}$$

4b. Find $\lim_{x\to 0}\frac{\sqrt{9+x}-3}{x}$.

As x approaches 0, the denominator of the expression approaches 0. Thus, the quotient property for limits cannot be used. We multiply the numerator and denominator by $\sqrt{9+x}+3$ to eliminate the radical in the numerator and then simplify.

$$\begin{aligned}\lim_{x\to 0}\frac{\sqrt{9+x}-3}{x} &= \lim_{x\to 0}\left[\frac{\sqrt{9+x}-3}{x}\cdot\frac{\sqrt{9+x}+3}{\sqrt{9+x}+3}\right]\\ &= \lim_{x\to 0}\frac{(\sqrt{9+x})^2-3^2}{x(\sqrt{9+x}+3)}\\ &= \lim_{x\to 0}\frac{9+x-9}{x(\sqrt{9+x}+3)}\\ &= \lim_{x\to 0}\frac{x}{x(\sqrt{9+x}+3)}\\ &= \lim_{x\to 0}\frac{\cancel{x}}{\cancel{x}(\sqrt{9+x}+3)}\\ &= \lim_{x\to 0}\frac{1}{\sqrt{9+x}+3}\\ &= \frac{1}{\sqrt{9+0}+3} = \frac{1}{3+3} = \frac{1}{6}\end{aligned}$$

Pencil Problem #4

4a. Find $\lim_{x\to 1}\frac{x^2-1}{x-1}$.

4b. Find $\lim_{x\to 0}\frac{\sqrt{1+x}-1}{x}$.

Answers for Pencil Problems *(Textbook Exercise references in parentheses)*:

1a. 8 *(11.2 #1)* **1b.** –6 *(11.2 #2)* **1c.** 2 *(11.2 #3)*

2a. 14 *(11.2 #5)* **2b.** 28 *(11.2 #7)* **2c.** 6 *(11.2 #9)* **2d.** 8 *(11.2 #11)*

2e. 5 *(11.2 #15)* **2f.** 3 *(11.2 #19)*

3. $\lim_{x\to 2^-} f(x) = 9$; $\lim_{x\to 2^+} f(x) = 9$; $\lim_{x\to 2} f(x) = 9$ *(11.2 #45)*

4a. 2 *(11.2 #21)* **4b.** $\frac{1}{2}$ *(11.2 #29)*

Section 11.3
Limits and Continuity

Taxing Your Limits

Did you know that as your income increases so does the amount that you pay in income taxes? But not only does the amount you pay in taxes increase but also the percentage of your income that you pay in taxes increases. So as you earn greater amounts of money, your tax liability increases at a greater rate.

In the Exercise Set, you will work with a piecewise function that describes the amount of taxes owed by a single taxpayer. You will explore whether or not there is a "big jump" in the amount of taxes owed when you move up to the next tax bracket.

Objective #1: Determine whether a function is continuous at a number.

✔ *Solved Problem #1*

1a. Determine whether the function $f(x)=\dfrac{x-2}{x^2-4}$ is continuous at 1.

Is $f(1)$ defined? Yes.

$$f(1)=\frac{1-2}{1^2-4}=\frac{-1}{-3}=\frac{1}{3}$$

Does $\lim\limits_{x\to 1} f(x)$ exist? Yes.

$$\lim_{x\to 1} f(x)=\lim_{x\to 1}\frac{x-2}{x^2-4}=\frac{\lim\limits_{x\to 1}(x-2)}{\lim\limits_{x\to 1}(x^2-4)}=\frac{1-2}{1^2-4}=\frac{-1}{-3}=\frac{1}{3}$$

Does $\lim\limits_{x\to 1} f(x)=f(1)$? Yes.

$f(1)$ and $\lim\limits_{x\to 1} f(x)$ both equal $\frac{1}{3}$.

Because the three conditions are satisfied, we conclude that f is continuous at 1.

1b. Determine whether the function $f(x)=\dfrac{x-2}{x^2-4}$ is continuous at 2.

Substituting 2 into the denominator of the expression results in 0. Since division by 0 is undefined, 2 is not in the domain of the function. Since f is not defined at 2, f is not continuous at 2.

Pencil Problem #1

1a. Determine whether the function $f(x)=\dfrac{x-5}{x+5}$ is continuous at 5.

1b. Determine whether the function $f(x)=\dfrac{x+5}{x-5}$ is continuous at 5.

Objective #2: Determine for what numbers a function is discontinuous.

✔ *Solved Problem #2*

2. Determine for what numbers, x, if any, the function f is discontinuous.

$$f(x)=\begin{cases}2x & \text{if } x\le 0\\ x^2+1 & \text{if } 0<x\le 2\\ 7-x & \text{if } x>2\end{cases}$$

Since each piece of the function is defined by a polynomial, the only possible discontinuities are at $x = 0$ and $x = 2$. We check for continuity at each of these x-values.

At $x = 0$:

$$\lim_{x\to 0^-} f(x)=\lim_{x\to 0^-} 2x = 2\cdot 0 = 0$$

$$\lim_{x\to 0^+} f(x)=\lim_{x\to 0^+} (x^2+1) = 0^2+1 = 1$$

Since $\lim_{x\to 0^-} f(x)\ne \lim_{x\to 0^+} f(x)$, $\lim_{x\to 0} f(x)$ does not exist.

Since the second condition for continuity is not satisfied, f is discontinuous at 0.

At $x = 2$:

$f(2)=2^2+1=5$, so f is defined at 2.

$$\lim_{x\to 2^-} f(x)=\lim_{x\to 2^-} (x^2+1) = 2^2+1 = 5$$

$$\lim_{x\to 2^+} f(x)=\lim_{x\to 2^+} (7-x) = 7-2 = 5$$

Since $\lim_{x\to 2^-} f(x)=\lim_{x\to 2^+} f(x)=5$, $\lim_{x\to 2} f(x)$ exists and $\lim_{x\to 2} f(x)=5$.

Finally, $\lim_{x\to 2} f(x)=5=f(2)$.

Since all three conditions for continuity are satisfied, f is continuous at 2.

The only value of x for which f is discontinuous is 0.

✎ *Pencil Problem #2* ✎

2. Determine for what numbers, x, if any, the function f is discontinuous.

$$f(x)=\begin{cases}x+6 & \text{if } x\le 0\\ 6 & \text{if } 0<x\le 2\\ x^2+1 & \text{if } x>2\end{cases}$$

Answers for Pencil Problems *(Textbook Exercise references in parentheses)*:

1a. continuous *(11.3 #9)* **1b.** not continuous *(11.3 #7)*

2. discontinuous at 2 *(11.3 #31)*

Section 11.4
Introduction to Derivatives

Measuring Change

We all know that things change. In this section, we combine two topics we've already studied, average rate of change and limits, to study how quickly things change. By applying limits to expressions representing average rate of change, we can define *instantaneous rate of change*.
If an object is thrown upward and we know its initial velocity and the height from which it is thrown, we can write a function that describes the height of the object until it hits the ground. When we apply the ideas in this section to that function, we can also determine the velocity of the object at each instant until it hits the ground.

Objective #1: Find slopes and equations of tangent lines.

✔ *Solved Problem #1*

1a. Find the slope of the tangent line to the graph of $f(x)=x^2-x$ at (4, 12).

We use the formula $m_{\tan}=\lim\limits_{h\to 0}\dfrac{f(a+h)-f(a)}{h}$, with $a=4$, the x-coordinate of the given point on the graph of f. After simplifying the expressions in the numerator, we factor h from the numerator and divide both the numerator and denominator by h to eliminate the factor of h in the denominator.

$$\begin{aligned}
m_{\tan} &= \lim_{h\to 0}\frac{f(4+h)-f(4)}{h}\\
&= \lim_{h\to 0}\frac{[(4+h)^2-(4+h)]-[4^2-4]}{h}\\
&= \lim_{h\to 0}\frac{[16+8h+h^2-4-h]-[16-4]}{h}\\
&= \lim_{h\to 0}\frac{[12+7h+h^2]-12}{h}\\
&= \lim_{h\to 0}\frac{7h+h^2}{h}\\
&= \lim_{h\to 0}\frac{h(7+h)}{h}\\
&= \lim_{h\to 0}(7+h)=7+0=7
\end{aligned}$$

Thus, the slope of the tangent line to the graph of $f(x)=x^2-x$ at (4, 12) is 7.

Pencil Problem #1

1a. Find the slope of the tangent line to the graph of $f(x)=x^2+4$ at (−1, 5).

1b. Find the slope-intercept equation of the tangent line to the graph of $f(x)=\sqrt{x}$ at (1, 1).

We begin by finding the slope of the tangent line. Note that we use the method of rationalizing the numerator in order to evaluate the limit. Along the way, we factor h from the numerator and divide both the numerator and denominator by h to eliminate the factor of h in the denominator.

$$\begin{aligned}
m_{\tan} &= \lim_{h\to 0}\frac{f(1+h)-f(1)}{h}\\
&= \lim_{h\to 0}\frac{\sqrt{1+h}-\sqrt{1}}{h}\\
&= \lim_{h\to 0}\left[\frac{\sqrt{1+h}-1}{h}\cdot\frac{\sqrt{1+h}+1}{\sqrt{1+h}+1}\right]\\
&= \lim_{h\to 0}\frac{(\sqrt{1+h})^2-1^2}{h(\sqrt{1+h}+1)}\\
&= \lim_{h\to 0}\frac{1+h-1}{h(\sqrt{1+h}+1)}\\
&= \lim_{h\to 0}\frac{h}{h(\sqrt{1+h}+1)}\\
&= \lim_{h\to 0}\frac{1}{\sqrt{1+h}+1}=\frac{1}{\sqrt{1+0}+1}=\frac{1}{1+1}=\frac{1}{2}
\end{aligned}$$

We now use the slope just found and the given point, (1, 1), to write the equation of the tangent line in point-slope form. Then we solve for y to write the equation in slope intercept form.

$$\begin{aligned}
y-y_1 &= m(x-x_1)\\
y-1 &= \frac{1}{2}(x-1)\\
y-1 &= \frac{1}{2}x-\frac{1}{2}\\
y &= \frac{1}{2}x+\frac{1}{2}
\end{aligned}$$

1b. Find the slope-intercept equation of the tangent line to the graph of $f(x)=\sqrt{x}$ at (9, 3).

Objective #2: Find the derivative of a function.

✔ *Solved Problem #2*

2a. Find the derivative of $f(x) = x^2 - 5x$ at x. That is, find $f'(x)$.

$$f'(x) = \lim_{h\to 0} \frac{f(x+h) - f(x)}{h}$$
$$= \lim_{h\to 0} \frac{[(x+h)^2 - 5(x+h)] - [x^2 - 5x]}{h}$$
$$= \lim_{h\to 0} \frac{[x^2 + 2xh + h^2 - 5x - 5h] - x^2 + 5x}{h}$$
$$= \lim_{h\to 0} \frac{2xh + h^2 - 5h}{h}$$
$$= \lim_{h\to 0} \frac{h(2x + h - 5)}{h}$$
$$= \lim_{h\to 0}(2x + h - 5) = 2x + 0 - 5 = 2x - 5$$

The derivative is $f'(x) = 2x - 5$.

2b. Find the slope of the tangent line to the graph of $f(x) = x^2 - 5x$ at $x = -1$.

We found $f'(x)$ in Solved Problem #2a. The slope of the tangent line is $f'(-1)$.

$f'(-1) = 2(-1) - 5 = -2 - 5 = -7$

The slope of the tangent line at $x = -1$ is -7.

Pencil Problem #2

2a. Find the derivative of $f(x) = x^2 - 3x + 5$ at x. That is, find $f'(x)$.

2b. Find the slope of the tangent line to the graph of $f(x) = x^2 - 3x + 5$ at $x = 2$.

Objective #3: Find average and instantaneous rates of change.

✔ *Solved Problem #3*

3a. The function $f(x) = x^3$ describes the volume of a cube, $f(x)$, in cubic inches, whose length, width, and height each measure x inches. If x is changing, find the average rate of change of the volume with respect to x as x changes from 4 inches to 4.1 inches and from 4 inches to 4.01 inches.

We use the difference quotient for both calculations.
From 4 inches to 4.1 inches:

$$\frac{f(4.1) - f(4)}{4.1 - 4} = \frac{4.1^3 - 4^3}{0.1}$$
$$= \frac{68.921 - 64}{0.1}$$
$$= \frac{4.921}{0.1}$$
$$= 49.21$$

(continued on next page)

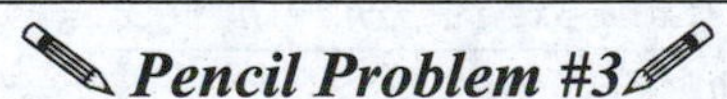

Pencil Problem #3

3a. The function $f(x) = x^2$ describes the area of a square, $f(x)$, in square inches, whose sides each measure x inches. If x is changing, find the average rate of change of the area with respect to x as x changes from 6 inches to 6.1 inches and from 6 inches to 6.01 inches.

The average rate of change is 49.21 cubic inches per inch as x changes from 4 inches to 4.1 inches.

From 4 inches to 4.01 inches:

$$\begin{aligned}\frac{f(4.01)-f(4)}{4.01-4} &= \frac{4.01^3-4^3}{0.01}\\ &= \frac{64.481201-64}{0.01}\\ &= \frac{0.481201}{0.01}\\ &= 48.1201\end{aligned}$$

The average rate of change is 48.1201 cubic inches per inch as x changes from 4 inches to 4.01 inches.

3b. The function $f(x)=x^3$ describes the volume of a cube, $f(x)$, in cubic inches, whose length, width, and height each measure x inches. If x is changing, find the instantaneous rate of change of the volume with respect to x when $x = 4$ inches.

The instantaneous rate of change is the derivative. We begin by finding $f'(x)$ and then find $f'(4)$.

$$\begin{aligned}f'(x) &= \lim_{h\to 0}\frac{f(x+h)-f(x)}{h}\\ &= \lim_{h\to 0}\frac{(x+h)^3-x^3}{h}\\ &= \lim_{h\to 0}\frac{[x^3+3x^2h+3xh^2+h^3]-x^3}{h}\\ &= \lim_{h\to 0}\frac{3x^2h+3xh^2+h^3}{h}\\ &= \lim_{h\to 0}\frac{h(3x^2+3xh+h^2)}{h}\\ &= \lim_{h\to 0}(3x^2+3xh+h^2) = 3x^2+3x\cdot 0+0^2 = 3x^2\end{aligned}$$

Since $f'(x)=3x^2$, we have $f'(4)=3(4)^2=48$.

The instantaneous rate of change of the volume with respect to x is 48 cubic inches per inch when $x = 4$ inches.

3b. The function $f(x)=x^2$ describes the area of a square, $f(x)$, in square inches, whose sides each measure x inches. If x is changing, find the instantaneous rate of change of the area with respect to x when $x = 6$ inches.

Objective #4: Find instantaneous velocity.

✔ *Solved Problem #4*

4a. A ball is thrown straight up from ground level with an initial velocity of 96 feet per second. The function $s(t) = -16t^2 + 96t$ describes the ball's height above the ground, $s(t)$, in feet, t seconds after it is thrown. What is the instantaneous velocity of the ball after 4 seconds?

The instantaneous velocity is the derivative.

$$\begin{aligned}
&s'(a)\\
&= \lim_{h\to 0} \frac{s(a+h)-s(a)}{h}\\
&= \lim_{h\to 0} \frac{[-16(a+h)^2 + 96(a+h)] - [-16a^2 + 96a]}{h}\\
&= \lim_{h\to 0} \frac{[-16(a^2 + 2ah + h^2) + 96(a+h)] - [-16a^2 + 96a]}{h}\\
&= \lim_{h\to 0} \frac{-16a^2 - 32ah - 16h^2 + 96a + 96h + 16a^2 - 96a}{h}\\
&= \lim_{h\to 0} \frac{-32ah - 16h^2 + 96h}{h}\\
&= \lim_{h\to 0} \frac{h(-32a - 16h + 96)}{h}\\
&= \lim_{h\to 0}(-32a - 16h + 96)\\
&= -32a - 16\cdot 0 + 96\\
&= -32a + 96
\end{aligned}$$

We now compute $s'(4)$.

$s'(4) = -32(4) + 96 = -128 + 96 = -32$

The instantaneous velocity of the ball after 4 seconds is -32 feet per second. The negative velocity indicates that the ball is moving downward at this point.

✎ *Pencil Problem #4* ✎

4a. A ball is thrown straight up from ground level with an initial velocity of 64 feet per second. The function $s(t) = -16t^2 + 64t$ describes the ball's height above the ground, $s(t)$, in feet, t seconds after it is thrown. What is the instantaneous velocity of the ball after 3 seconds?

4b. Use the information in Solved Problem #4a to determine the instantaneous velocity of the ball when it hits the ground.

We need to know the time when the ball hits the ground. Solve $s(t) = 0$.

$$-16t^2 + 96t = 0$$
$$-16t(t-6) = 0$$
$$-16t = 0 \quad \text{or} \quad t - 6 = 0$$
$$t = 0 \qquad\qquad t = 6$$

Since the ball is thrown at $t = 0$, we are not interested in this value of t. The ball hits the ground after 6 seconds. Use the derivative found in Solved Problem #4a to find $s'(6)$.

$$s'(6) = -32(6) + 96 = -192 + 96 = -96$$

The instantaneous velocity of the ball when it hits the ground is –96 feet per second.

4b. Use the information in Pencil Problem #4a to determine the instantaneous velocity of the ball when it hits the ground.

Answers for Pencil Problems *(Textbook Exercise references in parentheses)*:

1a. -2 *(11.4 #3)* **1b.** $y = \frac{1}{6}x + \frac{3}{2}$ *(11.4 #11)*

2a. $f'(x) = 2x - 3$ **2b.** 1 *(11.4 #19)*

3a. 12.1 square inches per inch; 12.01 square inches per inch *(11.4 #37a)*

3b. 12 square inches per inch *(11.4 #37b)*

4a. –32 feet per second **4b.** –64 feet per second *(11.4 #43)*

Classroom Activities to accompany

PRECALCULUS 6E

Classroom Activity: Applications of Scientific Notation 3

Classroom Activity: Reversing the Process of Multiplying 5

Classroom Activity: Identifying Common Factors; Least Common Denominator 7

Classroom Activity: Introduction to Solving Equations 9

Classroom Activity: Classifying Equations and Comparing Solution Methods 11

Classroom Activity: Applications of Absolute Value Inequalities 13

Classroom Activity: Applications of Linear Functions 15

Classroom Activity: Investigating Composition and Inverses of Functions 19

Classroom Activity: Investigating Zeros of Polynomial Functions 23

Classroom Activity: Exploring End Behavior of a Rational Function 25

Classroom Activity: Applications of Polynomial and Rational Functions 27

Classroom Activity: Applications of Exponential Functions 29

Classroom Activity: Investigating the Inverse Relationship 33

Classroom Activity: Changing Tire Size 35

Classroom Activity: Investigating Right Triangles 37

Classroom Activity: Fitting a Cosine Curve to Temperature Data 39

Classroom Activity: Using Inverse Functions to Find Angles 41

Classroom Activity: Using Identities to Find Exact Values 43

Classroom Activity: Applications of Trigonometric Equations 45

Classroom Activity: Law of Sines or Law of Cosines 47

Classroom Activity: Reviewing Angles, Distance, and Trigonometric Functions 49

Classroom Activity: More on Curve Fitting 51

Classroom Activity: Investigating Nonlinear Systems 55

Classroom Activity: Applications of Matrix Multiplication 57

Classroom Activity: Investigating Matrix Multiplication and Solving Matrix Equations ... 61

Classroom Activity: Investigating Second-Degree Equations 63

Classroom Activity: Applications of Ellipses and Parabolas 65

Classroom Activity: Applications of Sequences 67

Classroom Activity: Investigating Binomial Coefficients 69

Classroom Activity: Exploring Limits of a Rational Function 71

Classroom Activity: Applications of Scientific Notation

Complete this activity after Section P.2 is covered.

Group Size: 2–3
Material: Paper, Pencil, Calculator
Time: 20 minutes

In this activity, you will be working with small numbers expressed in scientific notation. We begin with two measurements from biology.

Diameter of a human hair: 99 micrometers (μm), where 1 μm = 10^{-6} meter (m)

Diameter of the Ebola virus: 80 nanometers (nm), where 1 nm = 10^{-9} meter (m)

Discuss each problem as a group and then perform any necessary calculations individually. Compare answers and check each others' work. Use a calculator as needed.

1. Express each of the measurements in meters using scientific notation.

2. Which has the smaller diameter, a human hair or the Ebola virus? Explain how to determine the smaller of two numbers expressed in scientific notation when both have negative exponents.

3. The associative property of multiplication states that $a(bc) = (ab)c$. Explain how this can be used to simplify $\frac{1}{2}(8 \times 10^{-8})$ and then simplify the number, expressing the result in scientific notation.

4. Use the fact that the radius of a circle is half its diameter to find the radius of a cross-section of a human hair and of the Ebola virus. Refer to #3. Express each of your answers in meters using scientific notation.

5. Explain how the exponent rules $(ab)^n = a^n b^n$ and $(a^m)^n = a^{mn}$ can be used to raise a number in scientific notation to a power. Then simplify $(4 \times 10^{-8})^2$ and express the result in scientific notation.

6. The cross-sectional areas, in square meters, of a human hair and of the Ebola virus are approximately $3.14(4.95 \times 10^{-5})^2$ and $3.14(4 \times 10^{-8})^2$, respectively. Simplify each expression by first squaring the number in scientific notation and then multiplying the result by 3.14. Refer to #3 and #5. Express each of your answers in square meters using scientific notation. Do not round.

7. Use your answers from #6. Divide the cross-sectional area of a human hair by the cross-sectional area of the Ebola virus. How many times greater is the cross-sectional area of the hair? Round your answer in a way that seems appropriate to the group.

8. Discuss how understanding properties of real numbers and exponents is critical in performing operations with numbers in scientific notation. Write a brief summary of your discussion.

Classroom Activity: Reversing the Process of Multiplying

Complete this activity before Section P.5 is covered.

Group Size: 2–3
Material: Paper, Pencil
Time: 15 minutes

In #1–6, as a group, discuss strategies for multiplying the given binomials. Then assign each problem to a group member who will then perform the multiplication. Group members should check each others' work.

1. $(x+2)(x+3)$

2. $(2x-1)(3x+5)$

3. $(x-6)(x+6)$

4. $(2x-7)(2x+7)$

5. $(x+9)(x+9)$ or $(x+9)^2$

6. $(3x-4)(3x-4)$ or $(3x-4)^2$

In #7–10, as a group, discuss ways in which the given polynomials are similar to your answers in #1–6. Then pair each polynomial in #7–10 with the answer in #1–6 that the group believes it most closely resembles.

7. $x^2+8x+16$

8. $49x^2-25$

9. $x^2+8x+12$

10. $6x^2+x-1$

In #11–14, the polynomials in #7–10 are repeated. Write each polynomial as the product of two binomials. To do this, consider how you might reverse the multiplications in the problems that you paired with each of these polynomials in #7–10. Some trial and error may be necessary. Check your answers by multiplying.

11. $x^2 + 8x + 16$

12. $49x^2 - 25$

13. $x^2 + 8x + 12$

14. $6x^2 + x - 1$

15. Describe the patterns in the multiplication problems that allowed you to write the polynomials as products of binomials in #11–14.

Classroom Activity: Identifying Common Factors; Least Common Denominator

Complete this activity before Section P.6 is covered.

Group Size: 2–3
Material: Paper, Pencil
Time: 20 minutes

In #1–6, as a group, discuss strategies for factoring each polynomial. Then assign each polynomial to a group member who will then factor the polynomial. After factoring their assigned polynomials, group members should check each others' work.

1. $10x^3 + 15x^2$

2. $10x^2 + 13x - 3$

3. $4x^2 - 9$

4. $4x^2 + 12x + 9$

5. $8x^3 + 27$

6. $8x^3 + 12x^2 - 18x - 27$

7. What binomial factor is common to all six factorizations in #1–6?

8. Which polynomials in #1–6 have a repeated binomial factor?

9. One of the factorizations in #1–6 includes all of the factors in two of the other factorizations. Identify these three factorizations.

In order to add or subtract two fractions with different denominators, we rewrite each fraction with a common denominator. A *common denominator* contains all factors of each of the original denominators. Refer to your work in #1–9 as you answer the remaining questions.

10. Which of the polynomials in #1–6 is a common denominator for $\frac{1}{4x^2-9}$ and $\frac{1}{4x^2+12x+9}$?

In factored form, your answer to #10 has three factors. If any of these factors are left out, the resulting polynomial is no longer a common denominator for both $\frac{1}{4x^2-9}$ and $\frac{1}{4x^2+12x+9}$. Consequently, your answer to #10 is the *least common denominator* for these fractions.

11. Compare the factorizations of $4x^2-9$ and $4x^2+12x+9$ to the factorization of the least common denominator in #10. Describe the relationships among the three factorizations.

12. Based on your work so far in this activity, devise a strategy for finding the least common denominator of fractions of the form $\frac{1}{(ax-b)(ax+b)}$ and $\frac{1}{(ax+b)^2}$. Explain why this strategy applies to the fractions $\frac{1}{25x^2-16}$ and $\frac{1}{25x^2+40x+16}$.

13. Find the least common denominator of fractions $\frac{1}{25x^2-16}$ and $\frac{1}{25x^2+40x+16}$.

Classroom Activity: Introduction to Solving Equations

Complete this activity before Section P.7 is covered.

Group Size: 1–2
Material: Paper, Pencil
Time: 15 minutes

Your success in solving equations in Section 1.2 and beyond will depend in part on your ability to simplify algebraic expressions. In this activity, you will practice a few of the necessary skills.

In #1–2, use the distributive property and then combine like terms to simplify each expression.

1. $5(3x-2)-8x$

2. $x(2x^2+1)-2x^3$

When you multiply a polynomial by a number, remember to multiply every term of the polynomial by that number. If you are asked to multiply $x+3$ by 4, write $4(x+3)$ and use the distributive property to write the answer as $4x+12$.

3. Multiply $x-6$ by 7.

4. Multiply $2x-8$ by 0.5.

In #5–6, you will need to multiply fractions and whole numbers. Recall that you can write the whole number as fraction and then divide out common factors in numerators and denominators.

5. $\frac{1}{3}(15x)$

6. $10\left(\frac{x}{2}+\frac{3}{5}\right)$

In #6, the number in front of the parentheses is the least common denominator (LCD) of the two fractions inside the parentheses. In #7–8, identify the LCD of the fractions in the expression. Then multiply each expression by this LCD.

7. $\frac{x}{3}-\frac{7}{4}$

8. $\frac{5}{6}x+\frac{8}{9}$

The remainder of this activity will focus on the equation $\frac{1}{x} = \frac{1}{x} + \frac{5}{3}$.

9. What number(s) must be excluded from the domain of $\frac{1}{x}$?

10. Find the LCD of the fractions $\frac{1}{x}$, $\frac{1}{x}$, and $\frac{5}{3}$.

11. Simplify the expression on each side of the equal sign: $3x\left(\frac{1}{x}\right) = 3x\left(\frac{1}{x} + \frac{5}{3}\right)$. What is the significance of $3x$?

12. After simplifying in #11, you should have obtained $3 = 3 + 5x$. Subtract 3 from both sides of this equation and then multiply both sides by $\frac{1}{5}$. What do you obtain?

13. Compare your answer in #12 to your answer in #9. What do you notice?

14. A solution to an equation is a number that makes a true statement when it is substituted for the variable. Explain why 0 cannot be a solution of the equation $\frac{1}{x} = \frac{1}{x} + \frac{5}{3}$.

15. The equation $\frac{1}{x} = \frac{1}{x} + \frac{5}{3}$ actually has no solution. What happens if you subtract $\frac{1}{x}$ from both sides of the equation? How might this demonstrate that the equation has no solution?

Classroom Activity: Classifying Equations and Comparing Solution Methods

Complete this activity after Section P.7 is covered.

Group Size: 2–3
Material: Paper, Pencil
Time: 20 minutes

In this chapter, you have studied several methods for solving equations. In order to choose an appropriate method to solve a particular equation, you first need to identify the type of equation or, in other words, *classify the equation*.

Refer to the equations below as you answer the questions in #1–10. Be sure to discuss each question as a group.

a. $3|x-5|+2=0$ **b.** $x^4-13x^2+36=0$ **c.** $x^3-2x^2=3x-6$

d. $4(x+5)^2-9=0$ **e.** $\sqrt{x-3}+x=3$ **f.** $(x-1)^{\frac{3}{2}}+8=0$

1. Which equations are polynomial equations? Identify the degree of each polynomial equation.

2. Which equations are *not* polynomial equations? Explain why they are not polynomial equations.

3. Which equations would you solve by first isolating an expression on one side? In each case, identify the expression you would isolate.

4. Which equations are solved using methods that potentially yield extraneous solutions? Explain why it is necessary to check your answers in this situation.

5. Which equation is a quadratic equation? Discuss three methods for solving this equation. Then have each group member solve the equation by a different method. Check each others' work. Does one method seem easier than others in this situation? Why?

6. Which equation is *not* quadratic but is quadratic in form? Explain how you identified this equation. Discuss a strategy for solving this equation and then solve it as a group.

7. There should be one equation that you haven't solved yet that can be solved by factoring. Identify this equation and solve it.

8. Solve each of the remaining three equations and check each others' work.

9. Which equations, if any, have no solution? For each equation with no solution, explain how you determined that it has no solution.

10. Which equations, if any, have an extraneous solution? Based on the equations you solved in this activity, does an equation with an extraneous solution necessarily have no solution? Explain.

Classroom Activity: Applications of Absolute Value Inequalities

Complete this activity after Section P.9 is covered.

Group Size: 1–2
Material: Paper, Pencil, Calculator
Time: 15 minutes

Polls abound, giving us some insight into people's preferences on a variety of issues. However, because polls are based on samples, it is unlikely that any poll will reflect people's opinions exactly. Therefore, polls are stated with a margin of error.

In manufacturing, parts are produced to design specifications, but again due to variation in the manufacturing process it is unlikely that any part is exactly the specified dimension. Thus, acceptable ranges are given.

Both of these situations may be described using absolute value inequalities.

Suppose that a poll reports that 63% of American voters support a plan to offer qualifying students two free years of community college. The margin of error is given as 2.5%. This means that the actual percentage of American voters who support the plan is believed to be between $63\% - 2.5\%$, or 60.5%, and $63\% + 2.5\%$, or 65.5%. If we let x represent the actual percentage of American voters who support the plan, we can describe this situation in three ways:

Absolute Value Inequality	Inequality Notation	Interval Notation
$\lvert x - 63 \rvert < 2.5$	$60.5 < x < 65.5$	$(60.5,\ 65.5)$.

1. Solve the absolute value equation above using the techniques in the text to confirm the solutions.

In #2–3, describe each situation using all three forms shown above: absolute value inequality, inequality notation, and interval notation.

2. According to one poll, 51% of American voters believe the driving age should be 18 or higher. The poll's margin of error is 3.6%. Let x be the percentage of American voters who believe the driving age should be 18 or higher.

3. The specifications for a food processor call for the blade to 2.875 inches long. The blade will function properly if it is within 0.015 inch of the specified length. Let x be the length of a blade that functions properly.

In #4–5, devise a strategy to work backwards and express each range of values as an absolute value inequality.

4. The acceptable diameter d of a DVD is between 11.695 cm and 11.705 cm.

5. The percentage p of college students who have a credit card is between 35.7% and 41.1%.

6. If the situation in #5 were given as the results of a poll, what number would be stated as the percentage of college students who have a credit card and what would be the margin of error?

In #7, refer back to #4.

7. If the acceptable diameter d of a DVD is between 11.695 cm and 11.705 cm, what values for d are unacceptable? Express the answer as an absolute value inequality, in inequality notation, and in interval notation.

8. Use your work in #4 and #7 to explain why it is usually the case that the solution to an absolute value inequality with $>$ or $\geq$ cannot be expressed as a single interval.

Classroom Activity: Applications of Linear Functions

Complete this activity after Section 1.4 is covered.

Group Size: 3–4
Material: Paper, Pencil, Calculator
Time: 20 minutes

In business and other applications, what has happened in the past is used to try to predict what might occur in the future. In this activity, you will work with several linear models and consider which might provide the most accurate predictions.

The manager of a café in a port city in the Caribbean schedules extra workers for times when cruise ships are docked nearby. The local government has an online calendar that she can consult to see how many ships will be docked and how many passengers will be on each ship each day. She plans to use the total number of passengers on all ships docked on a given day to predict the number of customers in her café and then use this to determine the number of workers she should schedule.

The manager has collected the following data from which to build a model.

Passengers	Customers
2400	340
2900	380
3100	410
5400	460
7200	490

1. The café had 340 customers on a day when there were 2400 passengers and 490 customers on a day when there were 7200 passengers. Use this information to build a linear model where the independent variable is the number of passengers. In other words, write the equation of the line passing through the points (2400, 340) and (7200, 490). Use function notation. Call this function f. Do not round.

2. The café had 380 customers on a day when there were 2900 passengers and 410 customers on a day when there were 3100 passengers. Use this information to build a linear model where the independent variable is the number of passengers. In other words, write the equation of the line passing through the points (2900, 380) and (3100, 410). Use function notation. Call this function g. Do not round.

In #3–4, you will use the models you obtained in #1–2. If you did not obtain the models shown in the tables, go back and check your work in #1–2. Fill in the tables by first evaluating the function for each number of passengers and then subtracting the number of customers from the predicted number. Round to the nearest whole number. The first row is already completed as an example. Answer the questions under the table.

3. Using f from #1:

Passengers (x)	Customers (y)	Predicted Number of Customers $f(x) = 0.03125x + 265$	Error: $f(x) - y$
2400	340	$f(2400) = 0.03125(2400) + 265 = 340$	$340 - 340 = 0$
2900	380		
3100	410		
5400	460		
7200	490		

a. For which numbers of passengers, x, does the model output, $f(x)$, match the number of customers, y, exactly? Explain why this is expected. Refer to #1, if necessary.

b. When the model output, $f(x)$, does not match the number of customers, y, exactly, did you find that the model consistently underestimates the number of customers? Would you trust this model to give accurate predictions? Explain.

4. Using g from #2:

Passengers (x)	Customers (y)	Predicted Number of Customers $g(x) = 0.15x - 55$	Error: $g(x) - y$
2400	340	$g(2400) = 0.15(2400) - 55 = 305$	$305 - 340 = -35$
2900	380		
3100	410		
5400	460		
7200	490		

a. For which numbers of passengers, x, does the model output, $g(x)$, match the number of customers, y, exactly? Explain why this is expected. Refer to #2, if necessary.

b. Did you find that the model prediction is more than double the number of customers for 7200 passengers? Would you trust this model to give accurate predictions? Explain.

5. Looking back at the error columns in #3–4, if you had to choose the better model between f and g, which would you choose and why?

The least-squares regression model for the passenger/customer data is $h(x) = 0.028x + 298$. If you are using a graphing utility, use the linear regression option to verify this equation.

6. Complete the table using the least-squares regression model. Round the predicted number of customers to the nearest whole number.

Passengers (x)	Customers (y)	Predicted Number of Customers $h(x) = 0.028x + 298$	Error: $h(x) - y$
2400	340	$h(2400) = 0.028(2400) + 298 \approx 365$	$365 - 340 = 25$
2900	380		
3100	410		
5400	460		
7200	490		

a. Do any of the predictions, $h(x)$, match the number of customers, y, exactly? If so, for which numbers(s) of passengers, x, does this occur?

b. In #3, we saw that the model f had a potential problem in that it consistently overestimated the number of customers when it was not exact. Does h have a similar problem in that it either consistently overestimates or underestimates the actual numbers of customers? Discuss any reservations you may have about the model h in this regard.

c. In #4, we saw that the model g had a potential problem in that it predicted more than double the number of passengers in one case. Does h have a similar problem in that it makes predictions that are vastly different from the actual numbers of customers? Discuss any reservations you may have about the model h in this regard.

7. Looking back at the tables in #3, #4, and #6, which model, f, g, or h, is the most accurate for 5400 passengers? Do you think that the most accurate prediction is a good estimate of the actual number of customers? Explain.

8. Looking back at the error columns in #3, #4, and #6, if you had to choose the best model from f, g, and h, which would you choose and why?

Refer to this partial table of the original data as you answer the questions in #9–11.

Passengers	Customers
3100	410
5400	460

9. Use f to predict the number of customers next Tuesday if there are a total of 4600 passengers on the cruise ships scheduled to be in port that day. Round to the nearest whole number. Does this result seem reasonable when compared to the numbers of customers for 3100 and 5400 passengers, shown in the table? Explain.

10. Use g to predict the number of customers next Tuesday if there are a total of 4600 passengers on the cruise ships scheduled to be in port that day. Does this result seem reasonable when compared to the numbers of customers for 3100 and 5400 passengers, shown in the table? Explain.

11. Use h to predict the number of customers next Tuesday if there are a total of 4600 passengers on the cruise ships scheduled to be in port that day. Round to the nearest whole number. Does this result seem reasonable when compared to the numbers of customers for 3100 and 5400 passengers, shown in the table? Explain.

12. Using what you consider to be the most reasonable predicted number of customers for 4600 passengers in your work from #9–11, how many workers should the manager schedule next Tuesday if she prefers to have one worker for every 30 customers. Round the number of workers up to the nearest whole number.

13. Considering the work you've done in this activity, discuss some criteria for choosing the best model from several options.

Classroom Activity: Investigating Composition and Inverses of Functions

Complete this activity before Section 1.7 is covered.

Group Size: 2–3
Material: Paper, Pencil, Calculator
Time: 20 minutes

In this activity, you will write functions for simple operations and then investigate how more complicated functions can be built from these simpler functions and how the more complicated functions can be undone.

Simple operations of addition, subtraction, multiplication, and division of two numbers are functions. When one of the two numbers in any of these operations is fixed, the operation can be expressed as a function in one variable. For example, adding 6 to a number x can be expressed as $f(x) = x + 6$.

In #1–6, write a function for each operation.

1. The function g adds a number x to 2.

2. The function h subtracts 5 from a number x.

3. The function j subtracts a number x from 5.

4. The function k multiplies a number x by 3.

5. The function F divides a number x by 4.

6. The function G divides 4 by a number x.

Some other operations involve only one number and can also be expressed as functions. In #7–12, write a function for each operation.

7. The function f squares a number x.

8. The function g cubes a number x.

9. The function h takes the square root a number x.

10. The function j takes the cube root of a number x.

11. The function F finds the absolute value of a number x.

12. The function G finds the reciprocal a number x.

Some functions combine two operations. For example, the function $h(x) = 3(x - 4)$ has two operations: The function first subtracts 4 from a number x and then multiplies the result by 3. The function h is a combination of two simpler functions, $f(x) = x - 4$ and $g(x) = 3x$.

In #13–16, for each function h, give a verbal description of what the function does to a number x and then write two simpler functions f and g representing the operations in h. Let f represent the operation that is performed first.

13. $h(x) = 4(x + 1)$

14. $h(x) = 5x + 2$

15. $h(x) = x^3 + 1$

16. $h(x) = |x - 5|$

In #17–20, a function h with two operations is described. Write an equation for h, like the equations in #13–16 above. Then write functions f and g representing the two operations. Let f represent the operation that is performed first.

17. The function h multiplies a number x by 7 and then subtracts 3.

18. The function h subtracts 3 from a number x and then multiplies by 7.

19. The function h adds 2 to a number x and squares the result.

20. The function h squares a number x and then adds 2.

In #21–24, refer back to #17–20.

21. Compare your answers in #17 and #18. Are the functions h that you found in each exercise the same? How are the functions f and g in #17 related to the functions f and g in #18?

22. Compare your answers in #19 and #20. Are the functions h that you found in each exercise the same? How are the functions f and g in #19 related to the functions f and g in #20?

23. The function in #18 is $h(x) = 7(x-3)$. Use the distributive property to simplify the right side of this equation and then give a verbal description of what the function h does to a number x based on this simplified form. How does this description differ from the one given in #17?

24. Based on you work in #17–23, does the order in which two operations are performed make a difference? Explain.

Many basic operations can be undone. For example, the function $f(x) = x + 5$, which adds 5 to a number x, is undone by the function $F(x) = x - 5$, which subtracts 5. If you add 5 to a number x and then subtract 5, $(x+5) - 5$, you come back to the original number, x.

In #25–28, write the function F that undoes the given function f.

25. $f(x) = x - 2$

26. $f(x) = 7x$

27. $f(x) = \dfrac{x}{9}$

28. $f(x) = x^3$

Look again at $h(x) = 3(x - 4)$; the function first subtracts 4 from a number x and then multiplies the result by 3. The function h is a combination of the functions $f(x) = x - 4$ and $g(x) = 3x$. Use this information in #29–32.

29. Write the function F that undoes f.

30. Write the function G that undoes g.

31. Undoing h should involve adding 4 and dividing by 3, but in which order? Find $h(10)$. Take the result and try adding 4 and then dividing by 3. Do you get 10? If not, trying taking the output of $h(10)$ and dividing by 3 and then adding 4. Do you get 10 now?

32. Using what you learned in #31, write the function H that undoes h. Comment on the order in which the operations are undone relative to the order of the operations in the original function h.

In #33–34, use what you learned in #29–32 to write the function H that undoes the given function h.

33. $h(x) = 3(x + 5)$

34. $h(x) = \frac{x}{8} - 3$

35. The function $j(x) = \frac{2x+1}{3}$ combines three operations. Using what you've learned in this activity, discuss the simpler functions representing these operations and how they are undone. Write a function J that undoes j.

Classroom Activity: Investigating Zeros of Polynomial Functions

Complete this activity before Section 2.5 is covered.

Group Size: 2–3
Material: Paper, Pencil, Calculator
Time: 15 minutes

Recall that a *zero* of a function is an input value that results in an output value of 0. Thus, if we want to check whether a given number is a zero of a function, we can evaluate the function at that number. If the resulting function value is 0, then the input value is a zero.

In #1–8, evaluate the polynomial function $f(x) = x^3 + 12x^2 + 44x + 48$ for the given value of x. Is the given value a zero of f? You may use a calculator or the Remainder Theorem to help with the evaluations.

1. $x = 1$ **2.** $x = 2$ **3.** $x = 3$ **4.** $x = 6$

5. $x = -1$ **6.** $x = -2$ **7.** $x = -4$ **8.** $x = -6$

You should have identified three zeros of $f(x) = x^3 + 12x^2 + 44x + 48$ in your work above. You will use these zeros in the rest of the activity.

9. Use the Factor Theorem to write the factor of f corresponding to each of the zeros. Then multiply the three factors. Since the leading coefficient of f is 1, the product of the three factors should be the original polynomial.

10. Use the degree and the leading coefficient of f to describe the end behavior and the maximum number of turning points of the graph of f. Then use this information and the zeros to sketch the graph of f. Recall that the zeros correspond to x-intercepts.

11. Use your work in #9–10 to argue that *f* has only the three zeros identified in #1–8. You might start by asking yourselves *What if there were an additional zero?* Then investigate how an additional zero would affect the factors in #9 and the maximum number of turning points in #10.

12. The coefficients of the terms in $f(x) = x^3 + 12x^2 + 44x + 48$ are all positive. Argue that the output value for a positive input value must be positive, and therefore not zero. Conclude that *f* has no positive zeros. Would the same conclusion be valid for any polynomial function with all positive coefficients?

13. Multiply the three zeros of *f* identified in #1–8. Compare the result to the constant term of *f*. Ignoring the sign, what do you notice? Using this and your knowledge of factoring, write a conjecture about the possible zeros of a polynomial function when the leading coefficient of *f* is 1.

14. Based on your work in this activity, do you think that any of −5, 3, or 16 could be a zero of $g(x) = x^4 + 3x^3 + 5x^2 + 7x + 12$? Explain.

When it comes to finding zeros of polynomial functions, there's a lot of information available to you. The challenge is keeping track of all of it and interpreting it in a consistent manner. Since trial-and-error is often applied, it is important to narrow the possible choices for zeros whenever possible. Understanding end behavior, the relationship between zeros and factors, and turning points will help. However, as you have seen in this activity, there's additional information about the zeros in the leading coefficient, the constant term, and the signs of the coefficients.

Classroom Activity: Exploring End Behavior of a Rational Function

Complete this activity before Section 2.6 is covered.

Group Size: 2–3
Material: Paper, Pencil, Calculator
Time: 20 minutes

The group should explore the functions $f(x)=\frac{1}{x+5}$, $g(x)=\frac{x}{x+5}$, and $h(x)=\frac{x^2}{x+5}$ by completing the following.

1. Use a calculator to fill in the following tables for each of the functions. Round to three decimal places, where necessary.

x	1	5	10	20	30	40	50	75
$f(x)$								

x	1	5	10	20	30	40	50	75
$g(x)$								

x	1	5	10	20	30	40	50	75
$h(x)$								

2. Which function has output values that get smaller as x gets larger? For this function, use trial and error and your calculator to find a value of x that has an output value less than 0.0001.

3. Which function has output values that increase rapidly as x gets larger? For this function, use trial and error and your calculator to find a value of x that has an output value greater than 1000.

4. Which function has output values that increase slowly as x gets larger? For this function, use trial and error and your calculator to find a value of x that has an output value greater than 0.995.

5. Match of each of the functions f, g, and h to the appropriate description.

a. As x increases, the function values decrease but are bounded below by 0.

b. As x increases, the function values increase but are bounded above by 1.

c. As x increases, the function values increase without bound.

6. The functions f, g, and h all have the same denominator. Can you explain the descriptions in #5 in terms of the numerators?

7. Write descriptions similar to the ones given in #5 for each of the following functions. Trial and error and your calculator may be helpful.

a. $R(x) = \dfrac{2x}{x+3}$

b. $F(x) = \dfrac{x^3}{x+3}$

c. $r(x) = \dfrac{5}{x+3}$

Classroom Activity: Applications of Polynomial and Rational Functions

Complete this activity after Section 2.6 is covered.

Group Size: 1–2
Material: Paper, Pencil, Calculator
Time: 15 minutes

Recently we've seen a trend where candidates for statewide elections campaign only in heavily populated metropolitan areas, believing that winning by large margins in these areas will offset any losses in the less-populated areas where they do not campaign.

Florida has three large metropolitan areas: Miami Metro, Tampa Bay, and the Orlando area. At the time of the 2016 general election, over half of Florida's population lived in one of these three areas.

The population, in millions, of Florida and each of its three largest metropolitan areas can be modeled by polynomial functions. In each of the functions below, the variable x represents decades after 1980: $x = 0$ represents 1980, $x = 1$ represents 1990, and so on.

Total Population of Florida: $F(x) = 0.19x^3 - 0.91x^2 + 4.12x + 9.7$

Population of Miami Metro Area: $M(x) = 0.06x^3 - 0.31x^2 + 1.25x + 3.2$

Population of Tampa Bay Area: $T(x) = 0.04x^3 - 0.21x^2 + 0.67x + 1.6$

Population of Orlando Area: $O(x) = 0.02x^3 - 0.05x^2 + 0.43x + 0.8$

Remember the input value, x, is measured in decades, and the output value is the population, in millions.

1. Use the functions above to find $F(3)$, $M(3)$, $T(3)$, and $O(3)$. Interpret each of these values in terms of the year and the population of each area in that year.

2. Let $U(x) = M(x) + T(x) + O(x)$. Find an equation for $U(x)$, writing the resulting polynomial in standard form. Explain what $U(x)$ represents.

3. Let $R(x) = F(x) - U(x)$. Find an equation for $R(x)$, writing the resulting polynomial in standard form. Explain what $R(x)$ represents.

4. While it is algebraically possible to multiply the functions $M(x)$ and $T(x)$, what would the product $M(x) \cdot T(x)$ represent? Is there a meaningful reason to multiply these functions? Explain.

5. Let $P(x) = \dfrac{U(x)}{F(x)}$. Find an equation for $P(x)$, leaving the result in fraction form. Explain what $P(x)$ represents.

6. Find an equation for the horizontal asymptote of $P(x)$. Explain what the horizontal asymptote represents, assuming that the model is valid over a long period of time.

7. In #5, you should have concluded that $P(x)$ represents the proportion of the population of Florida residing in the three largest metropolitan areas x decades after 1980. Use $P(x)$ to find the proportion of the population of Florida residing in the three largest metropolitan areas in 2020. Round to two decimal places.

8. You should have found approximately 0.57 of the population of Florida residing in the three largest metropolitan areas in 2020. Therefore, approximately $1 - 0.57 = 0.43$ of the population of Florida will reside outside these areas in 2020. Assume that voters are distributed in the same manner. If a candidate wins 65% of the vote in the metropolitan areas and only 30% of the vote outside these areas, the proportion of the vote received by this candidate is $0.65(.57) + 0.30(0.43)$. Evaluate this expression. Does the candidate receive a majority of the vote under these conditions?

9. In #6, you should have found that the proportion of the population of Florida residing in the three largest metropolitan areas will approach approximately 0.63 over time, if the trend continues. Use a calculation similar to the one in #8 to determine whether a candidate can eventually expect to win an election in Florida with 65% of the vote in the metropolitan areas and 30% of the vote outside these areas.

Classroom Activity: Applications of Exponential Functions

Complete this activity after Section 3.1 is covered.

Group Size: 3–4
Material: Paper, Pencil, Calculator
Time: 25 minutes

In this activity, you will compare two investments over a 15-year period.

Investment 1: $20,000 is placed in a fund that has an historical rate of return of 8% per year.

Investment 2: $20,000 is used as a 10% down payment on a 15-year mortgage for a rental property valued at $200,000. The rental property has monthly income in the form of rent and also appreciation on the real estate. However, the rental property also has a monthly mortgage payment and other expenses, such as taxes, insurance, and maintenance. The monthly mortgage payment is fixed at $1332. It is expected that the monthly rent, the value of the real estate, and the expenses will each increase by 3% per year.

1. Assuming an annual rate of return of 8% and no further deposits or withdrawals, the value of the first investment, in dollars, at the end of t years is expected to be $A = 20,000(1.08)^t$. Use this model to complete the second column of the table. (The table feature of a graphing utility might be helpful.) Then complete the third column by subtracting the value at the end of the previous year, as shown in the first two rows. Round all amounts to the nearest dollar.

Year	Value at End of Year	Increase in Value over Previous Year
1	$21,600	$21,600 – $20,000 = $1,600
2	$23,328	$23,328 – $21,600 = $1,728
3		
4		
5		
6		
7		
8		
9		
10		
11		
12		
13		
14		
15		

2. Referring to the table in #1, what is the value of the investment after 15 years?

3. Referring to the table in #1, describe what is happening to the annual increase in value over time. Explain how this shows that the growth in the value of the investment is not linear.

4. One component of the second investment is the value of the real estate, which is expected to increase by 3% per year. Use the model $A = 200{,}000(1.03)^t$ to complete the second column of the table. (The table feature of a graphing utility might be helpful.) Then complete the third column by subtracting the value at the end of the previous year, as shown in the first two rows. Round all amounts to the nearest dollar.

Year	Value at End of Year	Increase in Value over Previous Year
1	\$206,000	\$206,000 – \$200,000 = \$6,000
2	\$212,180	\$212,180 – \$206,000 = \$6,180
3		
4		
5		
6		
7		
8		
9		
10		
11		
12		
13		
14		
15		

5. After 15 years, the mortgage is paid off and the entire value of the real estate is an asset. Referring to the table in #4, what is the value of the real estate at this time? How do this compare with the value of the first investment at the same time?

6. Comparing the third columns of the tables in #1 and #4 on a year-by-year basis, make some observations about the relative sizes of the annual increases in value. Do you think the annual increase for the first investment will ever catch up or surpass the annual increase in the value of the real estate?

7. Compute the value of the first investment at the end of the 29th and 30th years. Then find the increase in value during the 30th year. Do the same for the value of the real estate. Which had the greater increase in value during the 30th year? Compare this to your observations in #6. Is this result consistent with #6?

Your work in #6–7 hints at an important property of exponential functions: The function with the larger base will eventually surpass the function with the smaller base. In this case, the value of the first investment will surpass the value of the real estate during the 49th year.

Of course, the value of the real estate is only part of the picture in the rental property investment. The rental property also produces income in the form of rent, but there is also the mortgage payment and other expenses. The goal is to have the rental income exceed the total of the mortgage payment and other expenses, although this does not always happen initially.

The monthly mortgage payment is fixed at $1332 for the entire 15-year mortgage. The monthly rent will be $1400 during the first year and will increase by 3% per year at the beginning of each subsequent year. The annual expenses are estimated to be $4800 during the first and are expected to increase by 3% per year each subsequent year.

8. Complete the following table. Use $A = 1400(1.03)^{t-1}$ to find the monthly rent income, in dollars, during each year t. Round to the nearest dollar and then multiply the rounded value by 12 to find the annual rent income for the year. Use $A = 4800(1.03)^{t-1}$ to find the annual expenses, in dollars, for each year t. Round to the nearest dollar. Finally subtract the annual sum of mortgage payments and the annual expenses from the annual rent income to find the annual profit or loss.

Year	Monthly Rent Income	Annual Rent Income	Monthly Mortgage Payment	Annual Sum of Mortgage Payments	Annual Expenses	Annual Profit or Loss
1	$1,400	$16,800	$1,332	$15,984	$4,800	–$3,984
2	$1,442	$17,304	$1,332	$15,984	$4,944	–$3,624
3			$1,332	$15,984		
4			$1,332	$15,984		
5			$1,332	$15,984		
6			$1,332	$15,984		
7			$1,332	$15,984		
8			$1,332	$15,984		
9			$1,332	$15,984		
10			$1,332	$15,984		
11			$1,332	$15,984		
12			$1,332	$15,984		
13			$1,332	$15,984		
14			$1,332	$15,984		
15			$1,332	$15,984		

9. Describe what is happening to the profit or loss in #8 over time. Then find the sum of the amounts in the last column of the table. Is there an overall profit or loss for the 15-year period?

Based on #9, the rental property may not seem like a good investment, but again this is only part of the picture. You should have found an overall operating loss of approximately $16,573 for the 15-year period. When you consider the initial investment of $20,000, the total cash investment is $36,573. However, the real estate is now paid off and is worth over $310,000.

There are also certain tax advantages along the way. For the first year, there is an operating loss of $3,984 but the real estate increased in value by $6,000, so there is a gain of $2,016. However, taxes are not paid on the appreciation of the real estate until it is sold. The loss of $3,984 may be used as a tax deduction.

In the 16th year and beyond, there is no longer a mortgage payment. Let's look at how this affects the bottom line.

10. Extend the table in #8 to include years 16–18. Look back at #8 if you need directions.

Year	Monthly Rent Income	Annual Rent Income	Monthly Mortgage Payment	Annual Sum of Mortgage Payments	Annual Expenses	Annual Profit or Loss
16			$0	$0		
17			$0	$0		
18			$0	$0		

11. Is there a profit or a loss when you consider the first 18 years? At this point, how do you feel about this investment? Remember the real estate, which is now paid off, is continuing to appreciate as well.

12. Looking back over your work in this activity, discuss some of the pro and cons of each type of investment, both in the short term and in the long term. Be sure to address risk in your discussion. Also, discuss the role of exponential functions and rate of growth.

As you finish your education and begin your career, you will probably want to invest some of your earnings for retirement or other purposes. Be sure to investigate a variety of investment options to see what is right for you.

Classroom Activity: Investigating the Inverse Relationship

Complete this activity after Section 3.5 is covered.

Group Size: 2–3
Material: Paper, Pencil, Calculator
Time: 15 minutes

You have seen that $f(x) = e^x$ and $f^{-1}(x) = \ln x$ are inverses of one another. You have also learned how to fit an exponential model to data that follow an exponential pattern. In this activity, you will use the inverse relationship between exponential and logarithmic functions to build an exponential model.

1. Complete the following table. Evaluate the exponential function $y = e^x$ for each value of x. Round to five decimal places. Then find $\ln y$ for each value of y. Round these values to the nearest whole number.

x	$y = e^x$	$\ln y$
0		
1		
2		
3		
4		

2. Form the ordered pairs $(x, \ln y)$ for all five values of x in the table in #1. What is the most obvious pattern in the ordered pairs? Why should this pattern not be surprising?

If graphed, the points (x, y) in the table in #1 would lie on the graph of the exponential function $y = e^x$. But what about the five ordered pairs that you wrote in #2? It should be obvious that these correspond to points on the graph of the linear function $y = x$.

For any data in the form (x, y) that follow an exponential pattern, the ordered pairs $(x, \ln y)$ will follow a linear pattern. You can use what you know about writing equations for lines to fit a linear model to the data $(x, \ln y)$ and then use the inverse relationship between exponential and logarithmic functions to rewrite the linear model as an exponential model.

First, you will write an exponential model using the techniques presented in the text so you will be able to compare models. You will use the two points in the form (t, A), $(0, 4.2)$ and $(100, 11.4)$. The first coordinate, t, in each ordered pair represents the number of years after 1900 and the second coordinate, A, represents the population of Ohio, in millions.

3. Use the ordered pair $(0, 4.2)$ to find the value of A_0 in the exponential growth model $A = A_0e^{kt}$. Then substitute the value of A_0 into $A = A_0e^{kt}$.

4. Use the ordered pair (100, 11.4) and your model from #3 to find the value of k. Round k to five decimal places. Then write your exponential model.

5. Now consider the ordered pairs (0, ln 4.2) and (100, ln 11.4). Use the slope formula, $m = \frac{x_2 - x_1}{y_2 - y_1}$, to find the slope of the line passing through these two points. Round to five decimal places. Does the slope look familiar?

6. Use the slope you found in #5 and the point (0, ln 4.2) to write the equation of a line in slope-intercept form, $y = mx + b$. You may round ln 4.2 to five decimal places or leave it as an exact number.

7. The original points, (0, 4.2) and (100, 11.4), are in the form (t, A). The points you used to write the linear equation in #6, (0, ln 4.2) and (100, ln 11.4), are in the form (t, ln A). In your equation from #6, replace x with t and y with ln A.

8. Solve the equation in #7 for A. [Hint: Your equation should be in the form $\ln A = mt + b$. Write $e^{\ln A} = e^{mt+b}$ and simplify the left side.]

9. Simplify the right side of the equation from #7 by rewriting e^{mt+b} as $e^{mt}e^{b}$, using the property of exponents $a^m a^n = a^{m+n}$ in reverse. Simplify e^b and round to the nearest tenth, if necessary. Write this number in front of e^{mt}. Did you get the same model as in #4?

While the process of finding the model in #3–4 is certainly shorter, the method in #5–9 demonstrates a more general problem-solving strategy: When confronted with data that don't follow a linear pattern, perhaps you can use an appropriate function to transform the data so that it does follow a linear pattern. Use the many tools you have available to you to write the equation of a linear function. Then use the properties of the inverse of the function used in the first step to transform back.

Classroom Activity: Changing Tire Size

Complete this activity after Section 4.1 is covered.

Group Size: 1–2
Material: Paper, Pencil, Calculator
Time: 10 minutes

When a car is manufactured, the speedometer is calibrated based on the angular speed of the tires, measured in rotations per unit of time, for the diameter of the tires that are standard for that car. The speedometer displays the linear speed of the car, in miles or kilometers per hour. Thus, a conversion from angular speed to linear speed is taking place.

1. Suppose that the tires on a car are 22 inches in diameter and are rotating at a rate of 500 revolutions per minute. Use the formula $v = r\omega$ to find the linear speed of the car in inches per minute. Use a calculator and round to the nearest tenth of an inch per minute.

2. Convert your answer from #1 to miles per hour to find the speed of the car. Round to the nearest tenth of a mile per hour.

3. Suppose that the tires on a car are 24 inches in diameter and are rotating at a rate of 500 revolutions per minute. Use the formula $v = r\omega$ to find the linear speed of the car in inches per minute. Use a calculator and round to the nearest tenth of an inch per minute.

4. Convert your answer from #3 to miles per hour to find the speed of the car. Round to the nearest tenth of a mile per hour.

5. Using your answers from #2 and #4, describe what happens to the linear speed of the car when the diameter of the tire is increased and the angular speed remains the same.

6. Suppose that the speedometer of a car is calibrated for standard tires that are 22 inches in diameter and that the car's owner replaces them with tires that are 24 inches in diameter but does not recalibrate the speedometer. Which of the linear speeds in #2 and #4 is displayed by the speedometer and which is the actual speed of the car when the tires are rotating at a rate of 500 revolutions per minute?

7. As in #6, suppose that the speedometer of a car is calibrated for standard tires that are 22 inches in diameter and that the car's owner replaces them with tires that are 24 inches in diameter but does not recalibrate the speedometer. How fast is the car actually traveling when the speedometer displays 70 miles per hour?

8. Referring to #7, suppose the owner is driving on a freeway where the posted speed limit is 70 miles per hour and tickets are issued for speeds over 75 miles per hour. If the speedometer displays 70 miles per hour, is the owner in danger of getting a ticket?

9. Suppose you own a business that customizes cars. A customer has requested larger wheels and tires on his car. Write a few sentences explaining the consequences of changing the tire size to the customer. How do you convince the customer that recalibrating the speedometer is really necessary?

Classroom Activity: Investigating Right Triangles

Complete this activity after Section 4.3 is covered.

Group Size: 2–3
Material: Paper, Pencil, Calculator
Time: 15 minutes

In each problem, begin by drawing and labeling a right triangle with the given measurements. Then answer the questions or find the required information.

1. The right triangle has one acute angle measuring 32° and the side opposite the 32° angle is 10 centimeters long.

a. As a group, discuss how to find the measure of the remaining angle. Does your strategy involve using any trigonometric functions? Find the measure of this angle.

b. As a group, discuss how to find the length of the side adjacent to the 32° angle. Does your strategy involve using any trigonometric functions? Find the length of this side. Use a calculator and round to the nearest tenth of a centimeter. If group members suggest more than one strategy, have some group members try different strategies. Do you get the same answers?

c. As a group, discuss how to find the length of the hypotenuse. Does your strategy involve using any trigonometric functions? Find this length. Use a calculator and round to the nearest tenth of a centimeter. If group members suggest more than one strategy, have some group members try different strategies. Do you get the same answers?

d. As a group, devise a strategy for finding the unknown sides and angle of a right triangle when the measurements of one side and one acute angle are given. Be sure to note when more than one option may be available.

2. The right triangle has one leg measuring 8 feet and the other leg measuring 13 feet.

a. As a group, discuss how to find the length of the hypotenuse. Does your strategy involve using any trigonometric functions? Find this length. Use a calculator and round to the nearest tenth of a foot.

b. As a group, discuss how to find the measure of the angle opposite the 8-foot side. Does your strategy involve using any trigonometric functions? Find the measure of this angle. Use a calculator and round to the nearest tenth of a degree. If group members suggest more than one strategy, have some group members try different strategies. Do you get the same answers?

c. As a group, discuss how to find the measure of the remaining acute angle. Does your strategy involve using any trigonometric functions? Find the measure of this angle. Use a calculator and round to the nearest tenth of a degree. If group members suggest more than one strategy, have some group members try different strategies. Do you get the same answers?

d. As a group, devise a strategy for finding the unknown side and angles of a right triangle when the measurements of two sides are given. Be sure to note when more than one option may be available.

Classroom Activity: Fitting a Cosine Curve to Temperature Data

Complete this activity after Section 4.5 is covered.

Group Size: 2–3
Material: Paper, Pencil, Calculator
Time: 15 minutes

In this activity, you will fit a function of the form $y = A\cos(Bt - C) + D$ to temperature data. Pay close attention to what information is needed to find each of the constants A, B, C, and D as we guide you through the process.

A town has a low average monthly temperature of 45°F in January ($t = 1$) and a high average monthly temperature of 75°F in July ($t = 7$). The average monthly temperatures for the town follow a sinusoidal pattern and repeat year after year.

1. As a group, discuss why the period of the function representing the average monthly temperature of the town is 12 months. Then solve the equation $\frac{2\pi}{B} = 12$ to find the value of B in $y = A\cos(Bt - C) + D$. Express B as a simplified fraction in terms of π.

2. The basic cosine curve $y = \cos x$ is at a maximum when $x = 0$. Our function has a maximum at $t = 7$. Thus, the phase shift is 7. Substitute your value of B from #1 into $\frac{C}{B} = 7$ and solve for C.

3. The amplitude, A, of the function is the difference between the high average monthly temperature and the low average monthly temperature, divided by 2. Find A.

4. The vertical shift, D, is the mean or average of the high average monthly temperature and the low average monthly temperature. Find D.

5. Use your answers from #1–4 to write the function $y = A\cos(Bt - C) + D$ that models the average monthly temperature of the town.

6. Suppose that we had fit a model of the form $y = A\sin(Bt - C) + D$ to the temperature data instead of $y = A\cos(Bt - C) + D$. Which of the constants *A*, *B*, *C*, and *D* that you found in #1–4 would be the same and which would be different? Explain.

7. Did you determine in #6 that only the value of *C* would be different? As a group, devise a strategy for determining the value of *C* and then write the function of the form $y = A\sin(Bt - C) + D$ that models the average monthly temperature of the town.

8. Now suppose that a different town has a low average monthly temperature of 41°F in January ($t = 1$) and a high average monthly temperature of 77°F in July ($t = 7$). The average monthly temperatures for the town follow a sinusoidal pattern and repeat year after year. Which of the constants *A*, *B*, *C*, and *D* that you found in #1–4 would be the same and which would be different? Explain.

9. Did you decide in #8 that the values of *A* and *D* would be different? As a group, devise a strategy for determining the values of *A* and *D* and then write the function of the form $y = A\cos(Bt - C) + D$ that models the average monthly temperature of the town in #8.

10. Write a brief summary of the information needed to fit a function of the form $y = A\cos(Bt - C) + D$ or $y = A\sin(Bt - C) + D$ to data that follow a sinusoidal pattern. Briefly describe how to find the values of the constants *A*, *B*, *C*, and *D*.

Classroom Activity: Using Inverse Functions to Find Angles

Complete this activity after Section 4.7 is covered.

Group Size: 1–2
Material: Paper, Pencil, Calculator
Time: 15 minutes

For this activity, we will measure all angles in degrees, rounding to the nearest tenth of a degree. Be sure your calculator is in degree mode.

1. Begin by writing down the domain and range for each of the inverse sine, inverse cosine, and inverse tangent functions. Write the range of each function using degrees. Use the text, if necessary.

$y = \sin^{-1} x$	$y = \cos^{-1} x$	$y = \tan^{-1} x$
Domain:	Domain:	Domain:
Range:	Range:	Range:

The values of the inverse trigonometric functions given by your calculator are based on these ranges.

2. Suppose that an angle θ lies in the first quadrant, $0° < \theta < 90°$, and $\tan\theta = 10$. Use your calculator to evaluate $\tan^{-1} 10$. Does your calculator give you an angle θ that satisfies $0° < \theta < 90°$? Explain this using the range of $y = \tan^{-1} x$ that you wrote down above.

3. Suppose that an angle θ lies in the second quadrant, $90° < \theta < 180°$, and $\tan\theta = -10$. Use your calculator to evaluate $\tan^{-1}(-10)$. Does your calculator give you an angle θ that satisfies $90° < \theta < 180°$? Explain this using the range of $y = \tan^{-1} x$ that you wrote down above.

4. Suppose that an angle θ lies in the second quadrant, $90° < \theta < 180°$, and $\cos\theta = -0.4$. Use your calculator to evaluate $\cos^{-1}(-0.4)$. Does your calculator give you an angle θ that satisfies $90° < \theta < 180°$? Explain this using the range of $y = \cos^{-1} x$ that you wrote down above.

5. Referring to your work from #3, describe how you can use the period of the tangent function to find the angle θ that satisfies $90° < \theta < 180°$ and $\tan\theta = -10$. Then find the angle.

6. Referring to your work from #3, describe how you can use reference angles to find the angle θ that satisfies $90° < \theta < 180°$ and $\tan\theta = -10$. Then find the angle.

In #7–12, use the inverse trigonometric functions on your calculator as well as periods and reference angles, as appropriate, to find the angle θ satisfying the given conditions.

7. $\tan\theta = 10,\ 180° < \theta < 270°$

8. $\cos\theta = -0.4,\ 180° < \theta < 270°$

9. $\sin\theta = 0.7,\ 0° < \theta < 90°$

10. $\sin\theta = 0.7,\ 90° < \theta < 180°$

11. $\sin\theta = -0.7,\ 180° < \theta < 270°$

12. $\sin\theta = -0.7,\ 270° < \theta < 360°$

Classroom Activity: Using Identities to Find Exact Values

Complete this activity after Section 5.3 is covered.

Group Size: 2–3
Material: Paper, Pencil, Calculator
Time: 15 minutes

You know that in mathematics there is often more than one way to solve a problem and that regardless of the method used the final answer will be the same. However, it isn't always easy to recognize that two different-looking answers are actually equal.

Let's consider two different ways to find an exact value for $\sin 75°$.

1. Write $\sin 75°$ as $\sin(45° + 30°)$ and use a sum formula to find the exact value. Write your answer as a single fraction in simplified form.

2. Write $\sin 75°$ as $\sin \frac{150°}{2}$ and use a half-angle formula to find the exact value. Write your answer as a single fraction in simplified form with a rational denominator.

3. Compare your answers in #1 and #2. Do you immediately recognize that the answers are equal? Explain.

4. Use a calculator to find approximations, to four decimal places, for your answers in #1 and #2. Compare these to a calculator approximation for $\sin 75°$. Are you convinced that your answers in #1 and #2 are equal? Explain.

5. Using algebraic techniques to try to rewrite your answer in #1 to look like your answer in #2, or vice-versa, turns out to be futile. However, try squaring each of your answers in #1 and #2 and simplifying. What happens?

6. Note that your squared expressions in #5 are both values for $\sin^2 75°$, not $\sin 75°$. As a group, discuss how showing that the values for $\sin^2 75°$ are equal allows you to conclude that your two values for $\sin 75°$ are equal. Briefly summarize your discussion.

7. Find an exact value for $\cos 105°$ first by using a sum or difference formula and then by using a half-angle formula. Verify that the two answers are equal using the technique discussed in #5–6.

8. In a homework exercise you are asked to find an exact value for $\sin 15°$. You work the problem and obtain $\frac{\sqrt{2-\sqrt{3}}}{2}$ as the value in simplified form. However, when you look in the answer section, the answer is given as $\frac{\sqrt{6}-\sqrt{2}}{4}$. Discuss what has occurred and how you might determine whether your answer is correct.

Classroom Activity: Applications of Trigonometric Equations

Complete this activity after Section 5.5 is covered.

Group Size: 2–3
Material: Paper, Pencil, Calculator
Time: 15 minutes

The number of hours of daylight in Boston is given by $y = 3\sin\left[\frac{2\pi}{365}(x-79)\right]+12,$ where *x* is the number of days after January 1. In this activity, you will use what you've learned about trigonometric functions to solve equations involving the number of daylight hours in Boston.

1. The function $y = 3\sin\left[\frac{2\pi}{365}(x-79)\right]+12$ is almost in the form $y = A\sin(Bx - C) + D.$ What must be done to write $y = 3\sin\left[\frac{2\pi}{365}(x-79)\right]+12$ in the form $y = A\sin(Bx - C) + D$? Rewrite the function in this form and identify the values of *A*, *B*, *C*, and *D*.

2. Describe the effect that each of the values of *A*, *B*, *C*, and *D*, that you found in #1, have on the graph of the function $y = 3\sin\left[\frac{2\pi}{365}(x-79)\right]+12.$ Identify the period and the range of the function.

3. Based on the range that you identified in #2, determine whether each of the following equations has solutions. Do not solve.

a. $3\sin\left[\frac{2\pi}{365}(x-79)\right]+12 = 12$

b. $3\sin\left[\frac{2\pi}{365}(x-79)\right]+12 = 15$

c. $3\sin\left[\frac{2\pi}{365}(x-79)\right]+12 = 10$

d. $3\sin\left[\frac{2\pi}{365}(x-79)\right]+12 = 8$

4. When solving the equation $3\sin\left[\frac{2\pi}{365}(x-79)\right]+12=12,$ explain why it is natural to ask for all solutions on the interval [0, 365).

5. Explain how solving the equation $3\sin\left[\frac{2\pi}{365}(x-79)\right]+12=12$ on the interval [0, 365) is related to solving the equation $\sin\theta=0$ on the interval $[0, 2\pi)$.

6. Solve each of the following equations on the interval [0, 365). Interpret the solutions in the context of the number of hours of daylight in Boston.

a. $3\sin\left[\frac{2\pi}{365}(x-79)\right]+12=12$

b. $3\sin\left[\frac{2\pi}{365}(x-79)\right]+12=9$

7. Describe ways in which understanding the properties of the trigonometric functions helps in solving trigonometric equations.

Classroom Activity: Law of Sines or Law of Cosines?

Complete this activity after Section 6.2 is covered.

Group Size: 2–3
Material: Paper, Pencil, Calculator
Time: 20 minutes

In this activity, you will practice identifying the appropriate method for solving a triangle.

1. As a group, discuss why neither the Law of Sines nor Law of Cosines is necessary for solving the triangle shown below. Determine a strategy for solving the triangle without the Law of Sines or Law of Cosines and then solve the triangle. Round lengths of sides to the nearest tenth of a centimeter.

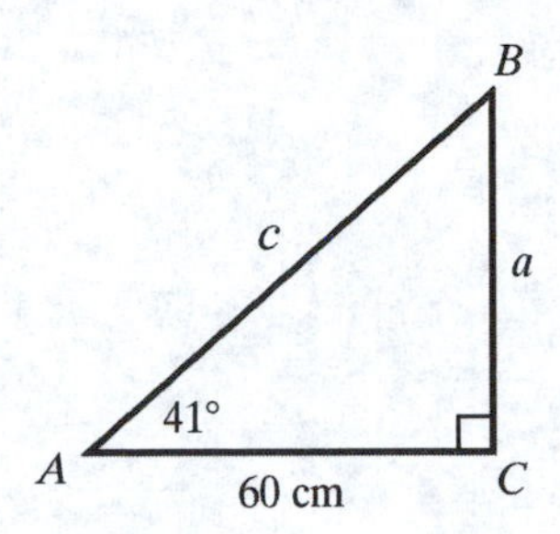

2. For the triangle in #1, you are given the measures of two angles and the length of the included side: ASA. Which method, Law of Sines or Law of Cosines, would be appropriate for solving this triangle? Solve the triangle by this method. Round lengths of sides to the nearest tenth of a centimeter.

3. Compare the work you did in solving the triangle in #1 and in #2. Did either method seem simpler? Explain.

4. Discuss a strategy for solving a triangle given two angles and one side. Does the position of the known side relative to the known angles matter? How can you tell whether the triangle is a right triangle before solving it? If you first identified the triangle as being a right triangle, would you change your strategy? Explain.

5. Discuss strategies for solving a triangle given two sides and one angle. Does the position of the known angle relative to the known sides matter? How can you tell whether the triangle is a right triangle before solving it? If you first identified the triangle as being a right triangle, would you change your strategy? Explain.

In #6–11, solve each triangle. Round angle measures to the nearest degree and lengths of sides to the nearest tenth. Side a is opposite angle A, and so on, as shown in the diagram.

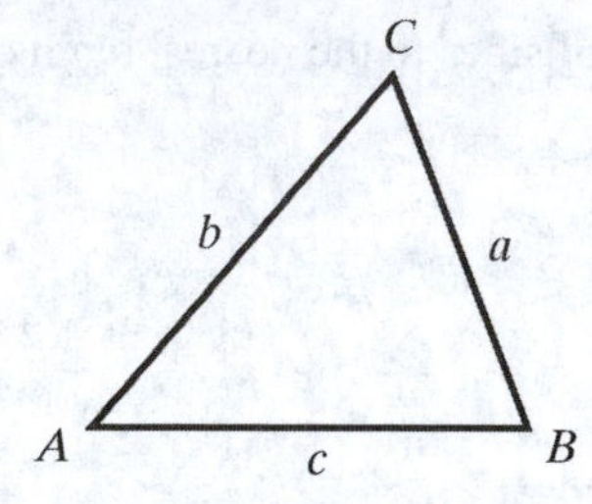

For each triangle, discuss possible strategies as a group before solving.

6. $a = 3, b = 4, c = 5$

7. $a = 7, b = 10, C = 110°$

8. $A = 25°, B = 65°, c = 15$

9. $A = 35°, B = 75°, b = 25$

10. $a = 3, b = 4, B = 50°$

11. $a = 9, b = 13, C = 90°$

12. Look back at #6–11. Which triangles are right triangles? Did you notice that these were right triangles before you solved them? For each right triangle, discuss how the given information can be used to conclude that it's a right triangle without first solving it.

Classroom Activity: Reviewing Angles, Distance, and Trigonometric Functions

Complete this activity before Section 6.3 is covered.

Group Size: 1–2
Material: Paper, Pencil, Calculator
Time: 15 minutes

In this activity, you will review important relationships among points in the plane, angles in standard position, and the trigonometric functions. These concepts will be used extensively in the next few sections of the text.

Recall that the distance between two points (x_1, y_1) and (x_2, y_2) in the rectangular coordinate system is given by $d = \sqrt{(x_2 - x_1)^2 + (y_2 - y_1)^2}$.

1. Use the formula above to find the distance of the point (x, y) in the rectangular coordinate system from the origin, (0, 0). Simplify.

Recall also that the six trigonometric functions of an angle θ in standard position in the rectangular coordinate system can be defined in terms of a point (x, y) on the terminal side of the angle.

2. Use x, y, and r to complete the definitions of the six trigonometric functions, where (x, y) is a point on the terminal side of the angle θ in standard position and r is the distance of the point (x, y) from the origin.

$\sin\theta =$ ___ $\cos\theta =$ ___ $\tan\theta =$ ___

$\csc\theta =$ ___ $\sec\theta =$ ___ $\cot\theta =$ ___

In #3–4, plot each point in the rectangular coordinate system. Let θ be an angle in standard position with terminal side passing through the given point. Draw θ. Then calculate r and find the values of the six trigonometric functions of θ. Use an exact value for r. Rationalize denominators, where necessary.

3. (3, 4)

$\sin\theta =$ ___ $\cos\theta =$ ___ $\tan\theta =$ ___

$\csc\theta =$ ___ $\sec\theta =$ ___ $\cot\theta =$ ___

4. (–2, –1)

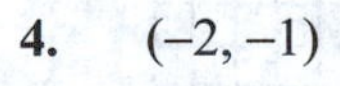

$\sin\theta =$ ___ $\cos\theta =$ ___ $\tan\theta =$ ___

$\csc\theta =$ ___ $\sec\theta =$ ___ $\cot\theta =$ ___

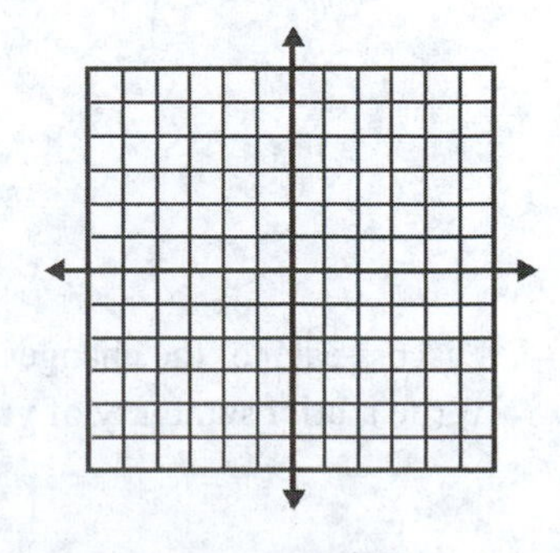

Given the coordinates of a point (x, y) on the terminal side of an angle θ in standard position, you may need to find the measure of the angle θ, $0° < \theta < 360°$? You can use the inverse tangent function and a calculator to find an angle $\alpha = \tan^{-1}\frac{y}{x}$. However, recall that the range of the inverse tangent function is $(-90°, 90°)$. The angle α, provided by your calculator, will need to be adjusted for angles θ in quadrants II, III, and IV.

In #5–8, plot each point in the rectangular coordinate system. Let θ be an angle in standard position, $0° < \theta < 360°$, with terminal side passing through the given point. Draw θ. Calculate $\alpha = \tan^{-1}\frac{y}{x}$; round to the nearest degree. Follow the instructions to find θ.

5. (1, 3); Since the angle is in quadrant I, $\theta = \alpha$.

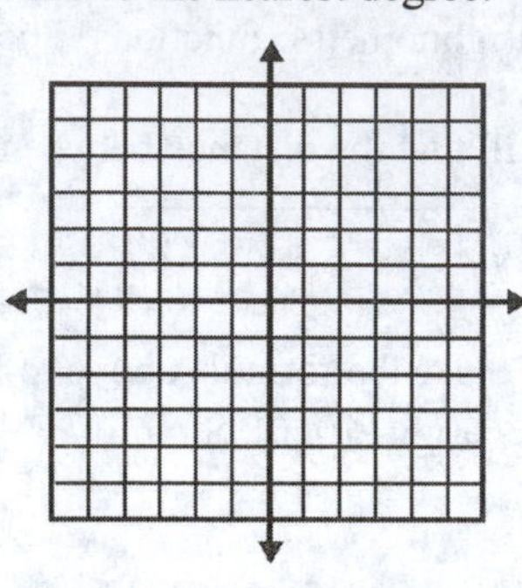

6. (−2, 3); Since the angle is in quadrant II, $\theta = \alpha + 180°$.

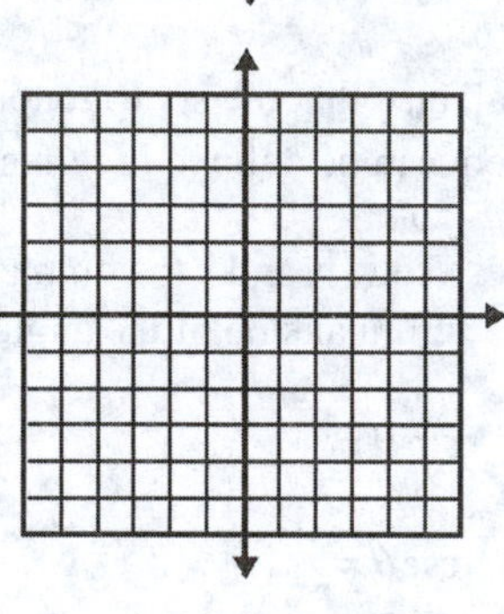

7. (−3, −4); Since the angle is in quadrant III, $\theta = \alpha + 180°$.

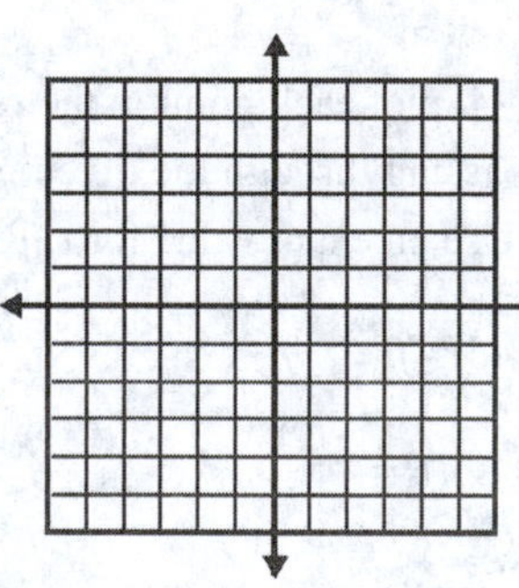

8. (1, −3); Since the angle is in quadrant IV, $\theta = \alpha + 360°$.

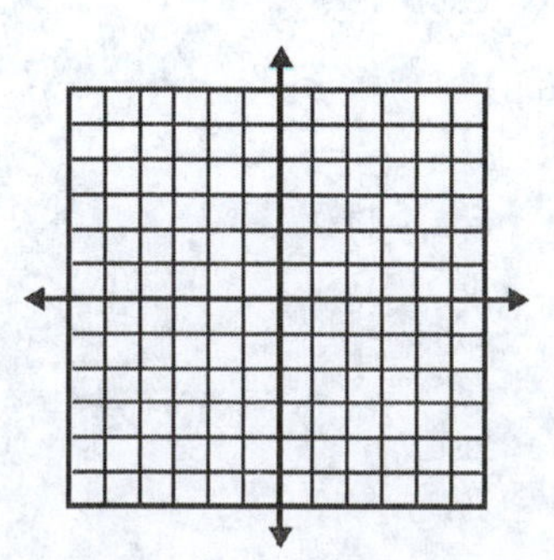

9. Discuss each of the changes made to α in #6–8 in order to obtain θ.
Write a brief summary of your discussion.

Classroom Activity: More on Curve Fitting

Complete this activity after Section 7.2 is covered.

Group Size: 2–3
Material: Paper, Pencil, Calculator
Time: 20 minutes

You have learned how to write an equation in the form $y = ax^2 + bx + c$ given three points on its graph. In this activity, you will look at this process in two special cases and then extend one of the cases to write an equation of the form $y = ax^3 + bx^2 + cx + d$. You will also look at how to solve the associated systems of equations more efficiently.

1. Use $y = ax^2 + bx + c$ and the points (−1, −8), (2, 1), and (4, 7) to write a system of three equations with the three unknowns a, b, and c. As a reminder, if you need one: For each ordered pair, substitute the first coordinate into $y = ax^2 + bx + c$ for x and the second coordinate for y and simplify. Do not solve yet.

2. Looking at your three equations in #1, what do all three equations have in common? Would this be true every time you use $y = ax^2 + bx + c$ to write a system of three equations? Based on this, which of a, b, and c would you recommend eliminating first? Explain.

3. Now solve the system you wrote in #1 for a, b, and c, using your recommendation from #2.

4. Substitute the values of a, b, and c that you found in #3 into $y = ax^2 + bx + c$ and simplify, if necessary. Is the graph of this equation a parabola? If not, what is it?

In #1–4, you fit an equation of the form $y = ax^2 + bx + c$ to three points. The resulting equation is linear, not quadratic. In this case, the three given points are all on the same line. In order to write a quadratic equation, you must start with three points that are not all on the same line. However, our method still produced the correct equation; there are just much more efficient ways to write a linear equation.

Now we look at a case where one of the given points contains the y-intercept.

5. Consider the three points $(-1, 1)$, $(0, 4)$, and $(2, -2)$. Quickly plot the points to confirm that they do not all lie on the same line. Which point contains the y-intercept?

6. Substitute the coordinates of the point containing the y-intercept into $y = ax^2 + bx + c$ and simplify. What happens?

7. In #6, you should have found that $c = 4$. Replacing c with 4 in $y = ax^2 + bx + c$, we obtain $y = ax^2 + bx + 4$. Use this form to write equations using the two remaining points given in #5. Simplify by isolating constant terms on one side in each equation. Do not solve yet.

8. You should now have a system of two linear equations with the unknowns a and b. As a group, discuss ways that you could solve this system. Then each group member should solve the system using different steps, if possible. Check each others' work. Does any solution seem more efficient than the others? Explain.

9. Now use your results to write the equation of the parabola that passes through the points $(-1, 1)$, $(0, 4)$, and $(2, -2)$. Discuss how you can check your work. Then verify that your equation is correct.

Now let's generalize what you've learned about writing equations of the form $y = ax^2 + bx + c$ to write equations of the form $y = ax^3 + bx^2 + cx + d$.

10. We need two distinct points to write a linear equation and three points not all on the same line to write a quadratic equation. How many points do you think you will need to write a cubic polynomial function of the form $y = ax^3 + bx^2 + cx + d$? What conditions do you think will be necessary to be sure that the resulting equation is actually cubic and not linear or quadratic?

11. Quickly plot the points (–1, 3), (0, 1), (1, 3), and (2, 15). Do these points satisfy your conditions from #10?

12. Generalizing from the situation in #6–7, discuss how knowing the *y*-intercept can be used to simplify the process of writing the equation $y = ax^3 + bx^2 + cx + d$ from the given points. Which value do you know? Rewrite $y = ax^3 + bx^2 + cx + d$, substituting the known value.

13. Use the form from #12 to write equations using the three remaining points given in #11. Simplify by isolating constant terms on one side in each equation. Do not solve yet.

14. As a group, discuss how to solve the system of equations you found in #13. Think about how to do this most efficiently. Can the work be split up among group members? Solve the system.

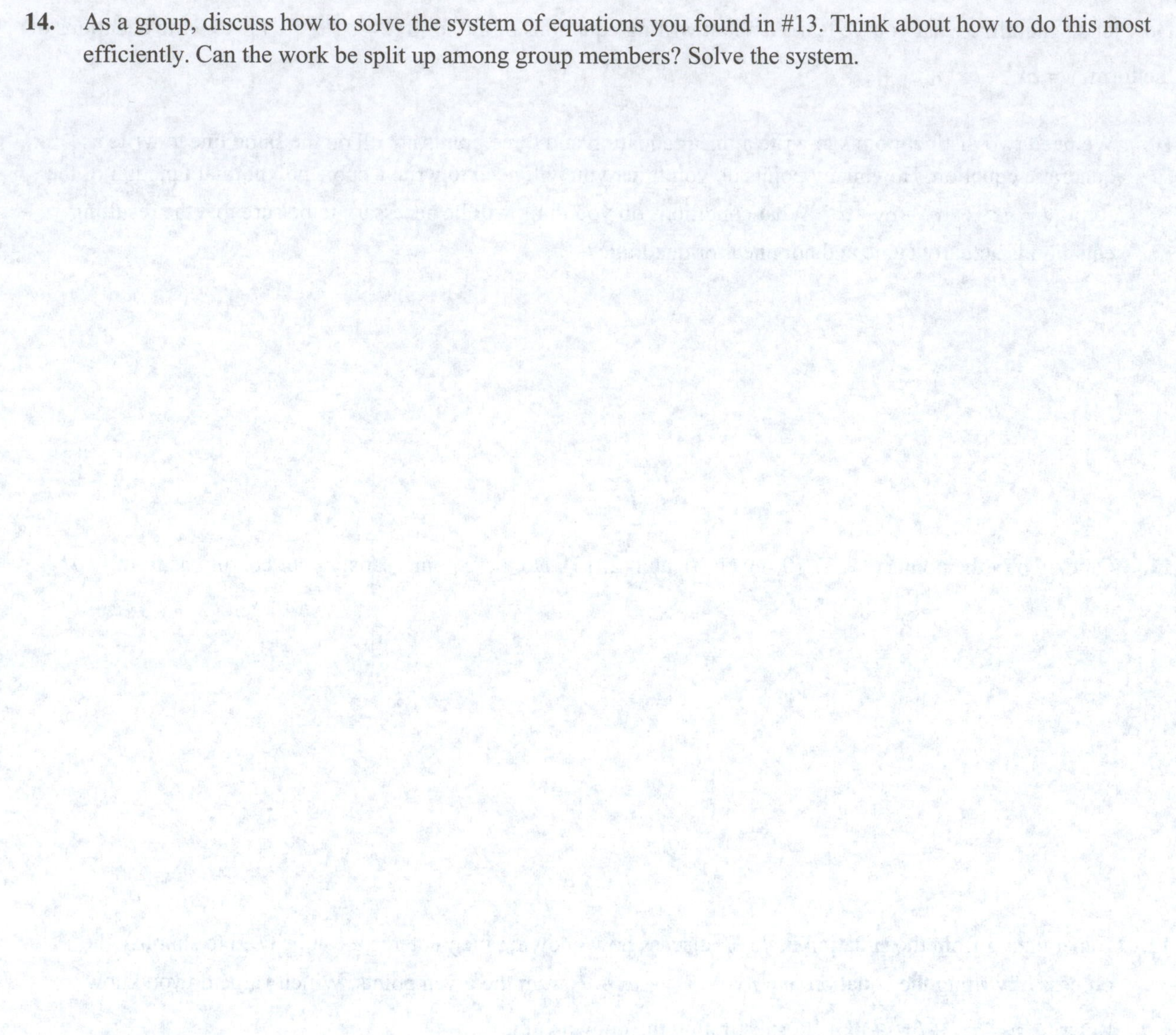

15. As in any problem you are solving, be sure your final answer is in the required form. Here you are looking for the cubic polynomial equation whose graph passes through the points $(-1, 3)$, $(0, 1)$, $(1, 3)$, and $(2, 15)$. Write this equation. Then verify that all four of the given points are solutions of your equation.

The methods for fitting higher-degree polynomial models to a set of data can be generalized from the methods discussed here and in the text. If you can fit a quadratic model to a set of three points not all on the same line, then you have the necessary skills and knowledge to fit a higher-degree polynomial model to a set of points, provided you have the right number of points and they satisfy certain conditions.

Classroom Activity: Investigating Nonlinear Systems

Complete this activity before Section 7.4 is covered.

Group Size: 1–2
Material: Paper, Pencil
Time: 15 minutes

The following graphs illustrate that two lines may intersect in exactly one point, be parallel, or coincide. This graphically confirms that a system of two linear equations in two variables, x and y, may have exactly one solution, no solution, or infinitely many solutions.

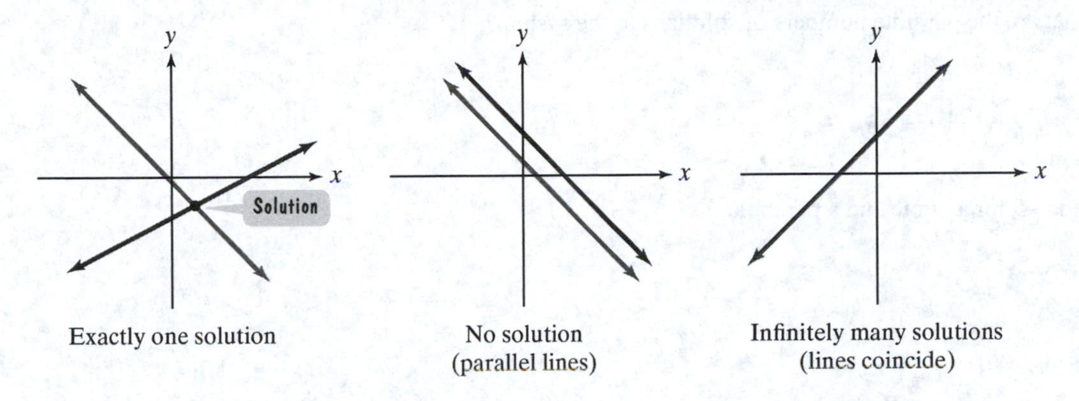

In this activity, you will explore the possible numbers of intersection points of two graphs, where at least one of the graphs is not a line. We start with a circle and a line.

1. In the space below, draw examples of a circle and a line where (a) the circle and line do not intersect, (b) the circle and the line intersect in exactly one point, and (c) the circle and the line intersect in exactly two points.

2. Do your graphs in #1 illustrate all possible numbers of intersection points of a circle and a line? In particular, can a circle and a line intersect in exactly three points or infinitely many points?

3. Based on your answers to #1 and #2, if a system of two equations consists of the equation of a circle and the equation of a line, what are the possible numbers of solutions to the system?

4. In the space below, draw combinations of a parabola and a line until you are convinced that you have illustrated all possible numbers of intersection points of a parabola and a line.

5. Based on your work in #4, if a system of two equations consists of the equation of a parabola and the equation of a line, what are the possible numbers of solutions to the system?

6. Repeat #4 and #5 for a circle and a parabola.

7. Repeat #4 and #5 for two circles.

8. Summarize any patterns you notice in your work in #1–7 and explain the benefit of classifying equations by their graphs in determining the number of possible solutions to a system of equations.

Classroom Activity: Applications of Matrix Multiplication

Complete this activity after Section 8.3 is covered.

Group Size: 3–4
Material: Paper, Pencil, Calculator
Time: 20 minutes

Each fall semester, Professor Smith teaches a graduate class with an enrollment of about five students. Each student is assigned a grade at the end of the course based on three exams and a course project. The grade book for this class last year is given below.

Student	Test 1	Test 2	Test 3	Project
Acosta, C.	100	95	90	79
Broussard, L.	85	84	87	100
Diaz, J.	86	83	85	70
Martin, S.	71	94	92	83
Walton, G.	60	85	75	60

Professor Smith wants to analyze the scores on each assignment and the final course grades in order to see whether she should make some adjustments before teaching the class again.

1. Perform the following matrix multiplication. Split up the calculations among the group members.

$$\begin{bmatrix} 1 & 1 & 1 & 1 & 1 \end{bmatrix} \begin{bmatrix} 100 & 95 & 90 & 79 \\ 85 & 84 & 87 & 100 \\ 86 & 83 & 85 & 70 \\ 71 & 94 & 92 & 83 \\ 60 & 85 & 75 & 60 \end{bmatrix}$$

2. In #1, your answer should be a 1×4 matrix. Did you notice that this multiplication simply sums the numbers in each column of the 5×4 matrix? Multiply your 1×4 matrix from #1 by the scalar $\frac{1}{5}$. Describe what each of the four numbers in this matrix represents.

3. In a graduate class, Professor Smith prefers that the class average on each assignment be at least 80 but no more than 90. Which assignments, if any, need to be adjusted to make this happen?

Last year Professor Smith weighted all four assignments equally: Each assignment comprised 25% or 0.25 of the final grade. The 4×1 matrix below represents these weights. The first row is the weight for Test 1, the second row Test 2, and so on.

$$\begin{bmatrix} 0.25 \\ 0.25 \\ 0.25 \\ 0.25 \end{bmatrix}$$

Note that the weights are nonnegative and sum to 1.

4. Perform the following matrix multiplication. Split up the calculations among the group members.

$$\begin{bmatrix} 100 & 95 & 90 & 79 \\ 85 & 84 & 87 & 100 \\ 86 & 83 & 85 & 70 \\ 71 & 94 & 92 & 83 \\ 60 & 85 & 75 & 60 \end{bmatrix} \begin{bmatrix} 0.25 \\ 0.25 \\ 0.25 \\ 0.25 \end{bmatrix}$$

5. In #4, your answer should be a 5×1 matrix. Describe what each of the five numbers in this matrix represents.

6. Assign a course grade to each student based on your answers in #4–5 and the following scale:

A: 90–100 B: 80–89 C: 70–79 D: 60–69 F: 59 or below.

Student	Course Grade
Acosta, C.	
Broussard, L.	
Diaz, J.	
Martin, S.	
Walton, G.	

7. Professor Smith's goal is to have at least one A in the class, but mostly Bs, and no Ds or Fs. Did she achieve this goal last year? If not, where did she fall short?

8. Describe what the following multiplication represents. As a group, devise a strategy for finding the product using work you've already done in this activity in order to minimize the number of calculations. Then find the product.

$$\frac{1}{5}\begin{bmatrix}1 & 1 & 1 & 1 & 1\end{bmatrix}\begin{bmatrix}100 & 95 & 90 & 79 \\ 85 & 84 & 87 & 100 \\ 86 & 83 & 85 & 70 \\ 71 & 94 & 92 & 83 \\ 60 & 85 & 75 & 60\end{bmatrix}\begin{bmatrix}0.25 \\ 0.25 \\ 0.25 \\ 0.25\end{bmatrix}$$

9. Professor Smith would like for her final class average to be within 2.5 points of 85. Did she achieve this goal last year? If not, discuss how far she is from meeting the goal and whether the class average is too high or too low.

10. Professor Smith believes that the project is the most important assignment in the class and is considering reweighting the assignments to reflect this. Each column in the following matrix represents a proposed set of new weights. Interpret the weights in each column.

$$\begin{bmatrix}0.2 & 0.1 \\ 0.2 & 0.1 \\ 0.2 & 0.3 \\ 0.4 & 0.5\end{bmatrix}$$

11. Perform the following matrix multiplication. Split up the calculations among the group members. Round entries to the nearest whole number.

$$\begin{bmatrix}100 & 95 & 90 & 79 \\ 85 & 84 & 87 & 100 \\ 86 & 83 & 85 & 70 \\ 71 & 94 & 92 & 83 \\ 60 & 85 & 75 & 60\end{bmatrix}\begin{bmatrix}0.2 & 0.1 \\ 0.2 & 0.1 \\ 0.2 & 0.3 \\ 0.4 & 0.5\end{bmatrix}$$

12. Assign a course grade to each student for each set of weights based on your work in #11, following the scale in #6.

Student	Course Grade (weights in column 1)	Course Grade (weights in column 2)
Acosta, C.		
Broussard, L.		
Diaz, J.		
Martin, S.		
Walton, G.		

13. Is there any practical difference between the two new proposed grading systems? Explain.

14. Is the goal stated in #7 achieved with either of these grading systems? Explain.

15. Compare the grades assigned in #12 to those assigned in #6. Did any of the students' grades change? If so, whose grades changed and in which direction?

16. Each group member should choose a set of weights different from those in the activity and different from each other. The project should be weighted more heavily than any other assignment. Remember the weights must be nonnegative and sum to 1. Assign grades to the five students based on these weights. Do any of the grade assignments in your group achieve the goals stated in #7? Explain.

Classroom Activity: Investigating Matrix Multiplication and Solving Matrix Equations

Complete this activity after Section 8.4 is covered.

Group Size: 2–3
Material: Paper, Pencil, Calculator
Time: 15 minutes

Although the way matrix multiplication is performed may seem odd to you, it is actually defined in a way that makes it extremely useful. As you have worked with more systems of linear equations, perhaps you have begun to look at the equations from a different perspective. Perhaps you have noticed that the left side of an equation, such as

$$3x-2y+5z=10,$$

is related to matrix multiplication. Let's begin by exploring this relationship.

Look at the matrix multiplication problem below. Note that the result is a 1×1 matrix. The single entry in the matrix, an algebraic expression with three terms, is the left side of the equation above.

$$\begin{bmatrix}3 & -2 & 5\end{bmatrix}\begin{bmatrix}x\\ y\\ z\end{bmatrix}=\begin{bmatrix}3x-2y+5z\end{bmatrix}$$

Thus, the original equation can be expressed as

$$\begin{bmatrix}3 & -2 & 5\end{bmatrix}\begin{bmatrix}x\\ y\\ z\end{bmatrix}=\begin{bmatrix}10\end{bmatrix}.$$

1. Write each equation in the system below in the matrix form shown above.

$$\begin{cases}2x-\ y-3z = 0\\ x+3y+\ z = 6\\ 4x+\ y-\ z = 8\end{cases}$$

The properties of matrices allow you to write your three matrix equations in #1 as a single matrix equation:

$$\begin{bmatrix}2x-\ y-3z\\ x+3y+\ z\\ 4x+\ y-\ z\end{bmatrix}=\begin{bmatrix}0\\ 6\\ 8\end{bmatrix}\quad\rightarrow\quad\begin{bmatrix}2 & -1 & -3\\ 1 & 3 & 1\\ 4 & 1 & -1\end{bmatrix}\begin{bmatrix}x\\ y\\ z\end{bmatrix}=\begin{bmatrix}0\\ 6\\ 8\end{bmatrix}.$$

The equation on the right above is in the form $AX=B,$ where A, X, and B are matrices. So, a system of three equations in three variables becomes a single matrix equation. The form of this equation is similar to the linear equation $ax=b,$ which can be solved by multiplying both sides by the multiplicative inverse (reciprocal) of a as long as a is not zero. Likewise, the matrix equation $AX=B$ can be solved by multiplying both sides by A^{-1}, provided A^{-1} exists.

2. The matrix A below is the coefficient matrix for the system of equations in #1. Find its inverse, A^{-1}, if it exists.

$$A = \begin{bmatrix} 2 & -1 & -3 \\ 1 & 3 & 1 \\ 4 & 1 & -1 \end{bmatrix}$$

3. The matrix B below contains the constants from the system of equations in #1. Using A^{-1} from #2, find, if possible, both $A^{-1}B$ and BA^{-1}.

$$B = \begin{bmatrix} 0 \\ 6 \\ 8 \end{bmatrix}$$

4. Are $A^{-1}B$ and BA^{-1} equal? What limitation on matrix multiplication is illustrated here? Which is the solution to the matrix equation $AX = B$ on the previous page?

5. Given a system of linear equations, describe how to write a matrix equation for the system. Then explain how to solve the system, provided that the appropriate inverse matrix exists. Include a warning about the multiplication involved.

6. Use your strategy described in #5 to solve the system of equations below. Use separate paper for your work.

$$\begin{cases} 2x - 3y = 12 \\ 3x + 4y = 1 \end{cases}$$

Classroom Activity: Investigating Second-Degree Equations

Complete this activity before Section 9.1 is covered.

Group Size: 2–3
Material: Paper, Pencil, Calculator
Time: 15 minutes

In this activity, you will investigate the graphs of certain second-degree equations in two variables using intercepts, symmetry, and point plotting. You are already familiar with second-degree equations of the form $x^2 + y^2 + Dx + Ey + F = 0$; you should recognize this as the general form of a circle's equation. Note that in this form the coefficients of the squared terms are both 1.

You will begin by reviewing some of the important properties of circles. Then you will look at what can happen if the coefficient of y^2 is changed while the coefficient of x^2 is kept fixed at 1.

1. The equation $x^2 + y^2 = 9$ represents a circle. Find the *x*-intercepts of its graph by letting $y = 0$ and solving for *x*. [Remember that equations of the form $u^2 = d,\ d > 0,$ have two solutions: $u = \pm\sqrt{d}$.] Then find the *y*-intercepts of the graph by letting $x = 0$ and solving for *y*.

2. Test the equation $x^2 + y^2 = 9$ for symmetry with respect to the *x*-axis, the *y*-axis, and the origin. Perform the substitutions indicated; if an equation equivalent to $x^2 + y^2 = 9$ results, the graph has the corresponding type of symmetry.

x-axis symmetry:
Replace *y* with $-y$.

y-axis symmetry:
Replace *x* with $-x$.

origin symmetry:
Replace *x* with $-x$ and *y* with $-y$.

3. Find additional points on the graph of $x^2 + y^2 = 9$ by letting $x = 1$ and $x = 2$ and solving for *y*. Round to one decimal place. Use these points, the intercepts, and symmetry to graph $x^2 + y^2 = 9$.

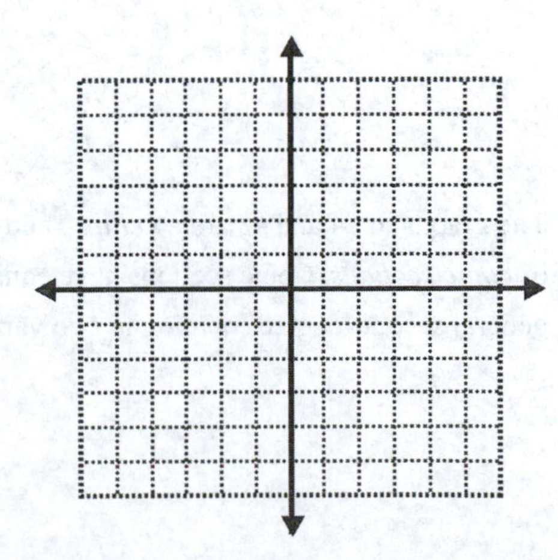

Now you will look at what happens when the coefficient of y^2 in $x^2 + y^2 = 9$ is changed to 9.

4. Repeat the steps in #1–3 for the equation $x^2 + 9y^2 = 9$. As a group, discuss how the graph of $x^2 + 9y^2 = 9$ compares to the graph of $x^2 + y^2 = 9$.

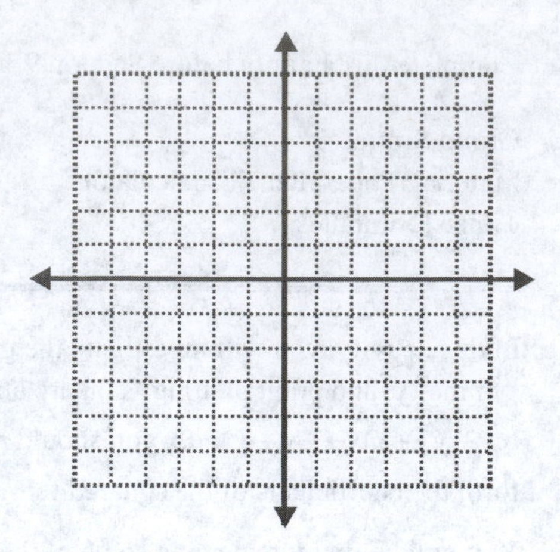

Now you will look at what happens when the coefficient of y^2 in $x^2 + y^2 = 9$ is changed to −1.

5. Repeat the steps in #1–3 for the equation $x^2 - y^2 = 9$. This time find additional points on the graph by letting $x = 4$ and $x = 5$ and solving for y. Discuss why values of x in the interval (−3, 3) are not in the domain of the relation. How does this graph compare to your graphs in #3 and #4.

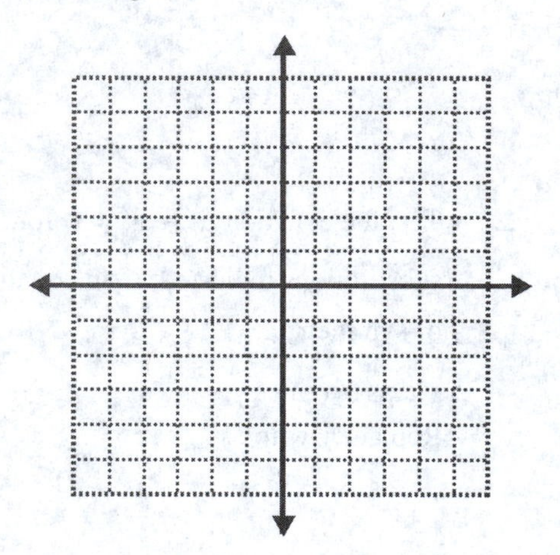

The graphs in #4 and #5 are an *ellipse* and a *hyperbola*, respectively. Circles, ellipses, and hyperbolas are three types of *conic sections*. Conic sections have equations of the form $Ax^2 + Bxy + Cy^2 + Dx + Ey + F = 0$, which is the general second-degree equation in two variables. In the text, you will study geometric definitions of conic sections.

Classroom Activity: Applications of Ellipses and Parabolas

Complete this activity after Section 9.3 is covered.

Group Size: 2–3
Material: Paper, Pencil, Calculator
Time: 15 minutes

Suppose that you are an engineer designing an arch bridge spanning the entrance to a marina. Small watercraft will be passing under the bridge, so it is important to allow for proper clearance.

The arch will be 120 feet wide at its base and 40 feet tall at its peak. In a rectangular coordinate system, the feet of the arch will be at (–60, 0) and (60, 0) and the peak will be at (0, 40).

In this activity, you will compare two different shapes for the arch described.

1. If the arch were parabolic in shape, it would have an equation of the form $y = ax^2 + c$. Use the coordinates given above to write an equation of a parabolic arch with the given dimensions. Write the value of a as a simplified fraction.

2. Use your equation from #1 to find the height of the arch at distances of 10 feet, 20 feet, 30 feet, 40 feet, and 50 feet from the center. Round to the nearest foot. Organize your results in the table. Then draw the arch in a rectangular coordinate system.

Distance from Center	Height
10 feet	
20 feet	
30 feet	
40 feet	
50 feet	

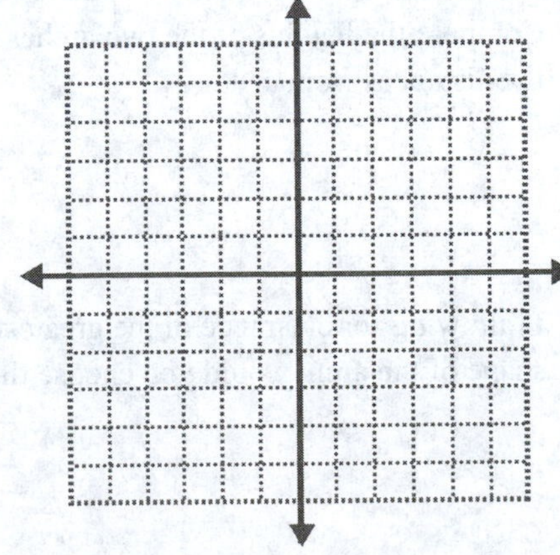

3. If the arch were semi-elliptical in shape, it would be the top half of an ellipse with an equation of the form $\frac{x^2}{a^2}+\frac{y^2}{b^2}=1$. Write an equation of the ellipse with vertices at (−60, 0) and (60, 0) and passing through (0, 40). Solve for y to find an equation for the semi-elliptical arch. You only want the nonnegative value of y.

4. Use your final equation from #3 to find the height of the arch at distances of 10 feet, 20 feet, 30 feet, 40 feet, and 50 feet from the center. Round to the nearest foot. Organize your results in the table. Then draw the arch in a rectangular coordinate system.

Distance from Center	Height
10 feet	
20 feet	
30 feet	
40 feet	
50 feet	

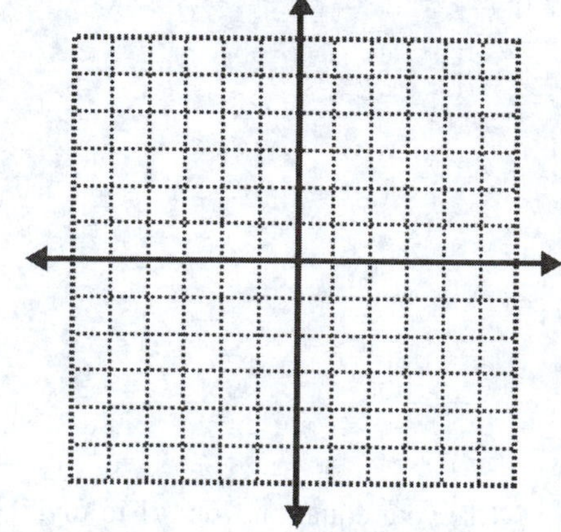

5. Compare the heights of the two arches in your tables in #2 and #4. Is either arch consistently higher at each of the distances from center?

6. If allowing for clearance of the greatest variety of watercraft were the most important criterion in selecting the shape of the arch, would you choose the parabolic or semi-elliptical arch? Explain.

Classroom Activity: Applications of Sequences

Complete this activity after Section 10.3 is covered.

Group Size: 2–3
Material: Paper, Pencil, Calculator
Time: 20 minutes

When you buy a home or a car, you may need to borrow money in the form of a loan and then repay the loan by making monthly payments over a number of years. A loan used to purchase a home is called a *mortgage.* In this activity, you will work with three sequences related to a mortgage: (1) the monthly interest, (2) the amount of the payment applied toward the principal, and (3) the balance after each payment.

Suppose you are purchasing a \$300,000 home. You have a \$60,000 down payment, so you are financing \$240,000. You obtain a 30-year mortgage with a fixed rate of 4.8%; the monthly payment will be \$1260.

Let's begin defining the three sequences with b_n, the balance of the loan. At this point, we know that at the beginning of the loan, when $n = 0$, the balance is \$240,000: $b_0 = 240{,}000$.

The interest for the first month, i_1, will be calculated using the original balance, b_0. The rate given, 4.8%, is an annual rate; the monthly rate is 0.4% or 0.004. Thus, $i_1 = 0.004b_0$, and in general, $i_n = 0.004b_{n-1}$.

The amount of the payment applied toward the principal for any month, p_n, is found by subtracting the interest from the payment: $p_n = 1260 - i_n$.

Finally, the balance after the nth payment, b_n, is found by subtracting the amount of the payment applied toward the principal from the previous balance: $b_n = b_{n-1} - p_n$.

The three sequences are completely defined by the following:

$$b_0 = 240{,}000 \qquad i_n = 0.004b_{n-1} \qquad p_n = 1260 - i_n \qquad b_n = b_{n-1} - p_n,$$

for $n = 1, 2, \ldots, 360$. Use this information to complete the table. Round to the nearest cent, where necessary.

Month, n	Interest, i_n	Principal, p_n	Balance, b_n
1			
2			
3			
4			
5			
6			
7			
8			
9			
10			
11			
12			

This table is an *amortization schedule* for the first year of the loan. An amortization schedule for all 30 years (360 months) could be constructed by continuing the process or using a spreadsheet.

1. Using your intuitive understanding of the terms, describe each of the three sequences represented in the amortization schedule using *increasing* or *decreasing*. Explain what is happening to the three quantities over time.

2. Find and interpret the sum: $\sum_{n=1}^{12} i_n$.

3. Mortgage interest is often deductible on your taxes. Assuming the first year of the mortgage coincides with a tax year and you are in the 15% tax bracket, your taxes may be reduced by 15% of the sum you found in #2. Find this amount.

4. Without calculating the sum $\sum_{n=13}^{24} i_n$, use your knowledge of the sequence i_n to compare this sum to $\sum_{n=1}^{12} i_n$.

What does this mean in terms of the tax benefit in the second year? What will happen to the tax benefit over the life of the loan?

5. Suppose you add an extra $100 to each payment and pay $1360 each month. Complete the table for the sequences defined as follows. Round to the nearest cent, where necessary.

$b_0 = 240{,}000$ $\qquad i_n = 0.004b_{n-1}$ $\qquad p_n = 1360 - i_n$ $\qquad b_n = b_{n-1} - p_n$

Month, n	Interest, i_n	Principal, p_n	Balance, b_n
1			
2			
3			

Compare the balance at the end of three months in this table with the balance at the end of three months in the previous table. Is the difference exactly $300? If not, explain the difference. Discuss possible benefits of paying extra on your mortgage payment each month.

Mortgages may have other tax benefits in addition to those mentioned in this activity. Be sure to consult tax, mortgage, and real estate professionals when you are ready to buy a home.

Classroom Activity: Investigating Binomial Coefficients

Complete this activity before Section 10.5 is covered.

Group Size: 1–2
Material: Paper, Pencil, Calculator
Time: 15 minutes

In the expression $(x+y)^3$, the exponent represents repeated multiplication, so the expression may be written as

$$(x+y)(x+y)(x+y).$$

One way to perform the multiplication is to multiply $(x+y)(x+y)$, simplify the result, and then multiply by $(x+y)$ again.

1. Use the steps described above to perform the multiplication: $(x+y)(x+y)(x+y)$.

While the steps above worked well for $(x+y)^3=(x+y)(x+y)(x+y)$ and can be extended to $(x+y)^4$, $(x+y)^5$, $(x+y)^6$, and so on, performing the multiplication in this manner is inefficient for these bigger exponents. We can, however, look for patterns that make the multiplications easier.

Another way to think about the multiplication is to consider the sum of all possible products where each product is formed by taking one term from each factor in parentheses. The expansion below should help to clarify what we mean by this. The numbers above the variables in $\overset{1}{x}\cdot\overset{2}{y}\cdot\overset{3}{x}$ mean that the first x comes from the first factor of $(x+y)$, the y comes from the second factor of $(x+y)$, and the second x comes from the third factor of $(x+y)$.

$$(x+y)^3=\underbrace{(x+y)}^{1^{\text{st}}\text{ factor}}\underbrace{(x+y)}^{2^{\text{nd}}\text{ factor}}\underbrace{(x+y)}^{3^{\text{rd}}\text{ factor}}$$

$$=\overset{1}{x}\cdot\overset{2}{x}\cdot\overset{3}{x}+\overset{1}{x}\cdot\overset{2}{x}\cdot\overset{3}{y}+\overset{1}{x}\cdot\overset{2}{y}\cdot\overset{3}{x}+\overset{1}{x}\cdot\overset{2}{y}\cdot\overset{3}{y}+\overset{1}{y}\cdot\overset{2}{x}\cdot\overset{3}{x}+\overset{1}{y}\cdot\overset{2}{x}\cdot\overset{3}{y}+\overset{1}{y}\cdot\overset{2}{y}\cdot\overset{3}{x}+\overset{1}{y}\cdot\overset{2}{y}\cdot\overset{3}{y}$$

2. Simplify the terms in the last line above, but for now do not combine line terms. Then fill in the table below by counting the number of times each of the distinct terms occurs.

Term	x^3	x^2y	xy^2	y^3
Number of Occurrences				

The numbers you wrote in the table above are called *binomial coefficients*. The binomial coefficients count the number of ways the distinct terms occur when expanding a binomial $(x+y)$ raised to a power.

Thinking about what the binomial coefficients mean, there are 3 ways that x^2y occurs in the expansion of $(x+y)^3$: $x \cdot x \cdot y$, $x \cdot y \cdot x$, and $y \cdot x \cdot x$. You may view this as the y in x^2y could be chosen from the third factor of $(x+y)$ or the y could be chosen from the second factor of $(x+y)$ or the y could be chosen from the first factor of $(x+y)$, so there are 3 different ways to choose the y in x^2y. Thus, there are three occurrences of x^2y.

3. As a group, discuss why there are no terms of the form x^4, x^2y^2, or xy in the expansion of $(x+y)^3$. Briefly summarize your discussion.

4. Using the method of forming all possible products by choosing either the x or the y from each factor, expand $(x+y)^4 = (x+y)(x+y)(x+y)(x+y)$. Note that you will have 16 terms before simplifying.

5. Complete the following table by listing the distinct terms from #4 in the first row and the number of times each term occurs in the second row. List the terms in order of descending powers of x.

Term					
Number of Occurrences					

Recall that the numbers in the second row of the table are binomial coefficients. Here is a similar table for $(x+y)^5$:

Term	x^5	x^4y	x^3y^2	x^2y^3	xy^4	y^5
Number of Occurrences	1	5	10	10	5	1

6. Compare the three tables in this activity. As a group, discuss patterns that you notice in both the distinct terms and the binomial coefficients. Then try to construct a similar table for $(x+y)^6$, by extending the patterns that you've observed. How far can you get? Do you have empty cells?

You will learn a formula for the binomial coefficients. However, an understanding of the patterns they follow will reduce the number of calculations you need to perform in order to expand an expression of the form $(x+y)^n$.

Classroom Activity: Exploring Limits of a Rational Function

Complete this activity before Section 11.2 is covered.

Group Size: 2–3
Material: Paper, Pencil, Calculator
Time: 15 minutes

In this activity, you will explore limits associated with the rational function $f(x)=\dfrac{x-2}{x^2+x-6}$. Refer to Section 2.6, as necessary, to review the properties of rational functions.

1. Find the domain of the rational function $f(x)=\dfrac{x-2}{x^2+x-6}$. Determine whether the graph of f has a vertical asymptote or a hole at each value excluded from its domain.

2. Does the graph of $f(x)=\dfrac{x-2}{x^2+x-6}$ have a horizontal asymptote? If not, does it have an oblique asymptote? If it has either of these types of asymptote, find an equation for the asymptote.

3. Use the domain and any asymptotes or holes that you identified in #1 and #2 to sketch a graph of the rational function $f(x)=\dfrac{x-2}{x^2+x-6}$. Plot additional points as necessary. Are there any x-intercepts? Is there a y-intercept?

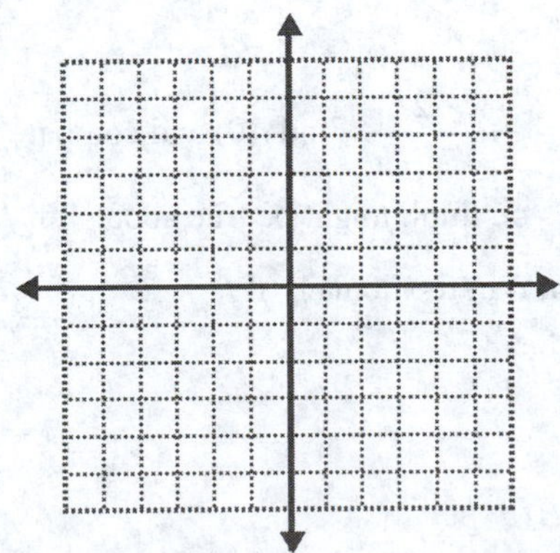

You will use your graph and additional numerical data to explore limits of the function f.

4. Use a calculator to fill in the following tables for the function $f(x)=\dfrac{x-2}{x^2+x-6}$. Round to three decimal places, where necessary.

X	−4	−3.5	−3.1	−3.01	−2.99	−2.9	−2.5	−2
$f(x)$								

X	1	1.5	1.9	1.99	2.01	40	50	75
$f(x)$								

X	−1	−0.5	−0.1	−0.01	0.01	0.1	0.5	1
$f(x)$								

5. Using your graph in #3 and your completed tables in #4, find the indicated limit or state that the limit does not exist.

a. $\lim_{x\to -3^-} f(x)$ **b.** $\lim_{x\to -3^+} f(x)$ **c.** $\lim_{x\to -3} f(x)$

c. $\lim_{x\to 2^-} f(x)$ **d.** $\lim_{x\to 2^+} f(x)$ **e.** $\lim_{x\to 2} f(x)$

f. $\lim_{x\to 0^-} f(x)$ **g.** $\lim_{x\to 0^+} f(x)$ **h.** $\lim_{x\to 0} f(x)$

6. Factor the denominator of $\dfrac{x-2}{x^2+x-6}$ and simplify the rational expression. Substitute 2 into the simplified expression. Compare the resulting value to your answer for #5e. What do you notice? Write a few sentences explaining the relationship that you think exists between holes in the graph and limits of a rational function.

7. Notice that 0 is in the domain of $f(x)=\dfrac{x-2}{x^2+x-6}$. Find $f(0)$ and compare the resulting value to your answer for #5h. What do you notice? What do you think might be true about $\lim_{x\to a} f(x)$ when a is in the domain of f? Test your conjecture for a few values of a in the domain of f.

Integrated Review Worksheets to accompany

PRECALCULUS 6E

R.1 Review of Algebraic Expressions and Real Numbers 1

Objective 1 – Recognize the sets that make up the set of real numbers and use set notation. 1

Objective 2 – Use the symbols for "is an element of" and "is not an element of." 3

Objective 3 – Use inequality symbols. 4

Objective 4 – Find a number's absolute value 5

Objective 5 – Use the order of operations. 6

Objective 6 – Evaluate algebraic expressions. 7

Objective 7 – Simplify algebraic expressions. 8

Objective 8 – Use commutative, associative, and distributive properties. 9

R.1 Answers 11

R.2 Review of Integer Exponents 12

Objective 1 – Use the product rule.. 12

Objective 2 – Use the quotient rule. 14

Objective 3 – Use the zero-exponent rule. 15

Objective 4 – Use the negative-exponent rule. 16

Objective 5 – Use the power rule. 18

Objective 6 – Find the power of a product. 19

Objective 7 – Find the power of a quotient. 20

Objective 8 – Evaluate exponential expressions. 22

Objective 9 – Simplify exponential expressions. 23

R.2 Answers 25

R.3 Review Radicals and Rational Exponents 26

Objective 1 – Evaluate square roots. 26

Objective 2 – Find even and odd roots. 28

Objective 3 – Rewrite expressions with rational exponents using radical notation. 30

Objective 4 – Simplify radical expressions using rational exponents. 32

Objective 5 – Use factoring and the product rule to simplify radicals. 33

Objective 6 – Rationalize denominators containing one term. 34

R.3 Answers 35

R.4 Review of Polynomials 36

Objective 1 – Use the vocabulary of polynomials. 36

Objective 2 – Add polynomials. 38

Objective 3 – Subtract polynomials. 39

Objective 4 – Multiply a monomial and a polynomial. 40

Objective 5 – Multiply polynomials when neither is a monomial. 41

Objective 6 – Use FOIL in polynomial multiplication. 42

Objective 7 – Square binomials. 44

Objective 8 – Multiply the sum and difference of two terms. 45

Objective 9 – Divide a polynomial by a monomial. 46

R.4 Answers 47

R.5 Review of Factoring 48
Objective 1 – Factor out the greatest common factor from a polynomial........ 48
Objective 2 – Factor by grouping........ 50
Objective 3 – Factor a trinomial whose leading coefficient is 1........ 51
Objective 4 – Factor a trinomial whose leading coefficient is not 1........ 53
Objective 5 – Factor difference of squares........ 55
Objective 6 – Factor polynomials completely........ 56
Objective 7 – Factor perfect square trinomials........ 57
Objective 8 – Factor the sum or difference of two cubes........ 59
Objective 9 – Complete the square of a binomial 61
R.5 Answers 63

R.6 Review of Rational Expressions 64
Objective 1 – Find the domain of a rational expression........ 64
Objective 2 – Simplify rational expressions........ 66
Objective 3 – Multiply rational expressions........ 67
Objective 4 – Divide rational expressions........ 69
Objective 5 – Find the least common denominator........ 71
Objective 6 – Add and subtract rational expressions with different denominators........ 73
Objective 7 – Simplify complex rational expressions by multiplying by 1........ 75
R.6 Answers 77

R.7 Review Solving Applications 78
Objective 1 – Translate English phrases into algebraic expressions........ 78
Objective 2 – Perimeter, Circumference, and Area........ 79
Objective 3 – Ratios and Proportions........ 81
Objective 4 – The Pythagorean Theorem and Its Converse........ 83
Objective 5 – Volumes of Prisms and Cylinders........ 85
Objective 6 – Solving applied problems using mathematical models........ 87
Objective 7 – Area of Parallelograms and Triangles........ 88
R.7 Answers 89

R.8 Review Solving Equations and Inequalities 90
Objective 1 – Solve linear equations........ 90
Objective 2 – Solve quadratic equations by factoring........ 92
Objective 3 – Solve quadratic equations using the quadratic formula........ 94
Objective 4 – Use interval notation........ 96
Objective 5 – Solve linear inequalities including inequalities with no solutions........ 98
R.8 Answers 100

R.1 Review of Algebraic Expressions and Real Numbers

Objective 1 – Recognize the sets that make up the set of real numbers and use set notation.
Objective 2 – Use the symbols for "is an element of" and "is not an element of."
Objective 3 – Use inequality symbols.
Objective 4 – Find a number's absolute value.
Objective 5 – Use the order of operations.
Objective 6 – Evaluate algebraic expressions.
Objective 7 – Simplify algebraic expressions.
Objective 8 – Use commutative, associative, and distributive properties.

Objective 1 – Recognize the sets that make up the set of real numbers and use set notation.

Key Terms
Set
Elements
Roster method
Set-builder notation
Natural numbers
Whole numbers
Integers
Rational numbers
Irrational numbers.

Summary
A **set** is a collection of objects whose contents can be clearly determined. The objects in a set are called the **elements** of the set. For example, the set of numbers used for counting can be represented by $\{1, 2, 3, 4, 5, \ldots\}$. The braces, $\{\ \}$, indicate that we are representing a set. This form of representation, called the **roster method**, uses commas to separate the elements of the set. The three dots after the 5, called an *ellipsis*, indicate that there is no final element and that the listing goes on forever. A set can also be written in **set-builder notation**. In this notation, the elements of the set are described, but not listed. Here is an example: $\{x \mid x \text{ is a natural number less than } 6\}$.

Three common sets of numbers are the *natural numbers*, the *whole numbers*, and the *integers*. **Natural numbers** are the numbers that we use for counting, $\{1, 2, 3, 4, 5, \ldots\}$. **Whole numbers** includes 0 and the natural numbers, $\{0, 1, 2, 3, 4, 5, \ldots\}$. **Integers** include the negatives of the natural numbers and the whole numbers, $\{\ldots, -5, -4, -3, -2, -1, 0, 1, 2, 3, 4, 5 \ldots\}$. Other common sets of numbers are the set of *rational numbers* and *irrational numbers*. The set of **rational numbers** is the set of all numbers that can be expressed as a quotient of two integers, with the denominator not 0. Three examples of rational numbers are $\frac{1}{4}, \frac{2}{3}$, and 5. The set of **irrational numbers** is the set of numbers whose decimal representations neither terminate nor repeat. Irrational numbers cannot be expressed as the quotient of integers. Examples of irrational numbers are $\sqrt{3}$ and π (pi).

Notes:

Guided Example	***Practice***
Use the roster method to list the elements in each set. 1. $\{x \mid x \text{ is a natural number less than } 6\}$ The set of natural numbers is $\{1, 2, 3, 4, 5, \ldots\}$. The set of natural numbers less than 6 is $\{1, 2, 3, 4, 5\}$. 2. $\{x \mid x \text{ is an odd whole number less than } 11\}$ The set of whole numbers is $\{0, 1, 2, 3, 4, 5, \ldots\}$. The set of odd whole number is $\{1, 3, 5, \ldots\}$, so the set of odd whole numbers less than 11 is $\{1, 3, 5, 7, 9\}$.	Use the roster method to list the elements in each set. 1. $\{x \mid x \text{ is an integer between } -8 \text{ and } -3\}$ 2. $\{x \mid x \text{ is a natural number greater than } 4\}$

Objective 2 – Use the symbols for "is an element of" and "is not an element of."

Summary

The symbol $\in$ is used to indicate that a number or object is in a particular set. The symbol $\in$ is read "is an element of." The symbol $\notin$ is used to indicate that a number or object is not an element in a particular set. The symbol $\notin$ is read "is not an element of."

Notes:

Guided Example	***Practice***
Use the meaning of the symbols $\in$ and $\notin$ to determine whether each statement is true of false. 1. $7 \in \{x \mid x \text{ is an integer}\}$ The set of integers is $\{\ldots, -5, -4, -3, -2, -1, 0, 1, 2, 3, 4, 5\ldots\}$, so 7 is an integer. The statement is true. 2. $\frac{1}{2} \notin \{x \mid x \text{ is a rational number}\}$ Rational numbers are numbers that can be expressed as the quotient of two integers, so $\frac{1}{2}$ is a rational number. The statement is false.	Use the meaning of the symbols $\in$ and $\notin$ to determine whether each statement is true of false. 1. $-6 \in \{x \mid x \text{ is an irrational number}\}$ 2. $\sqrt{2} \notin \{x \mid x \text{ is an integer}\}$

Objective 3 – Use inequality symbols.

Summary

Inequality symbols, such as < or >, are used to order real numbers. These symbols always point to the lesser of the two real numbers when the inequality statement is true. For example, $-4 < -1$ is read "-4 is less than -1. The symbol points to -4, the lesser number. $-1 > -4$ is read "-1 is greater than -4." The symbol still point to -4, the lesser number.

The symbols < and > may be combined with an equal sign, as shown in the following table:

Symbols	Meaning	Examples	Explanation
$a \le b$	a is less than or equal to b.	$2 \le 9$ $9 \le 9$	Because $2 < 9$ Because $9 = 9$
$b \ge a$	b is greater than or equal to a.	$9 \ge 2$ $2 \ge 2$	Because $9 > 2$ Because $2 = 2$

This inequality is true if either the < part or the = part is true.

This inequality is true if either the > part or the = part is true.

Notes:

Guided Example

Write out the meaning of each inequality and then determine whether the inequality is true or false.

1. $-7 < -3$

This says that -7 is less than -3. Since -7 is to the left of -3 on the number line, this is a true statement.

2. $-6 \ge -2$

This says that -6 is greater than or equal to -2. Since -6 is to the left of -2 on the number line, it is less than -2. This statement is false.

2. $5 \le 5$

This says that 5 is less than or equal to 5. Since 5 is equal to 5, this is a true statement.

Practice

Write out the meaning of each inequality and then determine whether the inequality is true or false.

1. $-8 \le -9$

2. $0 > -1$

3. $3 < \frac{1}{3}$

Objective 4 – Find a number's absolute value.

Summary

The absolute value of a real number a, denoted by $|a|$, is the distance from 0 to a on the number line. This distance is always taken to be nonnegative. Zero is the only number whose absolute value is 0. The absolute value of a real number is never negative.

Notes:

Guided Example	***Practice***
Find the absolute value. 1. $\|-4\|$ The absolute value of -4 is 4 because -4 is 4 units from 0. 2. $\|3.5\|$ The absolute value of 3.5 is 3.5 because 3.5 is 3.5 units from 0. 3. $-\|-5\|$ The absolute value of -5 is 5. Substitute 5 for the absolute value symbol and simplify. $-\|-5\| = -(5) = -5$	Find the absolute value. 1. $\left\|\frac{1}{2}\right\|$ 2. $\|-6.25\|$ 3. $-\|-\sqrt{7}\|$

Objective 5 – Use the order of operations.

Summary
The rules for order of operations can be applied to positive and negative real numbers. Recall the order of operations agreement.

- Perform operations within the innermost parentheses and work out.
- Evaluate exponential expressions.
- Multiply and divide, from left to right.
- Add and subtract, from left to right.

If an expression contains grouping symbols, we perform operations within these symbols first. Common grouping symbols are parentheses, brackets, and braces. Other grouping symbols include fraction bars, absolute value symbols, and radical symbols such as square root signs $\left(\sqrt{\ }\right)$.

Notes:

Guided Example	***Practice***
Simplify. 1. $4-7^2+8\div2(-3)^2$ $4-7^2+8\div2(-3)^2=4-49+8\div2(9)$ Evaluate exponents. $=4-49+4(9)$ Divide: $8\div2=4$. $=4-49+36$ Multiply: $4(9)=36$. $=-45+36$ Subtract: $4-49=-45$. $=-9$ Add. 2. $\dfrac{13-3(-2)^4}{3-(6-10)}$ $\dfrac{13-3(-2)^4}{3-(6-10)}=\dfrac{13-3(16)}{3-(-4)}$ Evaluate exponent. Subtract inside ( ). $=\dfrac{13-48}{3-(-4)}$ Multiply in numerator. $=\dfrac{-35}{7}$ Subtract in numerator and in denominator. $=-5$ Divide.	Simplify. 1. $10^2-100\div5^2\cdot2-3$ 2. $\dfrac{10\div2+3\cdot4}{(12-3\cdot2)^2}$

Objective 6 – Evaluate algebraic expressions.

Key Terms:
Variable
Algebraic expression
Evaluating

Summary
Algebra uses letters, such as x and y, to represent numbers. If a letter is used to represent various numbers, it is called a **variable**. A combination of variables and numbers using the operations of addition, subtraction, multiplication, or division, as well as powers or roots, is called an **algebraic expression**. Here are some examples of algebraic expressions:

$$x+6,\ x-6,\ 6x,\ \frac{x}{6},\ 3x+5,\ x^2-3,\ \sqrt{x}+7$$

Evaluating an algebraic expression means to find the value of the expression for a given value of the variable. Many algebraic expressions involve more than one operation. Evaluating an algebraic expression without a calculator involves carefully applying the order of operations agreement

Notes:

Guided Example	***Practice***
1. Evaluate $7+5(x-4)^3$ for $x=6$. $7+5(x-4)^3 = 7+5(6-4)^3$ Replace x with 6. $=7+5(2)^3$ Work inside parentheses: $6-4=2$. $=7+5(8)$ Evaluate the exponent: $2^3=8$. $=7+40$ Multiply: $5(8)=40$. $=47$ Add.	1. Evaluate $8+6(x-3)^2$ for $x=13$.
2. Evaluate $x^2-3(x-y)$ for $x=8$ and $y=2$. $x^2-3(x-y)=8^2-3(8-2)$ Replace x with 8 and y with 2. $=8^2-3(6)$ Work inside parentheses: $8-2=6$. $=64-3(6)$ Evaluate the exponent: $8^2=64$. $=64-18$ Multiply: $3(6)=18$. $=46$ Subtract.	2. Evaluate $x^2-4(y-x)$ for $x=6$ and $y=3$.

Objective 7 – Simplify algebraic expressions.

Key Terms:

Terms
Coefficient
Like terms
Simplified

Summary

The **terms** of an algebraic expression are those parts that are separated by addition. The numerical part of a term is called its **coefficient**. **Like terms** are terms that have exactly the same variable factors. To combine like terms, mentally add or subtract the coefficients of the terms. Use this result as the coefficient of the terms' variable factors. An algebraic expression is **simplified** when grouping symbols have been removed and like terms have been combined.

Notes:

Guided Example	***Practice***
1. Simplify. 1. $7x+12x^2+3x+x^2$ $7x+12x^2+3x+x^2=(12x^2+x^2)+(7x+3x)$ Group like terms. $=(12+1)x^2+(7+3)x$ Add coefficients. $=13x^2+10x$ Simplify. 2. $4(7x-3)-10x$ $4(7x-3)-10x=4\cdot 7x-4\cdot 3-10x$ Remove parentheses. $=28x-12-10x$ Multiply. $=18x-12$ Combine like terms.	Simplify. 1. $9x+5x^2-3x+4x^2$ 2. $5(3y-2)-(7y+2)$

Objective 8 – Use commutative, associative, and distributive properties.

Key Terms
Equivalent algebraic expressions

Summary
Below are some basic algebraic properties which enable us to write equivalent algebraic expressions. Two algebraic expressions that have the same value for all replacements are called **equivalent algebraic expressions**.

The Commutative Properties
Let a and b represent real numbers, variables, or algebraic expressions.

Addition: $a+b=b+a$

Multiplication: $ab=ba$

Changing order when adding or multiplying does not affect a sum or product.

The Associative Properties.
Let a and b represent real numbers, variables, or algebraic expressions.

Addition: $(a+b)+c=a+(b+c)$

Multiplication: $(ab)c=a(bc)$

Changing grouping when adding or subtracting does not affect a sum or product.

The Distributive Property
Let a and b represent real numbers, variables, or algebraic expressions.

$$a(b+c)=ab+ac$$

Multiplication distributes over addition.

Notes:

Guided Example	***Practice***
1. Write an equivalent algebraic expression using one of the commutative properties. $7x+5$ The commutative property of addition allows for reordering of the terms. $7x+5=5+7x$	1. Write an equivalent algebraic expression using one of the commutative properties. $-3(8)$
2. Write an equivalent algebraic expression using one of the associative properties. $-10(5x)$ The associative property of multiplication allows for regrouping of the terms. $-10(5x)=(-10\cdot 5)x$	2. Write an equivalent algebraic expression using one of the associative properties. $12+(3+x)$
3. Write an equivalent algebraic expression using the distributive property. $-3(2x+5)$ The distributive property eliminates the parentheses. Multiply -3 by both terms inside the parentheses. $-3(2x+5)=(-3)(2x)+(-3)(5)=-6x-15$	3. Write an equivalent algebraic expression using the distributive property. $-(x-3)$

R.1 Answers

R.1.1

1) $\{-7,-6,-5,-4\}$ 2) $\{5,6,7,\ldots\}$

R.1.2

1) False 2) True

R.1.3

1) -8 is less than or equal to -9 .; False 2) 0 is greater than -1 .; True 3) 3 is less than $\frac{1}{3}$.; False

R.1.4

1) $\frac{1}{2}$ 2) 6.25 3) $-\sqrt{7}$

R.1.5

1) 89 2) $\frac{17}{36}$

R.1.6

1) 608 2) 48

R.1.7

1) $9x^2+6x$ 2) $8y-12$

R.1.8

1) $8(-3)$ 2) $(12+3)+x$ 3) $-x+3$

R.2 Review of Integer Exponents

Learning Objectives
Objective 1 – Use the product rule.
Objective 2 – Use the quotient rule.
Objective 3 – Use the zero-exponent rule.
Objective 4 – Use the negative-exponent rule.
Objective 5 – Use the power rule.
Objective 6 – Find the power of a product.
Objective 7 – Find the power of a quotient.
Objective 8 – Evaluate exponential expressions.
Objective 9 – Simplify exponential expressions.

Objective 1 – Use the product rule.

Key Terms
Exponential expression
Base
Exponent
Product rule

Summary
The expression b^n is called an **exponential expression.** b^n is read "the n th power of b," or "b to the n th power." The n th power of b is defined as the product of n factors of b. In the expression b^n, b is called the **base** and n is the **exponent**. When multiplying exponential expressions with the same base, add the exponents. This is the **product rule.**

The Product Rule

$$b^m \cdot b^n = b^{m+n}$$

When multiplying exponential expressions with the same base, add the exponents. Use this sum as the exponent of the common base.

Notes:

Guided Example	***Practice***
Multiply each expression using the product rule. 1. $x^8 \cdot x^{10}$ $x^8 \cdot x^{10} = x^{8+10} = x^{18}$ 2. $(6a^4b^3)(5a^2b^7)$ $(6a^4b^3)(5a^2b^7) = 6 \cdot 5 \cdot a^4 \cdot a^2 \cdot b^3 \cdot b^7$ Reorder. $= 30a^{4+2}b^{3+7}$ Product rule $= 30a^6b^{10}$ Simplify.	Multiply each expression using the product rule. 1. $r^6 \cdot r^5$ 2. $(4u^3v^4)(10u^2v^6)$

Objective 2 – Use the quotient rule.

Key Terms
Quotient rule

Summary
When dividing exponential expressions with the same nonzero base, subtract the exponent in the denominator from the exponent in the numerator. Use this difference as the exponent of the common base. This is the **quotient rule**.

The Quotient Rule

$$\frac{b^m}{b^n} = b^{m-n}, \; b \neq 0$$

Notes:

Guided Example

Divide each expression using the quotient rule.

1. $\dfrac{(-2)^7}{(-2)^4}$

$\dfrac{(-2)^7}{(-2)^4} = (-2)^{7-4}$ Quotient rule

$= (-2)^3$ Subtract exponents.

$= -8$ Simplify.

2. $\dfrac{30m^{12}n^9}{5m^3n^7}$

$\dfrac{30m^{12}n^9}{5m^3n^7} = \dfrac{30}{5} \cdot \dfrac{m^{12}}{m^3} \cdot \dfrac{n^9}{n^7}$ Divide.

$= 6m^{12-3}n^{9-7}$ Quotient rule

$= 6m^9n^2$ Subtract exponents.

Practice

Divide each expression using the quotient rule.

1. $\dfrac{(-3)^6}{(-3)^3}$

2. $\dfrac{27x^{14}y^8}{3x^3y^5}$

Objective 3 – Use the zero-exponent rule.

Key Terms
Zero-exponent rule

Summary

The Zero-Exponent Rule

If b is any real number other than 0,

$$b^0 = 1.$$

Notes:

Guided Example	***Practice***
Use the zero-exponent rule to simplify each expression.	Use the zero-exponent rule to simplify each expression.
1. 8^0 $8^0 = 1$ Zero-exponent rule	1. 7^0
2. $(-6)^0$ $(-6)^0 = 1$ Zero-exponent rule	2. $(-5)^0$
3. -6^0 $-6^0 = -(6^0) = -1(6^0) = -1 \cdot 1 = -1$	3. -5^0
4. $(5x)^0$ $(5x)^0 = 1$	4. $10x^0$

Objective 4 – Use the negative-exponent rule.

Key Terms
Negative-exponent rule

Summary

The Negative-Exponent Rule

If b is any real number other than 0, then

$$b^{-n} = \frac{1}{b^n}.$$

Notes:

Guided Example

Use the negative-exponent rule to write each expression with a positive exponent. Simplify, if possible.

1. 9^{-2}

$$9^{-2}=\frac{1}{9^2}=\frac{1}{81}$$

2. $(-2)^{-5}$

$$(-2)^{-5}=\frac{1}{(-2)^5}=-\frac{1}{32}$$

3. $\frac{1}{6^{-2}}$

$$\frac{1}{6^{-2}}=\frac{1}{\frac{1}{6^2}}=1\cdot\frac{6^2}{1}=6^2=36$$

4. $7x^{-5}y^2$

$$7x^{-5}y^2=7\cdot\frac{1}{x^5}\cdot y^2=\frac{7y^2}{x^5}$$

Practice

Use the negative-exponent rule to write each expression with a positive exponent. Simplify, if possible

1. 5^{-2}

2. $(-3)^{-3}$

3. $\frac{1}{4^{-2}}$

4. $3x^{-6}y^4$

Objective 5 – Use the power rule.

Key Terms
Power rule

Summary
The **power rule** applies when an exponential expression is raised to a power.

The Power Rule (Powers to Powers)

$$\left(b^m\right)^n = b^{mn}$$

When an exponential expression is raised to a power, multiply the exponents. Place the product of the exponents on the base and remove the parentheses.

Notes:

Guided Example	***Practice***
Simplify using the power rule.	Simplify using the power rule.
1. $\left(x^6\right)^4$	1. $\left(5^3\right)^{-1}$
$\left(x^6\right)^4 = x^{6\cdot4} = x^{24}$	
2. $\left(y^5\right)^{-3}$	2. $\left(q^2\right)^4$
$\left(y^5\right)^{-3} = y^{5\cdot(-3)} = y^{-15} = \dfrac{1}{y^{15}}$	
3. $\left(b^{-4}\right)^{-2}$	3. $\left(6^{-1}\right)^{-3}$
$\left(b^{-4}\right)^{-2} = b^{(-4)(-2)} = b^8$	

Objective 6 – Find the power of a product.

Summary
When a product is raised to a power, raise each factor to that power.

The Products-to-Powers Rule

$$(ab)^n = a^n b^n.$$

Notes:

Guided Example	***Practice***
Simplify each expression using the products-to powers rule.	Simplify each expression using the products-to powers rule.
1. $(6x)^3$ $(6x)^3 = 6^3 \cdot x^3 = 216x^3$	1. $(2x)^4$
2. $(-2y^2)^4$ $(-2y^2)^4 = (-2)^4 \cdot (y^2)^4 = 16y^8$	2. $(-3y^2)^3$
3. $(-3x^{-1}y^3)^{-2}$ Raise each factor to the -2 power. $(-3x^{-1}y^3)^{-2} = (-3)^{-2}(x^{-1})^{-2}(y^3)^{-2}$ $= (-3)^{-2} x^{(-1)(-2)} y^{(3)(-2)}$ $(b^m)^n = b^{mn}$ $= (-3)^{-2} x^2 y^{-6}$ Simplify. $= \frac{1}{(-3)^2} \cdot x^2 \cdot \frac{1}{y^6}$ $b^{-n} = \frac{1}{b^n}$ $= \frac{x^2}{9y^6}$ Multiply.	3. $(-4x^5y^{-1})^{-2}$

Objective 7 Find the power of a quotient.

Summary
When a quotient is raised to a power, raise the numerator to that power and divide by the denominator to that power.

The Quotients-to-Powers Rule

$$\left(\frac{a}{b}\right)^n = \frac{a^n}{b^n}$$

Notes:

Guided Example

Simplify each expression using the quotients-to-powers rule.

1. $\left(\frac{x^2}{4}\right)^3$

$$\left(\frac{x^2}{4}\right)^3 = \frac{\left(x^2\right)^3}{4^3} = \frac{x^6}{64}$$

2. $\left(\frac{2x^3}{y^{-4}}\right)^5$

$$\left(\frac{2x^3}{y^{-4}}\right)^5 = \frac{\left(2x^3\right)^5}{\left(y^{-4}\right)^5} \quad \text{Quotient-to-power}$$

$$= \frac{2^5\left(x^3\right)^5}{\left(y^{-4}\right)^5} \quad \text{Product-to-power}$$

$$= \frac{32x^{15}}{y^{-20}} \quad \text{Power rule}$$

$$= 32x^{15}y^{20} \quad \text{Negative exponent rule}$$

3. $\left(\frac{x^3}{y^2}\right)^{-4}$

$$\left(\frac{x^3}{y^2}\right)^{-4} = \frac{\left(x^3\right)^{-4}}{\left(y^2\right)^{-4}} \quad \text{Quotient-to-power}$$

$$= \frac{x^{(3)(-4)}}{y^{(2)(-4)}} \quad \text{Power rule}$$

$$= \frac{x^{-12}}{y^{-8}} \quad \text{Simplify.}$$

$$= \frac{y^8}{x^{12}} \quad \text{Negative exponent rule}$$

Practice

Simplify each expression using the quotients-to-powers rule.

1. $\left(\frac{x^5}{4}\right)^3$

2. $\left(\frac{2x^{-3}}{y^2}\right)^4$

3. $\left(\frac{x^{-3}}{y^4}\right)^{-5}$

Objective 8 – Evaluate exponential expressions.

Summary

Because exponents indicate repeated multiplication, rules for multiplying real numbers can be used to evaluate exponential expressions. We will review these here. The product of two real numbers with different signs is found by multiplying their absolute values. The product is negative. The product of two real numbers with the same sign is found by multiplying their absolute values. The product is positive. If no number is 0, a product with an odd number of negative factors is found by multiplying their absolute values. The product is negative. If no number is 0, a product with an even number of negative factors is found by multiplying absolute values. The product is positive.

Notes:

Guided Example	***Practice***
Simplify.	Simplify.
1. $(-6)^2$ $(-6)^2 = (-6)(-6)$ Same signs $= 36$ Positive product	1. $(-5)^2$
2. -6^2 $-6^2 = -(6 \cdot 6)$ The base is 6, not -6. $= -36$ Apply negative to positive product.	2. -5^2
3. $(-5)^3$ $(-5)^3 = (-5)(-5)(-5)$ Odd number of negatives $= -(5 \cdot 5 \cdot 5)$ Negative product $= -125$ Simplify.	3. $(-4)^3$
4. $\left(-\frac{2}{3}\right)^4$ $\left(-\frac{2}{3}\right)^4 = \left(-\frac{2}{3}\right)\left(-\frac{2}{3}\right)\left(-\frac{2}{3}\right)\left(-\frac{2}{3}\right) = \frac{16}{81}$	4. $\left(-\frac{3}{5}\right)^4$

Objective 9 – Simplify exponential expressions.

Key Terms
Simplified

Summary
Properties of exponents are used to simplify exponential expressions. An exponential expression is **simplified** when

- No parentheses appear.
- No powers are raised to powers.
- Each base occurs only once.
- No negative or zero exponents appear.

Notes:

Guided Example	***Practice***
Evaluate. 1. $\left(-2xy^{-14}\right)\left(-3x^4y^5\right)^3$ Cube factors in the second parentheses. $\left(-2xy^{-14}\right)\left(-3x^4y^5\right)^3=\left(-2xy^{-14}\right)(-3)^3\left(x^4\right)^3\left(y^5\right)^3$ Use power-to-power rule. $=\left(-2xy^{-14}\right)(-27)x^{12}y^{15}$ Use the product rule. $=(-2)(-27)x^{1+12}y^{-14+15}$ $=54x^{13}y$ Simplify. 2. $\left(\dfrac{x^{-4}y^7}{2}\right)^{-5}$ $\left(\dfrac{x^{-4}y^7}{2}\right)^{-5}=\dfrac{\left(x^{-4}y^7\right)^{-5}}{2^{-5}}$ Quotient-to-power $=\dfrac{\left(x^{-4}\right)^{-5}\left(y^7\right)^{-5}}{2^{-5}}$ Product-to-power $=\dfrac{x^{20}y^{-35}}{2^{-5}}$ Power rule $=\dfrac{2^5x^{20}}{y^{35}}$ Negative-exponent rule $=\dfrac{32x^{20}}{y^{35}}$ Simplify.	Evaluate. 1. $\left(-3x^{-6}y\right)\left(-2x^3y^4\right)^2$ 2. $\left(\dfrac{x^3y^5}{4}\right)^{-3}$

R.2 Answers

R.2.1
1) r^{11} 2) $40u^5v^{10}$

R.2.2
1) -27 2) $9x^{11}y^3$

R.2.3
1) 1 2) 1 3) -1 4) 10

R.2.4
1) $\frac{1}{25}$ 2) $-\frac{1}{27}$ 3) 16 4) $\frac{3y^4}{x^6}$

R.2.5
1) $\frac{1}{125}$ 2) q^8 3) 216

R.2.6
1) $16x^4$ 2) $-27y^6$ 3) $\frac{y^2}{16x^{10}}$

R.2.7
1) $\frac{x^{15}}{64}$ 2) $\frac{16}{x^{12}y^8}$ 3) $x^{15}y^{20}$

R.2.8
1) 25 2) -25 3) -64 4) $\frac{81}{625}$

R.2.9
1) $-12y^9$ 2) $\frac{64}{x^9y^{15}}$

R.3 Review Radicals and Rational Exponents

Objective 1 – Evaluate square roots.
Objective 2 – Find even and odd roots.
Objective 3 – Rewrite expressions with rational exponents using radical notation.
Objective 4 – Simplify radical expressions using rational exponents.
Objective 5 – Use factoring and the product rule to simplify radicals
Objective 6 – Rationalize denominators containing one term.

Objective 1 – Evaluate square roots.

Key Terms
Radical sign
Principal square root
Radicand
Radical expression

Summary
In general, if $b^2 = a$, then b is the square root of a. The symbol $\sqrt{\ }$, called a **radical sign**, is used to denote the positive or **principal square root** of a number. The number under the radical sign is called the **radicand**. Together we refer to the radical sign and its radicand as a **radical expression**.

Definition of the Principal Square Root

If a is a nonnegative real number, the nonnegative number b such that $b^2 = a$, denoted by $b = \sqrt{a}$, is the **principal square root** of a.

Simplifying $\sqrt{a^2}$

For any real number a,

$$\sqrt{a^2} = |a|$$

In words, the principal square root of a^2 is the absolute value of a.

Notes:

Guided Example	***Practice***
Evaluate. 1. $\sqrt{81}$ $\sqrt{81}=9$ The principal square root of 81 is 9 because $9^2=81$. 2. $-\sqrt{9}$ $-\sqrt{9}=-3$ The negative square root of 9 is -3 because $(-3)^2=9$. 3. $\sqrt{\frac{4}{49}}$ $\sqrt{\frac{4}{49}}=\frac{2}{7}$ The principal square root of $\frac{4}{49}$ is $\frac{2}{7}$ because $\left(\frac{2}{7}\right)^2=\frac{4}{49}$. 4. $\sqrt{0.0064}$ $\sqrt{0.0064}=0.08$ The principal square root of $\sqrt{0.0064}$ is 0.08 because $(0.08)^2=0.0064$.	Evaluate. 1. $\sqrt{64}$ 2. $-\sqrt{49}$ 3. $\sqrt{\frac{16}{25}}$ 4. $\sqrt{0.0081}$

Objective 2 – Find even and odd roots.

Key terms

*n*th root

Index

Summary

The radical expression $\sqrt[n]{a}$ represents the ***n*th root** of a. The number n is called the **index**. An index of 2 represents a square root and is not written. An index of 3 represents a cube root. If the index n in $\sqrt[n]{a}$ is an odd number, a root is said to be an *odd root*. A cube root is an odd root. Other odd roots have the same characteristics as cube roots.

- Every real number has exactly one real root when n is odd. An odd root of a positive number is positive and an odd root of a negative number is negative.

If the index n in $\sqrt[n]{a}$ is an even number, a root is said to be an *even root*. A square root is an even root. Other even roots have the same characteristics as square roots.

- Every positive real number has two real roots when n is even. One root is positive and one is negative.

Simplifying $\sqrt[n]{a^n}$

For any real number a,

1. If n is even, $\sqrt[n]{a^n} = |a|$.
2. If n is odd, $\sqrt[n]{a^n} = a$.

Notes:

Guided Example	***Practice***
Simplify. 1. $\sqrt{(-3)^2}$ $\sqrt{(-3)^2} = \lvert -3 \rvert = 3$ $\quad\quad \sqrt[n]{a^n} = \lvert a \rvert$ if n is even. 2. $\sqrt[4]{(x-3)^4}$ $\sqrt[4]{(x-3)^4} = \lvert x-3 \rvert$ $\quad\quad \sqrt[n]{a^n} = \lvert a \rvert$ if n is even. 3. $\sqrt[5]{(2x+7)^5}$ $\sqrt[5]{(2x+7)^5} = 2x+7$ $\quad\quad \sqrt[n]{a^n} = a$ if n is odd. 4. $\sqrt[3]{(-6)^3}$ $\sqrt[3]{(-6)^3} = -6$ $\quad\quad \sqrt[n]{a^n} = a$ if n is odd.	Simplify. 1. $\sqrt{(-5)^2}$ 2. $\sqrt[6]{(x+5)^6}$ 3. $\sqrt[3]{(3x-1)^3}$ 4. $\sqrt[7]{(-4)^7}$

Objective 3 – Rewrite expressions with rational exponents using radical notation.

Summary

If $\sqrt[n]{a}$ represents a real number and $n \geq 2$ is an integer, then $a^{\frac{1}{n}} = \sqrt[n]{a}$. In words, the denominator of the rational exponent is the index of the radical. If n is even, a must be nonnegative. If n is odd, a can be any real number. We can generalize the definition of $a^{\frac{1}{n}}$ to define $a^{\frac{m}{n}}$.

If $\sqrt[n]{a}$ represents a real number, $\frac{m}{n}$ is a rational number reduced to lowest terms, and $n \geq 2$ is an integer, then

$$a^{\frac{m}{n}} = \left(\sqrt[n]{a}\right)^m \quad \text{(First take the } n\text{th root of } a\text{.)}$$

and

$$a^{\frac{m}{n}} = \sqrt[n]{a^m} \quad \text{(First raise } a \text{ to the } m \text{ power.)}$$

Notice in both forms that the numerator in the rational exponent is the exponent and the denominator is the radical's index.

Notes:

Guided Example	***Practice***
Use radical notation to rewrite each expression. Simplify, if possible.	Use radical notation to rewrite each expression. Simplify, if possible.
1. $64^{\frac{1}{2}}$ $64^{\frac{1}{2}} = \sqrt{64} = 8$	1. $25^{\frac{1}{2}}$
2. $(-125)^{\frac{1}{3}}$ $(-125)^{\frac{1}{3}} = \sqrt[3]{-125} = -5$	2. $(-8)^{\frac{1}{3}}$
3. $1000^{\frac{2}{3}}$ $1000^{\frac{2}{3}} = \left(\sqrt[3]{1000}\right)^2 = 10^2 = 100$	3. $8^{\frac{4}{3}}$
4. $-16^{\frac{3}{2}}$ The base is 16 and the negative sign is not affected by the exponent. $-16^{\frac{3}{2}} = -\left(\sqrt{16}\right)^3 = -4^3 = -64$	4. $-81^{\frac{3}{4}}$

Objective 4 – Simplify radical expressions using rational exponents.

Summary

Some radical expressions can be simplified using rational exponents. We will use the following procedure:

Simplifying Radical Expressions Using Rational Exponents

1. Rewrite each radical expression as an exponential expression with a rational exponent.
2. Simplify using properties of rational exponents.
3. Rewrite in radical notation if rational exponents still appear.

Notes:

Guided Example	***Practice***
Use rational exponents to simplify. Assume all variables represent nonnegative numbers. 1. $\sqrt[10]{x^5}$ $\sqrt[10]{x^5} = x^{\frac{5}{10}}$ Rewrite as an exponential expression. $= x^{\frac{1}{2}}$ Reduce exponent. $= \sqrt{x}$ Rewrite in radical notation. 2. $\sqrt[3]{27a^{15}}$ $\sqrt[3]{27a^{15}} = \left(27a^{15}\right)^{\frac{1}{3}}$ Rewrite as an exponential expression. $= 27^{\frac{1}{3}}\left(a^{15}\right)^{\frac{1}{3}}$ Product-to-power rule $= \sqrt[3]{27}\left(a^{15}\right)^{\frac{1}{3}}$ Rewrite in radical notation. $= 3a^5$ Simplify radical and use power rule on exponents.	Use rational exponents to simplify. Assume all variables represent nonnegative numbers. 1. $\sqrt[4]{x^6y^2}$ 2. $\sqrt[3]{8r^{12}}$

Objective 5 – Use factoring and the product rule to simplify radicals.

Summary

If $\sqrt[n]{a}$ and $\sqrt[n]{b}$ are real numbers, then the *product rule for radicals* says that $\sqrt[n]{a}\cdot\sqrt[n]{b}=\sqrt[n]{ab}$. In words this says that the product of two n th roots is the n th root of the product of the radicands. Sometimes we can use the product rule for radicals in the form $\sqrt[n]{ab}=\sqrt[n]{a}\cdot\sqrt[n]{b}$ and factoring to simplify a radical expression.

Simplifying Radical Expressions by Factoring

A radical expression whose index is n is simplified when its radicand has no factors that are perfect n th powers. To simplify, use the following procedure:

1. Write the radicand as the product of two factors, one of which is the greatest perfect n th power.
2. Use the product rule to take the n th root of each factor.
3. Find the n th root of the perfect n th power.

Notes:

Guided Example	***Practice***
Simplify.	Simplify.
1. $\sqrt{75}$ $\sqrt{75}=\sqrt{25\cdot 3}$ 25 is the greatest perfect square factor of 75. $=\sqrt{25}\cdot\sqrt{3}$ Take the square root of each factor. $=5\sqrt{3}$ Simplify $\sqrt{25}$.	1. $\sqrt[3]{40}$
2. $\sqrt[3]{54xy^3}$ $\sqrt[3]{54xy^3}=\sqrt[3]{27y^3\cdot 2x}$ $27y^3$ is the greatest perfect cube factor. $=\sqrt[3]{27y^3}\cdot\sqrt[3]{2x}$ Factor into two radicals. $=3y\sqrt[3]{2x}$ Take the cube root of $27y^3$.	2. $\sqrt{50r^4s}$

Objective 6 – Rationalizing denominators containing one term.

Key Terms
Rationalizing the denominator

Summary
The process of rewriting a radical expression as an equivalent expression in which the denominator no longer contains any radicals is called **rationalizing the denominator**. When the denominator contains a single radical with an *n*th root, multiply the numerator and the denominator by a radical of index *n* that produces a perfect *n*th power in the denominator's radicand.

Notes:

Guided Example	***Practice***
Simplify. 1. $\frac{\sqrt{5}}{\sqrt{6}}$ $\frac{\sqrt{5}}{\sqrt{6}}=\frac{\sqrt{5}}{\sqrt{6}}\cdot\frac{\sqrt{6}}{\sqrt{6}}$ Multiply the numerator and denominator by $\sqrt{6}$. $=\frac{\sqrt{30}}{\sqrt{36}}$ Multiply. 36 is a perfect square. $=\frac{\sqrt{30}}{6}$ Simplify $\sqrt{36}$. 2. $\sqrt[3]{\frac{7}{25}}$ $\sqrt[3]{\frac{7}{25}}=\frac{\sqrt[3]{7}}{\sqrt[3]{25}}$ Use the quotient rule. $=\frac{\sqrt[3]{7}}{\sqrt[3]{5^2}}$ Write the denominator's radicand with an exponent. $=\frac{\sqrt[3]{7}}{\sqrt[3]{5^2}}\cdot\frac{\sqrt[3]{5}}{\sqrt[3]{5}}$ Multiply the numerator and denominator by $\sqrt[3]{5}$. $=\frac{\sqrt[3]{35}}{\sqrt[3]{5^3}}$ Multiply. 5^3 is a perfect cube. $=\frac{\sqrt[3]{35}}{5}$ Simplify $\sqrt[3]{5^3}=5$. .	Simplify. 1. $\frac{\sqrt{3}}{\sqrt{7}}$ 2. $\sqrt[3]{\frac{2}{9}}$

R.3 Answers

R.3.1

1) 8 2) −7 3) $\frac{4}{5}$ 4) 0.09

R.3.2

1) 5 2) $|x+5|$ 3) $3x-1$ 4) −4

R.3.3

1) 5 2) −2 3) 16 4) −27

R.3.4

1) $\sqrt{x^3y}$ 2) $2r^4$

R.3.5

1) $2\sqrt[3]{5}$ 2) $5r^2\sqrt{2s}$

R.3.6

1) $\frac{\sqrt{21}}{7}$ 2) $\frac{\sqrt[3]{6}}{3}$

R.4 Review of Polynomials.

Objective 1 – Use the vocabulary of polynomials.
Objective 2 – Add polynomials.
Objective 3 – Subtract polynomials.
Objective 4 – Multiply a monomial and a polynomial.
Objective 5 – Multiply polynomials when neither is a monomial.
Objective 6 – Use FOIL in polynomial multiplication.
Objective 7 – Square binomials.
Objective 8 – Multiply the sum and difference of two terms.
Objective 9 – Divide a polynomial by a monomial.

Objective 1 – Use the vocabulary of polynomials.

Key Terms
Polynomial
Degree of term and of polynomial
Monomial
Binomial
Trinomial
Coefficient
Leading term
Leading Coefficient

Summary
A **polynomial** is a single term or the sum of two or more terms containing variables with whole-number exponents. Some polynomials contain only one variable. Each term of such a polynomial in x is of the form ax^n. If $a \neq 0$, the **degree** of ax^n is n. A polynomial is simplified when it contains no grouping symbols and no like terms. A simplified polynomial that has exactly one term is called a **monomial**. A **binomial** is a simplified polynomial that has two terms. A **trinomial** is a simplified polynomial with three terms. Simplified polynomials with four or more terms have no special names.

A polynomial in two variables, *x* and *y*, contains the sum of one or more monomials of the form $ax^n y^m$. The constant a is the **coefficient**. The exponents, n and m, represent whole numbers. The **degree of the term** $ax^n y^m$ is the sum of the exponents of the variables, $n+m$. The **degree of a polynomial** is the greatest degree of any term of the polynomial. If there is precisely one term of the greatest degree, it is called the **leading term**. Its coefficient is called the **leading coefficient**.

Notes:

Guided Example	***Practice***
Determine the coefficient of each term, the degree of each term, the degree of the polynomial, the leading term, and the leading coefficient of the polynomial.	Determine the coefficient of each term, the degree of each term, the degree of the polynomial, the leading term, and the leading coefficient of the polynomial.
1. $7x^2y^3-17x^4y^2+xy-6y^2+9$	1. $8x^4y^5-7x^3y^2-x^2y-5x+11$

Term	**Coefficient**	**Degree (Sum exponents)**
$7x^2y^3$	7	$2+3=5$
$-17x^4y^2$	-17	$4+2=6$
xy	1	$1+1=2$
$-6y^2$	-6	2
9	9	0

The degree of the polynomial is the greatest degree of any term. The degree is 6. The leading term is the term with the greatest degree. The leading term is $-17x^4y^2$. Its coefficient -17 is the leading coefficient.

Guided Example	***Practice***
2. Label the following polynomials as either a monomial, binomial, or trinomial.	2. Label the following polynomials as either a monomial, binomial, or trinomial.
a) x^2-9	a) $15x^6$
x^2-9 has two terms. It is a binomial.	b) y^3-3y^2+2y
b) $-3m^3n^2$	c) x^3-27
$-3m^3n^2$ has one term. It is a monomial.	
c) $3a^2+2ab-5b^2$	
$3a^2+2ab-5b^2$ has three terms. It is a trinomial.	

Objective 2 – Add polynomials.

Summary
Polynomials are added by adding like terms.

Notes:

Guided Example	***Practice***
Add. 1. $\left(-6x^3+5x^2+4\right)+\left(2x^3+7x^2-10\right)$ $\left(-6x^3+5x^2+4\right)+\left(2x^3+7x^2-10\right)$ $=-6x^3+5x^2+4+2x^3+7x^2-10$ Remove parentheses. $=-6x^3+2x^3+5x^2+7x^2+4-10$ Rearrange terms. $=-4x^3+12x^2-6$ Combine like terms. 2. $\left(5x^3y-4x^2y-7y\right)+\left(2x^3y+6x^2y-4y-5\right)$ $\left(5x^3y-4x^2y-7y\right)+\left(2x^3y+6x^2y-4y-5\right)$ Remove parentheses. $=5x^3y-4x^2y-7y+2x^3y+6x^2y-4y-5$ Rearrange terms. $=5x^3y+2x^3y-4x^2y+6x^2y-7y-4y-5$ $=7x^3y+2x^2y-11y-5$ Combine like terms.	1. Add. $\left(-6x^3+5x^2+4\right)+\left(2x^3+7x^2-10\right)$ 2. Add. $\left(7x^3y-5x^2y-3y\right)+\left(2x^3y+8x^2y-12y-9\right)$

Objective 3 – Subtract polynomials

Summary
To subtract one polynomial from another, add the opposite, or additive inverse, of the polynomial being subtracted.

Notes:

Guided Example

Subtract.

1. $\left(7x^3-8x^2+9x-6\right)-\left(2x^3-6x^2-3x+9\right)$

$\left(7x^3+8x^2+9x-6\right)-\left(2x^3-6x^2-3x+9\right)$

Add the opposite of the polynomial being subtracted.

$=\left(7x^3-8x^2+9x-6\right)+\left(-2x^3+6x^2+3x-9\right)$

Remove parentheses.

$=7x^3-8x^2+9x-6-2x^3+6x^2+3x-9$

$=5x^3-2x^2+12x-15$ Combine like terms.

2. Subtract $-2x^5-3x^3y+7$ from $3x^5-4x^3y-3$.

Set up the subtraction problem.

$\left(3x^5-4x^3y-3\right)-\left(-2x^5-3x^3y+7\right)$

$=\left(3x^5-4x^3y-3\right)+\left(2x^5+3x^3y-7\right)$ Add opposite.
$=3x^5-4x^3y-3+2x^5+3x^3y-7$ Remove parentheses.
$=3x^5+2x^5-4x^3y+3x^3y-3-7$ Rearrange terms.
$=5x^5-x^3y-10$ Combine like terms.

Practice

1. Subtract.

$\left(14x^3-5x^2+x-9\right)-\left(4x^3-3x^2-7x+1\right)$

2. Subtract $-7x^2y^5-4xy^3+2$ from $6x^2y^5-2xy^3-8$.

Objective 4 – Multiply a monomial and a polynomial.

Summary

We use the distributive property to multiply a monomial and a polynomial that is not a monomial. To multiply a monomial and a polynomial, multiply each term of the polynomial by the monomial. Once the monomial factor is distributed, we multiply the resulting monomials.

Notes:

Guided Example	***Practice***
Multiply. 1. $4x^3(6x^5-2x^2+3)$ Use the distributive property. $4x^3(6x^5-2x^2+3)=4x^3\cdot 6x^5-4x^3\cdot 2x^2+4x^3\cdot 3$ Multiply coefficients and add exponents. $=24x^8-8x^5+12x^3$ 2. $5x^3y^4(2x^7y-6x^4y^3-3)$ Use the distributive property. $5x^3y^4(2x^7y-6x^4y^3-3)=5x^3y^4\cdot 2x^7y-5x^3y^4\cdot 6x^4y^3-5x^3y^4\cdot 3$ Multiply coefficients and add exponents. $=10x^{10}y^5-30x^7y^7-15x^3y^4$	Multiply. 1. $6x^4(2x^5-3x^2+4)$ 2. $2x^4y^3(2xy^6-4x^3y^4-5)$

Objective 5 – Multiply polynomials when neither is a monomial.

Summary
To multiply polynomials when neither is a monomial, multiply each term of one polynomial by each term of the other polynomial. Then combine like terms. Another method is to use a vertical format similar to that used for multiplying whole numbers. This is the method we will use in the Guided Examples below.

Notes:

Guided Example	***Practice***
Multiply using the vertical method.	Multiply using the vertical method.
1. $(x+2)(x^2+4x-3)$	1. $(3x+7)(x^2+4x+5)$
x^2+4x-3 $x+2$ $2x^2+8x-6$ Multiply x^2+4x-3 by 2. x^3+4x^2-3x Multiply x^2+4x-3 by x. x^3+6x^2+5x-6 Add like terms.	
2. $(4x-3)(x^3-x+5)$	2. $(3x-2)(2x^3-2x+1)$
Add $0x^2$ to the trinomial. x^3+0x^2-x+5 $4x-3$ $-3x^3+0x^2+3x-15$ Multiply x^3-x+5 by -3. $4x^4+0x^3-4x^2+20x$ Multiply x^3-x+5 by $4x$. $4x^4-3x^3-4x^2+23x-15$ Add like terms.	

Objective 6 – Use FOIL in polynomial multiplication.

Summary
Frequently we need to find the product of two binomials. One way to perform this multiplication is to distribute each term in the first binomial throughout the second binomial. We can also find the product using a method called FOIL. Any two binomials can be quickly multiplied using the FOIL method, in which **F** represents the product of the **first** terms in each binomial, **O** represents the product of the **outside** terms, **I** represents the product of the two **inside** terms, and **L** represents the product of the **last**, or second, terms in each binomial.

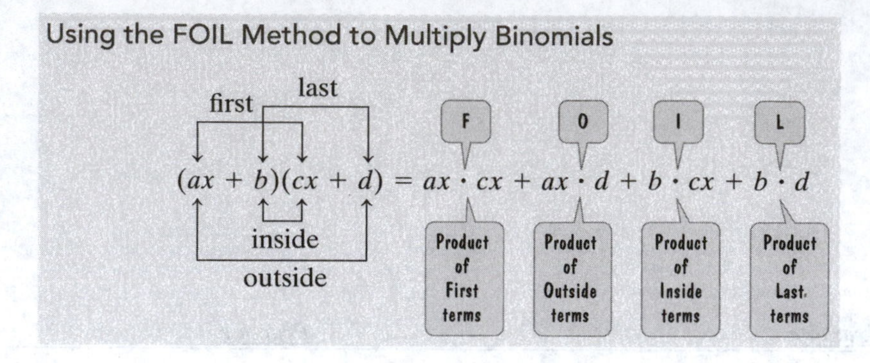

Notes:

Guided Example	*Practice*
Multiply using the FOIL method.	Multiply using the FOIL method.
1. $(8x-3)(2x-1)$	1. $(x-3)(2x+3)$
Use the FOIL method.	
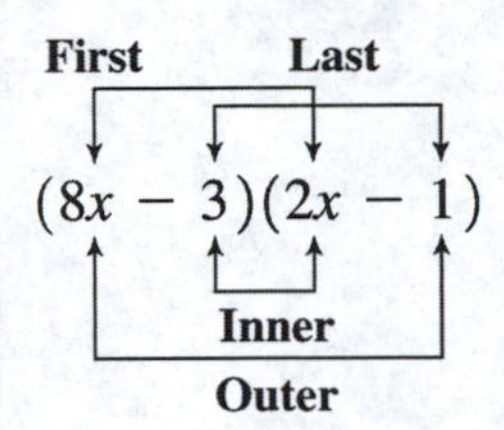	
$(8x-3)(2x-1) = \underbrace{8x(2x)}_{F} + \underbrace{8x(-1)}_{O} + \underbrace{(-3)(2x)}_{I} + \underbrace{(-3)(-1)}_{L}$	2. $(7x+4y)(2x-y)$
$= 16x^2 - 8x - 6x + 3$ Multiply.	
$= 16x^2 - 14x + 3$ Combine like terms.	
2. $(3x+5y)(x-2y)$	
Use the FOIL method.	
$(3x+5y)(x-2y) = \underbrace{3x(x)}_{F} + \underbrace{3x(-2y)}_{O} + \underbrace{(5y)(x)}_{I} + \underbrace{(5y)(-2y)}_{L}$	
$= 3x^2 - 6xy + 5xy - 10y^2$ Multiply.	
$= 3x^2 - xy - 10y^2$ Combine like terms.	

Objective 7 – Square binomials.

Summary

The square of a binomial, $(A+B)^2$, can be found using the FOIL method, but there is a general rule that can be followed. This general rule is a *special product formula* and is given below.

The Square of a Binomial Sum

$$(A+B)^2 = \underbrace{A^2}_{\text{first term squared}} \underbrace{+}_{\text{plus}} \underbrace{2AB}_{\text{2 times the product of the terms}} \underbrace{+}_{\text{plus}} \underbrace{B^2}_{\text{last term squared}}$$

The Square of a Binomial Difference

$$(A-B)^2 = \underbrace{A^2}_{\text{first term squared}} \underbrace{-}_{\text{minus}} \underbrace{2AB}_{\text{2 times the product of the terms}} \underbrace{+}_{\text{plus}} \underbrace{B^2}_{\text{last term squared}}$$

Notes:

Guided Example Practice

Multiply. 1. $(x+5)^2$ Use the special-product formula. $(x+5)^2 = \underbrace{x^2}_{\text{first term squared}} \underbrace{+}_{\text{plus}} \underbrace{2\cdot x\cdot 5}_{\text{2 times the product of the terms}} \underbrace{+}_{\text{plus}} \underbrace{5^2}_{\text{last term squared}}$ $= x^2 + 10x + 25$ Simplify. 2. $(2x-3y)^2$ Use the special-product formula. $(2x-3y)^2 = \underbrace{(2x)^2}_{\text{first term squared}} \underbrace{-}_{\text{minus}} \underbrace{2\cdot 2x\cdot 3y}_{\text{2 times the product of the terms}} \underbrace{+}_{\text{plus}} \underbrace{(3y)^2}_{\text{last term squared}}$ $= 4x^2 - 12xy + 9y^2$ Simplify.	Multiply. 1. $(r+3)^2$ 2. $(2x-5y)^2$

Objective 8 – Multiply the sum and difference of two terms.

Summary
We can use the FOIL method to multiply $A+B$ and $A-B$ as follow.

$$(A+B)(A-B) = A^2 - AB + AB - B^2 = A^2 - B^2$$

Notice that the outside and inside products have a sum of zero and the terms cancel. This leads us to another special-product formula.

Multiplying the Sum and Difference of Two Terms

$$(A+B)(A-B) = \underbrace{A^2}_{\text{square of the first term}} \underbrace{-}_{\text{minus}} \underbrace{B^2}_{\text{square of the second term}}$$

Notes:

Guided Example	***Practice***
Multiply. 1. $(x+8)(x-8)$ Use the special product formula. $(x+8)(x-8) = \underbrace{x^2}_{\text{square of the first term}} \underbrace{-}_{\text{minus}} \underbrace{8^2}_{\text{square of the second term}}$ $= x^2 - 64$ Simplify. 2. $(9x+5y)(9x-5y)$ Use the special product formula. $(9x+5y)(9x-5y) = \underbrace{(9x)^2}_{\text{square of the first term}} \underbrace{-}_{\text{minus}} \underbrace{(5y)^2}_{\text{square of the second term}}$ $= 81x^2 - 25y^2$ Simplify.	Multiply. 1. $(a+7)(a-7)$ 2. $(3r^2+4s)(3r^2-4s)$

Objective 9 – Divide a polynomial by a monomial.

Summary

When we divide a monomial by another monomial, we divide the coefficients and subtract the exponents when the bases are the same. To divide a polynomial by a monomial, divide each term of the polynomial by the monomial. Then simplify each quotient, if possible.

Notes:

Guided Example	***Practice***
Divide. 1. $\left(15x^3-5x^2+x+5\right)\div(5x)$ Rewrite the problem in a vertical format. $\left(15x^3-5x^2+x+5\right)\div(5x)=\dfrac{15x^3-5x^2+x+5}{5x}$ Divide the terms in the polynomial by the monomial. $=\dfrac{15x^3}{5x}-\dfrac{5x^2}{5x}+\dfrac{x}{5x}+\dfrac{5}{5x}$ Simplify each quotient. $=3x^2-x+\dfrac{1}{5}+\dfrac{1}{x}$ 2. $\left(8x^4y^5-10x^4y^3+12x^2y^3\right)\div\left(4x^3y^2\right)$ $=\dfrac{8x^4y^5-10x^4y^3+12x^2y^3}{4x^3y^2}$ Write vertically. $=\dfrac{8x^4y^5}{4x^3y^2}-\dfrac{10x^4y^3}{4x^3y^2}+\dfrac{12x^2y^3}{4x^3y^2}$ Divide each term. $=2xy^3-\dfrac{5}{2}xy+\dfrac{3y}{x}$ Simplify each quotient.	Divide. 1. $\left(16x^3-32x^2+2x+4\right)\div(4x)$ 2. $\left(15x^5y^4-5x^3y^5+10x^2y^2\right)\div\left(5x^2y^3\right)$

R.4 Answers

R.4.1

1)

Term	Coefficient	Degree (Sum exponents)
$8x^4y^5$	8	$4+5=9$
$-7x^3y^2$	-7	$3+2=5$
$-x^2y$	-1	$2+1=3$
$-5x$	-5	1
11	11	0

The degree is 9. The leading term is $8x^4y^5$ and the leading coefficient is 8.

2) *a)* monomial *b)* trinomial *c)* binomial

R.4.2

1) $-4x^3+12x^2-6$ 2) $9x^3y+3x^2y-15y-9$

R.4.3

1) $10x^3-2x^2+8x-10$ 2) $13x^2y^5+2xy^3-10$

R.4.4

1) $12x^9-18x^6+24x^4$ 2) $4x^5y^9-8x^7y^7-10x^4y^3$

R.4.5

1) $3x^3+19x^2+43x+35$ 2) $6x^4-4x^3-6x^2+7x-2$

R.4.6

1) $2x^2-3x-9$ 2) $14x^2+xy-4y^2$

R.4.7

1) r^2+6r+9 2) $4x^2-20xy+25y^2$

R.4.8

1) a^2-49 2) $9r^2-16s^2$

R.4.9

1) $4x^2-8x+\frac{1}{2}+\frac{1}{x}$ 2) $3x^3y-xy^2+\frac{2}{y}$

R.5 Review of Factoring

Objective 1 – Factor out the greatest common factor from a polynomial.
Objective 2 – Factor by grouping.
Objective 3 – Factor a trinomial whose leading coefficient is 1.
Objective 4 – Factor a trinomial whose leading coefficient is not 1.
Objective 5 – Factor difference of squares.
Objective 6 – Factor polynomials completely.
Objective 7 – Factor perfect square trinomials.
Objective 8 – Factor the sum or difference of two cubes.
Objective 9 – Complete the square of a binomial.

Objective 1 – Factor out the greatest common factor from a polynomial.

Key Terms
Factoring
Prime polynomials.
Greatest common factor (GCF)
Distributive property

Summary
Factoring a polynomial consisting of the sum of monomials means finding an equivalent expression that is a product. We will be factoring over the set of integers, meaning that the coefficients in the factors are integers. Polynomials that cannot be factored using integer coefficients are called **prime polynomials** over the set of integers. In any factoring problem, the first step is to look for the *greatest common factor*. The **greatest common factor**, abbreviated GCF, is an expression with the greatest coefficient and of the highest degree that divides each term of the polynomial. The variable part of the greatest common factor always contains the *smallest* power of a variable that appears in all terms of the polynomial.

When factoring a monomial from a polynomial, determine the greatest common factor of all terms in the polynomial. Then express each term as the product of the GCF and its other factor. Finally, use the distributive property to factor out the GCF.

Notes:

Guided Example	***Practice***
Factor. 1. $21x^2 + 28x$ The GCF is $7x$. Express each term as the product of the GCF and its other factor. $21x^2 + 28x = 7x(3x) + 7x(4)$ $= 7x(3x+4)$ Factor out the GCF. 2. $12x^5y^4 - 4x^4y^3 + 2x^3y^2$ The GCF is $2x^3y^2$. Express each term as the product of the GCF and its other factor. $12x^5y^4 - 4x^4y^3 + 2x^3y^2 = 2x^3y^2(6x^2y^2) + 2x^3y^2(-2xy) + 2x^3y^2(1)$ Factor out the GCF. $= 2x^3y^2(6x^2y^2 - 2xy + 1)$ 3. $2(x-7)+9a(x-7)$ The GCF, a binomial, is $x-7$. Express each term as the product of the GCF and its other factor. (This step is usually omitted when the GCF is a binomial.) $2(x-7)+9a(x-7) = (x-7)\cdot 2 + (x-7)\cdot 9a$ $= (x-7)(2+9a)$ Factor out the GCF.	Factor. 1. $20y^2 + 30y$ 2. $16x^4y^5 - 8x^3y^4 + 4x^2y^3$ 3. $3(x-4)+7a(x-4)$

Objective 2 – Factor by grouping.

Summary
Some polynomials have only a greatest common factor of 1. However, by a suitable grouping of the terms, it still may be possible to factor. This process is called ***factoring by grouping.***

Notes:

Guided Example	***Practice***
Factor. 1. $x^3 - 5x^2 + 3x - 15$ There is no common factor other than 1 common to all four terms. However, we can group the terms that have a common factor. $x^3 - 5x^2 + 3x - 15 = (x^3 - 5x^2) + (3x - 15)$ Factor out the GCF from the grouped terms. $= x^2(x-5) + 3(x-5)$ Factor out the common binomial factor. $= (x-5)(x^2+3)$ 2. $3x^2 + 12x - 2xy - 8y$ $3x^2 + 12x - 2xy - 8y = (3x^2 + 12x) + (-2xy - 8y)$ Group. Factor out the GCF from the grouped terms. Sometimes it is necessary to use a factor with a negative coefficient. $= 3x(x+4) - 2y(x+4)$ $= (x+4)(3x-2y)$ Factor.	Factor. 1. $x^3 - 4x^2 + 5x - 20$ 2. $4x^2 + 20x - 3xy - 15y$

Objective 3 – Factor a trinomial whose leading coefficient is 1.

Summary
We can use the FOIL method to multiply two binomial and often the product is a trinomial. Now we would like to start with the trinomial form $x^2 + bx + c$ and, assuming that it is factorable, return it to the factored form.

A Strategy for Factoring $x^2 + bx + c$

1. Enter x as the first term of each factor.

$$(x \quad)(x \quad) = x^2 + bx + c$$

2. List pairs of factors of the constant c.
3. Try various combinations of these factors. Select the combination in which the sum of the Outside and Inside products is equal to bx.

$$(x + \square)(x + \square) = x^2 + bx + c$$

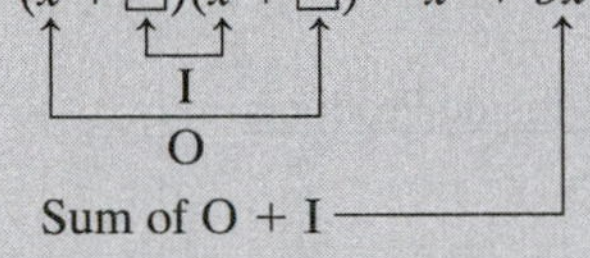

I

O

Sum of O + I

4. Check your work by multiplying the factors using the FOIL method. You should obtain the original trinomial.

If none of the possible combinations yield an Outside product and an Inside product whose sum is equal to bx, the trinomial cannot be factored using integers and is called **prime** over the set of integers.

Notes:

Guided Example	***Practice***
1. Factor x^2+5x+6.	Factor.

Enter x as the first term of each factor.

$x^2+5x+6=(x\quad)(x\quad)$

List pairs of factors of the constant 6.

Factors of 6	6,1	3,2	−6,−1	−3,−2

Try various combinations of these factors. The sum of the outside and inside products should equal $5x$.

Factors	Sum of Outside and Inside Products
$(x+6)(x+1)$	$x+6x=7x$
$(x+3)(x+2)$	$2x+3x=5x$ ←
$(x-6)(x-1)$	$-x-6x=-7x$
$(x-3)(x-2)$	$-2x-3x=-5x$

The arrow indicates that the factors 3 and 2 yield the required middle term, $5x$.

Thus, $x^2+5x+6=(x+2)(x+3)$.

2. Factor y^2-y-12.

Enter y as the first term of each factor.

$y^2-y-12=(y\quad)(y\quad)$

In this trinomial $b=-1$ and $c=-12$. We can speed things up by listing all of the factors of -12 and then finding the pair whose sum is -1.

Factors of 12	Sum of Factors
−1, 12	$-1+12=11$
1, −12	$1-12=-11$
−2, 6	$-2+6=4$
2, −6	$2-6=-4$
−3, 4	$-3+4=1$
3, −4	$3-4=-1$ ←

The factors 3 and −4 have a product of −12 and a sum of −1. Thus, $y^2-y-12=(y+3)(y-4)$.

Practice

Factor.

1. Factor $x^2+7x+10$.

2. Factor $b^2+2b-24$.

Objective 4 – Factor a trinomial whose leading coefficient is not 1.

Summary

To factor a trinomial of the form ax^2+bx+c, where $a \neq 1$, we will use a procedure called the *grouping method*. This method involves both trial and error, as well as grouping. The trial and error in factoring ax^2+bx+c depends upon finding two numbers, p and q, for which $p+q=b$. Then we factor $ax^2+px+qx+c$ using grouping.

Factoring ax^2+bx+c Using Grouping $a \neq 1$

1. Multiply the leading coefficient, a, and the constant, c.
2. Find the factors of ac whose sum is b.
3. Rewrite the middle term, bx, as a sum or difference using the factors from step 2.
4. Factor by grouping.

Notes:

Guided Example

Factor by grouping.

1. $12x^2 - 5x - 2$

$a = 12$ and $c = -2$, find the product ac.

$ac = 12(-2) = -24$

Find the factors of ac whose sum is $b = -5$.

$-8(3) = -24$ and $-8 + 3 = -5$

Rewrite the middle term $-5x$ as a sum or difference using the factors -8 and 3.

$$\begin{aligned} 12x^2 - 5x - 2 &= 12x^2 - 8x + 3x - 2 \\ &= (12x^2 - 8x) + (3x - 2) \quad \text{Group terms.} \\ &= 4x(3x - 2) + 1(3x - 2) \quad \text{Factor from each group.} \end{aligned}$$

Factor out the common binomial factor.

$$= (3x - 2)(4x + 1)$$

2. $3y^2 + 11y + 10$

$a = 3$ and $c = 10$, find the product ac.

$ac = 3(10) = 30$

Find the factors of ac whose sum is $b = 11$.

$6(5) = 30$ and $6 + 5 = 11$

Rewrite the middle term $11y$ as a sum or difference using the factors 6 and 5.

$$\begin{aligned} 3y^2 + 11y + 10 &= 3y^2 + 6y + 5y + 10 \\ &= (3y^2 + 6y) + (5y + 10) \quad \text{Group terms.} \\ &= 3y(y + 2) + 5(y + 2) \quad \text{Factor from each group.} \end{aligned}$$

Factor out the common binomial factor.

$$= (y + 2)(3y + 5)$$

Practice

Factor by grouping.

1. $5x^2 + 14x + 8$

2. $2r^2 - 7r - 4$

Objective 5 – Factor difference of squares.

Summary
A method for factoring the difference of two squares is obtained by reversing the special product for the sum and difference of two terms.

The Difference of Two Squares

If A and B are real numbers, variables, or algebraic expressions, then

$$A^2 - B^2 = (A+B)(A-B)$$

In words: the difference of the squares of two terms factors as the product of a sum and a difference of those terms.

Notes:

Guided Example	***Practice***
Factor. 1. $9x^2 - 100$ Express each term as the square of some monomial. Then use the formula for factoring $A^2 - B^2$. $9x^2 - 100 = \underbrace{(3x)^2}_{A^2} - \underbrace{(10)^2}_{B^2} = \underbrace{(3x+10)}_{(A+B)}\underbrace{(3x-10)}_{(A-B)}$ 2. $36y^6 - 49x^4$ Express each term as the square of some monomial. Then use the formula for factoring $A^2 - B^2$. $36y^6 - 49x^4 = \underbrace{(6y^3)^2}_{A^2} - \underbrace{(7x^2)^2}_{B^2} = \underbrace{(6y^3+7x^2)}_{(A+B)}\underbrace{(6y^3-7x^2)}_{(A-B)}$	Factor. 1. $16x^2 - 25$ 2. $100x^6 - 9y^4$

Objective 6 – Factor polynomials completely.

Summary

A polynomial is factored completely when it is written as the product of prime polynomials. To be sure that you have factored completely, check to see whether any factors with more than one term in the factored polynomial can be factored further. If so, continue factoring.

Notes

Guided Example	***Practice***
Factor. 1. $3y-3x^6y^5$ The GCF is $3y$. $3y-3x^6y^5=3y(1-x^6y^4)$ Factor out the GCF. Use the formula for the difference of two squares. $=3y\left[(1)^2-(x^3y^2)^2\right]=3y(1+x^3y^2)(1-x^3y^2)$ $\underbrace{}_{A^2}\ -\ \underbrace{}_{B^2}\ =\ \underbrace{(A+B)}\ \underbrace{(A-B)}$	Factor. 1. y^4-81
2. $x^3+5x^2-9x-45$ First factor by grouping. Group terms with common factors. $x^3+5x^2-9x-45=(x^3+5x^2)+(-9x-45)$ Factor out common factors from each group. $=x^2(x+5)-9(x+5)$ $=(x+5)(x^2-9)$ Factor out $x+5$. Factor $x^2-9=x^2-3^2$, the difference of two squares. $=(x+5)(x+3)(x-3)$	2. $x^3+7x^2-4x-28$

Objective 7 – Factor perfect square trinomial.

Key Terms
Perfect square trinomial

Summary
Certain trinomials can be factored by reversing the special products for squaring binomials. The trinomials that are factored using this technique are called **perfect square trinomials.**

Factoring Perfect Square Trinomials

Let A and B be real numbers, variables, or algebraic expressions.

1. $A^2 + 2AB + B^2 = (A + B)^2$
   ↑ Same sign ↑

2. $A^2 - 2AB + B^2 = (A - B)^2$
   ↑ Same sign ↑

The two items above show that perfect square trinomials, $A^2 + 2AB + B^2$ and $A^2 - 2AB + B^2$, come in two forms: one in which the coefficient of the middle term is positive and one in which the coefficient of the middle term is negative. Here's how to recognize a perfect square trinomial:

1. The first and last terms are squares of monomials or integers.
2. The middle term is twice the product of the expressions being squared in the first and last terms.

Notes

Guided Example

Factor.

1. $x^2 + 14x + 49$

This trinomial is a perfect square trinomial because the first and last terms are perfect squares (x^2 and $49 = 7^2$ are both perfect squares.), and the middle term $14x$ is twice the product of x and 7.

Identify A and B and apply the formula.

$A = x$ and $B = 7$

$$x^2 + 14x + 49 = \underbrace{x^2}_{A^2} + \underbrace{2(x)(7)}_{2AB} + \underbrace{(7)^2}_{B^2} = \underbrace{(x+7)^2}_{(A+B)^2}$$

2. $4x^2 - 12xy + 9y^2$

This trinomial is a perfect square trinomial because the first and last terms are perfect squares ($4x^2 = (2x)^2$ and $9y^2 = (3y)^2$ are both perfect squares.), and the middle term $12xy$ is twice the product of $2x$ and $3y$.

Identify A and B and apply the formula.

$A = 2x$ and $B = 3y$

$$4x^2 - 12xy + 9y^2 = \underbrace{(2x)^2}_{A^2} - \underbrace{2(2x)(3y)}_{2AB} + \underbrace{(3y)^2}_{B^2} = \underbrace{(2x-3y)^2}_{(A-B)^2}$$

Practice

Factor.

1. $x^2 + 6x + 9$

2. $16x^2 - 40xy + 25y^2$

Objective 8 – Factor the sum or difference of two cubes.

Summary

There are special formulas that let us factor the sum or difference of two cubes. These are given below.

Factoring the Sum or Difference of Two Cubes

1. Factoring the Sum of Two Cubes

$$A^3 + B^3 = (A + B)(A^2 - AB + B^2)$$

Same signs | Opposite signs | Always positive

2. Factoring the Difference of Two Cubes

$$A^3 - B^3 = (A - B)(A^2 + AB + B^2)$$

Same signs | Opposite signs | Always positive

Notes

Guided Example	***Practice***
Factor. 1. $x^3 + 125$ This trinomial is the sum of two cubes because both terms are perfect cubes (x^3 and $125 = 5^3$). Identify A and B and apply the formula. $A = x$ and $B = 5$ $x^3 + 125 = x^3 + 5^3 = \overbrace{(x+5)}^{\uparrow}\overbrace{(x^2 - x \cdot 5 + 5^2)}^{\uparrow}$ $(A+B)(A^2 - AB + B^2)$ $= (x+5)(x^2 - 5x + 25)$ Simplify. 2. $x^3 - 216$ This trinomial is the difference of two cubes because both terms are perfect cubes (x^3 and $216 = 6^3$). Identify A and B and apply the formula. $A = x$ and $B = 6$ $x^3 - 216 = x^3 - 6^3 = \overbrace{(x-6)}^{\uparrow}\overbrace{(x^2 + x \cdot 6 + 6^2)}^{\uparrow}$ $(A-B)(A^2 + AB + B^2)$ $= (x-6)(x^2 + 6x + 36)$ Simplify.	Factor. 1. $x^3 + 27$ 2. $t^3 - 8$

Objective 9 – Complete the square of a binomial.

Key Terms

Completing the square

Summary

If $x^2 + bx$ is a binomial, then adding $\left(\frac{b}{2}\right)^2$, which is the square of half the coefficient of x, a perfect square trinomial will result.

$$x^2 + bx + \left(\frac{b}{2}\right)^2 = \left(x + \frac{b}{2}\right)^2$$

This process is called **completing the square.** Note the coefficient of x^2 must be 1 to complete the square.

Notes:

Guided Example

What term should be added to each binomial so that it becomes a perfect square trinomial? Write and factor the trinomial.

1. x^2+8x

$\left(\frac{b}{2}\right)^2$ must be added. First identify b.

b is the coefficient of x, so $b=8$.

$$\left(\frac{b}{2}\right)^2=\left(\frac{8}{2}\right)^2$$

$=(4)^2$ Divide.

$=16$ Evaluate exponent.

Add 16 to complete the square and then factor.

$x^2+8x+16=(x+4)^2$

2. x^2-7x

$\left(\frac{b}{2}\right)^2$ must be added. First identify b.

b is the coefficient of x, so $b=-7$.

$$\left(\frac{b}{2}\right)^2=\left(\frac{-7}{2}\right)^2$$

$=\frac{49}{4}$ Evaluate exponent.

Add $\frac{49}{4}$ to complete the square and then factor.

$x^2-7x+\frac{49}{4}=\left(x-\frac{7}{2}\right)^2$

Practice

What term should be added to each binomial so that it becomes a perfect square trinomial? Write and factor the trinomial.

1. x^2+10x

2. $x^2-\frac{3}{4}x$

R.5 Answers

R.5.1
1) $10y(2y+3)$ 2) $4x^2y^3(4x^2y^2-2xy+1)$ 3) $(x-4)(3+7a)$

R.5.2
1) $(x-4)(x^2+5)$ 2) $(x+5)(4x-3y)$

R.5.3
1) $(x+2)(x+5)$ 2) $(b+6)(b-4)$

R.5.4
1) $(5x+4)(x+2)$ 2) $(2r+1)(r-4)$

R.5.5
1) $(4x+5)(4x-5)$ 2) $(10x^3+3y^2)(10x^3-3y^2)$

R.5.6
1) $(y^2+9)(y+3)(y-3)$ 2) $(x+7)(x+2)(x-2)$

R.5.7
1) $(x+3)^2$ 2) $(4x-5y)^2$

R.5.8
1) $(x+3)(x^2-3x+9)$ 2) $(t-2)(t^2+2t+4)$

R.5.9
1) $x^2+10x+25=(x+5)^2$ 2) $x^2-\frac{3}{4}x+\frac{9}{64}=\left(x-\frac{3}{8}\right)^2$

R.6 Review of Rational Expressions

Objective 1 – Find the domain of a rational function.

Objective 2 – Simplify rational expressions.

Objective 3 – Multiply rational expressions.

Objective 4 – Divide rational expressions.

Objective 5 – Find the least common denominator.

Objective 6 – Add and subtract rational expressions with different denominators.

Objective 7 – Simplify complex rational expressions by multiplying by 1.

Objective 1 – Find the domain of a rational expression.

Key Terms

Rational expression
Rational function
Domain

Summary

A **rational expression** consists of a polynomial divided by a nonzero polynomial. A **rational function** is a function defined by a formula that is a rational expression. The **domain** of a rational function is the set of all real numbers except those for which the denominator is zero. We can find the domain by determining when the denominator is zero.

Notes:

Guided Example	***Practice***
Find the domain of the function. 1. $f(x)=\frac{2}{x-5}$ Set the denominator equal to 0 and then solve for x. $x-5=0$ $x=5$ Because 5 makes the denominator zero, the domain is all real numbers, except 5. 2. $f(x)=\frac{2x+1}{2x^2-x-1}$ Set the denominator equal to 0 and then solve for x. $2x^2-x-1=0$ $(2x+1)(x-1)=0$ Factor. $2x+1=0$ or $x-1=0$ Set factors equal to 0. $x=-\frac{1}{2}$ or $x=1$ Solve for x. Because $-\frac{1}{2}$ and 1 make the denominator zero, the domain is all real numbers, except $-\frac{1}{2}$ and 1.	Find the domain of the function. 1. $f(x)=\frac{4}{x+2}$ 2. $f(x)=\frac{x-5}{2x^2+5x-3}$

Objective 2 – Simplify rational expressions.

Key Terms
Simplified

Summary
A rational expression is **simplified** if its numerator and denominator have no common factors other than 1 or -1. The following procedure can be used to simplify rational expressions.

Simplifying Rational Expressions

1. Factor the numerator and the denominator completely.
2. Divide both the numerator and the denominator by any common factors.

For simplicity we will assume that a simplified rational expression is equal to the original rational expression for all real numbers, except those for which the denominator is 0.

Notes:

Guided Example	***Practice***
Write in simplest form.	Write in simplest form.
1. $\frac{x^2+4x+3}{x+1}$	1. $\frac{x^2+7x+10}{x+2}$
Factor the numerator and the denominator.	
$\frac{x^2+4x+3}{x+1}=\frac{(x+3)(x+1)}{x+1}$	
Divide out the common factor $x+1$.	
$=\frac{(x+3)\cancel{(x+1)}}{\cancel{(x+1)}}$	
$=x+3$	2. $\frac{x^2-2x-15}{3x^2+8x-3}$
2. $\frac{x^2-7x-18}{2x^2+3x-2}$	
$\frac{x^2-7x-18}{2x^2+3x-2}=\frac{(x-9)(x+2)}{(2x-1)(x+2)}$ Factor.	
$=\frac{(x-9)\cancel{(x+2)}}{(2x-1)\cancel{(x+2)}}$ Divide out $x+2$.	
$=\frac{x-9}{2x-1}$	

Objective 3 – Multiply rational expressions.

Summary

The product of two rational expressions is the product of their numerators divided by the product of their denominators. Below is a step-by-step procedure for multiplying rational expressions.

Multiplying Rational Expressions

1. Factor all numerators and denominators completely.
2. Divide numerators and denominators by common factors.
3. Multiply the remaining factors in the numerators and multiply the remaining factors in the denominators.

Notes:

Guided Example	***Practice***
Multiply. 1. $\frac{x+3}{x-4}\cdot\frac{x^2-2x-8}{x^2-9}$ Factor all numerators and denominators completely. $\frac{x+3}{x-4}\cdot\frac{x^2-2x-8}{x^2-9}=\frac{x+3}{x-4}\cdot\frac{(x-4)(x+2)}{(x+3)(x-3)}$ Divide numerators and denominators by common factors. $=\frac{1(x+3)}{1(x-4)}\cdot\frac{(x-4)(x+2)}{(x+3)(x-3)}$ Multiply the remaining factors in the numerators and denominators. $=\frac{x+2}{x-3}$ 2. $\frac{5x+5}{7x-7x^2}\cdot\frac{2x^2+x-3}{4x^2-9}$ Factor all numerators and denominators completely. $\frac{5x+5}{7x-7x^2}\cdot\frac{2x^2+x-3}{4x^2-9}=\frac{5(x+1)}{7x(1-x)}\cdot\frac{(2x+3)(x-1)}{(2x+3)(2x-3)}$ Divide numerators and denominators by common factors. Because $1-x$ and $x-1$ are opposites their quotient is -1. . $=\frac{5(x+1)}{7x(1-x)}\cdot\frac{(2x+3)\overset{(-1)}{(x-1)}}{(2x+3)(2x-3)}$ Multiply the remaining factors in the numerators and denominators. $=-\frac{5(x+1)}{7x(2x-3)}$	Multiply. 1. $\frac{x+4}{x-7}\cdot\frac{x^2-4x-21}{x^2-16}$ 2. $\frac{4x+8}{6x-3x^2}\cdot\frac{3x^2-4x-4}{9x^2-4}$

Objective 4 – Divide rational expressions.

Key Terms
Reciprocal

Summary
The quotient of two rational expressions is the product of the first expression and the multiplicative inverse, or **reciprocal**, of the second expression. The reciprocal is found by interchanging the numerator and the denominator of the expression.

Dividing Rational Expressions

If P, Q, R, and S are polynomials, where $Q \neq 0$, $R \neq 0$, and $S \neq 0$, then

$$\frac{P}{Q} \div \frac{R}{S} = \frac{P}{Q} \cdot \frac{S}{R} = \frac{PS}{QR}$$

Thus, **we find the quotient of two rational expressions by inverting the divisor and multiplying**.

Notes:

Guided Example

Divide.

1. $\left(4x^2-25\right)\div\dfrac{2x+5}{14}$

Write $4x^2-25$ with a denominator of 1. Then invert the divisor and multiply.

$$\left(4x^2-25\right)\div\frac{2x+5}{14}=\frac{4x^2-25}{1}\cdot\frac{14}{2x+5}$$
$$=\frac{(2x+5)(2x-5)}{1}\cdot\frac{14}{(2x+5)}\quad\text{Factor.}$$

Divide the numerator and denominator by the common factor $2x+5$.

$$=\frac{\cancel{(2x+5)}(2x-5)}{1}\cdot\frac{14}{1\cancel{(2x+5)}}$$

Multiply the remaining factors in the numerators and denominators.

$$=14(2x-5)$$

2. $\dfrac{x^2+3x-10}{2x}\div\dfrac{x^2-5x+6}{x^2-3x}$

Invert the divisor and multiply.

$$\frac{x^2+3x-10}{2x}\div\frac{x^2-5x+6}{x^2-3x}=\frac{x^2+3x-10}{2x}\cdot\frac{x^2-3x}{x^2-5x+6}$$

Factor.

$$=\frac{(x+5)(x-2)}{2x}\cdot\frac{x(x-3)}{(x-3)(x-2)}$$

Divide the numerators and denominators by common factors.

$$=\frac{(x+5)\cancel{(x-2)}}{2\cancel{x}}\cdot\frac{\cancel{x}\cancel{(x-3)}}{\cancel{(x-3)}\cancel{(x-2)}}$$

Multiply the remaining factors in the numerators and denominators.

$$=\frac{x+5}{2}$$

Practice

Divide.

1. $\left(9x^2-49\right)\div\dfrac{3x-7}{9}$

2. $\dfrac{x^2-x-12}{5x}\div\dfrac{x^2-10x+24}{x^2-6x}$

Objective 5 – Find the least common denominator (LCD).

Key Terms

Least common denominator

Summary

To add or subtract rational expressions, the rational expressions must have the same denominators. To combine rational expressions with unlike denominators, we must first find the least common denominator (LCD). The **least common denominator** of several rational expressions is a polynomial consisting of the product of all prime factors in the denominators, with each factor raised to the greatest power of its occurrence in any denominator.

Finding the Least Common Denominator

1. Factor each denominator completely.
2. List the factors of the first denominator.
3. Add to the list in step 2 any factors of the second denominator that do not appear in the list.
4. Form the product of each different factor from the list in step 3. This product is the LCD.

Notes:

Guided Example	***Practice***
Find the least common denominator. 1. $\frac{3}{10x^2}$ and $\frac{7}{15x}$ Factor each denominator completely. $10x^2 = 5 \cdot 2x^2$ $15x = 5 \cdot 3x$ List the factors of the first denominator. 5, 2, x^2 Add any unlisted factors from the second denominator. 3, 5, 2, x^2 Find the product of all factors in the final list. $\text{LCD} = 3 \cdot 2 \cdot 5 \cdot x^2 = 30x^2$ 2. $\frac{9}{7x^2+28x}$ and $\frac{11}{x^2+8x+16}$ Factor each denominator completely. $7x^2+28x = 7x(x+4)$ $x^2+8x+16 = (x+4)^2$ List the factors of the first denominator. 7, x, $(x+4)$ Add any unlisted factors from the second denominator. We add a second $x+4$ factor because the second denominator was $(x+4)^2$. 7, x, $(x+4)$, $(x+4)$ Find the product of all factors in the final list. $\text{LCD} = 7x(x+4)(x+4) = 7x(x+4)^2$	Find the least common denominator. 1. $\frac{7}{6x^2}$ and $\frac{2}{9x}$ 2. $\frac{7}{5x^2+15x}$ and $\frac{9}{x^2+6x+9}$

Objective 6 – Add and subtract rational expressions with different denominators.

Summary
Below are the steps for adding or subtracting rational expressions with different denominators.

Adding or Subtracting Rational Expressions That Have Denominators

1. Find the LCD of the rational expressions.
2. Rewrite each rational expression as an equivalent expression whose denominator is the LCD. To do so, multiply the numerator and the denominator of each rational expression by any factor(s) needed to convert the denominator into the LCD.
3. Add or subtract numerators, placing the resulting expression over the LCD.
4. If possible, simplify the resulting rational expression.

Notes:

Guided Example	***Practice***
Perform the indicated operation.	Perform the indicated operation.
1. $\frac{3}{10x^2}+\frac{7}{15x}$	1. $\frac{7}{6x^2}+\frac{2}{9x}$
Find the LCD. In Guided Example #1 in Objective 5, we found that the LCD for these rational expressions is $30x^2$.	
Write an equivalent expressions with the LCD as the denominator.	
$\frac{3}{10x^2}+\frac{7}{15x}=\frac{3}{10x^2}\cdot\frac{3}{3}+\frac{7}{15x}\cdot\frac{2x}{2x}$	
$=\frac{9}{30x^2}+\frac{14x}{30x^2}$ Perform the required multiplication.	
$=\frac{9+14x}{30x^2}$ Add numerators. Put sum over the LCD.	
2. $\frac{x-1}{x^2+x-6}-\frac{x-2}{x^2+4x+3}$	2. $\frac{2x-3}{x^2-5x+6}-\frac{x+4}{x^2-2x-3}$
Factor the denominators.	
$\frac{x-1}{x^2+x-6}-\frac{x-2}{x^2+4x+3}=\frac{x-1}{(x+3)(x-2)}-\frac{x-2}{(x+3)(x+1)}$	
The LCD is $(x+3)(x-2)(x+1)$.	
Multiply each numerator and denominator by the extra factor required to form the LCD.	
$=\frac{(x-1)(x+1)}{(x+3)(x-2)(x+1)}-\frac{(x-2)(x-2)}{(x+3)(x+1)(x-2)}$	
Subtract the numerators and put the difference over the LCD.	
$=\frac{x^2-1-(x^2-4x+4)}{(x+3)(x-2)(x+1)}$ Multiply in numerators.	
Remove parentheses and change the signs of the terms in the second parentheses.	
$=\frac{x^2-1-x^2+4x-4}{(x+3)(x-2)(x+1)}$	
$=\frac{4x-5}{(x+3)(x-2)(x+1)}$ Combine like terms.	

Objective 7 – Simplify complex rational expressions by multiplying by 1.

Key Terms
Complex rational expression
Complex fractions

Summary
Complex rational expressions, also called **complex fractions**, have numerators or denominators containing one or more rational expressions. One method for simplifying a complex rational expression is to find the least common denominator of all the rational expressions in its numerator and denominator. Then multiply each term in its numerator and denominator by this least common denominator. Because we are multiplying by a form of 1, we will obtain an equivalent expression that does not contain fractions in the numerator or denominator.

Simplifying a Complex Rational Expression by Multiplying by 1 in the Form $\frac{\text{LCD}}{\text{LCD}}$

1. Find the LCD of all rational expressions within the complex rational expression.
2. Multiply both the numerator and the denominator of the complex rational expression by this LCD.
3. Use the distributive property and multiply each term in the numerator and denominator by this LCD. Simplify each term. No fractional expressions should remain within the numerator or denominator of the main fraction.
4. If possible, factor and simplify.

Notes:

Guided Example	***Practice***
1. Simplify $\dfrac{\frac{1}{x}+\frac{y}{x^2}}{\frac{1}{y}+\frac{x}{y^2}}$. Find the LCD. The LCD of $\frac{1}{x}$, $\frac{1}{y}$, $\frac{y}{x^2}$, and $\frac{x}{y^2}$ is x^2y^2. Multiply by x^2y^2 in the numerator and denominator. $\dfrac{\frac{1}{x}+\frac{y}{x^2}}{\frac{1}{y}+\frac{x}{y^2}}=\dfrac{x^2y^2}{x^2y^2}\cdot\dfrac{\left(\frac{1}{x}+\frac{y}{x^2}\right)}{\left(\frac{1}{y}+\frac{x}{y^2}\right)}$ $=\dfrac{x^2y^2\cdot\frac{1}{x}+x^2y^2\cdot\frac{y}{x^2}}{x^2y^2\cdot\frac{1}{y}+x^2y^2\cdot\frac{x}{y^2}}$ Use the distributive property. $=\dfrac{xy^2+y^3}{x^2y+x^3}$ Simplify. $=\dfrac{y^2(x+y)}{x^2(y+x)}=\dfrac{y^2}{x^2}$ Factor and simplify. 2. Simplify $\dfrac{\frac{1}{x+h}-\frac{1}{x}}{h}$. The LCD of $x+h$ and x is $x(x+h)$. Multiply by $x(x+h)$ in the numerator and denominator. $\dfrac{\frac{1}{x+h}-\frac{1}{x}}{h}=\dfrac{x(x+h)}{x(x+h)}\cdot\dfrac{\left(\frac{1}{x+h}-\frac{1}{x}\right)}{h}$ $=\dfrac{x(x+h)\cdot\frac{1}{x+h}-x(x+h)\cdot\frac{1}{x}}{x(x+h)h}$ Distributive property $=\dfrac{x-(x+h)}{x(x+h)h}$ Simplify Remove parentheses, simplify, and divide by h. $=\dfrac{x-x-h}{x(x+h)h}=\dfrac{-h}{xh(x+h)}=-\dfrac{1}{x(x+h)}$	1. Simplify $\dfrac{\frac{x}{y}-1}{\frac{x^2}{y^2}-1}$. 2. Simplify $\dfrac{\frac{1}{x+7}-\frac{1}{x}}{7}$.

R.6 Answers

R.6.1

1) Because -2 makes the denominator zero, the domain is all real numbers, except -2.

2) Because $\frac{1}{2}$ and -3 makes the denominator zero, the domain is all real numbers, except $\frac{1}{2}$ and -3.

R.6.2

1) $x+5$ 2) $\frac{x-5}{3x-1}$

R.6.3

1) $\frac{x+3}{x-4}$ 2) $-\frac{4(x+2)}{3x(3x-2)}$

R.6.4

1) $9(3x+7)$ 2) $\frac{x+3}{5}$

R.6.5

1) $18x^2$ 2) $5(x+3)^2$

R.6.6

1) $\frac{21+4x}{18x^2}$ 2) $\frac{x^2-3x+5}{(x-3)(x+1)(x-2)}$

R.6.7

1) $\frac{y}{x+y}$ 2) $-\frac{1}{x(x+7)}$

R.7 Review Solving Applications

Objective 1 – Translate English phrases into algebraic expressions.
Objective 2 – Perimeter, Circumference, and Area
Objective 3 – Ratios and Proportions
Objective 4 – The Pythagorean Theorem and Its Converse
Objective 5 – Volumes of Prisms and Cylinders
Objective 6 – Solving applied problems using mathematical models.
Objective 7 – Area of Parallelograms and Triangles.

Objective 1 – Translate English phrases into algebraic expressions.

Summary
Problem solving in algebra involves translating English phrases into algebraic expressions. Here is a list of words and phrases for the four basic operations.

Addition	Subtraction	Multiplication	Division
sum	difference	product	quotient
plus	minus	times	divide
increased by	decreased by	of (used with fractions)	per
more than	less than	twice	ratio

Notes:

Guided Example

Write each English phrase as an algebraic expression. Let x represent the number.

1. Nine less than six times a number

$$\overbrace{6x}^{\text{six times a number}} - \underbrace{9}_{\text{nine less than}}$$

2. The quotient of five and a number increased by twice the number

$$\overbrace{\frac{5}{x}}^{\text{the quotient of five and a number}} \underbrace{+}_{\text{increased by}} \overbrace{2x}^{\text{twice the number}}$$

Practice

Write each English phrase as an algebraic expression. Let x represent the number.

1. Five more than eight times a number

2. Two less than the quotient of four and twice a number

Objective 2 – Perimeter, Circumference, and Area

Key Terms
Perimeter
Circumference

Summary
The **perimeter** of a polygon is the sum of the length of its sides. Perimeter is measured in linear units, such as feet or meters. The formula for the perimeter of a square is $P = s+s+s+s = 4s$, where s is the length of the side of the square. The formula for the perimeter of a rectangle is $P = l+l+w+w = 2l+2w$, where l is the length and w is the width of the rectangle.

The distance around a circle is called its **circumference**. The formula for the circumference of a circle is $C = 2\pi r$, where r is the radius of the circle, or $C = \pi d$, d is the diameter of the circle. The circumference is also a linear measure.

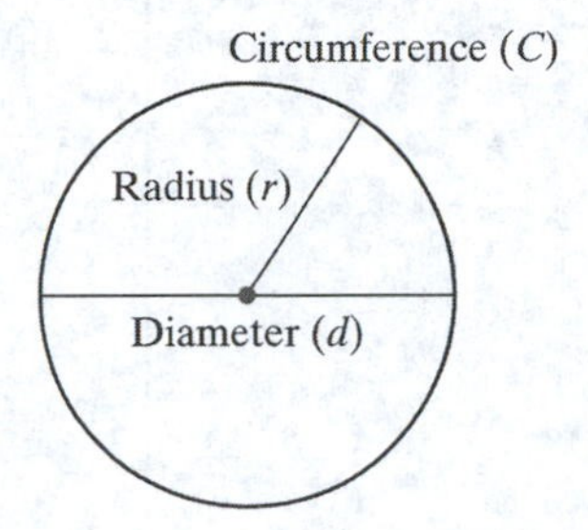

Area is the measure of the size of a plane geometric figure. Areas are measured in square units, such as square feet or square meters. The formula for the area of a square is $A = s^2$, the formula for the area of a rectangle is $A = l \times w$, and the formula for the area of a circle is $A = \pi r^2$.

Notes:

Guided Example	***Practice***
1. Find the area of a square with one side of length 16.8 inches. $A = s^2$ $A = (16.8)^2$ $A = 282.24\text{ in.}^2$	1. Find the area of the circle below. Use $\pi = 3.14$. 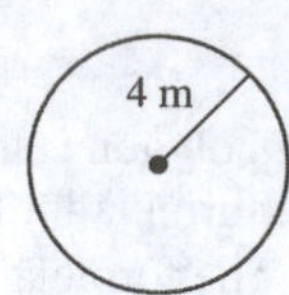
2. Find the circumference of the circle below. Use $\pi = 3.14$. 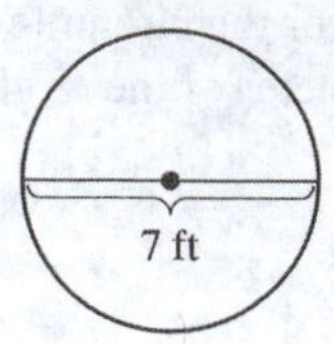 $C = \pi d$ $C = \pi(7)$ $C = 7\pi$ $C = 7(3.14)$ $C = 21.98\text{ ft}$	2. Find the perimeter of the square below. 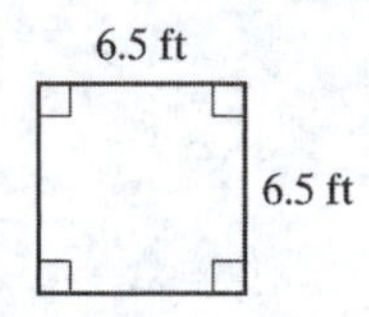
3. Find the perimeter of a rectangle below. 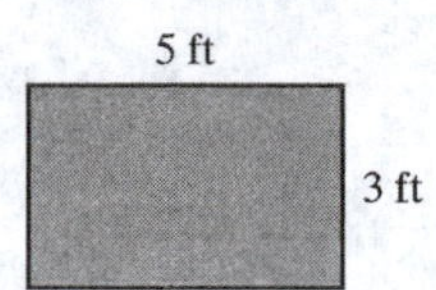 $P = 2l + 2w$ $P = 2(5) + 2(3)$ $P = 10 + 6$ $P = 16\text{ ft}$	3. Find the area of a rectangle whose length is 12 meters and whose width is 9 meters.

Objective 3 – Ratios and Proportions

Key Terms

Ratio
Proportion
Cross-product Property

Summary

A **ratio** is the comparison of two quantities by division. The ratio of two numbers a and b, where $b \neq 0$, can be written as $\frac{a}{b}$, $a:b$, or a to b. A **proportion** states that two ratios are equal. If $\frac{a}{b}$ and $\frac{c}{d}$ are two equal ratios, then $\frac{a}{b} = \frac{c}{d}$ is a proportion. The **cross-product property** for a proportion states that if $\frac{a}{b} = \frac{c}{d}$, where $b \neq 0$ and $d \neq 0$, then $ad = bc$.

Notes:

<table>
<tr><th><u>Guided Example</u></th><th><u>Practice</u></th></tr>
<tr><td>
1. Write the ratio of the first measurement to the second measurement.

Diameter of ball A: 5 centimeters
Diameter of ball B: 15 centimeters

The ratio of the diameter of ball A to the diameter of ball B can be expressed as

$\frac{\text{diameter of A}}{\text{diameter of B}}$

Now substitute the measurements.

$\frac{\text{diameter of A}}{\text{diameter of B}} = \frac{5 \text{ cm}}{15 \text{ cm}}$
$= \frac{1}{3}$ Reduce the fraction.

The ratio of the diameter of ball A to ball B is $\frac{1}{3}$ or $1:3$ or 1 to 3.

2. Solve the proportion for x.

$\frac{8}{3} = \frac{x}{12}$

Use the cross-product property.

$3 \cdot x = 8 \cdot 12$
$3x = 96$ Multiply.
$x = 32$ Divide both sides by 3.
</td><td>
1. Write the ratio of the length to the width of a rectangle.

Length of a rectangle: 36 feet
Width of a rectangle: 8 feet

2. Solve the proportion for x.

$\frac{3}{4} = \frac{9}{x}$
</td></tr>
</table>

Objective 4 – The Pythagorean Theorem and Its Converse

Key Terms

Right angle
Right triangle
Hypotenuse
Legs
Pythagorean Theorem

Summary

A **right angle** is an angle that measures $90°$. A triangle that has an angle measuring $90°$ is called a **right triangle**. The side of the triangle opposite the $90°$ angle is called the **hypotenuse**. The other two sides of the triangle are called the **legs**. The figure below shows a right triangle with legs of length a and b and hypotenuse of length c. The **Pythagorean Theorem** says that the sum of the squares of the lengths of the legs of a right triangle equals the square of the length of the hypotenuse. For the triangle below, the Pythagorean Theorem says that $c^2 = a^2 + b^2$. The converse of the Pythagorean Theorem states that if $c^2 = a^2 + b^2$, then the triangle is a right triangle.

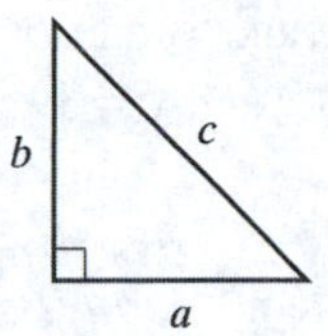

Notes:

Guided Example	*Practice*
1. A right triangle has one leg of length 8 and one leg of length 15. Find the length of the hypotenuse. Let c represent the length of the hypotenuse and a and b represent the lengths of the legs. Then $a=8$ and $b=15$. $c^2=a^2+b^2$ Pythagorean Theorem $c^2=8^2+15^2$ Substitute. $c^2=64+225$ Evaluate exponents. $c^2=289$ Add. $c=17$ Take the square root of both sides.	1. If $a=7$ and $b=24$ are the legs of a right triangle, find the hypotenuse.
2. A right triangle has one leg of length 15 and a hypotenuse of length 39. Find the length of the other leg. Let c represent the length of the hypotenuse and a and b represent the lengths of the legs. Then $a=15$ and $c=39$. $c^2=a^2+b^2$ Pythagorean Theorem $39^2=15^2+b^2$ Substitute. $1521=225+b^2$ Evaluate exponents. $1521-225=b^2$ Subtract 225 from both sides. $1296=b^2$ Simplify. $b=36$ Square root of both sides	2. If one leg $b=14$ and the hypotenuse $c=50$, find the length of the other leg in the right triangle.

Objective 5 – Volumes of Prisms and Cylinders

Summary
Solving geometry problems usually requires a knowledge of basic geometric ideas and formulas. The formulas for the volume of a rectangular solid (prism) and a cylinder are given below. In words, to find the volume of a rectangular solid, we find the product of its length, width, and height. To find the volume of a cylinder, we find the product of pi, the square of its radius, and its height.

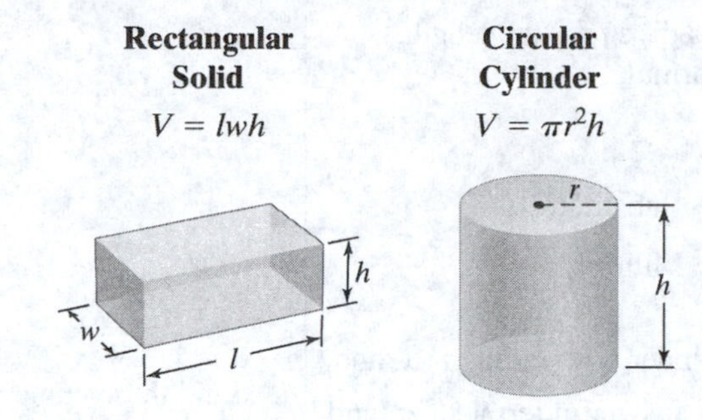

Notes:

Guided Example	***Practice***
1. Find the volume of a rectangular solid with length 9 cm, width 4 cm, and height 5 cm. The formula for the volume of a rectangular solid is $V = lwh$ Substitute the measurements given for the length, width, and height into the formula. $V = lwh$ $V = (9\text{cm})(4\text{cm})(5\text{cm})$ Substitute. $V = 180 \text{ cm}^3$ Multiply.	1. Find the volume of a rectangular solid with length 12 m, width 6 m, and height 4 m.
2. Find the volume of a cylinder with radius 8 ft and height 10 ft. Give an exact answer in terms of π and an approximate answer, using $\pi = 3.14$, rounded to the nearest tenth. The formula for the volume of a cylinder is $V = \pi r^2 h$ Substitute the measurements given for the radius and height into the formula. $V = \pi r^2 h$ $V = \pi(8\text{ft})^2(10\text{ft})$ Substitute. $V = \pi(64\text{ft}^2)(10\text{ft})$ Evaluate the exponent. $V = 640\pi \text{ ft}^3 \approx 2009.6 \text{ ft}^3$ Multiply	2. Find the volume of a cylinder with radius 7 in. and height 5 in.. Give an exact answer in terms of π and an approximate answer, using $\pi = 3.14$, rounded to the nearest tenth.

Objective 6 – Solving applied problems using mathematical models.

Key Terms
Model

Summary
A **model** is a mathematical representation of a real-world situation. Many times we must obtain models by translating from the ordinary language of English into the language of algebraic equations. When we have an equation that "models" the verbal conditions of a problem, frequently we can use the equation to project what might occur in the future.

Notes:

Guided Example	***Practice***
1. The formula, $T = 394x + 3123$ models the average cost of tuition and fees, T, at public four-year colleges for the school year ending x years after 2000. Use the formula in the problem above to predict the tuition in 2015. To find the tuition, we must know the value of x. $x = 2015 - 2000 = 15$ $T = 394x + 3123$ Substitute 15 into the formula for x. $T = 394(15) + 3123$ Substitute. $T = 5910 + 3123$ Multiply. $T = 9033$ Add. The tuition will be \$9033 in 2015.	1. The formula $T = 394x + 3123$ models the average cost of tuition and fees, T, at public four-year colleges for the school year ending x years after 2000. Use the formula to predict the tuition in 2017.

Objective 7 – Area of Parallelograms and Triangles.

Summary

The area of a triangle is half the product of a base b and a corresponding height h. A base of a triangle can be any of its sides. The corresponding height is the length of the altitude containing the corresponding height. The formula for the area of a triangle can be written as $A=\frac{1}{2}bh$.

Notes:

Guided Example	***Practice***
1. Find the area of the triangle below. 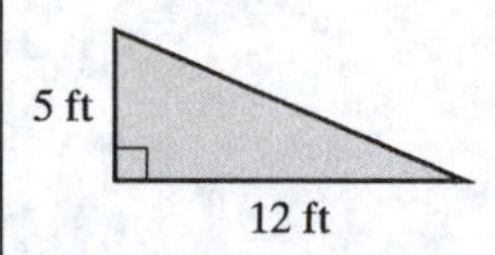 The formula for the area of a triangle is $A=\frac{1}{2}bh$. Identify the base, b, and the height h. $b=12,\ h=5$ $A=\frac{1}{2}bh=\frac{1}{2}(12)(5)$ Substitute for b and h. $=30\text{ ft}^2$ Multiply and add units.	1. Find the area of the triangle below. 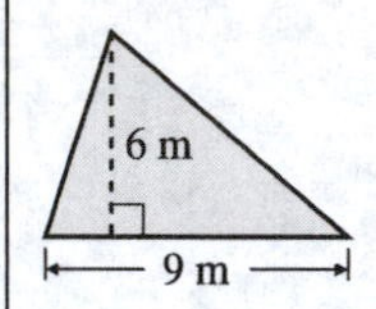
2. Find the area of the triangle below. 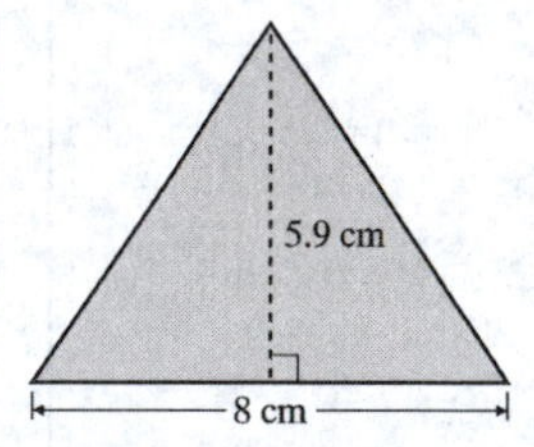 The formula for the area of a triangle is $A=\frac{1}{2}bh$. Identify the base, b, and the height h. $b=8,\ h=5.9$ $A=\frac{1}{2}bh=\frac{1}{2}(8)(5.9)$ Substitute for b and h. $=23.6\text{ cm}^2$ Multiply and add units.	2. Find the area of the triangle below. 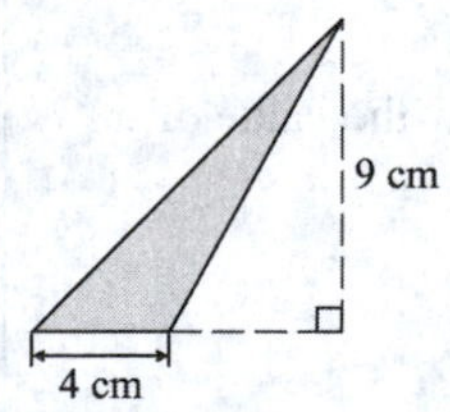

R.7 Answers

R.7.1

1) $5x+8$ 2) $\frac{4}{2x}-2$

R.7.2

1) $50.24\,\text{m}^2$ 2) $26\,\text{ft}$ 3) $108\,\text{m}^2$

R.7.3

1) $\frac{9}{2}$ or 9:2 or 9 to 2 2) $x=12$

R.7.4

1) 25 2) 48

R.7.5

1) $288\,\text{m}^2$ 2) $245\pi\,\text{in.}^3 \approx 769.3\,\text{in.}^3$

R.7.6

1) $9821

R.7.7

1) $A=27\,\text{m}^2$ 2) $A=18\,\text{cm}^2$

R.8 Review Solving Equations and Inequalities

Objective 1 – Solve linear equations.
Objective 2 – Solve quadratic equations by factoring.
Objective 3 – Solve quadratic equations using the quadratic formula.
Objective 4 – Use interval notation.
Objective 5 – Solve linear inequalities including inequalities with no solutions.

Objective 1 – Solve linear equations.

Key Terms

Linear equation
Solving an equation
Solutions or roots
Equivalent equations

Summary

A linear equation in one variable x is an equation that can be written in the form $ax+b=0$, where a and b are real numbers, and $a \neq 0$. **Solving an equation** in x involves determining all values of x that result in a true statement when substituted into the equation. Such values are **solutions**, or **roots**, of the equation. Two or more equations that have the same solution set are called **equivalent equations**. To solve a linear equation in x, we transform the equation into an equivalent equation one or more times. To generate equivalent equations, we will use the following properties:

The Addition and Multiplication Properties of Equality

The Addition Property of Equality

The same real number or algebraic expression may be added to both sides of an equation without changing the equation's solution set.

$a = b$ and $a + c = b + c$ are equivalent equations.

The Multiplication Property of Equality

The same nonzero real number may multiply both sides of an equation without changing the equation's solution set.

$a = b$ and $ac = bc$ are equivalent equations as long as $c \neq 0$.

Here is a step-by-step procedure for solving a linear equation in one variable. Not all of these steps are necessary to solve every equation.

Solving a Linear Equation

1. Simplify the algebraic expression on each side by removing grouping symbols and combining like terms.
2. Collect all the variable terms on one side and all the numbers, or constant terms, on the other side.
3. Isolate the variable and solve.
4. Check the proposed solution in the original equation.

Notes

Guided Example	***Practice***
1. Solve $2x+3=17$ $2x+3=17$ $2x+3-3=17-3$ Subtract 3 from both sides. $2x=14$ Simplify. $\frac{2x}{2}=\frac{14}{2}$ Divide both sides by 2. $x=7$ Simplify. The solution set is $\{7\}$.	1. Solve $4x+5=29$
2. Solve $8y-3=11y+8$ $8y-3=11y+8$ $8y-11y-3=11y-11y+8$ Subtract $11y$ from both sides. $-3y-3=8$ Simplify. $-3y-3+3=8+3$ Add 3 to both sides. $-3y=11$ Simplify. $\frac{-3y}{-3}=\frac{11}{-3}$ Divide both sides by -3. $y=-\frac{11}{3}$ Simplify. The solution set is $\left\{\frac{11}{3}\right\}$.	2. Solve $5y-2=9y+2$
3. $4(2x+1)-29=3(2x-5)$ $4(2x+1)-29=3(2x-5)$ $8x+4-29=6x-15$ Use the distributive property. $8x-25=6x-15$ Simplify. Subtract $6x$ from both sides. $8x-6x-25=6x-6x-15$ $2x-25=-15$ Simplify. $2x-25+25=-15+25$ Add 25 to both sides. $2x=10$ Simplify. $\frac{2x}{2}=\frac{10}{2}$ Divide both sides by 2. $x=5$ Simplify The solution set is $\{5\}$.	3. $2(x-3)-17=13-(3x+2)$

Objective 2 – Solve quadratic equations by factoring.

Key terms
Quadratic equation
Standard form
Second-degree-polynomial
Zero-product property

Summary
A **quadratic equation** in x is an equation that can be written in the **standard form** $ax^2 + bx + c = 0$, where a, b, and c are real numbers and $a \neq 0$. A quadratic equation in x is also called a **second-degree polynomial** in x. If a quadratic equation has zero on one side and a factored expression on the other side, it can be solved using the **zero-product property**.

The Zero-Product Property

If the product of two algebraic expressions is zero, then at least one of the factors is equal to zero.

If $AB = 0$, then $A = 0$ or $B = 0$.

Here is a set of steps for solving a quadratic equation by factoring.

Solving a Quadratic Equation by Factoring

1. If necessary, rewrite the equation in the standard form $ax^2 + bx + c = 0$, moving all terms to one side, thereby obtaining zero on the other side.
2. Factor completely.
3. Apply the zero-product principle, setting each factor containing a variable equal to zero.
4. Solve the equations in step 3.
5. Check the solutions in the original equation.

Notes

Guided Example	***Practice***
Solve by factoring. 1. $2x^2-5x=12$. $2x^2-5x=12$ $2x^2-5x-12=0$ Obtain zero on one side. $(2x+3)(x-4)=0$ Factor. $2x+3=0$ or $x-4=0$ Set each factor equal to zero. $2x=-3$ $x=4$ Solve resulting equations. $x=-\frac{3}{2}$ The solution set is $\left\{-\frac{3}{2}, 4\right\}$.	Solve by factoring. 1. $2x^2-9x=5$.
2. $5x^2=20x$. $5x^2=20x$ $5x^2-20x=0$ Obtain zero on one side. $5x(x-4)=0$ Factor. $5x=0$ or $x-4=0$ Set each factor equal to zero. $x=0$ $x=4$ Solve resulting equations. The solution set is $\{0, 4\}$. .	2. $3x^2=2x$.

Objective 3 – Solve quadratic equations using the quadratic formula.

Summary

The quadratic formula is a formula that can be used to solve quadratic equations.

The Quadratic Formula

The solutions of a quadratic equation in standard form $ax^2 + bx + c = 0$, with $a \neq 0$, are given by the quadratic formula:

$$x = \frac{-b \pm \sqrt{b^2 - 4ac}}{2a}$$

To use the quadratic formula, write the quadratic equation in standard form if necessary. Then determine the numerical values for a (the coefficient of the x^2-term), b (the coefficient of the x-term), and c (the constant term). Substitute the values of a, b, and c into the quadratic formula and evaluate the expression. The $\pm$ sign indicates that there are two (not necessarily distinct) solutions of the equation.

Notes

Guided Example	***Practice***
1. Solve using the quadratic formula: $8x^2+2x-1=0$. The equation is in standard form. Identify the values of a, b, and c. $a=8, b=2, c=-1$ $x=\frac{-b\pm\sqrt{b^2-4ac}}{2a}$ The quadratic formula $x=\frac{-2\pm\sqrt{2^2-4(8)(-1)}}{2(8)}$ Substitute a, b, and c. $x=\frac{-2\pm\sqrt{4+32}}{16}$ $2^2-4(8)(-1)=4+32$ $x=\frac{-2\pm\sqrt{36}}{16}=\frac{-2\pm 6}{16}$ $\sqrt{4+32}=\sqrt{36}=6$ Evaluate to obtain two solutions. $x=\frac{-2+6}{16}$ or $x=\frac{-2-6}{16}$ $x=\frac{4}{16}=\frac{1}{4}$ or $x=\frac{-8}{16}=-\frac{1}{2}$ The solution set is $\left\{\frac{1}{4},-\frac{1}{2}\right\}$.	1. Solve using the quadratic formula: $2x^2+9x-5=0$.
2. Solve using the quadratic formula: $2x^2=4x+1$. , $2x^2-4x-1=0$, so $a=2, b=-4, c=-1$. $x=\frac{-(-4)\pm\sqrt{(-4)^2-4(2)(-1)}}{2(2)}$ Substitute. $x=\frac{4\pm\sqrt{16-(-8)}}{4}$ $(-4)^2-4(2)(-1)=16-(-8)$ $x=\frac{4\pm\sqrt{24}}{4}$ $16-(-8)=16+8=24$ $x=\frac{4\pm 2\sqrt{6}}{4}$ $\sqrt{24}=\sqrt{4\cdot 6}=\sqrt{4}\cdot\sqrt{6}=2\sqrt{6}$ $x=\frac{2\left(2\pm\sqrt{6}\right)}{4}$ Factor 2 out from the numerator. $x=\frac{2\pm\sqrt{6}}{2}$ Divide numerator and denominator by 2. The solution set is $\left\{\frac{2+\sqrt{6}}{2},\frac{2-\sqrt{6}}{2}\right\}$.	2. Solve using the quadratic formula $2x^2=6x-1$.

Objective 4 – Use interval notation.

Key terms

Interval notation

Summary

Some sets of real numbers can be represented using **interval notation**. Below is a table that lists nine possible types of intervals used to describe real numbers. Keep in mind that parentheses indicate endpoints that are not included in an interval. Square brackets indicate endpoints that are included in an interval. Parentheses are always used with ∞ or $-\infty$.

Table 1.1 Intervals on the Real Number Line

Let a and b be real numbers such that $a < b$.

Interval Notation	Set-Builder Notation	Graph
(a, b)	$\{x \mid a < x < b\}$	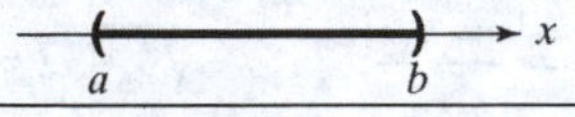
$[a, b]$	$\{x \mid a \le x \le b\}$	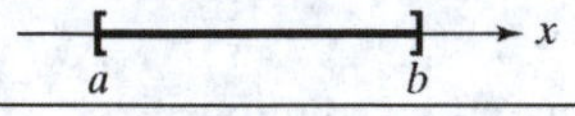
$[a, b)$	$\{x \mid a \le x < b\}$	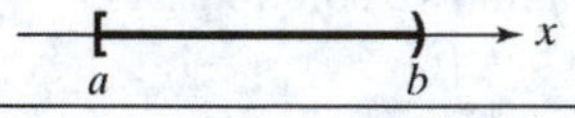
$(a, b]$	$\{x \mid a < x \le b\}$	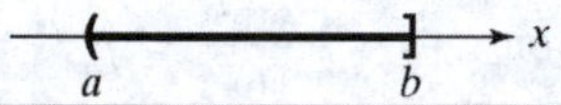
(a, ∞)	$\{x \mid x > a\}$	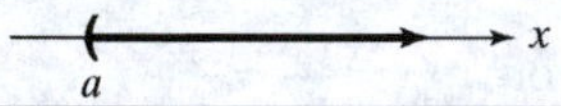
$[a, \infty)$	$\{x \mid x \ge a\}$	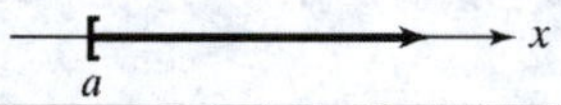
$(-\infty, b)$	$\{x \mid x < b\}$	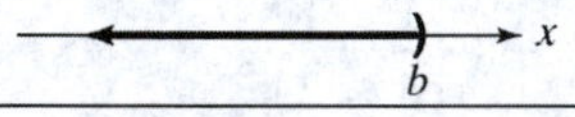
$(-\infty, b]$	$\{x \mid x \le b\}$	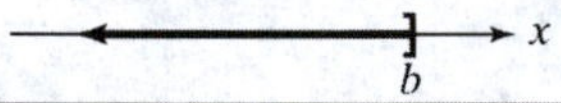
$(-\infty, \infty)$	$\{x \mid x$ is a real number$\}$ or $\mathbb{R}$ (set of real numbers)	

Notes:

Guided Example

Express each interval notation in set builder notation and graph.

1. $(-1,4]$

$(-1,4]=\{x\mid -1<x\leq 4\}$

−4 −3 −2 −1 0 1 2 3 4 x

2. $(-4,\infty)$

$(-4,\infty)=\{x\mid x>-4\}$

−4 −3 −2 −1 0 1 2 3 4 x

Practice

Express each interval notation in set builder notation and graph.

1. $[2.5,4]$

2. $[-2,\infty)$

Objective 5 – Solve linear inequalities including inequalities with no solutions.

Key Terms
Linear inequality
Solving a linear inequality
Solutions
Solution set

Summary
Placing an inequality symbol between a linear expression ($mx + b$) and a constant results in a **linear inequality** *in one variable*. **Solving an inequality** is the process of finding the set of numbers that make the inequality a true statement. These numbers are called the **solutions** of the inequality and we say that they *satisfy* the inequality. The set of all solutions is called the **solution set** of the inequality. We will use interval notation to represent these solution sets. A **linear inequality in** x can be written in one of the following forms: $ax+b<0$, $ax+b\le 0$, $ax+b>0$, $ax+b\ge 0$. In each form, $a\ne 0$. We can isolate a variable in a linear inequality the same way we can isolate a variable in a linear equation. The following properties are used to create equivalent inequalities:

The Addition Property of Inequality
If $a<b$, then $a+c<b+c$.
If $a<b$, then $a-c<b-c$.

The Positive Multiplication Property of Inequality
If $a<b$ and c is positive, then $a\cdot c<b\cdot c$.
If $a<b$ and c is positive, then $\frac{a}{c}<\frac{b}{c}$.

The Negative Multiplication Property of Inequality
If $a<b$ and c is negative, then $a\cdot c>b\cdot c$.
If $a<b$ and c is negative, then $\frac{a}{c}>\frac{b}{c}$.

Below is a set of step for solving a linear inequality

Solving a Linear Inequality

1. Simplify the algebraic expression on each side.
2. Use the addition property of inequality to collect all the variable terms on one side and all the constant terms on the other side.
3. Use the multiplication property of inequality to isolate the variable and solve. Change the sense of the inequality when multiplying or dividing both sides by a negative number.
4. Express the solution set in interval notation and graph the solution set on a number line.

Notes:

Guided Example	***Practice***
Solve. Write the answer in interval notation.	Solve. Write the answer in interval notation.
1. $3x-5>-17$ $3x-5>-17$ $3x-5+5>-17+5$ Add 5 to both sides. $3x>-12$ Simplify. $\frac{3x}{3}>\frac{-12}{3}$ Divide by 3. $x>-4$ Simplify. The solution set is $(-4,\infty)$.	1. $4x-3\leq-23$
2. $-2x-4\geq x+5$ $-2x-4\geq x+5$ $-2x-x-4\geq x-x+5$ Subtract x from both sides. $-3x-4\geq5$ Simplify. $-3x-4+4\geq5+4$ Add 4 to both sides. $-3x\geq9$ Simplify. Divide both sides by -3. Change $\geq$ to $\leq$. $\frac{-3x}{-3}\leq\frac{9}{-3}$ $x\leq-3$ Simplify. The solution set is $(-\infty,-3]$.	2. $3x+1>7x-15$

R.8 Answers

R.8.1

1) $\{6\}$ 2) $\{-1\}$ 3) $\left\{\frac{34}{5}\right\}$

R.8.2

1) $\left\{-\frac{1}{2}, 5\right\}$ 2) $\left\{0, \frac{2}{3}\right\}$

R.8.3

1) $\left\{\frac{1}{2}, -5\right\}$ 2) $\left\{\frac{3 \pm \sqrt{7}}{2}\right\}$

R.8.4

1) $\{x \mid 2.5 \leq x \leq 4\}$

−4 −3 −2 −1 0 1 2 3 4 x

2) $\{x \mid x \geq -2\}$

−4 −3 −2 −1 0 1 2 3 4 x

R.8.5

1) $(-\infty, -5]$ 2) $(-\infty, 4)$